Communications and Control Engineering

Springer
London
Berlin
Heidelberg
New York
Barcelona
Hong Kong
Milan
Paris
Singapore
Tokyo

Published titles include:

Nonlinear Control Systems (3rd edition)
Alberto Isidori

Theory of Robot Control
C. Canudas de Wit, B. Siciliano and G. Bastin (Eds)

Fundamental Limitations in Filtering and Control
María M. Seron, Julio Braslavsky and Graham C. Goodwin

Constructive Nonlinear Control
R. Sepulchre, M. Janković and P.V. Kokotović

A Theory of Learning and Generalization
M. Vidyasagar

Adaptive Control
I.D. Landau, R. Lozano and M.M'Saad

Stabilization of Nonlinear Uncertain Systems
Miroslav Krstić and Hua Deng

Passivity-based Control of Euler-Lagrange Systems
Romeo Ortega, Antonio Loría, Per Johan Nicklasson and Hebertt Sira-Ramírez

Stability and Stabilization of Infinite Dimensional Systems with Applications
Zheng-Hua Luo, Bao-Zhu Guo and Omer Morgul

Nonsmooth Mechanics (2nd edition)
Bernard Brogliato

Nonlinear Control Systems II
Alberto Isidori

L_2-Gain and Passivity Techniques in Nonlinear Control
Arjan van der Schaft

Control of Linear Systems with Regulation and Input Constraints
Ali Saberi, Anton A. Stoorvogel and Peddapullaiah Sannuti

Robust and $H\infty$ Control
Ben M. Chen

Computer Controlled Systems
Efim N. Rosenwasser and Bernhard P. Lampe

Dissipative Systems Analysis and Control
Rogelio Lozano, Bernard Brogliato, Olav Egeland and Bernhard Maschke

Control of Complex and Uncertain Systems
Stanislav V. Emelyanov and Sergey K. Korovin

Ian R. Petersen, Valery A. Ugrinovskii
and Andrey V. Savkin

Robust Control Design Using H-$^{\infty}$ Methods

With 53 Figures

Springer

Ian R. Petersen, BE, MSEE, PhD
School of Electrical Engineering, ADFA, Canberra ACT 2600, Australia

Valery A. Ugrinovskii, PhD
School of Electrical Engineering, ADFA, Canberra ACT 2600, Australia

Andrey V. Savkin, PhD
Department of Electrical and Electonic Engineering, University of Western Australia, Nedlands,WA6907 Australia

Series Editors

E.D. Sontag • M. Thoma

ISBN 1-85233-171-2 Springer-Verlag London Berlin Heidelberg

British Library Cataloguing in Publication Data
Petersen, Ian R.
Robust control design using H- [infinity symbol] methods. -
(Communication and control engineering)
1.Automatic control 2. H [infinity symbol] control
I.Title II. Ugrinovskii, Valery A. III.Savkin, Audrey V.
629.8'312
ISBN 1852331712

Library of Congress Cataloging-in-Publication Data
Petersen, Ian Richard, 1956-
Robust control design using H-È methods / Ian R. Petersen, Valery A. Ugrinovskii, and Audrey V. Savkin.
p. cm. -- (Communication and control engineering series)
ISBN 1-85233-171-2 (alk. paper)
1. Robust control. I. Ugrinovskii, Valery A., 1960- II. Savkin, Audrey V. III. Tile. IV. Series.
TJ217.2P48 2000
629.8'312--dc21 99-056302

Typesetting: Camera ready by authors
Printed and bound by Athenæum Press Ltd., Gateshead, Tyne & Wear
69/3830-543210 Printed on acid-free paper SPIN 10713809

Preface

One of the key ideas to emerge in the field of modern control is the use of optimization and optimal control theory to give a systematic procedure for the design of feedback control systems. For example, in the case of linear systems with full state measurements, the linear quadratic regulator (LQR) approach provides one of the most useful techniques for designing state feedback controllers; e.g., see [3]. Also, in the case of linear systems with partial information, the solution to the linear quadratic Gaussian (LQG) stochastic optimal control problem provides a useful technique for the design of multivariable output feedback controllers; see [3]. However, the above optimal control techniques suffer from a major disadvantage in that they do not provide a systematic means for addressing the issue of robustness. The robustness of a control system is a fundamental requirement in designing any feedback control system. This property of the control system reflects an ability of the system to maintain adequate performance and in particular, stability in the face of variations in the plant dynamics and errors in the plant model which is used for controller design. The enhancement of robustness is one of the main reasons for using feedback; e.g., see [87].

The lack of robustness which may result when designing a control system via standard optimal control methods, such as those mentioned above[1], has been a major motivation for research in the area of robust control. Research in this area has resulted in a large number of approaches being available to deal with the issue of robustness in control system design; e.g., see [47]. These methods include those

[1] It is known for example, that in the partial information case, an LQG controller may lead to very poor robustness properties; e.g., see [50].

based on Kharitonov's Theorem (see [104, 7]), H^∞ control theory (e.g., see [52, 9]) and the quadratic stabilizability approach (e.g., see [6, 102]). Also, the theory of absolute stability should be mentioned; e.g., see [120, 2, 137, 162]. For the purposes of this book, it is worth mentioning two important features of absolute stability theory. These features lead us to consider absolute stability theory as a forerunner of contemporary robust control theory. Absolute stability theory was apparently one of the first areas of control theory that introduced and actively used the notion of an uncertain system. Also, absolute stability theory introduced and made use of matrix equations that were termed Lur'e equations in the Russian literature and are known as Riccati equations in the West; e.g., see [119, 95, 256, 23].

In this book, we present the reader with some of the latest developments in robust control theory for uncertain systems. In particular, we will consider the quadratic stabilizability approach to the control of uncertain systems with *norm bounded-uncertainty* and a more general absolute stabilizability approach to the control of uncertain systems described by *integral quadratic constraints (IQCs)*. The reader will see that the integral quadratic constraint concept and its generalizations leads to an uncertain system framework in which the standard LQR and LQG optimal controller design methodologies can be extended into minimax optimal control and guaranteed cost control methodologies. Such an uncertain system framework allows the designer to retain existing methods of LQR/LQG design while the issue of robustness can be addressed by an appropriate choice of the uncertainty structure in the uncertain system model. In the book, this approach is illustrated by the two practical problems, a missile autopilot design problem and an active noise control problem.[2]

Note that for the most part, the robust control problems considered in this book are closely related to corresponding H^∞ control problems. Furthermore, throughout the book, the solutions to these robust control problems will be formulated in terms of Riccati equations of the form arising in H^∞ control theory. These facts motivate the title of this book.

The book focuses only on robust control system design problems. Other related topics such as robust filtering and robust model validation are outside the scope of the book. However, the methods presented here have been applied successfully to these and other related problems; e.g., see [196, 186, 192, 130, 199, 201, 203, 132]. In particular, robust filtering problems were carefully studied in the research monograph [159].

Another area in which the methods and ideas developed in this book have been successfully applied is in the analysis and synthesis of hybrid control systems; see [204, 175, 205, 214]. A survey of results on hybrid control systems together with a bibliography relating to this area can be found in the book [122].

---

[2]The reader who is interested in the practical aspects of these examples can find some additional details on the book website http://routh.ee.adfa.edu.au/~irp/RCD/index.html.

It is worth noting that in this book, we deal only with continuous-time systems. This is in no way due to limitations in the theory. The reader who is interested in robust control of discrete-time uncertain systems can use this book as a conceptual guideline to the analogous discrete-time theory. Moreover, discrete-time counterparts to the majority of the results in this book have already been published. We will point out the relevant references in the text.

In the production of this book, the authors wish to acknowledge the support they have received from the Australian Research Council. Also the authors wish to acknowledge the major contributions made by their colleagues Duncan McFarlane, Matthew James, Paul Dupuis, Uffe Thygesen and Hemanshu Pota in the research that underlies much of the material presented in this book. Furthermore, the first author is grateful for the enormous support he has received from his wife Lynn and children Charlotte and Edward. Also, the second author is truly thankful to Grisha, Anna and Elena for their love, encouragement and emotional support.

Contents

Frequently used notation

$\mathbf{R}^n$ is the n-dimensional Euclidean space of real vectors.

$\mathbf{R}^k_+$ is the positive orthant of $\mathbf{R}^n$. That is, $\mathbf{R}^k_+$ consists of all vectors in $\mathbf{R}^n$ such that each component is positive.

$\mathrm{cl}\, M$ denotes the closure of a set $M \subset \mathbf{R}^n$.

$\|\cdot\|$ is the standard Euclidean norm.

$\|A\|$ where A is a matrix, is the induced operator norm of A.

A' denotes the transpose of the matrix A.

$\rho(A)$ denotes the spectral radius of a matrix A.

$\lambda_{max}(N)$ denotes the maximum eigenvalue of a symmetric matrix N.

$\mathrm{tr}\, N$ denotes the trace of a matrix N.

$M \geq N$ ($M > N$) where M and N are square matrices, denotes the fact that the matrix $M - N$ is non-negative (positive) definite.

$C(\mathcal{J}, X)$ is the set of continuous functions $f: \mathcal{J} \to X$.

$C_b(\mathbf{R}^n)$ denotes the set of bounded continuous functions from $\mathbf{R}^n$ to $\mathbf{R}$.

$\mathbf{L}_2[T_1, T_2)$ is the Lebesgue space of square integrable functions, defined on $[T_1, T_2)$, $T_1 < T_2 \leq \infty$.

$\|x(\cdot)\|_2^2$ is the norm in $\mathbf{L}_2[T_1, T_2)$, that is $\|x(\cdot)\|_2^2 = \int_{T_1}^{T_2} \|x(t)\|^2 dt$, $T_1 < T_2 \leq \infty$.

$\langle x, y \rangle$ is the inner product in a Hilbert space.

G^* denotes the adjoint of an operator G.

H^∞ is the Hardy space of complex functions $\boldsymbol{f}(s)$ analytic and bounded in the open right half of the complex plane.

$\|\boldsymbol{f}(s)\|_\infty$ is the norm in the Hardy space H^∞, $\|\boldsymbol{f}(s)\|_\infty := \sup_{\mathrm{Re}\, s > 0} |\boldsymbol{f}(s)|$.

$\mathcal{P}$ is a set of uncertainty inputs

Ξ is a set of admissible uncertainty inputs.

$\delta \downarrow 0$ denotes the fact that δ approaches zero monotonically from above. Similarly, $T \uparrow \infty$ means that T approaches ∞ monotonically.

$(\Omega, \mathcal{F})$ is a measurable space.

$W(t)$ is a Wiener process.

$\mathbf{E}^Q$ is the expectation operator with respect to a probability measure Q. Also, $\mathbf{E}$ is the expectation with respect to a reference probability measure.

a.s. is an acronym for *almost surely*. In probability theory, this refers to a property which is true for almost all $\omega \in \Omega$, *i.e.*, the set $\{\omega : \text{the property is false}\}$ has probability zero.

$Q \ll P$ denotes the fact that the measure Q is absolutely continuous with respect to the measure P where Q and P are Lebesgue measures defined on a measurable space.

$\mathbf{L}_p(\Omega, \mathcal{F}, Q)$ is the Lebesgue space of $\mathcal{F}$-measurable functions $f \colon \Omega \to X$ such that $\mathbf{E}^Q \|f\|_X^p < \infty$; here X is a Banach space.

$\mathbf{L}_2(s, T; \mathbf{R}^n)$ is the Hilbert space generated by the (t, ω)-measurable non-anticipating random processes $x(t, \omega) : [s, T] \times \Omega \to \mathbf{R}^n$.

$|||\cdot|||$ is the norm on the space $\mathbf{L}_2(s, T; \mathbf{R}^n)$ defined by $|||\cdot||| = \left(\int_s^T \mathbf{E}\|\cdot\|^2 dt\right)^{1/2}$.

$a \wedge b$ denotes the minimum of the two real numbers a and b.

$\mathcal{M}(\mathbf{R}^n)$ denotes the set of probability measures on the set $\mathbf{R}^n$.

1. Introduction

1.1 The concept of an uncertain system

In designing a robust control system, one must specify the class of uncertainties the control system is to be robust against. Within the modern control framework, one approach to designing robust control systems is to begin with a plant model which not only models the nominal plant behavior but also models the type of uncertainties which are expected. Such a plant model is referred to as an uncertain system.

There are many different types of uncertain system model and the form of model to be used depends on type of uncertainty expected and the tractability of robust control problem corresponding to this uncertain system model. In many cases, it is useful to enlarge the class of uncertainties in the uncertain system model in order to obtain a tractable control system design problem. This process may however lead to a conservative control system design. Thus, much of robust control theory can be related to a trade off between the conservatism of the uncertain system model used and the tractability of the corresponding robustness analysis and robust controller synthesis problems.

Some commonly occurring uncertainty descriptions are as follows:

(i) A constant or time varying real parameter representing uncertainty in the value of a parameter in the system model; e.g., uncertainty in a resistance value in an electrical circuit.

(ii) A transfer function representing the uncertainty which might arise from neglecting some of the system dynamics; e.g., the effect of neglecting parasitic capacitances in an electrical circuit.

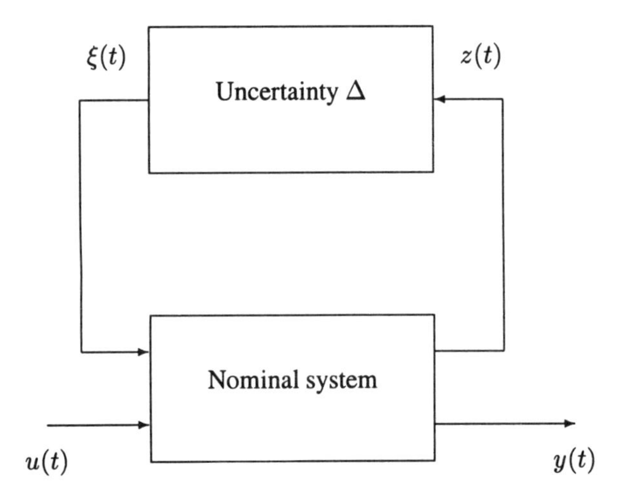

Figure 1.1.1. Uncertain system model block diagram

(iii) A nonlinear mapping which represents the uncertainty due to neglected nonlinearities.

An important class of uncertain system models involves separating the nominal system model from the uncertainty in the system in a feedback interconnection; see Figure 1.1.1. Such a feedback interconnection between the nominal model and the uncertainty is sometimes referred to as an linear fractional transformation (LFT); e.g., see [48].

Whatever the form of the uncertainty Δ in an uncertain system, it is typically a quantity which is unknown but bounded in magnitude in some way. That is, we do not know the value of the uncertainty but we know how big it can be. Some typical uncertainty bounds are as follows:

(i) A real time-varying uncertain parameter $\Delta(t)$, can be bounded in absolute value:

$$|\Delta(t)| \leq \mu \quad \text{for all } t \geq 0. \tag{1.1.1}$$

This includes time-varying norm-bounded uncertainty blocks; e.g., see [102].

(ii) If $\Delta(s)$ is an uncertain transfer function, we could bound its magnitude at all frequencies:

$$|\Delta(j\omega)| \leq \mu \quad \text{for all } \omega. \tag{1.1.2}$$

This amounts to a bound on the H^∞ norm of the transfer function $\Delta(s)$; e.g., see [51].

Also, note that each of the above uncertainty bounds can be extended in a straightforward way to a matrix of uncertainties.

Uncertain systems of the types described above have been widely studied. Furthermore, useful solutions have been obtained to various robust control problems for these classes of uncertain systems; e.g., see [150, 142, 102, 156, 157, 146, 51, 58]. However, these classes of uncertain systems do have some deficiencies.

Consider the class of systems with norm bounded uncertainties of the form (1.1.1). Although this class of uncertain systems allows for time-varying uncertain parameters and also for cone-bounded memoryless nonlinearities, it does not allow for dynamic uncertainties such as might arise from unmodeled dynamics. Another problem concerns the results which can be obtained for this class of uncertainties. Necessary and sufficient conditions for quadratic stabilizability of such systems can be obtained. However, the notion of quadratic stabilizability does not directly relate to the behavior of the system. Also, although tight state-feedback guaranteed cost results can be obtained (see [156, 157] and Section 5.2 of this book), the generalization to the output-feedback case is less satisfactory and only leads to upper bounds. All of these facts motivate the integral quadratic constraint (IQC) uncertainty description which will be described in the sequel.

H^∞ type uncertainty constraints of the form (1.1.2) provide further motivation for an uncertainty description in terms of integral quadratic constraints. First consider the transfer function uncertainty block in Figure 1.1.1. From the definition of the norm on the Hardy space H^∞, it follows that the frequency domain bound (1.1.2) is equivalent to the *frequency domain* integral quadratic constraint

$$\int_{-\infty}^{\infty} |\hat{\xi}(j\omega)|^2 d\omega \leq \int_{-\infty}^{\infty} |\hat{z}(j\omega)|^2 d\omega \tag{1.1.3}$$

which must be satisfied for all signals $z(t)$ provided these integrals exist. Here, $\hat{\xi}(s)$ is the Laplace transforms of the uncertainty input $\xi(t)$. Also, $\hat{z}(s)$ is the Laplace transform of the uncertainty output $z(t)$. Furthermore, $z(t)$ is the input to the uncertainty block Δ and $\xi(t)$ is the output of this uncertainty block.

It is well known (e.g., see [124]) that integral quadratic constraints of the form of (1.1.3) may be applied in the analysis of nonlinear systems. In particular, a number of the system properties such as passivity, structural information about perturbations, energy constraints on exogenous perturbations, saturations, uncertain delays etc., allow for a description in terms of suitable constraints of the form (1.1.3). For further details, we refer the reader to the frequency domain integral quadratic constraints given in [124]. However as mentioned in [124], there is an evident problem in using this form of uncertainty description. This is because the condition (1.1.3) makes sense only if the signals $\xi(t)$ and $z(t)$ are square integrable. Reference [124] points out how this problem can be resolved in certain situations. For example, in the case where the nominal system is stable, this problem can be easily overcome. However in stabilization problems, it is possible that the uncontrolled system is *unstable* and then its inputs and outputs will not be square summable unless the system is controlled using a stabilizing controller. In this book, we show that the problems arising from uncertain systems with an unstable nominal can be overcome to a large extent by using *time domain versions* of the integral quadratic constraint uncertainty description.

An uncertain system model must be chosen so that it captures the essential features of the real system and the uncertainty in that system. Also, the uncertainty class must be chosen so that it leads to a tractable solution to the control problem under consideration. These facts have led us to use the time domain integral quadratic constraint uncertainty descriptions presented in Section 2. Since this form of uncertainty description prevails throughout the book, from now on we will usually drop the words "time domain". That is, we will use the expression *Integral Quadratic Constraint (IQC)* to denote time domain integral quadratic constraints.

There are a number of advantages in dealing with uncertain systems defined by IQCs. The class of uncertainties satisfying an IQC is richer than the class of uncertainties satisfying the corresponding norm bound condition. Also, a time domain IQC can be applied to model uncertainty in finite-horizon problems. Among other advantages of the time domain IQC uncertainty description, we note that this uncertainty description allows us to model structured uncertain dynamics in systems subject to stochastic noise processes. Our motivation for considering uncertain systems which are subject to stochastic noise process disturbances is twofold. In the first instance, many engineering control problems involve dealing with systems which are subject to disturbances and measurement noise which can be well modeled by stochastic processes. A second motivation is that in going from the state-feedback linear quadratic regular (LQR) optimal control problem to the measurement feedback linear quadratic Gaussian (LQG) optimal control problem, a critical change in the model is the introduction of noise disturbances. Hence, it would be expected that in order to obtain a reasonable generalization of LQG control for uncertain systems, it would be necessary to consider uncertain systems which are subject to disturbances in the form of stochastic noise processes.

Note that stochastic extensions to the integral quadratic constraint uncertainty description provide a possible approach to the problem of non-worst case robust controller design. Although the worst case design methodology has proved its efficacy in various engineering problems, it suffers from the disadvantage that the designer lacks the opportunity to discriminate between "expected" uncertainties and those uncertainties which seldom occur. Indeed, the standard deterministic worst case approach to robust controller design presumes that all uncertainties are equally likely. That is, one can think of the uncertainty as arising from a uniformly distributed random variable taking its values in the space of uncertainties. However, this may not accurately represent the uncertainty in the system under consideration. For example, it may have been determined that the uncertainties have a probability distribution other than uniform and the designer may wish to make use of this *à priori* information about the distribution of the uncertainties. In this case, it is useful to consider a stochastic uncertain system in which the uncertainty has a stochastic nature. Hence, a suitable description of stochastic uncertainty is required. To this end, in this book we consider stochastic integral quadratic constraints or more generally, stochastic relative entropy constraints.

1.2 Overview of the book

A fundamental idea recurring in this book is the connection between the Riccati equation approach to H^∞ control and the problem of stabilizing an uncertain system. For systems with norm-bounded uncertainty, this idea is central in the problem of quadratic stabilization; e.g., see [143, 102]. However, it is often required to design a control system which is not only stable but also ensures the satisfaction of certain performance conditions. This additional requirement leads us to consider a number of control problems which address the issue of robustness together with that of optimal or guaranteed performance. These problems are as follows:

- The infinite-horizon and finite-horizon optimal guaranteed cost control problems;
- The problems of absolute stability, absolute stabilizability and structured dissipativity for nonlinear systems;
- Robust LQR and LQG control of stochastic systems; and
- The issue of nonlinear versus linear control in robust controller design.

Each of these problems will require a suitable definition of the underlying uncertain system. The essential features which we will try to capture when formulating the corresponding uncertain system definitions are as follows. First, the nominal system should be continuous time, linear (or in some cases nonlinear) and time-invariant (or in some cases time-varying). Second, the uncertainty in the system should be structured (or in some cases unstructured). Here the term 'structured uncertainty' refers to the fact that the uncertainty can be broken up into a number of independent uncertainty blocks. Also, it is required that each uncertainty block be subject to some sort of induced norm bound. Provided these conditions are satisfied, the uncertainty class will be then chosen so that it will lead to a tractable solution of the corresponding robust control problem. The definitions of the uncertainty classes to be considered, as well as further discussion of these definitions constitutes the subject of Chapter 2. This collection of uncertainty models includes the traditional definition of norm-bounded uncertainty as well as definitions exploiting an Integral Quadratic Constraint to bound the uncertainty in the system.

The class of uncertain systems satisfying integral quadratic constraints was developed by Yakubovich and has been extensively studied in the Russian literature; e.g., see [256, 258, 259, 260, 173, 241, 106, 212]. Associated with this class of uncertain systems is a corresponding stability notion referred to as absolute stability. One of the aims of this book is to bring this class of uncertain systems and the corresponding stability notion to the attention of readers of the Western literature. Also, we extend this class of uncertain systems to systems containing a control input. This enables us give controller synthesis results by using H^∞ control theory (a theory which has developed exclusively in the Western literature).

Although this book focuses on continuous-time uncertain systems, it is worth noting that many ideas of this book have already been applied to discrete-time systems. For example, the papers [157, 131, 129] provide discrete time versions of the results presented in Sections 5.2, 5.3 and 3.5, respectively. Also, the results of references [153, 236] are discrete-time versions of the results presented in Sections 8.4 and 8.5.

Some of the advantages in dealing with the class of uncertain systems defined by an integral quadratic constraint have already been discussed above. Also, it should be noted that the notions of absolute stability and absolute stabilizability associated with IQCs are more natural stability concepts than those of quadratic stability and quadratic stabilizability. Indeed, the notion of quadratic stability is defined in terms of the existence of a single quadratic Lyapunov function for the uncertain system in question. However, the requirement of a single quadratic Lyapunov function is not directly motivated in terms of the behavior of the uncertain system solutions. In contrast, the definition of absolute stability is directly related to the stability properties of the solutions to the uncertain system. Furthermore, the reader will see from the definitions that the absolute stability of an uncertain system also implies common stability properties such as asymptotic stability, uniformly bounded $\mathbf{L}_2$ gain, etc. That is why in this book, we shall be mainly interested in absolutely stabilizing controllers whenever the control problem being considered involves stabilization as a goal.

In all of the above mentioned problems, our method of controller design will rely on the interpretation of a worst case design in the presence of uncertainty as a game type minimax optimization problem. In this game type problem, the designer is considered as the "minimizing player" who endeavors to find an optimal control strategy to maintain a certain guaranteed level of performance for the closed loop system in the face of uncertainty. In contrast, the uncertainty in the underlying plant impairs the performance of the closed loop system and may be considered as the "maximizing player" in the game problem. The set of admissible uncertainty inputs described by either bounds on the norm of the uncertainty or by integral quadratic constraints (see Chapter 2), restricts the choice of the maximizing strategies in this game. This motivates us to think of the control problems considered in this book as constrained optimization problems. The advantage of this approach is that it allows one to readily convert problems of robust control into mathematically tractable game type minimax optimization problems. These problems can then be solved using H^∞ control methods.

In Chapter 3, we present some of the background theoretical results required in proving our main robust control results. Section 3.2 familiarizes the reader with the standard H^∞ control problem and with the game theoretic approach to H^∞ control. We discuss infinite-horizon and finite-horizon state-feedback and output-feedback H^∞ control problems. These problems are considered for linear time-invariant systems with zero initial conditions. We also present some extensions to the standard H^∞ control results involving time-varying systems and H^∞ problems with transients. In Section 3.3, we collect some preliminary results concerning risk-sensitive control which will be exploited in Chapter 8. Results concerning Riccati equations associated with H^∞ control theory and risk-sensitive control theory are given in Section 3.1. In presenting these results on H^∞ control and risk-sensitive control, we do not aim to present the most general available results in this area but rather present results as needed for the robust control techniques contained in this book.

The objective of Section 3.5 of Chapter 3 is quite specific. The main result of this section establishes a connection between output-feedback H^∞ control and absolute stabilizability via nonlinear and linear output-feedback control. Although this result deserves particular attention as it solves the problem of robust output-feedback control for uncertain systems with unstructured uncertainty, for the moment it is more important for the reader to note that the approach developed in this section is one which will be exploited throughout the book.

The availability of rigorous mathematical tools for converting a constrained optimization problem into an intermediate unconstrained minimax optimization problem is a critical technical issue in the many of the robust control problems considered in this book. It is well known in the theory of convex optimization that this procedure can be achieved by the use of Lagrange multipliers; e.g., see [118]. A similar technical result which allows one to convert the constrained optimization problems under consideration into corresponding unconstrained game type problems is used in this book. Following the Russian historical tradition, we refer to this result as the S-procedure. This result is related to Finsler's Theorem concerning pairs of quadratic forms; e.g., see [239] and also [255]. In 1960s and 1970s, the S-procedure was extensively used for solving absolute stability problems. Megretski and Treil [127] and Yakubovich [261] pointed out connections between the S-procedure and H^∞ robustness analysis.

Note that almost all of the problems considered in this book require a corresponding version of the S-procedure which is consistent with the chosen uncertainty description. This fact has motivated a number of extensions to the S-procedure results of [127, 261]. Indeed, the reader will find a collection of the S-procedure results available in the Russian and Western literature presented in Chapter 4. This collection includes the earliest versions of the S-procedure, a result due to Yakubovich on the S-procedure in Hilbert space which encompasses the corresponding result of Megretski and Treil as a special case, and also recent results by authors. As noted above, the S-procedure is analogous to the Lagrange multiplier technique in convex optimization. In particular, it leads to an unconstrained minimax optimization problem involving a number of "scaling" parameters which are analogous to Lagrange multipliers. However unlike Lagrange multipliers, these parameters are restricted to those which give rise to a suitable solution to an associated (algebraic or differential) game type Riccati equation. This game type Riccati equation arises from treating the unconstrained minimax optimization problem as an H^∞ control problem in which the uncertainty inputs are correspondingly scaled using the multiplier parameters. In fact, this connection between absolute stabilizability and the H^∞ control of a scaled system exhibits a common feature found in much of the robust control theory presented in this book.

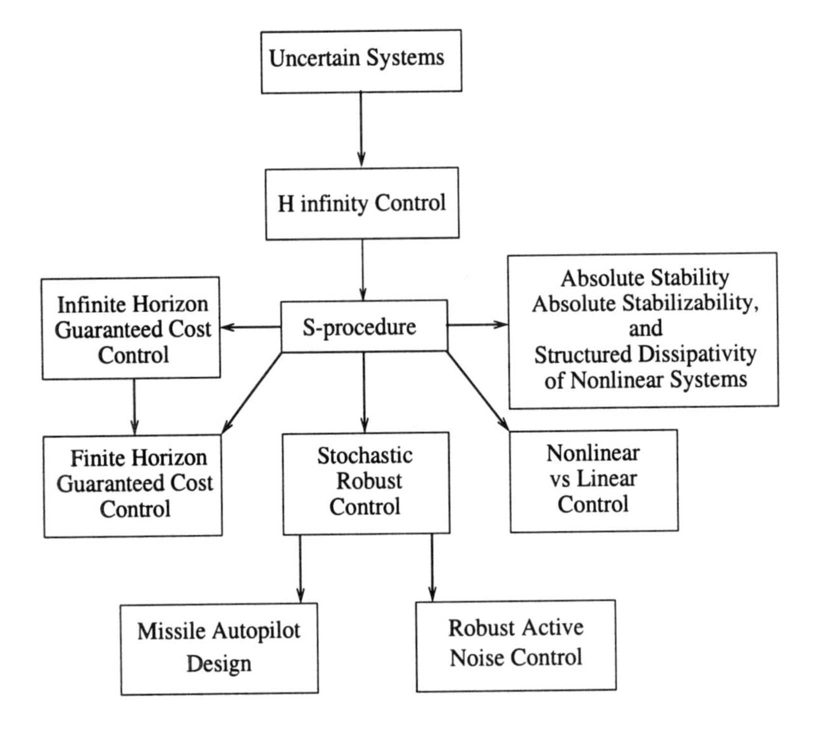

Figure 1.2.1. The organization of the book

We now summarize three key features of the robust control methods described in this book :

- A description of the uncertainty is given in terms of a norm bound or terms of integral quadratic constraints;
- The connection between stabilizability and H^∞ control is exploited to allow for the derivation of a stabilizing controller from a corresponding Riccati equation;
- The S-procedure is applied to convert constrained minimax optimization problems into mathematically tractable game type H^∞ optimization problems.

The organization of the book is shown in Figure 1.2.1. The book is written in such a way that after the reader is familiar with the concepts of admissible uncertainty and the S-procedure in presented in Chapters 2 and 4, any of the remaining chapters in the book should be accessible.

We now are in a position to overview the main problems considered in the book.

Optimal guaranteed cost control of uncertain systems

In this problem, one is concerned with designing a control system which is not only stable but also guarantees an adequate level of performance. The approach presented in this book involves extending the state-feedback linear quadratic regulator to the case in which the underlying system is uncertain.

It is known that the linear quadratic regulator provides good robustness in terms of gain margin and phase margin; e.g., see [3]. However, as shown in the paper [208], the linear quadratic regulator is unable to provide robustness against more general types of uncertainty such as considered in the robust control literature; e.g., see [128, 110, 5, 150, 143, 102, 138, 252, 253, 41]. In particular, these papers consider uncertain system models involving parameter uncertainty.

In Chapter 5, a number of approaches are presented for extending the linear quadratic regulator to the case in which the underlying plant is modeled as an uncertain system. We discuss two versions of this problem. In the first case, the class of uncertainties in the system are those termed as norm-bounded uncertainties; see Section 5.2. The main result which is presented on this problem is a Riccati equation approach to the construction of an "optimal" state-feedback quadratic guaranteed cost controller which minimizes a certain upper bound on the closed loop value of a quadratic cost function. This result originally appeared in the papers [154, 155, 156]. In Section 5.3, we extend the results of Section 5.2 in that structured uncertainty is considered. That is, we allow for multiple uncertainties in the underlying system as shown in Figure 1.2.2. However unlike Section 5.2, each uncertainty block in Figure 1.2.2 is assumed to satisfy a certain IQC. It is shown that for the class of uncertain systems considered in Section 5.3, a state-feedback controller obtained from a correspondingly scaled Riccati equation actually solves a minimax optimal control problem. This result originally appeared in the paper [184]. In Section 5.4, the results of Section 5.3 are extended to the output-feedback case. The main result presented in this section gives a necessary and sufficient condition for existence of an output-feedback guaranteed cost controller. This result is given in terms of an algebraic Riccati equation and a Riccati *differential* equation. Note that the results of Section 5.4 apply only to uncertain systems with a single uncertainty block. The main result presented in Section 5.4 originally appeared in the papers [158, 179, 200].

Note that both the quadratic stability and absolute stability approaches to guaranteed cost control use a fixed quadratic Lyapunov function to give an upper bound on a quadratic cost functional. However, the question remains as to whether a better controller could be obtained if one goes beyond the class of quadratic Lyapunov functions. In Section 5.5, we give a positive answer to this question. Specifically, we show that for a certain class of uncertainties, a guaranteed cost controller can be constructed by solving a modified Riccati equation which depends on certain scaling parameters, additional to those considered in the parameter-dependent Riccati equation of Section 5.3.

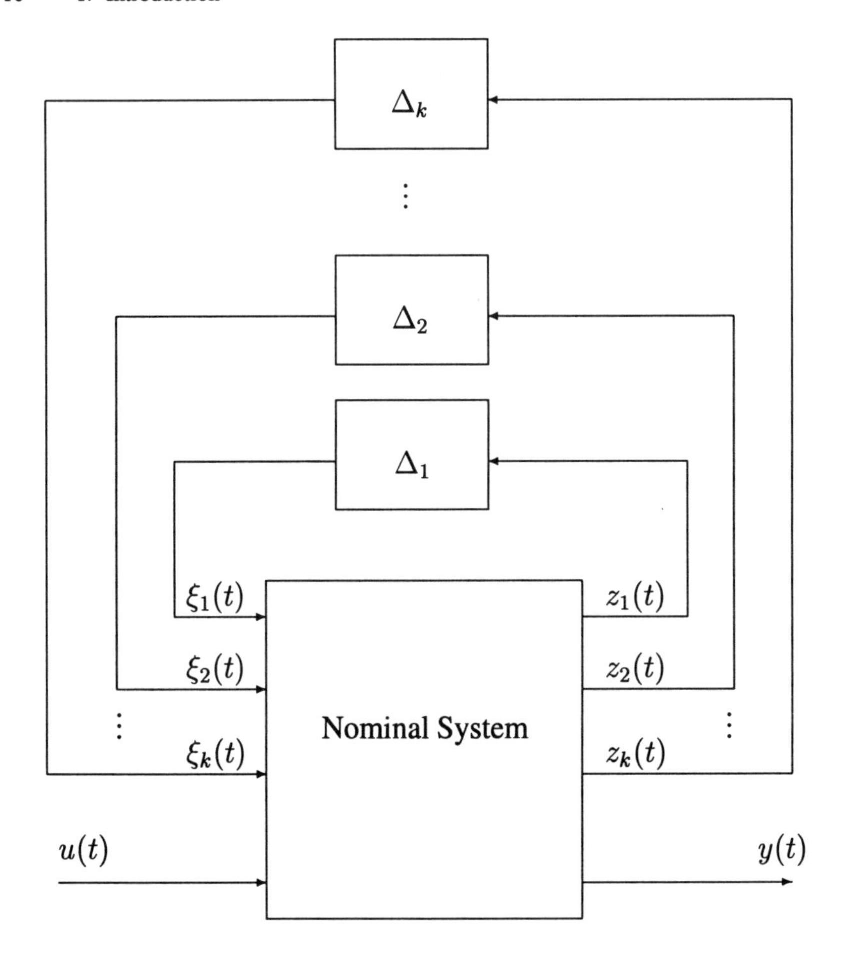

Figure 1.2.2. Block diagram of an uncertain system with structured uncertainty.

Finite-horizon guaranteed cost control

The results of Chapter 5 apply only to guaranteed cost control problems in which the underlying nominal system is time-invariant and the optimal control problem is considered on an infinite time horizon. Unfortunately, the problem of directly extending the results of Chapter 5 to a finite-horizon or time-varying state-feedback guaranteed cost control problem appears to be mathematically intractable. The results presented in Chapter 6, overcome this difficulty by considering a new class of uncertain systems with structured uncertainty. These uncertain systems are referred to as uncertain systems with an "Averaged Integral Quadratic Constraint". Within the framework of this class of uncertain systems, it is possible to obtain a mathematically tractable solution to the finite-horizon time-varying minimax optimal control problem. A state-feedback version of this result is presented in Section 6.2 and an output-feedback version of this result is presented in Section 6.3.

The main results of Sections 6.2 and 6.3 are necessary and sufficient conditions for the existence of a guaranteed cost controller for an uncertain system with structured uncertainties satisfying an averaged integral quadratic constraint. These results are given in terms of the existence of suitable positive-definite solutions to a pair of parameterized Riccati differential equations. If such solutions exist, it is shown that they can be used to construct a corresponding guaranteed cost controller. In the state-feedback case, this is a linear time-varying static state-feedback controller and in the output-feedback case this is a time-varying dynamic output-feedback controller. The main result presented in Section 6.2 was originally published in the papers [181, 191]. Also, the main result presented in Section 6.3 was originally published in the papers [182, 198].

In section 6.4, we consider a new robust control problem referred to as "robust control with a terminal state constraint." Given an uncertain linear system, it is desired to construct an output-feedback controller which steers a given initial state to the vicinity of the origin in a specified time. Also, it is required that the control energy required to achieve this is bounded. Again, the class of uncertain systems considered in this section are uncertain time-varying linear systems in which the uncertainty is described by an integral quadratic constraint. For this class of uncertain systems, we consider the robust control problem with a terminal state constraint described above. Such a robust control problem may arise in a number of practical situations. For example, it may be desired to steer a robotic arm from a given initial state to an equilibrium point with a specified degree of accuracy and within a specified time; e.g., see [34]. Furthermore, in such a situation, the mathematical model describing the robotic arm may be uncertain due to variations in load mass and the truncation of high frequency flexible modes. Thus, it is desirable to find a robust controller to steer the initial state into the terminal bounding set.

The main result presented in Section 6.4 gives a necessary and sufficient condition for an uncertain linear system to be robustly controllable with a given terminal state constraint; see Section 6.4 for a rigorous definition of the notion of robust controllability. This result is established by converting the problem into a finite-horizon H^∞ control problem which can be solved using the results presented in Chapter 3; see also [100] where this solution to the finite-horizon H^∞ problem was originally published. As in the majority of the problems in this book, a key step in converting the robust control problem into a corresponding H^∞ control problem is the use of the S-procedure. The results presented in Section 6.4 were originally published in the papers [187, 195].

Another problem considered in Chapter 6 is a problem of robust control with rejection of harmonic disturbances; see Section 6.5. In many control problems, there is a requirement to reject harmonic disturbances of a known frequency as well as a requirement for robustness against uncertainty. Such harmonic disturbances may arise, for example, from vibrations in rotating machinery with a known angular velocity, or as a result of electrical interference at line frequency.

Also, this problem finds applications in the control of oil drilling; e.g., see [207]. A standard approach to dealing with the presence of harmonic disturbances of a known frequency is to augment the plant with dynamics corresponding to the known frequencies. This approach is referred to as the internal model principle; e.g., [40]. This approach has been extended in [1] to uncertain linear systems with uncertainty satisfying a H^∞ norm bound.

In Section 6.5, we consider a finite-horizon control problem in which a quadratic cost function is to be minimized for a system subject to harmonic disturbances. Furthermore, the system is also subject to uncertainties satisfying an integral quadratic constraint. The results of Section 6.5 are related to those of [1]. However, we construct a controller which minimizes the bound on a quadratic cost functional in the presence of harmonic disturbances. In contrast, no quadratic cost functional was considered in [1]. It should also be noted that the papers [262, 115] also consider a control problem with a quadratic cost function for a system subject to harmonic disturbances. However, in these papers, the system is not uncertain. The results presented in Section 6.5 were originally published in the papers [188, 189].

Absolute stability, absolute stabilizability and structured dissipativity of nonlinear systems

Recent advances in H^∞ control (e.g., see the results presented in Chapter 3) and in theory of linear matrix inequalities (LMIs) (e.g., see [26]) have allowed a revisiting of the Lur'e-Postnikov approach to absolute stability of nonlinear control systems. The systems of interest include both the Lur'e type systems with sector bounded nonlinearity and also systems with a more general type of uncertainty [15, 17, 16, 76, 28]. Instead of the Kalman-Yakubovich-Popov lemma, the modern approach exploits positive-definite solutions to certain algebraic Lyapunov or Riccati equations.

In Section 7.2, we present one such extension of the approach of Lur'e and Postnikov to the problem of robust stabilization. This involves a new notion of "absolute stabilizability with a Lyapunov function of the Lur'e-Postnikov form". This notion extends the absolute stability notion of Lur'e and Popov (e.g., see [137]) to a corresponding stabilizability notion for output feedback control. The main result of this section is a necessary and sufficient condition for absolute stabilizability with a Lyapunov function of the Lur'e-Postnikov form. This condition is computationally tractable and involves the solution of a parameter dependent pair of algebraic Riccati equations. Previous results in this area such as in [73, 74, 75] have presented sufficient conditions for the existence of a solution to this problem. The main results presented in Section 7.2 originally appeared in the papers [178, 183, 232]. In Section 7.3, we consider connections between the areas of robust stability of nonlinear uncertain systems with structured uncertainty, the absolute stability of nonlinear uncertain systems, and the notion of dissipativity for nonlinear systems. Note that much research on the problem of robust stability for uncertain systems containing structured uncertainty can be traced to the papers [171, 51, 98, 209, 125]. In all of these papers, the underlying uncertain system is required to be linear. In Section 7.3, it is shown that these ideas can be extended into the realm of nonlinear systems. A feature of the uncertainty model considered in Section 7.3 and the corresponding definition of absolute stability is that it is closely related to the notion of dissipativity which arises in the modern theory of nonlinear systems; e.g., see [248, 79, 80, 81, 33]. Specifically, it is shown that there is a direct connection between the absolute stability of an uncertain nonlinear system containing a number of uncertain nonlinearities and the notion of *structured dissipativity* which is introduced in Section 7.3. The main results presented in Section 7.3 were originally published in the papers [180, 190].

Robust control of stochastic systems

Chapter 8 extends the ideas of minimax optimal control and absolute stabilization to stochastic uncertain systems. In Sections 2.4.1 and 2.4.2, a number of motivating examples are presented which give insight into how such stochastic robust control problems may arise.

In Section 8.3, we address a minimax optimal control problem for stochastic systems with multiplicative noise. We introduce a rather general model of stochastic uncertainty which naturally extends the deterministic structured uncertainty model [102, 184, 252] by allowing for stochastic perturbations. The main problem considered in Section 8.3 is to find a linear static state-feedback controller yielding a prescribed level of performance in the face of stochastic structured uncertainty. The class of controllers considered are static state-feedback controllers, which absolutely stabilize the system in a stochastic sense. However, the key property of the controller derived here is that its construction is based on the stabilizing solution of a *generalized* Riccati equation related to a stochastic H^∞ control problem and to a stochastic differential game. A solution to the stochastic H^∞ control problem which extends H^∞ control ideas to stochastic systems with multiplicative noise is provided in Section 8.2; see also [227, 234]. The main result presented in Section 8.3 originally appeared in the papers [228, 234].

The problem considered in Section 8.4 is motivated by the desire to develop a robust linear quadratic Gaussian optimal control methodology. We address the problem of finding an output-feedback minimax optimal controller for the case of partially observed stochastic systems. This problem is considered in the finite-horizon case. In fact, the results of this section provide a complete solution to the problem of robust LQG control on finite time intervals.

Note that to a certain extent, the issue of robustness of LQG control has been addressed within the framework of mixed H_2/H^∞ control [206, 49, 134, 72]. Specifically, the mixed H_2/H^∞ control methodology allows one to minimize a *nominal* performance measure subject to a robust stability constraint. In this sense, the solution to the robust LQG control problem obtained within the mixed H_2/H^∞ control framework may be considered as incomplete since it does not address the *robust* performance of the proposed controller. Another potential drawback of the H_2/H^∞ control methodology is that the mixed H_2/H^∞ control technique assumes a Gaussian model of noise disturbances. However, in many practical control problems, a Gaussian noise description may not be realistic. A more suitable way of describing noise disturbances may be to treat the disturbances as uncertain stochastic processes. Formalizing this idea leads to the concept of a *stochastic uncertain system* introduced in the papers [151, 152, 229]. The stochastic uncertain system framework introduced in these references involves finding a controller which minimizes the worst case performance in the face of uncertainty which satisfies a certain stochastic *relative entropy uncertainty constraint*. This constraint describes restrictions on the probability distribution of the stochastic uncertainty input. The definition and further comments concerning the relative entropy uncertainty constraint are given in Chapter 2.

The approach used in Section 8.4 involves a version of S-procedure theorem which is used to convert a constrained optimization problem into an unconstrained dynamic game type problem. To solve this stochastic game problem, we refer to certain duality connections between risk-sensitive stochastic control problems and stochastic dynamic games [39, 53, 170]. In Section 8.4, this duality is used together with the known solutions of corresponding risk-sensitive stochastic control problems (e.g., see Section 3.3 and [12, 139]). In fact, the risk-sensitive control problem that arises in Section 8.4 plays the same role in our analysis as H^∞ control plays in deterministic problems. The results presented in Section 8.4 originally appeared in the papers [230, 235]. These results are in turn based on the corresponding discrete time results originally published in the paper [152].

The infinite-horizon version of the problem considered in Section 8.4 is presented in Section 8.5. It is worth noting that as we proceed from the finite-horizon case to the infinite-horizon case, the fact that the systems under consideration are those with additive noise becomes important. The solutions of such systems do not necessarily belong to $\mathbf{L}_2[0,\infty)$. Hence, we shall be interested in the time-averaged properties of the system trajectories. In particular, we will need a definition of absolute stabilizability which properly accounts for this feature of the systems under consideration.

The main result presented in Section 8.5 is a robust LQG control synthesis procedure based on a pair of algebraic Riccati equations arising in risk-sensitive optimal control; see Section 3.3 and [139]. It is shown that solutions to a certain scaled risk-sensitive control problem lead to a controller which guarantees a certain upper bound on the time-averaged performance of the closed loop stochastic uncertain system. As in Section 8.4, we employ the relative entropy between the perturbation probability measure and the reference probability measure to define the class of admissible uncertainties. The main result of Section 8.5 originally appeared in the paper [233].

Nonlinear versus linear control

In the area of robust control, there has been a significant interest in the question of whether any advantage can be obtained via the use of nonlinear control as opposed to using linear control. In particular, it has been shown that in a number of problems of robust stabilization, an advantage can be obtained by using nonlinear or time-varying control as opposed to a linear time-invariant controller; e.g., see [140, 219]. However, in some special cases, it has been shown that no advantage can be obtained by using nonlinear or time-varying control; see [160, 103, 99, 168, 97]. For the particular case where the uncertainty entering into the system is subject to a single norm bound constraint, Khargonekar and Poolla formulated the following *Plant Uncertainty Principle*: *"In robust control problems for linear time-invariant plants, nonlinear time-varying controllers yield no advantage over linear time-invariant controllers if the plant uncertainty is unstructured"*; see [103, 168]. The connection between uncertain plants with unstructured uncertainty and the use of linear controllers was further reinforced by a counterexample given by Petersen in [140]; see also [218, 24]. The main result of [140] was an example of a system with structured uncertainty which could be quadratically stabilized via a nonlinear state-feedback controller but could not be quadratically stabilized via a linear state-feedback controller. Thus, from this result and some other existing results in the literature, it would appear that uncertainty structure is the most important issue relating to the question of nonlinear versus linear control of uncertain linear systems. However, the results presented in Section 3.5 indicate that the most important issue is not the uncertainty structure but rather the "richness" of the class of uncertainties allowed in the uncertain system. That is, if the uncertainty is defined by an integral quadratic constraint, the main result of Section 3.5 shows that absolute stabilizability via nonlinear output-feedback control for an uncertain system implies that the system can be absolutely stabilized via linear output-feedback control. Furthermore, it is shown that under certain conditions, the absolutely stabilizing controllers are those that solve an associated H^∞ control problem.

In Chapter 9, the connection between the Riccati equation approach to H^∞ control and the problem of stabilizing an uncertain system is further investigated from the point of view of nonlinear versus linear control. In Section 9.2, we discuss an extension of this problem to the case where the uncertainty in the underlying system is structured as shown in Figure 1.2.2. The paper [140] shows that the above plant uncertainty principle does not hold in the case of structured norm-bounded uncertainty in the underlying system. The result presented in Section 9.2 however shows that the above plant uncertainty principle remains true if the class of admissible uncertainty is extended to include structured uncertainty defined by a structured integral quadratic constraint uncertainty description. Specifically, it seems that it would be appropriate to replace the plant uncertainty principle of Khargonekar and Poolla with the following: *In robust control problems for linear time-invariant plants, nonlinear controllers yield no advantage over linear controllers if the class of plant uncertainties is sufficiently rich so as to include nonlinear time-varying dynamic uncertainties satisfying specific IQCs*. The main result presented in Section 9.2 originally appeared in the papers [176, 185].

As an illustration of the ideas of Section 9.2, we apply the results of this section to the problem of decentralized control. Although ideas of decentralized control for large-scale systems have attracted much attention in the literature (e.g., see [213]), the problem of designing decentralized control schemes which are able to cope with a wide range of uncertain perturbations still remains challenging. A number of attempts have been made to solve this problem using advances in H^∞ control theory and quadratic stabilization; e.g., see [37, 38, 243, 263, 269, 270]. A number of sufficient conditions for decentralized quadratic stabilization have been obtained in these and other papers using a Riccati equation approach. We will review these results in Section 9.3. From the existing literature and our previous experience of solving non-decentralized robust control problems, one may conclude that in order to obtain a necessary and sufficient condition for robust decentralized stabilizability with disturbance attenuation via a Riccati equation approach, one needs to broaden the class of uncertainty. The results presented in Section 9.3, show that by allowing for a broader class of uncertainty in which the perturbations and interaction between subsystems satisfy certain integral quadratic constraints, such a necessary and sufficient condition can be obtained. This result is also given in terms of a set of algebraic Riccati equations. The main results presented Section 9.3 were originally published in the paper [238].

The results of Sections 3.5 and 9.2 are concerned with the case of full order dynamic output-feedback controllers or full state feedback controllers. Section 9.4 presents a further advance concerning the nonlinear versus linear control problem. In this section, we present a corresponding result for the case of static output-feedback controllers and fixed-order dynamic output-feedback controllers. To this end, we define a notion of "robust stabilizability with a quadratic storage function". This notion is a slightly stronger notion than that of quadratic stabilizability such as defined in Section 3.4 and [168] and closely related to the notion of dissipativity such as considered in [248, 79]. Indeed, if attention were restricted to linear controllers, then it can be shown that our notion of robust stabilizability with a quadratic storage function is actually equivalent to the notion of quadratic stabilizability. Within this framework, the main result of Section 9.4 shows that if an uncertain system is robustly stabilizable with a quadratic storage function via a nonlinear time-varying controller, then it will also be robustly stabilizable with a quadratic storage function via a linear time-invariant controller of the same order. The main result presented in Section 9.4 originally appeared in the papers [193, 194].

Another problem in which the use of integral quadratic constraints may give an advantage is the problem of simultaneous H^∞ control of a finite collection of linear time-invariant systems. This problem, which is considered in Section 9.5, has attracted a considerable amount of interest; e.g., see [25]. In particular, if one restricts attention to the case of linear time-invariant controllers, the problem of finding a useful necessary and sufficient condition for simultaneous stabilization has remained unsolved except in the case of two plants. The motivation for the simultaneous stabilization problem is derived from the fact that many problems of robust control with parameter uncertainty can be approximated by simultaneous stabilization problems. Also, the fact that the simultaneous stabilization problem is tantalizingly simple to state and yet apparently very difficult to solve has attracted many researchers. It is clear that the problem of simultaneous stabilization can be interpreted as a particular case of the problem of simultaneous H^∞ control. Therefore, the problem of simultaneous H^∞ control is more general and difficult than the problem of simultaneous stabilization. However, unlike many papers on simultaneous stabilization (e.g., see [225, 240, 266]), the results presented in Section 9.5 do not restrict attention to linear time-invariant controllers. The class of compensators being considered contains nonlinear digital controllers which update the control signal $u(t)$ at discrete times, with $u(t)$ constant between updates.

The main result presented in Section 9.5 shows that if the simultaneous H^∞ control problem for k linear time-invariant plants of orders $n_1, n_2, \ldots, n_k$ can be solved, then this problem can be solved via a nonlinear digital controller of order $n = n_1 + n_2 + \cdots + n_k + k$. This conclusion is surprising because in the problem of simultaneous stabilization of a finite collection of linear time-invariant plants, the order of the required linear time-invariant compensator may be arbitrarily large even in the case of two plants; e.g. see [225]. Therefore, the main result of this section shows that in the problem of simultaneous H^∞ control via reduced-order output feedback, the use of nonlinear control may give an advantage. The main result presented in Section 9.5 originally appeared in the paper [174].

Applications

Throughout the book, the reader will find a number of examples illustrating ideas and methods of robust controller design for uncertain systems defined by integral quadratic constraints. However, to a large extent these examples are mostly of academic interest. In Chapters 10 and 11, we address two more practical control system design problems: the problem of missile autopilot design and the problem of control of acoustic noise in ducts. In the both problems, we illustrate how the theoretical robust LQG control design ideas developed in the book may be applied in a particular practical situation. Note that the uncertain system model considered in Chapter 10 can be regarded as a physical model whereas in Chapter 11, the uncertain system model is developed using experimental data.

The missile autopilot problem considered in Chapter 10 is taken from a collection of examples for robust controller design [216]. A feature of this robust control problem is that a single controller is required to give adequate tracking performance over a range of flight conditions corresponding to different values of mass, speed and altitude. The given data in this problem is a finite collection of plant models for the missile corresponding to different flight conditions. A total least squares approach is taken to fit the data to a norm-bounded uncertain system model. Then the robust LQG control approach to controller design developed in Section 8.5, is applied to allow for the presence of uncertainty.

Chapter 11 considers the problem of robust active noise control of an acoustic duct. This approach to active noise control involves designing an LQG controller which minimizes the duct noise level at the microphone. It is known that significant uncertainties in the model used to design the controller can arise from variations in the temperature and pressure of the air in the duct, nonlinearities in the speaker behavior and inaccuracies in constructing a suitable model for controller design. In this chapter, we will concentrate on robustness with respect to the uncertainty which arises from neglecting the higher order modes of the duct. In particular, an acoustic duct is a distributed parameter system with an infinite dimensional mathematical model. We consider a finite dimensional model which gives a good approximation to the duct behavior over a finite frequency range. The neglected "spillover" dynamics are then treated as uncertainty. Our approach to modeling this uncertainty is a frequency weighted multiplicative uncertainty approach. This uncertainty is then overbounded by a stochastic uncertain system of the type considered in Section 8.5. This enables us to design a controller which is robust with respect to the specified uncertainties, using the minimax LQG approach developed in Section 8.5.

2. Uncertain systems

2.1 Introduction

As mentioned in Chapter 1, the uncertain system models we will deal with in this book originate from the linear fractional transformation shown in Figure 1.1.1 on page 2 and Figure 1.2.2 on page 10. In Chapter 1, we discussed two typical bounds on the uncertainty Δ in the uncertain systems shown in these figures. This allows one to describe uncertainties arising either in the form of time-varying norm-bounded blocks satisfying (1.1.1) or in the form of transfer function blocks in which the H^∞ norm is bounded as in condition (1.1.2). The H^∞-norm bound uncertainty description is closely related to the frequency domain integral quadratic constraint (IQC) uncertainty description. The limitations of these two forms of uncertainty description were discussed briefly in Chapter 1.

In this chapter, we introduce mathematically rigorous definitions of the uncertainty classes referred to in Chapter 1 as norm-bounded uncertainties and uncertainties satisfying integral quadratic constraints. Also, further generalizations of these uncertainty classes will be presented. In the sequel, it will be seen that the class of uncertainties defined using an IQC uncertainty description is normally more general and encompasses those uncertainties which are defined in terms of a corresponding norm-bound constraint. However, for the sake of completeness, we begin with the definition of norm-bounded uncertainty given in Section 2.2.

2.2 Uncertain systems with norm-bounded uncertainty

We now formally define a class of uncertain systems with norm-bounded time-varying uncertainty. In the continuous-time case, this class of uncertain systems is described by the following state equations (e.g., see [150, 142, 102, 156, 157, 146, 252, 253]):

$$\begin{aligned}\dot{x}(t) &= [A + B_2\Delta(t)C_1]x(t) + [B_1 + B_2\Delta(t)D_1]u(t);\\ y(t) &= [C_2 + D_2\Delta(t)C_1]x(t) + D_2\Delta(t)D_1u(t)\end{aligned} \tag{2.2.1}$$

where $x(t) \in \mathbf{R}^n$ is the *state*, $u(t) \in \mathbf{R}^m$ is the *control input*, $y(t) \in \mathbf{R}^l$ is the *measured output*, and $\Delta(t) \in \mathbf{R}^{p\times q}$ is a *time varying matrix of uncertain parameters* satisfying the bound

$$\Delta(t)'\Delta(t) \leq I. \tag{2.2.2}$$

Also in the above equations, A, B_1, B_2, C_1, C_2, D_1 and D_2 are matrices of suitable dimensions.

In the sequel, it will be convenient to use another form of equation (2.2.1). Let $z(t) = C_1x(t) + D_1u(t)$ denote the *uncertainty output* of the system (2.2.1) and let

$$\xi(t) = \Delta(t)z(t) \tag{2.2.3}$$

be the *uncertainty input*. Note that for the uncertainty block in the block-diagram in Figure 1.1.1, $z(\cdot)$ is the input and $\xi(\cdot)$ is the output. Using this notation, one can re-write the system (2.2.1) as follows:

$$\begin{aligned}\dot{x}(t) &= Ax(t) + B_1u(t) + B_2\xi(t);\\ z(t) &= C_1x(t) + D_1u(t);\\ y(t) &= C_2x(t) \qquad\qquad + D_2\xi(t).\end{aligned} \tag{2.2.4}$$

The bound (2.2.2) on the uncertainty in this system then becomes the following:

$$\|\xi(t)\| \leq \|z(t)\|. \tag{2.2.5}$$

From the above definitions, it follows that although this class of uncertain systems allows for time-varying uncertain parameters and also for cone bounded memoryless nonlinearities, it does not allow for dynamic uncertainties such as might arise from unmodeled dynamics. This fact, along with some other deficiencies of this class of uncertain systems that were mentioned in Section 1.1, motivate the IQC uncertainty description presented in Section 2.3.

2.2.1 *Special case: sector-bounded nonlinearities*

This class of uncertainties arose from the celebrated theory of absolute stability; e.g., see [137]. Consider the time-invariant uncertain system (2.2.4) with scalar uncertainty input ξ and uncertainty output z in which $D_1 = 0$, $D_2 = 0$, and the uncertainty is described by the equation

$$\xi(t) = \phi(z(t)) \tag{2.2.6}$$

where $\phi(\cdot) : \mathbf{R} \to \mathbf{R}$ is an uncertain nonlinear mapping. This system is represented in the block diagram shown in Figure 2.2.1.

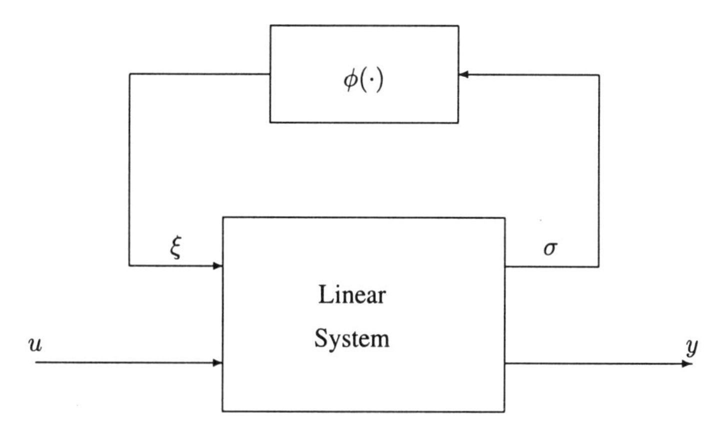

Figure 2.2.1. Uncertain system with a single nonlinear uncertainty.

We will suppose that the uncertain nonlinearity $\phi(\cdot)$ satisfies the following sector bound (e.g., see [137])

$$0 \leq \frac{\phi(z)}{z} \leq k \tag{2.2.7}$$

where $0 < k \leq \infty$ is a given constant associated with the system. Using the change of variables $\tilde{z} = k/2z$, $\tilde{\xi} = \phi(2/k\tilde{z}) - \tilde{z}$, this system is transformed to the system

$$\begin{aligned}
\dot{x}(t) &= (A + \frac{k}{2}B_2C_1)x(t) + B_1u(t) + B_2\tilde{\xi}(t);\\
\tilde{z}(t) &= \frac{k}{2}C_1x(t);\\
y(t) &= C_2x(t).
\end{aligned}$$

The bound (2.2.7) on the uncertainty in this system then becomes a standard bound on the norm of the uncertainty input:

$$|\tilde{\xi}(t)| \leq |\tilde{z}(t)|. \tag{2.2.8}$$

This observation motivates us to think of sector bounded uncertainty as a special case of norm-bounded uncertainty.

2.3 Uncertain systems with integral quadratic constraints

2.3.1 *Integral quadratic constraints*

In order to motivate this uncertainty description, first consider a transfer function uncertainty block as shown in Figure 2.3.1. Assuming that the transfer function

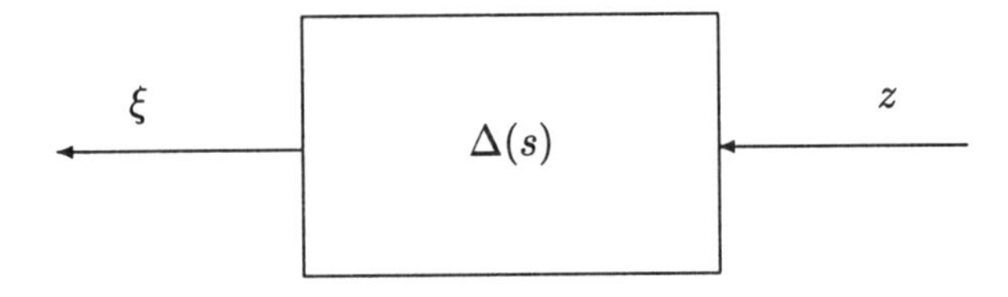

Figure 2.3.1. Transfer function uncertainty

$\Delta(s)$ is stable and using Parseval's Theorem, it follows that the H^∞ norm bound:

$$\|\Delta(j\omega)\| \leq 1 \quad \text{for all } \omega \in \mathbf{R}$$

is equivalent to the time domain bound

$$\int_0^\infty \|\xi(t)\|^2 dt \leq \int_0^\infty \|z(t)\|^2 dt \tag{2.3.1}$$

for all signals $z(\cdot)$ (provided these integrals exist); see Section 3.1 for a more detailed discussion of the H^∞ norm. Here $\|\cdot\|$ denotes the standard Euclidean norm. The time domain uncertainty bound (2.3.1) is an example of an integral quadratic constraint (IQC). This time domain uncertainty bound applies equally well in the case of a time-varying real uncertainty parameter $\Delta(t)$ or a nonlinear mapping. Note that by applying Laplace transforms, it can be seen that an IQC of the form (2.3.1) is equivalent to the frequency domain integral quadratic constraint (1.1.3).

The integral quadratic constraint uncertainty bound (2.3.1) can easily be extended to model the noise acting on the system as well as uncertainty in the system dynamics. This situation is illustrated in Figure 2.3.2. To model this situation, we would modify the integral quadratic constraint (2.3.1) to

$$\int_0^\infty \|\xi(t)\|^2 dt \leq d + \int_0^\infty \|z(t)\|^2 dt \tag{2.3.2}$$

where $d > 0$ is a constant which determines the bound on the size of the noise (again assuming that the integrals exist). If the signal $z(t)$ is zero, the uncertainty block Δ makes no contribution to the signal $\xi(t)$ (assuming zero initial condition on the dynamics of the uncertainty block). However, $\xi(t)$ can still be non-zero due to the presence of the noise signal. This IQC modeling of noise corresponds to an

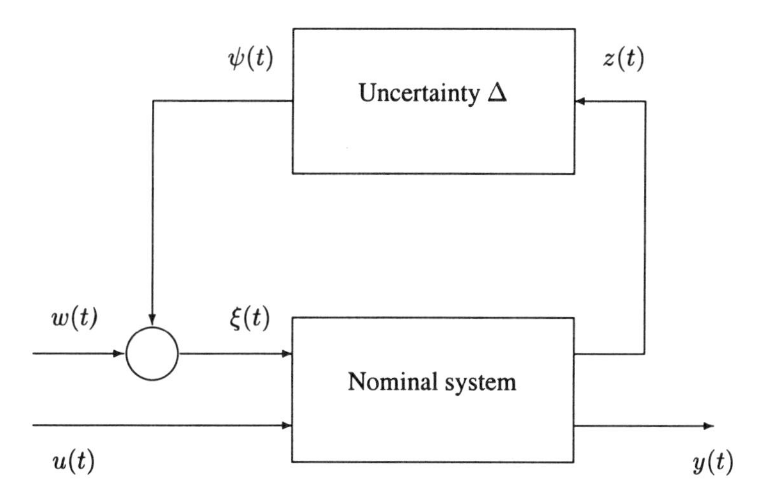

Figure 2.3.2. Uncertain system with noise inputs

energy bound on the noise rather than a stochastic white noise description. Also note that the presence of the d term in the IQC (2.3.2) can allow for a nonzero initial condition on the uncertainty dynamics.

We now present our formal definition of a continuous-time uncertain system with an IQC uncertainty description. We present this definition for a general case that allows for a structured uncertainty description in which multiple uncertainty blocks are admitted as illustrated in Figure 1.2.2. This corresponds to the case in which the process being modeled has a number of sources of uncertainty each acting "independently". Such a general uncertain system is described by the following state equations:

$$\begin{aligned} \dot{x}(t) &= Ax(t) + B_1 u(t) + \sum_{j=1}^{k} B_{2j}\xi_j(t); \\ z_1(t) &= C_{11}x(t) + D_{11}u(t); \\ z_2(t) &= C_{12}x(t) + D_{12}u(t); \\ &\vdots \\ z_k(t) &= C_{1k}x(t) + D_{1k}u(t); \\ y(t) &= C_2 x(t) + \sum_{j=1}^{k} D_{2j}\xi_j(t) \end{aligned} \tag{2.3.3}$$

where $x(t) \in \mathbf{R}^n$ is the *state*, $u(t) \in \mathbf{R}^m$ is the *control input*, $z_1(t) \in \mathbf{R}^{q_1}$, $z_2(t) \in \mathbf{R}^{q_2}, \ldots, z_k(t) \in \mathbf{R}^{q_k}$ are the *uncertainty outputs* and $\xi_1(t) \in \mathbf{R}^{p_1}$, $\xi_2(t) \in \mathbf{R}^{p_2}, \ldots, \xi_k(t) \in \mathbf{R}^{p_k}$ are the *uncertainty inputs*, $y(t) \in \mathbf{R}^l$ is the *measured* output. Also in the above equations, A, B_1, B_2, $C_{11}, \ldots, C_{1k}$, C_2, D_{11}, D_{1k}, and D_2 are constant matrices of suitable dimensions.

The uncertainty in the above system can be described by equations of the form:

$$\begin{aligned}
\xi_1(t) &= \phi_1(t, x(\cdot), u(\cdot));\\
\xi_2(t) &= \phi_2(t, x(\cdot), u(\cdot));\\
&\ \ \vdots\\
\xi_k(t) &= \phi_k(t, x(\cdot), u(\cdot)).
\end{aligned} \tag{2.3.4}$$

This uncertainty description allows for nonlinear, time-varying, dynamic uncertainties. An important feature of dynamic uncertainty as opposed to an exogenous disturbance input is the functional dependence of the uncertainty input on the control input as described by equations (2.3.4). The corresponding set of admissible uncertainties will therefore depend on the control input $u(\cdot)$. In minimax optimal control problems such as those which will be considered in Sections 5.3, 8.4 and 8.5, this leads to the observation that the minimizing player imposes restrictions on the choice of strategy available to the maximizing player. This observation reveals a significant difference between standard game type minimax optimization problems and minimax optimal control problems related to worst case controller design. In this book, we note this difference by referring to the minimax optimization problems considered in Sections 5.3, 8.4 and 8.5 as constrained minimax optimization problems.

For the uncertain system (2.3.3) and uncertainties (2.3.4), a bound on the uncertainty is determined by the following integral quadratic constraint condition.

Definition 2.3.1 *An uncertainty of the form (2.3.4) is an admissible uncertainty for the system (2.3.3) if the following conditions hold: Given any locally square integrable control input $u(\cdot)$ and any corresponding solution to equations (2.3.3), (2.3.4) defined on an existence interval[1] $(0, t_*)$, then there exists a sequence $\{t_i\}_{i=1}^{\infty}$ and constants $d_1 \geq 0, \ldots, d_k \geq 0$ such that $t_i \to t_*$, $t_i \geq 0$ and*

$$\int_0^{t_i} \|\xi_j(t)\|^2 dt \leq d_j + \int_0^{t_i} \|z_j(t)\|^2 dt \tag{2.3.5}$$

for all i and for $j = 1, 2, \ldots, k$. Also, note that t_* and t_i may be equal to infinity.

As mentioned in Chapter 1, the above definition of admissible uncertainty originates in the work of Yakubovich [254, 256, 257][2] and has been extensively studied in the Russian literature; e.g., see [256, 258, 259, 260, 173, 241, 106, 212]. In particular, references [241] and [106], give a number of examples of physical systems in which the uncertainty naturally fits into the above framework; see also [124]. The above definition extends this class of uncertain systems to systems containing a control input. This enables us to use H^∞ control theory in the problem of controller synthesis for such uncertain systems.

[1] That is, t_* is the upper limit of the time interval over which the solution exists.

[2] The reader may notice a connection between the integral quadratic constraint uncertainty description and Popov's notion of hyperstability; see [162].

Note that in the above definition of the IQC, we do not require any assumption that the uncertain system is stable or even that its solutions exist on an infinite time interval. The IQC (2.3.5) involves a sequence of times $\{t_i\}_{i=1}^{\infty}$. In the particular case where $u(\cdot)$ is defined by a *stabilizing* controller which guarantees that $x(\cdot) \in \mathbf{L}_2[0,\infty)$ for any $\xi(\cdot) \in \mathbf{L}_2[0,\infty)$, there is no need to introduce the sequence $\{t_j\}_{i=1}^{\infty}$ to describe the uncertainty. Indeed, given a constraint in the form (2.3.5) and a stabilizing control input $u(\cdot)$, by passing to the limit in (2.3.5) as $t_j \to \infty$, one can replace the integral over $[0, t_j]$ in (2.3.5) by an integral over the infinite interval $[0,\infty)$. Furthermore, by making use of Parseval's Theorem, it follows that this IQC is equivalent to the frequency domain integral quadratic constraint:

$$\int_{-\infty}^{\infty} \begin{bmatrix} \hat{z}_i(j\omega) \\ \hat{\xi}_(j\omega) \end{bmatrix}^* \begin{bmatrix} I & 0 \\ 0 & -I \end{bmatrix} \begin{bmatrix} \hat{z}_i(j\omega) \\ \hat{\xi}_i(j\omega) \end{bmatrix} d\omega \geq -d_i; \tag{2.3.6}$$

c.f. [124]. Here $\hat{z}_i(s)$ and $\hat{\xi}_i(s)$ denote the Laplace transforms of the signals $z_i(t)$ and $\xi_i(t)$) respectively. However in this book, we wish to define the class of admissible uncertainties for a generic control input and avoid referring to any particular stabilizing properties of the control input when the constraints on the uncertainty are being defined. This is achieved in Definition 2.3.1 by considering control inputs, uncertainty inputs and the corresponding solutions defined on a sequence of expanding finite intervals $[0, t_j]$.

Observation 2.3.1 It is important to note that the class of uncertainties satisfying an IQC of the form (2.3.5) includes norm-bounded uncertainties as a particular case. Indeed, from condition (2.2.5) it follows that any norm-bounded uncertainty input $\xi(\cdot)$ of the form (2.2.3) also satisfies the integral constraint

$$\int_0^T \|\xi(t)\|^2 dt \leq \int_0^T \|z(t)\|^2 dt$$

for all $T > 0$. Thus, any norm-bounded uncertainty of the form (2.2.3) satisfies the IQC (2.3.5) with any $d > 0$ and an arbitrarily chosen sequence $\{t_j\}_{j=1}^{\infty}$.

2.3.2 *Integral quadratic constraints with weighting coefficients*

It is often useful to introduce weighting matrices into the definition of an IQC. For instance, such a need may arise in the case where some of the entries of the vector $\xi(\cdot)$ represent noise signals acting on the system; e.g., sample paths of a sufficiently smooth stationary stochastic process. In this case, a matrix of weighting coefficients Q can be chosen to be the reciprocal of the covariance matrix of the noise. For simplicity, we will illustrate this idea for the system (2.2.4) in which the uncertainty is unstructured and given in the form

$$\xi(t) = \phi(t, x(\cdot), u(\cdot)). \tag{2.3.7}$$

In this case, the uncertainty can be defined by a single integral quadratic constraint. An integral quadratic constraint with weighting coefficients is defined in a similar fashion as that of Definition 2.3.1.

Definition 2.3.2 *Let $Q = Q'$ be a given weighting matrix associated with the system (2.2.4), (2.3.7). An uncertainty of the form (2.3.7) is an admissible uncertainty for the system (2.2.4) if the following conditions hold: Given any locally square integrable control input $u(\cdot)$, then any corresponding solution to equations (2.2.4), (2.3.7) is locally square integrable and there exists a sequence $\{t_i\}_{i=1}^{\infty}$ and constant $d > 0$ such that $t_i \to t_\star$, $t_i \geq 0$ and*

$$\int_0^{t_i} \xi(t)'Q\xi(t)dt \leq d + \int_0^{t_i} \|z(t)\|^2 dt. \tag{2.3.8}$$

Here $t_\star$ is the upper limit of the time interval over which the solution exists.

2.3.3 Integral uncertainty constraints for nonlinear uncertain systems

In this section, we present a nonlinear extension of the IQC uncertainty description introduced in Section 2.3.2. To this end, we replace the quadratic forms in Definition 2.3.1 by nonlinear functionals and consider a nonlinear nominal system. For simplicity we will consider the case of uncertain nonlinear systems without a control input.

Consider the nonlinear system

$$\dot{x}(t) = g(x(t), \xi(t)) \tag{2.3.9}$$

where $x(t) \in \mathbf{R}^n$ is the *state* and $\xi(t) \in \mathbf{R}^p$ is the *uncertainty input*. The uncertainty input is contained in the set Ξ of all locally integrable vector functions from $\mathbf{R}$ to $\mathbf{R}^p$. Associated with the system (2.3.9) is the following set of functions referred to as *supply rates*

$$w_1(x(t), \xi(t)); \quad w_2(x(t), \xi(t)); \quad \ldots; \quad w_k(x(t), \xi(t)). \tag{2.3.10}$$

System Uncertainty

The uncertainty in the system (2.3.9) is described by an equation of the form:

$$\xi(t) = \phi(t, x(\cdot)) \tag{2.3.11}$$

where the following integral uncertainty constraint is satisfied.

Definition 2.3.3 *An uncertainty of the form (2.3.11) is an admissible uncertainty for the system (2.3.9) if the following condition holds: Given any solution to equations (2.3.9), (2.3.11) with an interval of existence $[0, t_*)$, then there exist constants $d_1 \geq 0, d_2 \geq 0, \ldots, d_k \geq 0$ and sequence $\{t_i\}_{i=1}^{\infty}$ such that $t_i \to t_*$, $t_i \geq 0$ and*

$$\int_0^{t_i} w_j(x(t), \xi(t))dt \leq d_j \tag{2.3.12}$$

for all i and for $j = 1, 2, \ldots, k$. Note that t_* and t_i may be equal to infinity.

Remark 2.3.1 Note that the uncertain system (2.3.9), (2.3.11) allows for structured uncertainty satisfying norm bound conditions. In this case, the uncertain system would be described by the state equations

$$\dot{x}(t) = A(x(t)) + \sum_{j=1}^{k} B_j(x(t))\Delta_j(t)C_j(x(t)); \quad \|\Delta_j(t)\| \leq 1$$

where $\Delta_j(t)$ are uncertainty matrices and $\|\cdot\|$ denotes the standard induced matrix norm. This nonlinear uncertain system with norm-bounded uncertainty is a nonlinear version of the uncertain system (2.2.1) (without a control input). To verify that such uncertainty is admissible for the uncertain system (2.3.9), (2.3.11), let

$$\xi_j(t) := \Delta_j(t)C_j(x(t))$$

where $\|\Delta_j(t)\| \leq 1$ for all $t \geq 0$ and

$$\xi(t) := \left[\xi_1'(t) \ \xi_2'(t) \ \ldots \ \xi_k'(t) \right]'.$$

Then $\xi(\cdot)$ satisfies condition (2.3.12) with

$$w_j(x(t), \xi(t)) = \|\xi_j(t)\|^2 - \|C_j(x(t))\|^2,$$

$d_j = 0$ and any t_i.

2.3.4 Averaged integral uncertainty constraints

The *averaged integral quadratic constraint* uncertainty description is an extension of the IQC uncertainty description which was introduced by Savkin and Petersen in references [191, 198]. The motivation for introducing this uncertainty description is that it enables the extension of results on state-feedback infinite-horizon optimal guaranteed cost control to the finite-horizon and output-feedback cases. The uncertainty class defined by the averaged integral quadratic constraint, is as rich as the uncertainty class defined via the standard integral quadratic constraint in Section 2.3.1. It allows for nonlinear time-varying dynamic structured uncertainties but with a different characterization of the initial conditions on the uncertainty dynamics. Within the framework of this uncertainty description, it is possible to derive mathematically tractable solutions to problems of output-feedback finite-horizon time-varying minimax optimal control.

Roughly speaking, the difference between the averaged integral quadratic constraint uncertainty description and most other structured uncertainty descriptions (including those presented in previous sections) can be related to the initial conditions on the uncertainty dynamics. In the μ type uncertainty description of [51, 171], the initial condition on the uncertain dynamics is implicitly assumed to be zero. In the standard integral quadratic constraint uncertainty description such as occurs in Section 2.3.1, the initial conditions on the uncertainty dynamics are implicitly assumed to be unknown but bounded. One interpretation of the averaged integral quadratic constraint uncertainty description is that the initial condition on the uncertainty dynamics has an elementary probabilistic characterization.

In most applications, the description of the initial conditions on the uncertainty dynamics is not an important consideration and thus it is reasonable to use the problem formulation which is most tractable. This is the main motivation behind the introduction of the uncertainty averaging uncertainty description. Also, it should be noted that in situations where the initial conditions on the uncertainty dynamics are important, the uncertainty averaging approach is often the most reasonable approach. For example, it may be the case that the bounds on the size of the initial conditions of the uncertainty dynamics are to be estimated via a series of measurements and the use of averaging. In this case, an averaging description would be a natural choice in describing the resulting bound on the uncertainty initial conditions.

Consider the following uncertain time-varying system defined on the finite time interval $[0, T]$:

$$\begin{aligned} \dot{x}(t) &= A(t)x(t) + B_1(t)u(t) + \sum_{j=1}^{k} B_{2j}(t)\xi_j(t); \qquad (2.3.13) \\ z_1(t) &= C_{11}(t)x(t) + D_{11}(t)u(t); \\ &\vdots \\ z_k(t) &= C_{1k}(t)x(t) + D_{1k}(t)u(t); \\ y(t) &= C_2(t)x(t) + \sum_{s=1}^{k} D_{2j}(t)\xi_j(t) \qquad (2.3.14) \end{aligned}$$

where $x(t) \in \mathbf{R}^n$ is the *state*, $u(t) \in \mathbf{R}^m$ is the *control input*, $z_1(t) \in \mathbf{R}^{q_1}$, $z_2(t) \in \mathbf{R}^{q_2}$, ..., $z_k(t) \in \mathbf{R}^{q_k}$ are the *uncertainty outputs*, $\xi_1(t) \in \mathbf{R}^{p_1}$, $\xi_2(t) \in \mathbf{R}^{p_2}$, ..., $\xi_k(t) \in \mathbf{R}^{p_k}$ are the *uncertainty inputs*, and $y(t) \in \mathbf{R}^l$ is the *measured* output. Also, $A(\cdot)$, $B_1(\cdot)$, $B_{21}(\cdot)$, ..., $B_{2k}(\cdot)$, $C_{11}(\cdot)$, ..., $C_{1k}(\cdot)$, $D_{11}(\cdot)$, ..., $D_{1k}(\cdot)$, $C_2(\cdot)$, $D_{21}(\cdot)$, ..., $D_{2k}(\cdot)$ are bounded piecewise continuous matrix functions defined on $[0, T]$.

The uncertainty in the above system is described by a set of equations of the form (2.3.4). Alternatively, the uncertainty inputs and uncertainty outputs may be assembled together into two vectors. That is, we define

$$\xi(t) := \begin{bmatrix} \xi_1(t) \\ \xi_2(t) \\ \vdots \\ \xi_k(t) \end{bmatrix}; \quad z(t) := \begin{bmatrix} z_1(t) \\ z_2(t) \\ \vdots \\ z_k(t) \end{bmatrix}.$$

Then (2.3.4) can be re-written in the more compact form

$$\xi(t) = \Phi(t, x(\cdot), u(\cdot)); \qquad \Phi(t, x(\cdot), u(\cdot)) := \begin{bmatrix} \phi_1(t, x(\cdot), u(\cdot)) \\ \phi_2(t, x(\cdot), u(\cdot)) \\ \vdots \\ \phi_k(t, x(\cdot), u(\cdot)) \end{bmatrix}. \quad (2.3.15)$$

We consider finite sequences of uncertainty functions of the form (2.3.15) such that the following constraint is satisfied.

Definition 2.3.4 *Let $d_1 > 0$, $d_2 > 0$, ..., $d_k > 0$, be given positive constants associated with the system (2.3.13). We will consider sequences of uncertainty functions $\mathcal{S} = \{\Phi^1(\cdot), \Phi^2(\cdot), \ldots \Phi^q(\cdot)\}$ of arbitrary length q. A sequence of uncertainty functions $\mathcal{S}$ is an admissible* uncertainty sequence *for the system (2.3.13) if the following conditions hold: Given any $\Phi^i(\cdot) \in \mathcal{S}$, any control input $u^i(\cdot) \in \mathbf{L}_2[0, T]$, and any corresponding solution $\{x^i(\cdot), \xi^i(\cdot)\}$ to equations (2.3.13), (2.3.15) defined on $[0, T]$, then $\xi^i(\cdot) \in \mathbf{L}_2[0, T]$ and*

$$\frac{1}{q} \sum_{i=1}^{q} \int_0^T (\|\xi_j^i(t)\|^2 - \|z_j^i(t)\|^2) dt \leq d_j \quad (2.3.16)$$

for $j = 1, 2, \ldots, k$. The class of all such admissible uncertainty sequences is denoted Ξ.

Remark 2.3.2 The above definition extends the definition of the integral quadratic constraint given in Section 2.3.1; see also, references [260, 259, 173, 184, 177, 185]. In Definition 2.3.1, only individual uncertainty functions are considered rather than sequences of uncertainty functions. The uncertainty description given in Definition 2.3.4 allows for a large class of nonlinear time-varying structured uncertainties. Furthermore in this case, each constant d_j can be regarded as providing a bound on the size of the initial condition on the corresponding uncertainty dynamics. In the uncertainty averaging description of structured uncertainty, we introduce sequences of uncertainty functions and it is the *averaged* value of the size of the initial condition on the uncertainty dynamics which is bounded by the constant d_j. This uncertainty description can be given a probabilistic interpretation as follows: Given any admissible uncertainty sequence $\mathcal{S} \in \Xi$, each uncertainty function $\Phi^i(\cdot) \in \mathcal{S}$ is assigned an equal probability. Then condition (2.3.16) amounts to a bound on the expected value of the quantity

$$\int_0^T (\|\xi_j(t)\|^2 - \|z_j(t)\|^2) dt.$$

As mentioned above, this quantity can be regarded as providing a measure of the size of the initial conditions on the uncertainty dynamics.

The above averaged uncertainty description lends itself to a procedure for constructing the constants d_j based on a series of measurements on a physical system. That is, if the signals $\xi(\cdot)$ and $z(\cdot)$ are available for measurement, then the uncertainty description would be determined so that condition (2.3.16) is satisfied for the available measurements. If the signals $\xi(\cdot)$ and $z(\cdot)$ are not directly available for measurement, it still may be possible to construct a suitable uncertainty description based on input and output measurements. This might be achieved by using similar ideas to those contained in recent results on model validation theory; e.g., see [215, 192, 159].

2.4 Stochastic uncertain systems

Stochastic processes and deterministic uncertainties described by integral quadratic constraints are often regarded as two completing methodologies for modeling uncertainty in dynamical systems. However, there are situations in which it may be useful to model uncertainty via a combination of these two methodologies. This leads to the concept of a stochastic uncertain system. In the robust control literature, one can find a number of different definitions for stochastic uncertain systems. In this book, we consider two different definitions for a stochastic uncertain system. The difference between these two definitions relates to the manner in which the random perturbations effect the system. Essentially, we shall consider two alternative cases commonly referred to as the "multiplicative noise" case and the "additive noise" case. In the first case, random perturbations effecting the system give rise to random fluctuations in the system parameters; e.g., see the examples given in Section 2.4.1. In this case, the system in Figure 2.3.2 referred to as the nominal system, is a stochastic system even if there is no external noise input $w(t)$ entering into the system. A peculiar feature of such systems is that the noise perturbations cease as the system reaches its equilibrium.

In the case of systems with additive noise, the plant is subject to exogenous noise inputs such as, for example, sensor noises. Since this type of noise has an external source, there is no particular reason why its effect should cease when the plant reaches its equilibrium. If the system does not include any uncertainty in its dynamics, this class of stochastic systems is covered within the standard LQG framework. The stochastic uncertain system framework introduced in references [151, 152, 229] allows one to extend the standard LQG stochastic optimal controller design methodology into a minimax stochastic optimal control methodology for this class of stochastic uncertain systems. The papers [151, 152] consider discrete-time systems, while [229] deals with continuous-time systems.

For both classes of stochastic uncertain systems considered in this book, the corresponding robust control problems involve characterizing a controller which minimizes worst case performance in the face of uncertainty which satisfies certain stochastic uncertainty constraints. These constraints can be thought of as stochastic counterparts of the deterministic integral quadratic uncertainty constraints which have been discussed previously. As in the standard IQC uncertainty description, this uncertainty description allows for the stochastic uncertainty inputs to depend dynamically on the uncertainty outputs.

2.4.1 Stochastic uncertain systems with multiplicative noise

Stochastic systems with multiplicative noise naturally arise in many practical problems. An example illustrating how this uncertainty description may arise is given by the electric circuit shown in Figure 2.4.1.

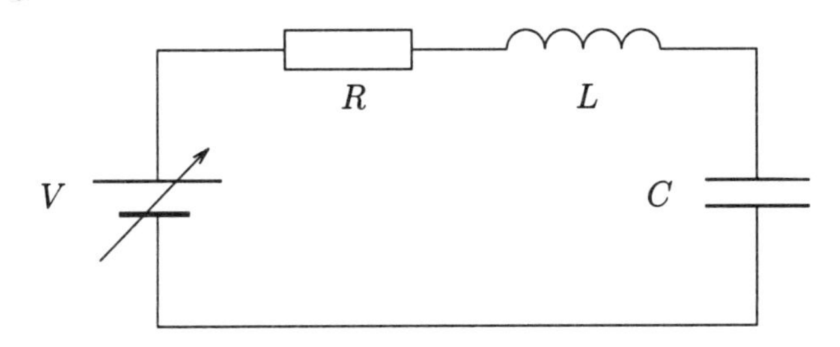

Figure 2.4.1. An uncertain electric circuit.

Example 2.4.1 The differential equations describing the current and voltage dynamics for the circuit shown in Figure 2.4.1 are as follows:

$$\begin{aligned}\frac{di}{dt} &= -\frac{R}{L}i + \frac{1}{L}(V - V_C);\\ \frac{dV_C}{dt} &= \frac{1}{C}i\end{aligned}$$

where i is the current in the circuit and V_C is the capacitor voltage. One can control this system by applying an appropriate voltage V to the circuit. In this circuit, the resistance R and the inductance L vary as follows.

It is known that the resistance R varies slowly between R^- and R^+, due to temperature variations. That is,

$$R = R(t) = R_0 + R_1 \Delta R(t),$$

where $R_0 = (R^+ + R^-)/2$ is the nominal value of the resistor, $R_1 = (R^+ - R^-)/2$, and $\Delta R(t)$ satisfies the following standard norm bound constraint:

$$|\Delta R(t)| \leq 1.$$

Also, suppose that movements of nearby magnetic materials may induce rapid changes in the inductance value. That is, one may suppose that the inductance $L = L(t)$ is given by

$$L^{-1} = L_0^{-1} + L_1^{-1}\zeta(t),$$

where $\zeta(t)$ is Gaussian white noise with the mean zero and unity covariance and L_0 is the mean inductance. The value of L_1 can be determined by estimating the covariance of the reciprocal inductance.

Letting $x = [i \ V_C]' \in \mathbf{R}^2$, $z_1 = i$, $z_2 = V_C - V$, one can write a formal Langevin equation for the system shown in Figure 2.4.1:

$$
\begin{aligned}
\dot{x} &= \begin{bmatrix} -R_0/L_0 & -1/L_0 \\ -1/C & 0 \end{bmatrix} x + \begin{bmatrix} 1/L_0 \\ 0 \end{bmatrix} V + \begin{bmatrix} -R_1/L_0 \\ 0 \end{bmatrix} \xi_1 \\
&\quad + \left(\begin{bmatrix} -R_0/L_1 & -1/L_1 \\ 0 & 0 \end{bmatrix} x + \begin{bmatrix} 1/L_1 \\ 0 \end{bmatrix} V + \begin{bmatrix} -R_1/L_1 \\ 0 \end{bmatrix} \xi_1 \right) \zeta(t); \\
z &= x + \begin{bmatrix} 0 \\ -1 \end{bmatrix} V; \qquad \xi_1 = \Delta R(t) z_1.
\end{aligned}
$$

A rigorous mathematical description of the system is then given by the following Ito stochastic differential equation in which $W(t)$ is a scalar Wiener process:

$$
\begin{aligned}
dx &= \left(\begin{bmatrix} -\frac{R_0}{L_0} + \frac{R_0^2}{2L_1^2} & -\frac{1}{L_0} + \frac{R_0}{2L_1^2} \\ -1/C & 0 \end{bmatrix} x + \begin{bmatrix} \frac{1}{L_0} - \frac{R_0}{2L_1^2} \\ 0 \end{bmatrix} V \right. \\
&\quad \left. + \begin{bmatrix} -\frac{R_1}{L_0} + \frac{R_0 R_1}{L_1^2} & \frac{R_1}{2L_1^2} & \frac{R_1^2}{2L_1^2} \\ 0 & 0 & 0 \end{bmatrix} \xi \right) dt \\
&\quad + \left(\begin{bmatrix} -R_0/L_1 & -1/L_1 \\ 0 & 0 \end{bmatrix} x + \begin{bmatrix} 1/L_1 \\ 0 \end{bmatrix} V \right. \\
&\quad \left. + \begin{bmatrix} -R_1/L_1 & 0 & 0 \\ 0 & 0 & 0 \end{bmatrix} \xi \right) dW(t); \\
z &= x + \begin{bmatrix} 0 \\ -1 \end{bmatrix} V; \qquad \xi = \Delta(t) z; \\
\Delta(t) &= \begin{bmatrix} \Delta R(t) & 0 \\ 0 & \Delta R(t) \\ (\Delta R(t))^2 & 0 \end{bmatrix}.
\end{aligned}
$$

It is easy to see that $\Delta'\Delta \leq 2I$. As mentioned above (see Observation 2.3.1), this constraint can be converted into a certain integral quadratic constraint. It is also worth noting that in this example, the uncertainty is structured. Hence, integral quadratic constraints on structured uncertainty can be used to describe the uncertainty in this example. This motivates the following integral quadratic constraint stochastic uncertainty description for systems with multiplicative noise.

Let $\{\Omega, \mathcal{F}, \mathcal{P}\}$ be a complete probability space and let $W_1(t)$ and $W_2(t)$ be two independent Wiener processes in $\mathbf{R}^{r_1}$ and $\mathbf{R}^{r_2}$ with covariance matrices Q_1 and Q_2, respectively. Let $\mathcal{F}_t$ denote the increasing sequence of Borel sub-σ-fields of $\mathcal{F}$, generated by $\{W_{1,2}(s), 0 \leq s < t\}$. Also, let $\mathbf{E}$ and $\mathbf{E}\{\cdot|\mathcal{F}_t\}$ be the corresponding unconditional and conditional expectation operators, respectively, where the latter is the expectation with respect to $\mathcal{F}_t$.

For $T \leq \infty$, let $\mathbf{L}_2(s, T; \mathbf{R}^n)$ denote the Hilbert space generated by the (t, ω)-measurable $\{\mathcal{F}_t, t \geq 0\}$-non-anticipating processes $x(t, \omega) : [s, T] \times \Omega \to \mathbf{R}^n$ and complete with respect to the norm

$$|||\cdot||| = \left(\int_s^T \mathbf{E}\|\cdot\|^2 dt \right)^{1/2}.$$

Given an $r \times r$ positive-definite symmetric matrix Q, let $\mathbf{R}_Q^{n \times r}$ denote the Hilbert space of $n \times r$ matrices with the inner product $\operatorname{tr} \Theta_1 Q \Theta_2'$.

We consider a stochastic uncertain system described by the following Ito stochastic differential equation:

$$\begin{aligned}
dx(t) &= (Ax(t) + B_1 u(t) + \sum_{i=1}^{k} B_{2i}\xi_i(t))dt \\
&\quad + (Hx(t) + P_1 u(t))dW_1(t) + \left(\sum_{i=1}^{k} P_{2i}\xi_i(t) \right) dW_2(t); \\
z_1(t) &= C_{11}x(t) + D_{11}u(t); \\
z_2(t) &= C_{12}x(t) + D_{12}u(t); \\
&\vdots \\
z_k(t) &= C_{1k}x(t) + D_{1k}u(t)
\end{aligned} \tag{2.4.1}$$

where $x(t) \in \mathbf{R}^n$ is the *state*, $u(t) \in \mathbf{R}^m$ is the *control input*, $z_1(t) \in \mathbf{R}^{q_1}$, $z_2(t) \in \mathbf{R}^{q_2}, \ldots, z_k(t) \in \mathbf{R}^{q_k}$ are the *uncertainty outputs* and $\xi_1(t) \in \mathbf{R}^{p_1}$, $\xi_2(t) \in \mathbf{R}^{p_2}, \ldots, \xi_k(t) \in \mathbf{R}^{p_k}$ are the *uncertainty inputs*. In equation (2.4.1), H, $P_1, P_{21}, \ldots, P_{2k}$ are bounded linear operators $\mathbf{R}^n \to \mathbf{R}_{Q_1}^{n \times r_1}$, $\mathbf{R}^m \to \mathbf{R}_{Q_1}^{n \times r_1}$, $\mathbf{R}^{p_1} \to \mathbf{R}_{Q_2}^{n \times r_2}, \ldots, \mathbf{R}^{p_k} \to \mathbf{R}_{Q_2}^{n \times r_2}$, respectively.

We suppose that the uncertainty inputs to the system (2.4.1) are defined by equation (2.3.7) and satisfy the following *Stochastic Integral Quadratic Constraint.*

Definition 2.4.1 *Let $s > 0$ be given and let $d_1 > 0, \ldots, d_k > 0$ be given $\mathcal{F}_s$-measurable random variables. The uncertainty inputs to the system (2.4.1) define an admissible uncertainty for this system if the following conditions hold: There exists a sequence $\{t_i\}_{i=1}^{\infty}$ such that $t_i > s$, $t_j \to \infty$ as $j \to \infty$ and the following condition is satisfied. If $u(\cdot) \in \mathbf{L}_2([s, t_i]; R^m)$, then $x(\cdot) \in \mathbf{L}_2([s, t_i]; R^n)$, and*

$$\mathbf{E}\int_s^{t_i} \|\xi_j(t)\|^2 dt \leq \mathbf{E}d_j + \mathbf{E}\int_s^{t_i} \|z_j(t)\|^2 dt \tag{2.4.2}$$

for all $j = 1, 2, \ldots, k$.

The stochastic uncertainty constraint (2.4.2) extends the integral quadratic constraint definition such as given in Section 2.3 to the case of stochastic systems with multiplicative noise. As in these references, this stochastic uncertainty description allows for the uncertainty input ξ to depend dynamically on the uncertainty outputs. Also note that the uncertainty description corresponding to the constraint (2.4.2) encompasses the standard structured norm-bounded uncertainty description such as in Example 2.4.1.

Remark 2.4.1 Contrary to Definition 2.3.1 in Section 2.3, here we assume that $d_1, \ldots d_k$ are random variables. To motivate this assumption, we note that one of the purposes of including these constants in the definition of admissible uncertainty was to allow for non-zero initial conditions on the uncertain dynamics. Since we are dealing with a stochastic system in this case, it is natural to assume that the uncertain dynamics are also stochastic with random initial conditions. This naturally leads to the assumption that the quantities $d_1, \ldots, d_k$ are random.

Remark 2.4.2 In contrast to the definitions given in Section 2.3, we have assumed that for any admissible uncertainty input, a solution to the stochastic differential equation (2.4.1) exists on any finite interval provided the control input is summable on any finite interval. This is done to avoid a rather technical and involved discussion concerning the existence of a "regular" (not having a finite escape time) solution to the stochastic differential equation. One possible modification to Definition 2.4.1 allowing for solutions to stochastic differential equations that exist only on an interval $[s, t_*]$ would be to consider t_* as a stopping time for the stochastic process $(x(t), \xi(t))$. However in this case, one would have to deal with the fact that t_* is a random variable $t_* = t_*(\omega)$ rather than with a constant as in the deterministic case. An example of this will be given in Section 2.4.3. The reader who wishes to study this issue in more detail is referred to references [55, 105].

Remark 2.4.3 Stochastic extensions to the integral quadratic constraint uncertainty description may provide a possible approach to the problem of non-worst case robust controller design. Recall that the standard deterministic worst case robust control design presumes that all uncertainties have an equal chance of occurring so that one does not expect that certain uncertainty inputs are more or less likely than others. Although the worst case design methodology has proved its efficacy in various engineering problems, it suffers from the disadvantage that the designer lacks the opportunity to discriminate between "expected" uncertainties and those uncertainties which seldom occur. In other words, the standard worst case design methodology proceeds from the assumption that the values which the uncertainty may take are all equally likely; *i.e.*, one can think of the uncertainty as arising from a uniformly distributed random variable taking its values in the space of uncertainty inputs. However, this may not accurately represent the uncertainty in the system under consideration. For example, it may have been determined that the uncertainty inputs have a distribution other than the uniform distribution and the designer may wish to make use of the *á priori* information about the distribution of the uncertainty values. The following example illustrates a situation in which this idea can be applied to go beyond worst case robust controller design in the case where the distribution of uncertainties is Gaussian. In this case, we arrive at an underlying system of the form (2.4.1).

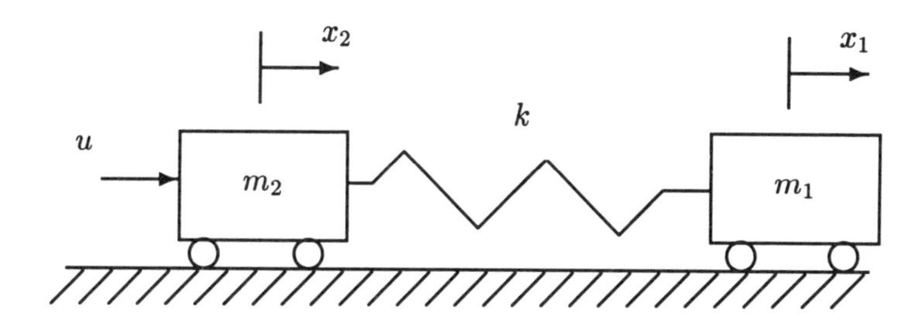

Figure 2.4.2. Two mass spring system

Suppose that the system to be controlled consists of two carts connected by a spring as shown in Figure 2.4.2. There is an uncertainty input force ξ acting on the first cart. This uncertainty input is required to satisfy a certain IQC and represents unmodeled dynamics relating to the motion of the first cart. A control force u drives the second cart. The spring constant k has a nominal value of $k_0 = 1.25$, but may vary and is considered uncertain. A series of experiments is undertaken to determine value of the spring constant in various conditions. This reveals that for each time instant t, the histogram of observed spring constant values is consistent with a stationary Gaussian distribution with the mean k_0. It is also found in these experiments that k ranges over the interval $[0.5, 2]$. Assume, that the masses of the carts are $m_1 = m_2 = 1$. Then, the system is described by the equation

$$\dot{x} = (A + F\Delta(t)C)x + B_1 u + B_2 \xi$$

where $x = [x_1\ x_2\ \dot{x}_1\ \dot{x}_2]' \in \mathbf{R}^4$, $\Delta(t) = k(t) - k_0$, and

$$A = \begin{bmatrix} 0 & 0 & 1 & 0 \\ 0 & 0 & 0 & 1 \\ -1.25 & 1.25 & 0 & 0 \\ 1.25 & -1.25 & 0 & 0 \end{bmatrix}; \quad B_1 = \begin{bmatrix} 0 \\ 0 \\ 0 \\ 1 \end{bmatrix}; \quad B_2 = \begin{bmatrix} 0 \\ 0 \\ 1 \\ 0 \end{bmatrix};$$

$$F = \begin{bmatrix} 0 \\ 0 \\ -1 \\ 1 \end{bmatrix}; \quad C = \begin{bmatrix} 1 & -1 & 0 & 0 \end{bmatrix}. \tag{2.4.3}$$

Assume that $\Delta(t)$ is the Gaussian white noise process with zero mean and $\mathbf{E}\Delta^2(t) = \sigma^2$. We can then choose the value of the parameter σ such that $k(t) = k_0 + \Delta(t)$ satisfies the bound $0.5 \leq k(t) \leq 2$ with a sufficiently high probability. For example, for $\sigma = 0.25$, we have $P(|k(t) - k_0| \leq 0.75) \geq 0.997$. This model of spring constant variations leads us to a stochastic uncertain system[3] of the form (2.4.1)

$$\begin{aligned} dx &= (Ax + B_1 u + B_2 \xi)dt + HxdW_1(t); \\ z &= Cx + Du \end{aligned}$$

where $H = \sigma FC$ and $W_1(\cdot)$ is the scalar Wiener process. Note that in this model, the probability $P(|k(t) - k_0| \leq 0.75)$ is increasing as $\sigma^2 \downarrow 0$. However, we will always have $P(|k(t) - k_0| \leq 0.75) < 1$. That is, the value of the spring constant $k(t)$ may exceed the presumed bounds on uncertainty $[0.5, 2]$ with a nonzero probability. This phenomenon indicates a "soft" norm bound on the uncertainty.

2.4.2 *Stochastic uncertain systems with additive noise: Finite-horizon relative entropy constraints*

In the IQC uncertainty descriptions considered in Section 2.3.1, noise signals were allowed but they were required to be $\mathbf{L}_2$ norm-bounded. In many applications, it would be more appropriate to consider noise signals which are stochastic white noise signals. This is particularly true if it is desired to develop a robust version of LQG control. Our approach to treating this class of uncertain systems involves the concept of relative entropy between two probability measures; e.g., see [53].

We first consider the finite-horizon case. Technically, this case is much simpler than the infinite-horizon case because in the finite-horizon case, we do not face difficulties arising as time approaches infinity. Let $T > 0$ be a constant which will denote the finite time horizon. Also, let $(\Omega, \mathcal{F}, P)$ be a complete probability space on which a p-dimensional standard Wiener process $W(\cdot)$ and a Gaussian random variable $x_0 : \Omega \to \mathbf{R}^n$ are defined, $p = r + l$. The first r entries of the vector process $W(\cdot)$ correspond to the system noise, while the last l entries correspond to the measurement noise. The space Ω can be thought of as the noise space $\mathbf{R}^n \times \mathbf{R}^l \times C([0,T], \mathbf{R}^p)$ [39]. The probability measure P is defined as the product of the probability measure

$$\mu(dx \times dy) = \frac{1}{(2\pi)^{n/2}|Y_0|^{1/2}} e^{-\frac{1}{2}(x-\check{x}_0)' Y_0^{-1} (x-\check{x}_0)} dx \times \delta(y) dy \qquad (2.4.4)$$

on $\mathbf{R}^n \times \mathbf{R}^l$ and the standard Wiener measure $\mathfrak{W}$ on $C([0,T], \mathbf{R}^p)$. Here $\check{x}_0$, $Y_0 > 0$ denote the mean and variance of the Gaussian variable x_0, and $\delta(y)$ denotes the delta-function on $\mathbf{R}^l$. Also as in [39], we endow the space Ω with the filtration $\{\mathcal{F}_t, t \geq 0\}$ generated by the mappings $\{\Pi_t, t \geq 0\}$ where

$$\Pi_0(x, \eta, W(\cdot)) = (x, \eta)$$

and

$$\Pi_t(x, \eta, W(\cdot)) = W(t)$$

[3] In this example, one must take into account the fact that $(FC)^2 = 0$ when proceeding from the formal Langevin equation to a corresponding mathematically rigorous Ito equation.

for $t > 0$. This filtration is completed by including all corresponding sets of P-probability zero. The random variable x_0 and the Wiener process $W(\cdot)$ are stochastically independent in $(\Omega, \mathcal{F}, P)$. In this and subsequent sections, we will only consider random variables which are measurable functionals of the Wiener process $W(\cdot)$ and the random variable x_0. Thus without loss of generality, we assume that $\mathcal{F}$ is the minimum σ-algebra which contains all of the σ-algebras $\mathcal{F}_t$, $t \geq 0$. This fact is denoted by

$$\mathcal{F} = \bigvee_{t \geq 0} \mathcal{F}_t. \tag{2.4.5}$$

Stochastic nominal system

On the probability space defined above, we consider the system and measurement dynamics driven by the noise input $W(\cdot)$ and a control input $u(\cdot)$, as described by the following stochastic differential equations:

$$\begin{aligned} dx(t) &= (A(t)x(t) + B_1(t)u(t))dt + B_2(t)dW(t); \\ &\qquad\qquad\qquad\qquad\qquad\qquad\qquad x(0) = x_0; \\ z(t) &= C_1(t)x(t) + D_1(t)u(t); \\ dy(t) &= C_2(t)x(t)dt + D_2(t)dW(t); \qquad y(0) = 0. \end{aligned} \tag{2.4.6}$$

In the above equations, $x(t) \in \mathbf{R}^n$ is the state, $u(t) \in \mathbf{R}^m$ is the control input, $z(t) \in \mathbf{R}^q$ is the uncertainty output, and $y(t) \in \mathbf{R}^l$ is the measured output. All coefficients in equations (2.4.6) are assumed to be deterministic sufficiently smooth matrix valued functions mapping $[0, T]$ into the spaces of matrices of corresponding dimensions.

We will consider control inputs $u(\cdot)$ adapted to the filtration $\{\mathcal{Y}_t, t \geq 0\}$, generated by the observation process y,

$$\mathcal{Y}_t = \sigma\{y(s), s \leq t\}.$$

This assumption means that attention will be restricted to output-feedback controllers of the form

$$u(t) = \mathcal{K}(t, y(\cdot)|_0^t) \tag{2.4.7}$$

where $\mathcal{K}(\cdot)$ is a non-anticipative function. That is, for any $t \geq 0$, the function $\mathcal{K}(t, \cdot) : C([0, T], \mathbf{R}^q) \to \mathbf{R}^m$ is $\mathcal{Y}_t$-measurable. Also, we assume that the function $\mathcal{K}(t, y)$ is piecewise continuous in t and Lipschitz continuous in y.

It is assumed that the stochastic differential equation defined by (2.4.6), (2.4.7) has a unique strong solution.

Stochastic uncertain system

An important feature of the stochastic uncertain systems considered in [151, 152, 153, 229, 235] is the description of stochastic uncertainty in terms of a given set

$\mathcal{P}$ of probability measures. To define the class of uncertain stochastic systems in which the uncertainty satisfies a relative entropy constraint, we begin with a set $\mathcal{P}$ of all the probability measures Q such that $Q(\Lambda) = P(\Lambda)$ for $\Lambda \in \mathcal{F}_0$ and

$$h(Q\|P) < +\infty. \tag{2.4.8}$$

In equation (2.4.8), $h(Q\|P)$ denotes the *relative entropy* between a probability measure Q and the given reference probability measure P; e.g., see Definition A.2 in Appendix A and also [42, 53]. It is worth mentioning that the relative entropy $h(Q\|P)$ can be regarded as a measure of the "distance" between the probability measure Q and the reference probability measure P. Note that the relative entropy is a convex, lower semicontinuous functional of Q; e.g., see [53, 44]. This fact implies that the set $\mathcal{P}$ defined above is convex.

We now define the uncertainty entering into the system in terms of a certain relative entropy constraint. This uncertainty description can be thought of as extending the integral quadratic constraint uncertainty description given in Section 2.3.1 to the case of stochastic uncertain systems. Also, this uncertainty description represents a continuous-time version of the discrete-time relative entropy constraint uncertainty description introduced in [152, 153].

In order to formally introduce our definition of the relative entropy constraint, we first let a symmetric nonnegative definite weighting matrix M be given. Also, let d be a given positive $\mathcal{F}_0$-measurable random variable and let $\mathbf{E}^Q$ denote the expectation with respect to the probability measure Q.

Definition 2.4.2 *A probability measure $Q \in \mathcal{P}$ is said to define an admissible uncertainty if the following* stochastic uncertainty constraint *is satisfied:*

$$h(Q\|P) \leq \frac{1}{2}\mathbf{E}^Q \left[d + x'(T)Mx(T) + \int_0^T \|z(t)\|^2 dt \right]. \tag{2.4.9}$$

In (2.4.9), $x(\cdot)$, $z(\cdot)$ are defined by equations (2.4.6).

We denote the set of probability measures defining the admissible uncertainties by Ξ. Elements of Ξ are also called admissible probability measures.

A representation of relative entropy

It is known (see [39]), that each probability measure $Q \in \mathcal{P}$ can be characterized in terms of a progressively measurable stochastic process $(\xi(t), \mathcal{F}_t)$, $t \in [0,T]$, such that the stochastic process $\tilde{W} = (\tilde{W}(t), \mathcal{F}_t)$, $t \in [0,T]$, as defined by

$$\tilde{W}(t) := W(t) - \int_0^t \xi(t)dt, \tag{2.4.10}$$

is a Wiener process on $(\Omega, \mathcal{F}, Q)$. Indeed, the Radon-Nikodym Theorem (see [117]) implies that each probability measure $Q \in \mathcal{P}$ can be characterized in terms of a P-a.s. unique nonnegative random variable $\zeta(\omega)$ defined as follows:

$$dQ(\omega) = \zeta(\omega)dP(\omega).$$

It is evident that

$$\mathbf{E}\zeta < \infty.$$

We now define a stochastic process

$$\zeta(t) = \mathbf{E}(\zeta|\mathcal{F}_t)$$

for $t \in [0,T]$, where the conditional expectations are chosen so that the process $\zeta(t)$ has right continuous trajectories. This can be done, since the sub-σ-field $\mathcal{F}_t$ has been completed with respect to the probability measure P; see Theorem 3.1 of [117]. Then $(\zeta(t),\mathcal{F}_t)$ is a martingale with right continuous trajectories such that

$$\sup_{t\in[0,T]} \mathbf{E}|\zeta(t)| < \infty$$

and

$$\mathbf{E}\zeta(0) = 1.$$

Hence, it follows from Theorem 5.7 of [117] that for each martingale satisfying these conditions, there exists a stochastic process $(\hat{\xi}(t),\mathcal{F}_t)$ with $\hat{\xi}(t) \in \mathbf{R}^p$ such that

$$P\left(\int_0^T \|\hat{\xi}(s,\omega)\|^2 ds < \infty\right) = 1 \tag{2.4.11}$$

and

$$\zeta(t) = 1 + \int_0^t \hat{\xi}'(s,\omega)dW(s). \tag{2.4.12}$$

The representation given by (2.4.12) is unique modulo sets of probability zero. Furthermore, it follows from Theorem 6.2 of [117] that the stochastic process $\tilde{W} = (\tilde{W}(t),\mathcal{F}_t)$ defined by (2.4.10) where

$$\xi(t,\omega) = \begin{cases} \zeta^{-1}(t,\omega)\hat{\xi}(t,\omega) & \text{if } \zeta(t,\omega) > 0; \\ 0 & \text{if } \zeta(t,\omega) = 0 \end{cases}$$

is a Wiener process on $(\Omega,\mathcal{F},Q)$. Note that it follows from Lemma 6.5 of [117] that

$$\xi(t) = \zeta^{-1}(t)\hat{\xi}(t) \ Q\text{-a.s.}$$

Also, it is proved on page 228 of [117] that

$$Q\left(\int_0^T \|\xi(s,\omega)\|^2 ds < \infty\right) = 1 \tag{2.4.13}$$

and the representation given by (2.4.12) is unique. Thus, each probability measure $Q \in \mathcal{P}$ generates a unique process $(\xi(t),\mathcal{F}_t)$ as required. Furthermore,

$$h(Q\|P) = \frac{1}{2}\mathbf{E}^Q \int_0^T \|\xi(t)\|^2 dt; \tag{2.4.14}$$

see [53, 39].

Conversely, the following lemma establishes the fact that under certain conditions, the uncertainty input $\xi(\cdot)$ can be characterized in terms of a probability measure $Q \in \mathcal{P}$. The existence of such a probability measure follows from Girsanov's Theorem; see [117]. This result is summarized in the following lemma.

Lemma 2.4.1 *Let $(\xi(t), \mathcal{F}_t)$, $0 \le t \le T$, be a random process such that*

$$P\left(\int_0^T \|\xi(s)\|^2 ds < \infty\right) = 1; \tag{2.4.15}$$

$$\mathbf{E} \exp\left(\frac{1}{2}\int_0^T \|\xi(s)\|^2 ds\right) < \infty. \tag{2.4.16}$$

Then the equation

$$\zeta(t) = 1 + \int_0^t \zeta(s)\xi(s)' dW(s). \tag{2.4.17}$$

defines a continuous positive martingale $\zeta(t)$, $t \in [0, T]$. Equivalently, martingale $\zeta(t)$ is given by the equation

$$\zeta(t) = \exp\left(\int_0^t \xi'(s) dW(s) - \frac{1}{2}\|\xi(s)\|^2 ds\right) \tag{2.4.18}$$

Furthermore, the random process $\tilde{W}(\cdot)$ defined by equation (2.4.10) is a Wiener process with respect to the system $\{\mathcal{F}_t\}$, $0 \le t \le T$, and the probability measure Q such that $Q(d\omega) = \zeta(T, \omega)P(d\omega)$.

Proof: Conditions (2.4.15) are the conditions of Novikov's Theorem (e.g., see Theorem 6.1 on page 216 of [117]). It follows from this theorem that the random process $(\zeta(t), \mathcal{F}_t)$, $0 \le t \le T$, defined by equation (2.4.18) is a martingale and in particular, $\mathbf{E}\zeta(t) = 1$ for all $0 \le t \le T$. The statement of the lemma now follows from Girsanov's Theorem; e.g., see Theorem 6.3 on page 232 of [117].

■

Lemma 2.4.1 establishes that the probability measure Q is absolutely continuous with respect to the reference probability measure P, $Q \ll P$. If in addition

$$\mathbf{E}^Q \int_0^T \|\xi(s)\|^2 ds < \infty,$$

then from (2.4.14), $h(Q\|P) < \infty$ and hence $Q \in \mathcal{P}$.

An equivalent representation of the stochastic uncertain system

With the above representation of the uncertain probability measure, the system (2.4.6) becomes a system of the following form on the probability space

$(\Omega, \mathcal{F}, Q)$:

$$\begin{aligned} dx(t) &= (A(t)x(t) + B_1(t)u(t) + B_2(t)\xi(t))dt + B_2(t)d\tilde{W}(t); \\ z(t) &= C_1(t)x(t) + D_1(t)u(t); \\ dy(t) &= (C_2(t)x(t) + D_2(t)\xi(t))dt + D_2(t)d\tilde{W}(t). \end{aligned} \tag{2.4.19}$$

This is the standard form for a stochastic uncertain system driven by an uncertainty input $\xi(t)$ and by the additive noise input described by the Wiener process $\tilde{W}(t)$. Equations (2.4.19) can be viewed as an equivalent representation of the original stochastic uncertain system (2.4.6). Also, the stochastic uncertainty constraint (2.4.9) becomes a constraint of the form

$$\mathbf{E}^Q \int_0^T \|\xi(t)\|^2 dt \leq \mathbf{E}^Q \left[d + x'(T)Mx(T) + \int_0^T \|z(t)\|^2 dt \right]. \tag{2.4.20}$$

In (2.4.20), $x(\cdot)$, $z(\cdot)$ are defined by equations (2.4.6), or equivalently by equations (2.4.19). The process $\xi(\cdot)$ corresponds to the probability measure Q as described above. In particular, $\xi(\cdot)$ is an $\mathcal{F}_t$-progressively measurable process for which equation (2.4.19) has a unique weak solution and the right-hand side of equation (2.4.14) is finite; see [39]. Also note that the function $\xi(\cdot) \equiv 0$ corresponds to the case $Q = P$. As mentioned above, the constraint (2.4.9) and equivalently, the constraint (2.4.20), are satisfied strictly in this case.

Remark 2.4.4 The uncertainty description (2.4.20) suggests an analogy between the stochastic uncertainty constraint description (2.4.9) and its deterministic IQC counterpart considered in this chapter. First, it should be noted that the uncertainty description derived from the stochastic uncertainty constraint (2.4.20) allows for the stochastic uncertainty inputs to depend dynamically on the uncertainty outputs. This may be the case in many realistic situations such as shown in Figure 2.4.4; see the example given below.

Motivating example

We now consider an example which leads to a stochastic uncertain system of the form described above. Consider a linear system driven by a standard vector Gaussian white noise process $(\alpha(t), \beta(t))'$:

$$\begin{aligned} \dot{x} &= Ax + B_1u + \tilde{B}_2\alpha(t); \\ \dot{y} &= C_2x + \tilde{D}_2\beta(t); \\ z &= C_1x + D_1u. \end{aligned} \tag{2.4.21}$$

A block diagram of this system is shown in Figure 2.4.3. A formal mathematical description of the system (2.4.21) can be given by equation (2.4.6) with $B_2 = \left[\tilde{B}_2 \ 0 \right]$ and $D_2 = \left[0 \ \tilde{D}_2 \right]$. We suppose that the uncertainties in this system can

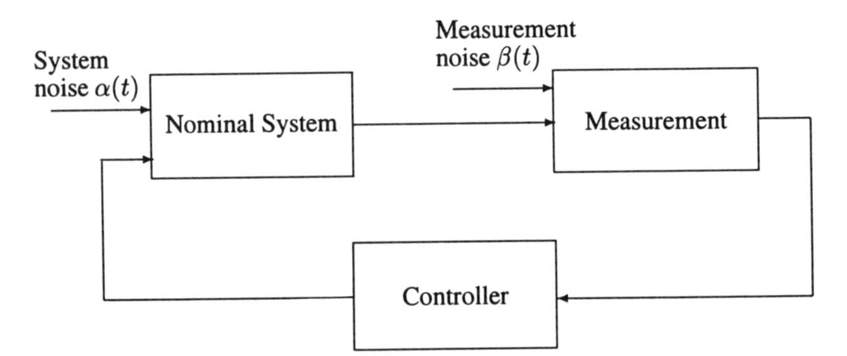

Figure 2.4.3. Nominal system block diagram

be described by introducing the following perturbations of the noise inputs:

$$\alpha(t) = \alpha^{\phi}(t) + \phi(t, z(\cdot)|_0^t); \qquad \beta(t) = \beta^{\psi}(t) + \psi(t, z(\cdot)|_0^t). \tag{2.4.22}$$

In the above equations, $\alpha^{\phi}(\cdot)$ and $\beta^{\psi}(\cdot)$ are uncertain noise inputs, and the functions $\phi(\cdot,\cdot)$ and $\psi(\cdot,\cdot)$ describe the uncertain dynamics. A block diagram of the corresponding perturbed system is shown in Figure 2.4.4. In the presence of per-

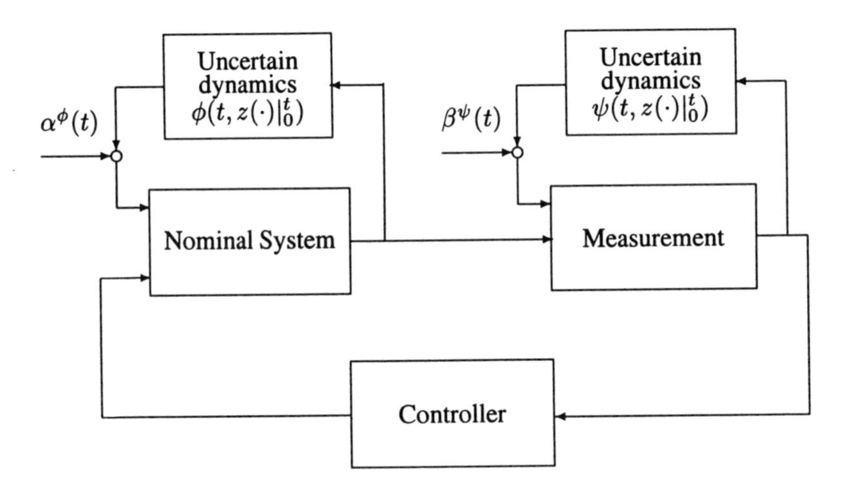

Figure 2.4.4. Uncertain system block diagram

turbations, the inputs $\alpha(\cdot)$ and $\beta(\cdot)$ in the system (2.4.21) cease to be Gaussian white noise processes when considered on the probability space $(\Omega, \mathcal{F}, P)$. However, it is still possible to give a mathematically tractable description of the system (2.4.21) in terms of a stochastic differential equation. Indeed, note that by the substitution of (2.4.22) into equation (2.4.21), it follows that the perturbed system is described by the equations

$$\begin{aligned} \dot{x} &= Ax + B_1 u + \tilde{B}_2 \phi(t, z(\cdot)|_0^t) + \tilde{B}_2 \alpha^{\phi}(t); \\ \dot{y} &= C_2 x + \tilde{D}_2 \psi(t, z(\cdot)|_0^t) + \tilde{D}_2 \beta^{\psi}(t); \\ z &= C_1 x + D_1 u. \end{aligned} \quad (2.4.23)$$

Also, suppose the control input $u(\cdot)$ and perturbation functions $\phi(\cdot)$ and $\psi(\cdot)$ are such that the uncertainty input

$$\xi(t) = \left[\phi(t, z(\cdot)|_0^t) \;\; \psi(t, z(\cdot)|_0^t) \right]$$

of the closed loop system is adapted to the filtration $\{\mathcal{F}_t^W, t \in [0, T]\}$ and square integrable. Consider the process $\tilde{W}(t)$ defined by equation (2.4.10). In order to refer to $\tilde{W}(t)$ as a Wiener process, we introduce a change of probability measure using Girsanov's Theorem as described above. With this change of probability measure, equation (2.4.19) arises naturally as a formal mathematical description of the perturbed system (2.4.23).

In this case, the uncertainty constraint (2.4.9) has the following interpretation. Condition (2.4.9) in this case restricts the $\mathbf{L}_2$-norm of the uncertainty dynamic processes $\phi(t, z(\cdot)|_0^t)$, $\psi(t, z(\cdot)|_0^t)$ as follows:

$$\begin{aligned} &\mathbf{E}^Q \int_0^T \left[\|\phi(t, z(\cdot)|_0^t)\|^2 + \|\psi(t, z(\cdot)|_0^t)\|^2 \right] dt \\ &\quad \leq \mathbf{E}^Q \left[d + x'(T) M x(T) + \int_0^T \|z(t)\|^2 dt \right] \end{aligned} \quad (2.4.24)$$

where $M \geq 0$ is a given weighting matrix and $d > 0$ is a given constant. Note that the uncertainty inputs $\phi(t, z(\cdot)|_0^t)$, $\psi(t, z(\cdot)|_0^t)$ depend on the uncertainty output $z(\cdot)$ and hence depend on the applied control input $u(\cdot)$. As mentioned above, this is in contrast to standard differential games where minimizing and maximizing strategies are chosen independently; e.g., see [9].

The constraints (2.4.24) and (2.4.20), can be thought of as extensions of the standard norm-bounded uncertainty constraint. Indeed, in the case of norm-bounded uncertainty, the uncertain dynamics are described by the equations

$$\phi(t, z) = \Delta_1(t) z, \qquad \psi(t, z) = \Delta_2(t) z,$$

where the matrix functions $\Delta_1(\cdot)$ and $\Delta_2(\cdot)$ satisfy the constraint

$$\|[\Delta_1'(t) \;\; \Delta_2'(t)]\| \leq 1.$$

It is easy to see that condition (2.4.24) is satisfied in this case with $M = 0$, and any constant $d > 0$.

2.4.3 *Stochastic uncertain systems with additive noise: Infinite-horizon relative entropy constraints*

In this section, we introduce an uncertainty description for stochastic uncertain systems with additive noise which can be regarded as the extension of the uncertainty description considered in Section 2.4.2 to the case of an infinite time horizon. As in Section 2.4.2, the stochastic uncertain systems to be considered are described by a nominal system which is driven by a Wiener process on a reference probability space, and also by a set of perturbation probability measures satisfying a certain uncertainty constraint. This uncertainty constraint restricts the relative entropy between an admissible perturbation probability measure and the reference probability measure.

As we proceed from the finite-horizon case to the infinite-horizon case, the fact that the systems under consideration are those with additive noise becomes important. The solutions to such systems do not necessarily belong to $\mathbf{L}_2[0, \infty)$. Hence, we will be interested in the time averaged properties of the system. Accordingly, the class of admissible uncertainty inputs will include bounded power processes rather than bounded energy processes. In order for such an uncertainty input to generate an uncertainty probability measure on the measurable space of sample functions defined on an infinite time interval, the random process $\zeta(t)$ defined in equation (2.4.18) must be a uniformly integrable martingale. This requirement severely limits the class of uncertainties which can be considered. For example, this requirement rules out standard H^∞ norm-bounded uncertainties. In order to remove this limitation in the infinite-horizon case, we make use of an uncertainty model in which the uncertainty is characterized using a family of probability measures on the measurable spaces $(\Omega, \mathcal{F}_T)$ rather than a probability measure on the measurable space $(\Omega, \mathcal{F})$.

Except for the feature mentioned above, we adopt a set up similar to that of Section 2.4.2. Let $(\Omega, \mathcal{F}, P)$ be a complete probability space. As in Section 2.4.2, the space Ω can be thought of as the noise space $\mathbf{R}^n \times \mathbf{R}^l \times C([0, \infty), \mathbf{R}^p)$. The probability measure P can then be defined as the product of a given probability measure on $\mathbf{R}^n \times \mathbf{R}^l$ and the standard Wiener measure on $C([0, \infty), \mathbf{R}^p)$. Also, let a p-dimensional standard Wiener process $W(\cdot)$ and a Gaussian random variable $x_0 : \Omega \to \mathbf{R}^n$ with mean $\check{x}_0$ and non-singular covariance matrix Y_0 be defined as in Section 2.4.2, $p = r + l$. The space Ω is endowed with a complete filtration $\{\mathcal{F}_t, t \geq 0\}$ constructed as in Section 2.4.2. The random variable x_0 and the Wiener process $W(\cdot)$ are stochastically independent on $(\Omega, \mathcal{F}, P)$.

As in Section 2.4.2, we will consider only random variables which are measurable functionals of the Wiener process $W(\cdot)$ and the random variable x_0. Thus, without loss of generality, we assume that the σ-field $\mathcal{F}$ is the minimum σ-algebra which contains all of the σ-algebras $\mathcal{F}_t$, $t \geq 0$; see equation (2.4.5) of Section 2.4.2.

Stochastic nominal system

On the probability space defined above, we consider the system and measurement dynamics driven by the noise input $W(\cdot)$ and a control input $u(\cdot)$, described by the following stochastic differential equation:

$$\begin{aligned} dx(t) &= (Ax(t) + B_1u(t))dt + B_2dW(t); \quad x(0) = x_0; \\ z(t) &= C_1x(t) + D_1u(t); \\ dy(t) &= C_2x(t)dt + D_2dW(t); \quad y(0) = 0. \end{aligned} \tag{2.4.25}$$

In the above equations, $x(t) \in \mathbf{R}^n$ is the state, $u(t) \in \mathbf{R}^m$ is the control input, $z(t) \in \mathbf{R}^q$ is the uncertainty output, and $y(t) \in \mathbf{R}^l$ is the measured output. All coefficients in equations (2.4.25) are assumed to be constant matrices of corresponding dimensions. Thus, the system (2.4.25) is a time-invariant version of the system (2.4.6) considered in the previous section.

In the infinite-horizon case, we consider linear output-feedback controllers $u(\cdot)$ of the form

$$\begin{aligned} d\hat{x} &= A_c\hat{x} + B_c dy(t); \\ u &= K\hat{x} \end{aligned} \tag{2.4.26}$$

where $\hat{x} \in \mathbf{R}^{\hat{n}}$ is the state of the controller and $A_c \in \mathbf{R}^{\hat{n}\times\hat{n}}$, $K \in \mathbf{R}^{m\times\hat{n}}$, and $B_c \in \mathbf{R}^{\hat{n}\times q}$. Note that the controller (2.4.26) is adapted to the filtration $\{\mathcal{Y}_t, t \geq 0\}$ generated by the observation process y.

Stochastic uncertain system

As in Section 2.4.2, the stochastic uncertain systems to be considered are described by the nominal system (2.4.25) considered over the probability space $(\Omega, \mathcal{F}, P)$, and also by a set of perturbations of the reference probability measure P. These perturbations are defined as follows. Consider the set $\mathcal{M}$ of continuous positive martingales $(\zeta(t), \mathcal{F}_t, t \geq 0)$ such that for each $T \geq 0$, $\mathbf{E}\zeta(T) = 1$. Note that the set $\mathcal{M}$ is convex.

Every martingale $\zeta(\cdot) \in \mathcal{M}$ gives rise to a probability measure Q^T on the measurable space $(\Omega, \mathcal{F}_T)$ defined by the equation

$$Q^T(d\omega) = \zeta(T)P^T(d\omega). \tag{2.4.27}$$

Here, P^T denotes the restriction of the reference probability measure P to $(\Omega, \mathcal{F}_T)$. From this definition, for every $T > 0$, the probability measure Q^T is absolutely continuous with respect to the probability measure P^T, $Q^T \ll P^T$. The uncertain system is described by state equation (2.4.25) considered over the probability spaces $(\Omega, \mathcal{F}_T, Q^T)$ for every $T > 0$.

A connection between the disturbance signal uncertainty model considered in the previous sections of this chapter and the perturbation martingale uncertainty model is based on Novikov's Theorem; see Lemma 2.4.1. Using the result of Lemma 2.4.1, a given disturbance input $\xi(\cdot)$ (satisfying the conditions of this lemma on every finite interval $[0, T]$) can be associated with a martingale $\zeta(\cdot) \in \mathcal{M}$ which is defined by equation (2.4.17) or equivalently, by equation (2.4.18). Furthermore, for each $T > 0$, a probability measure Q^T and a Wiener process $\tilde{W}(\cdot)$ can be defined on the measurable space $(\Omega, \mathcal{F}_T)$ as described in Lemma 2.4.1:

$$\tilde{W}(t) := W(t) - \int_0^t \xi(t)dt. \tag{2.4.28}$$

On the probability space $(\Omega, \mathcal{F}_T, Q^T)$, the state equation

$$\begin{aligned} dx &= (Ax + B_1 u + B_2 \xi)dt + B_2 d\tilde{W}(t); \\ z &= C_1 x + D_1 u; \\ dy &= (C_2 x + D_2 \xi)dt + D_2 d\tilde{W}(t) \end{aligned} \tag{2.4.29}$$

is a time-invariant version of equation (2.4.6). Therefore, the system (2.4.29) provides a representation of the above stochastic uncertain system. Note that the case $\xi(\cdot) \equiv 0$ corresponds to $\zeta(t) \equiv 1$ and $Q^T = P^T$ for all $T \geq 0$.

The stochastic uncertainty constraint

We now present our infinite-horizon uncertainty description for stochastic uncertain systems with additive noise. This uncertainty description may be regarded as an extension of the uncertainty description considered in Section 2.4.2 and references [152, 153, 235] to the infinite-horizon case. Also, this uncertainty description can be thought of as an extension of the deterministic integral quadratic constraint uncertainty description to the case of stochastic uncertain systems with additive noise. Recall that the integral quadratic constraints arising in the deterministic case (see Section 2.3) and the stochastic multiplicative noise case (see Section 2.4.1 and the paper [234]) exploit a sequence of times $\{t_i\}_{i=1}^{\infty}$ to "localize" the uncertainty inputs and uncertainty outputs to time intervals $[0, t_i]$. The consideration of the system dynamics on these finite time intervals then allows one to deal with bounded energy processes. However in this section, the systems under consideration are those with additive noise. For this class of stochastic systems, it is natural to consider bounded power processes rather than bounded energy processes. This motivates us to modify the integral quadratic constraint uncertainty description to accommodate bounded power processes.

In contrast to the case of deterministic integral quadratic constraints, the uncertainty description considered in this subsection exploits a sequence of continuous positive martingales $\{\zeta_i(t), \mathcal{F}_t, t \geq 0\}_{i=1}^{\infty}$ which converges to a limiting martingale $\zeta(\cdot)$ in the following sense: For any $T > 0$, the sequence $\{\zeta_i(T)\}_{i=1}^{\infty}$ converges weakly to $\zeta(T)$ in $\mathbf{L}_1(\Omega, \mathcal{F}_T, P^T)$. Using the martingales $\zeta_i(t)$, we define a sequence of probability measures $\{Q_i^T\}_{i=1}^{\infty}$ as follows:

$$Q_i^T(d\omega) = \zeta_i(T) P^T(d\omega). \tag{2.4.30}$$

From the definition of the martingales $\zeta_i(t)$, it follows that for each $T > 0$, the sequence $\{Q_i^T\}_{i=1}^{\infty}$ converges to the probability measure Q^T corresponding to a limiting martingale $\zeta(\cdot)$ in the following sense: For any bounded $\mathcal{F}_T$-measurable random variable η,

$$\lim_{i\to\infty} \int_{\Omega} \eta Q_i^T(d\omega) = \int_{\Omega} \eta Q^T(d\omega). \tag{2.4.31}$$

We denote this fact by $Q_i^T \Rightarrow Q^T$ as $i \to \infty$.

Remark 2.4.5 The property $Q_i^T \Rightarrow Q^T$ implies that the sequence of probability measures Q_i^T converges weakly to the probability measure Q^T. Indeed, consider the Polish space of probability measures on the measurable space $(\Omega, \mathcal{F}_T)$ endowed with the topology of weak convergence of probability measures. Note that Ω is a metric space. Hence, such a topology can be defined; e.g., see [53]. For the sequence $\{Q_i^T\}$ to converge weakly to Q^T, it is required that equation (2.4.31) holds for all bounded continuous random variables η. Obviously, this requirement is satisfied if $Q_i^T \Rightarrow Q^T$.

As in Section 2.4.2, we describe the class of admissible uncertainties in terms of the relative entropy functional $h(\cdot\|\cdot)$; for the definition and properties of the functional $h(\cdot\|\cdot)$, see Appendix A and also [53].

Definition 2.4.3 *Let d be a given positive constant. A martingale $\zeta(\cdot) \in \mathcal{M}$ is said to define an admissible uncertainty if there exists a sequence of continuous positive martingales $\{\zeta_i(t), \mathcal{F}_t, t \geq 0\}_{i=1}^{\infty}$ which satisfies the following conditions:*

(i) For each i, $h(Q_i^T\|P^T) < \infty$ for all $T > 0$;

(ii) For all $T > 0$, $Q_i^T \Rightarrow Q^T$ as $i \to \infty$;

(iii) The following stochastic uncertainty constraint *is satisfied: For any sufficiently large $T > 0$ there exists a constant $\delta(T)$ such that $\lim_{T\to\infty} \delta(T) = 0$ and*

$$\inf_{T'>T} \frac{1}{T'} \left[\frac{1}{2} \mathbf{E}^{Q_i^{T'}} \int_0^{T'} \|z(t)\|^2 dt - h(Q_i^{T'}\|P^{T'}) \right] \geq -\frac{d}{2} + \delta(T) \tag{2.4.32}$$

for all $i = 1, 2, \ldots$. In (2.4.32), the uncertainty output $z(\cdot)$ is defined by equation (2.4.25) considered on the probability space $(\Omega, \mathcal{F}_T, Q_i^T)$.

In the above conditions, Q_i^T is the probability measure defined by (2.4.30) corresponding to the martingale $\zeta_i(t)$ and time $T > 0$. We let Ξ denote the set of martingales $\zeta(\cdot) \in \mathcal{M}$ corresponding to admissible uncertainties. Elements of Ξ are also called admissible martingales.

Observe that the reference probability measure P corresponds to the admissible martingale $\zeta(t) \equiv 1$. Hence, the set Ξ is not empty. Indeed, choose $\zeta_i(t) = 1$ for all i and t. Then, $Q_i^T = P^T$ for all i. It follows from the identity $h(P^T \| P^T) = 0$ that

$$\inf_{T'>T} \frac{1}{T'} \left[\frac{1}{2} \mathbf{E}^{Q_i^{T'}} \int_0^{T'} \|z(t)\|^2 dt - h(Q_i^{T'} \| P^{T'}) \right]$$
$$= \inf_{T'>T} \frac{1}{2T'} \mathbf{E} \int_0^{T'} \|z(t)\|^2 dt.$$

Note that the expectations are well defined. Also, the infimum on the right hand side of the above equation is nonnegative for any $T > 0$. Therefore, for any constant $d > 0$, one can find a sufficiently small $\delta = \delta(T)$ such that $\lim_{T\to\infty} \delta(T) = 0$ and the constraint (2.4.32) is satisfied strictly in this case.

Remark 2.4.6 Note that condition (2.4.32) implies that

$$\liminf_{T\to\infty} \frac{1}{T} \left[\frac{1}{2} \mathbf{E}^{Q_i^T} \int_0^T \|z(t)\|^2 dt - h(Q_i^T \| P^T) \right] \geq -\frac{d}{2}$$

for all $i = 1, 2, \ldots$.

The definition of admissible uncertainties given above involves a collection of martingales $\{\zeta_i(\cdot)\}_{i=1}^{\infty}$ which has a given uncertainty martingale $\zeta(\cdot)$ as its limit point. In the deterministic case and the multiplicative noise case, similar approximations were defined by restricting uncertainty inputs to finite time intervals and then extending the restricted processes by zero beyond these intervals; *c.f.* Definitions 2.3.1 and 2.4.1. In the case of a stochastic uncertain system with additive noise considered on an infinite time interval, we apply a similar idea. However, in contrast to the deterministic and multiplicative noise cases, we use a sequence of martingales and corresponding probability measures in Definition 2.4.3. This procedure may be thought of as involving a spatial restriction rather than the temporal restriction used previously. A natural way to define the required sequence of martingales and corresponding probability measures is to consider martingales corresponding to the uncertainty inputs being "truncated" at certain Markov times $\mathfrak{t}_i$. For example, this can be achieved by choosing an expanding sequence of compact sets K_i in the uncertainty space and letting $\mathfrak{t}_i$ be the Markov time when the disturbance input reaches the boundary of the set K_i. In this case, we focus on spatial domains rather than time intervals on which the uncertainty inputs and uncertainty outputs are then constrained. An illustration of this idea is given below.

Connection to H^∞ norm-bounded uncertainty

The description of stochastic uncertainty presented in Definition 2.4.3 encompasses many important classes of uncertainty arising in control systems. As an example illustrating this fact, we will show that the standard H^∞ norm-bounded uncertainty description can be incorporated into the framework of Definition 2.4.3.

In a similar fashion, one can also show that the cone-bounded uncertainty description defines an admissible class of uncertainties according to Definition 2.4.3.

In the sequel, we will use the following well known property of linear stochastic systems. On the probability space $(\Omega, \mathcal{F}, \tilde{P})$, consider the following linear system driven by the Wiener process $\tilde{W}(\cdot)$ and a disturbance input $\xi(t)$, $t \in [0, T]$:

$$d\bar{x} = (\bar{A}\bar{x} + \bar{B}\xi(t))dt + \bar{B}d\tilde{W}(t). \tag{2.4.33}$$

Proposition 1 *If for some constant $\rho > 0$,*

$$\tilde{\mathbf{E}} \int_0^T \|\xi(t)\|^2 dt \leq \rho, \tag{2.4.34}$$

then the corresponding solution to the equation (2.4.33) is mean square bounded on the interval $[0, T]$. Here $\tilde{\mathbf{E}}$ denotes the expectation with respect to the probability measure $\tilde{P}$.

Proof: We choose a positive-definite symmetric matrix H and a positive constant α such that

$$\bar{A}'H + H\bar{A} + H\bar{B}\bar{B}'H \leq \alpha H.$$

This condition can always be satisfied by choosing a sufficiently large $\alpha > 0$. Then, using Ito's formula along with the above condition, one obtains the following inequality:

$$\begin{aligned}
&\tilde{\mathbf{E}}\bar{x}(t)'H\bar{x}(t) - \tilde{\mathbf{E}}\bar{x}(0)'H\bar{x}(0) \\
&= 2\tilde{\mathbf{E}} \int_0^t \left(\bar{A}\bar{x}(s) + \bar{B}\xi(s)\right)' H\bar{x}(s)ds + \int_0^t \operatorname{tr} \bar{B}\bar{B}'H ds \\
&\leq \tilde{\mathbf{E}} \int_0^t \bar{x}(s)' \left(\bar{A}'H + H\bar{A} + H\bar{B}\bar{B}'H\right) \bar{x}(s)dt \\
&\quad + \tilde{\mathbf{E}} \int_0^T \|\xi(s)\|^2 ds + \operatorname{tr} \bar{B}\bar{B}'H \cdot T \\
&\leq \alpha\tilde{\mathbf{E}} \int_0^t \bar{x}(s)'H\bar{x}(s)dt + \rho + \operatorname{tr} \bar{B}\bar{B}'H \cdot T
\end{aligned}$$

Using Gronwall's Lemma, it follows from the above inequality that

$$\tilde{\mathbf{E}}\bar{x}(t)'H\bar{x}(t) \leq \left(\tilde{\mathbf{E}}\bar{x}(0)'H\bar{x}(0) + \rho + \operatorname{tr} \bar{B}\bar{B}'H \cdot T\right) e^{\alpha t}.$$

Thus, the proposition follows.

Consider an uncertain system of the form (2.4.25) on the probability space $(\Omega, \mathcal{F}, P)$, driven by a controller (2.4.26). Associated with the system (2.4.25) and controller (2.4.26), consider the disturbance input $\xi(\cdot)$ defined by the convolution operator

$$\xi(t) = \int_0^t g(t-\theta)z(\theta)d\theta \tag{2.4.35}$$

corresponding to a given causal uncertainty transfer function $\Delta(s)$ which belongs to the Hardy space H^∞. In equation (2.4.35), $z(\cdot)$ is the output of the closed loop system corresponding to the system (2.4.25) and a given controller (2.4.26).

Lemma 2.4.2 *Let an uncertainty transfer function $\Delta(s) \in H^\infty$ be given which satisfies the norm bound condition*

$$||\Delta(s)||_\infty \leq 1. \tag{2.4.36}$$

Also, suppose the random process $(\zeta(t), \mathcal{F}_t)$, defined by equation (2.4.17) is a martingale; here $\xi(\cdot)$ is the disturbance input generated by the operator (2.4.35). Then, this martingale satisfies the conditions of Definition 2.4.3. That is, there exists a sequence of martingales $\{\zeta_i(t)\}_{i=1}^\infty$ satisfying the requirements of Definition 2.4.3.

Remark 2.4.7 The requirements of Lemma 2.4.2 are satisfied if $\Delta(s)$ is a stable rational transfer function satisfying condition (2.4.36). Indeed, in this case one can show that the augmented dynamics $[x'(\cdot), \hat{x}'(\cdot), \eta'(\cdot), z'(\cdot), \xi'(\cdot)]'$ are described by a linear system driven by a Wiener process, with Gaussian initial condition; here η denotes the state of the uncertainty. Hence for any $T > 0$, there exists a constant δ_T such that

$$\sup_{t \leq T} \mathbf{E} \exp(\delta_T ||\xi(t)||^2) < \infty;$$

see the remark on page 138 of [117]. This implies that $\zeta(t)$ is a martingale; see Example 3 on page 220 of [117]. Hence, any uncertainty described by a stable rational transfer function satisfying condition (2.4.36) will belong to the class Ξ of uncertainties admissible for the system (2.4.25) controlled by a linear output-feedback controller of the form (2.4.26).

Proof of Lemma 2.4.2: Since the random process $(\zeta(t), \mathcal{F}_t), 0 \leq t \leq T$, defined by equation (2.4.17) is a martingale and $\mathbf{E}\zeta(T) = \mathbf{E}\zeta(0) = 1$, it follows from Girsanov's Theorem that the random process $\tilde{W}(\cdot)$ defined by equation (2.4.28) is a Wiener process with respect to the filtration $\{\mathcal{F}_t, 0 \leq t \leq T\}$, and the probability measure Q^T defined as in equation (2.4.27); see [117]. Note that on the probability space $(\Omega, \mathcal{F}_T, Q^T)$, the system (2.4.25) becomes the system of the form (2.4.29).

To verify that the martingale $\zeta(t)$ corresponding to the H^∞ norm-bounded uncertainty under consideration defines an admissible uncertainty, we need to prove the existence of a sequence of martingales $\{\zeta_i(t)\}_{i=1}^\infty$ satisfying the conditions of

Definition 2.4.3. To construct such a sequence, consider the following family of Markov stopping times[4] $\{\mathfrak{t}_\rho, \rho > 0\}$. For any $\rho > 0$, define

$$\mathfrak{t}_\rho := \begin{cases} \inf\{t \geq 0 : \int_0^t \|\xi(s)\|^2 ds > \rho\} & \text{if } \int_0^\infty \|\xi(s)\|^2 ds > \rho; \\ \infty & \text{if } \int_0^\infty \|\xi(s)\|^2 ds \leq \rho. \end{cases}$$

The family $\{\mathfrak{t}_\rho\}$ is monotonically increasing and $\mathfrak{t}_\rho \to \infty$ P-a.s.

We now are in a position to construct an approximating sequence of martingales $\{\zeta_i(t)\}_{i=1}^\infty$ using the above sequence of Markov stopping times. First, note that the stochastic integral $\mu(t) := \int_0^t \xi(s) dW(s)$ defines a local continuous martingale; see Definition 6 on page 69 of [117]. Also, for any stopping time $\mathfrak{t}_\rho$ defined above,

$$\mu(t \wedge \mathfrak{t}_\rho) = \int_0^{t \wedge \mathfrak{t}_\rho} \xi(s)' dW(s) = \int_0^t \xi_\rho(s)' dW(s) = \int_0^t \xi(s)' dW(s \wedge \mathfrak{t}_\rho)$$

where the process $\xi_\rho(\cdot)$ is defined as follows:

$$\xi_\rho(t) = \xi(t) \chi_{\{\mathfrak{t}_\rho \geq t\}}. \tag{2.4.37}$$

Here, χ_Λ denotes the indicator function of a set $\Lambda \subseteq \Omega$. In the above definitions, the notation $\mathfrak{t} \wedge t := \min\{t, \mathfrak{t}\}$ is used. The above properties are fundamental properties of Ito's integral.

Associated with the positive continuous martingale $\zeta(t)$ and the family of stopping times $\{\mathfrak{t}_\rho, \rho > 0\}$ defined above, consider the stopped process

$$\zeta_\rho(t) = \zeta(t \wedge \mathfrak{t}_\rho).$$

From this definition, $\zeta_\rho(t)$ is a continuous martingale; e.g., see Lemma 3.3 on page 69 of [117]. Furthermore, using the representation (2.4.17) of the martingale $\zeta(t)$, it follows that $\zeta_\rho(t)$ is an Ito process with the stochastic differential

$$d\zeta_\rho(t) = \zeta_\rho(t) \xi_\rho'(t) dW(t) = \zeta_\rho(t) d\mu(t \wedge \mathfrak{t}_\rho); \quad \zeta_\rho(0) = 1. \tag{2.4.38}$$

From (2.4.38), the martingale $\zeta_\rho(t)$ admits the following representation

$$\zeta_\rho(t) = \exp\left(\int_0^t \xi_\rho' dW(s) - \frac{1}{2} \int_0^t \|\xi_\rho(s)\|^2 ds \right). \tag{2.4.39}$$

Using the martingale $\zeta_\rho(t)$ defined above, we define probability measures Q_ρ^T on $(\Omega, \mathcal{F}_T)$ as follows:

$$Q_\rho^T(d\omega) = \zeta(T \wedge \mathfrak{t}_\rho) P^T(d\omega).$$

[4] A random variable $\mathfrak{t}$ taking values in $[0, \infty) \cup \{\infty\}$ is said to be the Markov stopping time (with respect to the filtration $\{\mathcal{F}_t, t \geq 0\}$) if for all $t \geq 0$, $\{\omega : \mathfrak{t}(\omega) \leq t\} \in \mathcal{F}_t$; see [117].

From (2.4.39), the relative entropy between the probability measures Q_ρ^T and P^T is given by

$$h(Q_\rho^T\|P^T) = \frac{1}{2}\mathbf{E}^{Q_\rho}\int_0^T \|\xi_\rho(s)\|^2 ds = \frac{1}{2}\mathbf{E}^{Q_\rho}\int_0^{\mathfrak{t}_\rho\wedge T} \|\xi(s)\|^2 ds. \quad (2.4.40)$$

From this equation and from (2.4.37), it follows that $h(Q_\rho^T\|P^T) \leq (1/2)\rho < \infty$ for all $T > 0$. Thus, the condition (i) of Definition 2.4.3 is satisfied.

Also, using part 1 of Theorem 3.7 on page 62 of [117], we observe that for every $T > 0$, the family $\{\zeta(\mathfrak{t}_\rho \wedge T), \rho > 0\}$ is uniformly integrable. Also, since $\mathfrak{t}_\rho \to \infty$ with probability one as $\rho \to \infty$, then $\zeta_\rho(T) \to \zeta(T)$ with probability one. This fact together with the property of uniform integrability of the family $\{\zeta_\rho(T), \rho > 0\}$ implies that

$$\lim_{\rho\to\infty} \mathbf{E}(|\zeta(T\wedge \mathfrak{t}_\rho) - \zeta(T)||\mathcal{G}) = 0 \qquad P\text{-a.s.}, \quad (2.4.41)$$

for any σ-algebra $\mathcal{G} \subset \mathcal{F}_T$; see the Corollary on page 16 of [117]. We now observe that for any $\mathcal{F}_T$-measurable bounded random variable η with values in $\mathbf{R}$,

$$\mathbf{E}|\eta\zeta(T\wedge \mathfrak{t}_\rho) - \eta\zeta(T)| \leq \sup_\omega |\eta| \cdot \mathbf{E}|\zeta(T\wedge \mathfrak{t}_\rho) - \zeta(T)|.$$

Therefore, it follows from the definition of the probability measures Q_ρ^T and Q^T and from equation (2.4.41) that

$$\lim_{\rho\to\infty}\int_\Omega \eta Q_\rho^T(d\omega) = \int_\Omega \eta Q^T(d\omega)$$

for any bounded $\mathcal{F}_T$-measurable random variable η. That is, $Q_\rho^T \Rightarrow Q^T$ as $\rho \to \infty$ for all $T > 0$. Thus, we have verified that the family of martingales $\zeta_\rho(t)$ satisfies condition (ii) of Definition 2.4.3.

We now consider the system (2.4.25) on the probability space $(\Omega, \mathcal{F}_T, Q_\rho^T)$. Equivalently, we consider the system (2.4.29) driven by the uncertainty input $\xi_\rho(t)$ on the probability space $(\Omega, \mathcal{F}_T, Q_\rho^T)$. Note that since $\int_0^T \|\xi_\rho(t)\|^2 \leq \rho$ P-a.s., Proposition 1 implies that the corresponding output $z(\cdot)$ of the system (2.4.25) satisfies the conditions

$$\mathbf{E}^{Q_\rho^T}\int_0^T \|z(s)\|^2 ds < \infty, \qquad \int_0^T \|z(s)\|^2 ds < \infty \quad Q_\rho^T\text{-a.s.} \quad (2.4.42)$$

for any $T > 0$. We now use the fact that condition (2.4.36) implies that for any pair $(\tilde{z}(\cdot), \tilde{\xi}(\cdot))$, $\tilde{z}(\cdot) \in \mathbf{L}_2[0,T], T > 0$, related by equation (2.4.35),

$$\int_0^T \|\tilde{\xi}(t)\|^2 dt \leq \int_0^T \|\tilde{z}(t)\|^2 dt;$$

e.g., see [271]. Hence from this observation and from (2.4.42), it follows that the pair $(z(\cdot), \xi(\cdot))$ where $z(\cdot)$ and $\xi(\cdot)$ are defined by the system (2.4.25) and the operator (2.4.35), satisfies the condition

$$\int_0^T \|\xi(t)\|^2 dt \le \int_0^T \|z(t)\|^2 dt \quad Q_\rho^T\text{-a.s.} \tag{2.4.43}$$

Then, the definition of the uncertainty input ξ_ρ and condition (2.4.43) imply that for each $T > 0$,

$$\frac{1}{T}\int_0^T \left[\|z(s)\|^2 - \|\xi_\rho(s)\|^2\right] ds \ge 0 \quad Q_\rho^T\text{-a.s.} \tag{2.4.44}$$

From the above condition, it follows that for each $\rho > 0$,

$$\inf_{T'>T} \frac{1}{T'} \int_0^{T'} \mathbf{E}^{Q_\rho^{T'}} \left[\|z(s)\|^2 - \|\xi_\rho(s)\|^2\right] ds \ge 0.$$

Note that the expectation on the left hand side of the above inequality exists by virtue of (2.4.42). Obviously in this case, one can find a constant $d > 0$ and a variable $\delta(T)$ which is independent of ρ and such that $\lim_{T\to\infty} \delta(T) = 0$ and

$$\inf_{T'>T} \frac{1}{T'} \mathbf{E}^{Q_\rho^{T'}} \int_0^{T'} \left[\|z(s)\|^2 - \|\xi_\rho(s)\|^2\right] ds \ge -d + \delta(T).$$

This, along with the representation of the relative entropy between the probability measure Q_ρ^T and the reference probability measure P^T given in equation (2.4.40), leads us to the conclusion that for the H^∞ norm-bounded uncertainty under consideration, the martingale $\zeta_\rho(t)$, $\rho > 0$, satisfies the constraint (2.4.32). This completes the proof of the lemma.

■

Remark 2.4.8 In the special case where the uncertainty is modeled by the operator (2.4.35) with $\mathbf{L}_2$ induced norm less then one, and where the uncertainty output $z(\cdot)$ of the closed loop system is known to be Q^T mean square integrable on any interval $[0, T]$, the above proof shows that such an uncertainty can be characterized directly in terms of the martingale $\zeta(t)$ and the associated probability measures Q^T. That is, one can choose $\zeta_i(t) = \zeta(t)$ and $Q_i^T = Q^T$ in Definition 8.5.1. This will be true for example, if the chosen controller is a stabilizing controller; see Definition 8.5.2 of Section 8.5. However, in a general case the connection between the uncertainty output $z(\cdot)$ and the uncertainty input $\xi(\cdot)$ can be of a more complex nature than that described by equation (2.4.35). In this case, the Q^T mean square integrability of the uncertainty output $z(\cdot)$ is not known *à priori*. Hence, one cannot guarantee that $h(Q^T\|P^T) < \infty$ for all $T > 0$. Also, the expectation

$$\frac{1}{T}\left[\mathbf{E}^{Q^T} \int_0^T \|z(t)\|^2 dt - h(Q^T\|P^T)\right]$$

may not exist for all $T > 0$ unless it has already been proved that the controller (8.5.2) is a stabilizing controller. In this case, the approximations of the martingale $\zeta(t)$ allow us to avoid the difficulties arising when defining an admissible uncertainty for the uncertain system (8.5.1) controlled by a generic linear output-feedback controller.

3.

H^∞ control and related preliminary results

This chapter contains a number of important results related to H^∞ control theory, risk sensitive control theory, algebraic Riccati equations and uncertain systems with norm-bounded uncertainty. These results will be useful when addressing the problems of optimal guaranteed cost control and minimax optimal control.

3.1 Riccati equations

Most of the results presented in this book make use of nonnegative-definite or positive-definite solutions to algebraic or differential Riccati equations. In this section, we review some of the properties of these equations. We first introduce some standard notation regarding Riccati equation solutions.

Let A, M and N be $n \times n$ matrices, with $M = M'$ and $N = N' \geq 0$.

Definition 3.1.1 *A symmetric matrix P^+ is said to be a* stabilizing solution *to the algebraic Riccati equation*

$$A'P + PA - PMP + N = 0$$

if it satisfies the Riccati equation and the matrix $A - MP^+$ is stable; i.e., *all of its eigenvalues lie in the open left half of the complex plane. Similarly, a symmetric matrix P^+ is said to be a* strong solution *to this Riccati equation if it satisfies the Riccati equation and the matrix $A - MP^+$ has all of its eigenvalues in the closed left half of the complex plane. Note that any stabilizing solution to the Riccati equation will also be a strong solution.*

In a similar fashion to the above, one can also define the notion of a stabilizing solution to a differential Riccati equation. Given a time-varying system of the form

$$\begin{aligned} \dot{x} &= A(t)x + B(t)u; \\ y &= C(t)x + D(t)u; \end{aligned} \tag{3.1.1}$$

let $\mathcal{R}(t,s)$ denote the state transition matrix of the homogeneous part of (3.1.1). We first define the notion of an exponentially stable time-varying matrix function.

Definition 3.1.2 *The system (3.1.1) with $u(t) \equiv 0$ and corresponding matrix function $A(t)$ are said to be* exponentially stable *if there exist constants $a > 0$ and $\alpha > 0$ such that*

$$\|\mathcal{R}(t,s)\| \leq ae^{-\alpha(t-s)}$$

for all $t \geq s$.

Definition 3.1.3 *Let $M(t) = M(t)'$ and $N(t) = N(t)' \geq 0$ be symmetric matrix functions. A symmetric matrix function $P^+(t)$ is said to be a* stabilizing solution *to the differential Riccati equation*

$$-\dot{P} = A(t)'P + PA(t) - PM(t)P + N(t)$$

if it satisfies this differential Riccati equation and the matrix $A(t) - M(t)P^+(t)$ is exponentially stable.

The stability notion introduced in Definition 3.1.2 can be used to define notions of stabilizability and detectability for linear time-varying systems of the form (3.1.1); e.g., see [165]. That is, the system (3.1.1) is said to be *stabilizable* if there exists a bounded matrix function $K(t)$ such that the system

$$\dot{x} = (A(t) + B(t)K(t))x$$

is exponentially stable. In this case, we say the pair $(A(\cdot), B(\cdot))$ is stabilizable.

Similarly, the system (3.1.1) is said to be *detectable* if there exists a bounded matrix function $L(t)$ such that the system

$$\dot{x} = (A(t) - L(t)C(t))x$$

is exponentially stable. In this case, we say the pair $(A(\cdot), C(\cdot))$ is detectable.

The following comparison result for stabilizing solutions to the algebraic Riccati equation was first presented in [163]. It is frequently used in the theory of H^∞ control and quadratic stabilization.

Lemma 3.1.1 *Consider the algebraic Riccati equation*

$$\tilde{A}'P + P\tilde{A} - P\tilde{M}P + \tilde{N} = 0 \tag{3.1.2}$$

where $\tilde{M} \geq 0$ and the pair $(\tilde{A}, \tilde{M})$ is stabilizable. Suppose this Riccati equation has a symmetric solution $\tilde{P}$ and define the Hamiltonian matrix

$$\tilde{H} = \begin{bmatrix} \tilde{N} & \tilde{A}' \\ \tilde{A} & -\tilde{M} \end{bmatrix}.$$

Also consider the Riccati equation

$$A'P + PA - PMP + N = 0 \tag{3.1.3}$$

where $M \geq 0$ and define the Hamiltonian matrix

$$H = \begin{bmatrix} N & A' \\ A & -M \end{bmatrix}.$$

If $H \geq \tilde{H}$, then Riccati equation (3.1.3) will have a unique strong solution P^+ which satisfies $P^+ \geq \tilde{P}$.

Proof: See Theorems 2.1 and 2.2 of [163].

■

The following corollary to this lemma is used in the proof of the Strict Bounded Real Lemma given below.

Corollary 3.1.1 *Suppose the matrix A is stable and the Riccati equation*

$$A'\tilde{P} + \tilde{P}A + \tilde{P}BB'\tilde{P} + \tilde{Q} = 0 \tag{3.1.4}$$

has a symmetric solution $\tilde{P}$. Furthermore, suppose $\tilde{Q} \geq Q \geq 0$. Then the Riccati equation

$$A'P + PA + PBB'P + Q = 0 \tag{3.1.5}$$

will have a unique strong solution P and moreover, $0 \leq P \leq \tilde{P}$.

Proof: Let $\tilde{X} = -\tilde{P}$. Hence, Riccati equation (3.1.4) can be re-written as

$$A'\tilde{X} + \tilde{X}A - \tilde{X}BB'\tilde{X} - \tilde{Q} = 0.$$

Furthermore, since A is stable, the pair (A, B) must be stabilizable. Hence, using Lemma 3.1.1, it follows that the Riccati equation

$$A'X + XA - XBB'X - Q = 0$$

will have a unique strong solution $X \geq \tilde{X}$. Now let $P = -X$. It follows immediately that $P \leq \tilde{P}$ is the unique strong solution to (3.1.5). Moreover, using a standard result on Lyapunov equations, it follows from (3.1.5) that $P \geq 0$; e.g., see Lemma 12.1 of [251].

■

Many of the results to be presented in this book make extensive use of the following Strict Bounded Real Lemma; see [147]. In order to present this result, we must first introduce some notation. Consider the Hardy space H^∞ which consists of complex vector-valued functions $f(s)$ which are analytic and bounded in the open right half of the complex plane [271]. The norm on this space is defined as follows:

$$\|f(s)\|_\infty := \sup_{\mathrm{Re}\, s>0} \|f(s)\|.$$

Note that if $f(s)$ is a transfer function which is analytic and bounded in the closed right half-plane, by making use of Plancherel's Theorem along with the Maximum-Modulus Theorem, one can establish that

$$\|f(s)\|_\infty = \sup_{\omega \in \mathbf{R}} \|f(j\omega)\|.$$

In a similar fashion, the H^∞ norm of the transfer function for a linear time-invariant system can be defined as follows. Consider a linear time-invariant system

$$\begin{aligned} \dot{x} &= Ax + Bw, \\ z &= Cx, \end{aligned} \tag{3.1.6}$$

where A is a stable matrix and B, C are real constant matrices of appropriate dimensions. Also, $x \in \mathbf{R}^n$ is the state, $w \in \mathbf{R}^p$ is the disturbance input and $z \in \mathbf{R}^q$ is the uncertainty output. Suppose that the disturbance input satisfies $w(\cdot) \in \mathbf{L}_2[0,\infty)$. Let $\mathfrak{W}_{wz}$ denote the linear operator which maps $w(\cdot)$ to $z(\cdot)$ for this system. Furthermore, let $\|\mathfrak{W}_{wz}\|_\infty$ denote the induced norm of this linear operator:

$$\|\mathfrak{W}_{wz}\|_\infty := \sup_{\substack{w(\cdot)\in \mathbf{L}_2[0,\infty) \\ \|w(\cdot)\|_2 \neq 0}} \frac{\|z(\cdot)\|_2}{\|w(\cdot)\|_2} \tag{3.1.7}$$

where $z(\cdot)$ is the output of the system (3.1.6) corresponding to the disturbance input $w(\cdot)$ and the initial condition $x(0) = 0$. The H^∞ norm of the corresponding transfer function $H_{wz}(s) = C(sI - A)^{-1}B$ is defined as follows:

$$\|H_{wz}(s)\|_\infty := \sup_{\omega \in \mathbf{R}} \sigma\left(C(j\omega I - A)^{-1}B\right)$$

where $\sigma(\cdot)$ denotes the maximum singular value of a matrix. It then follows that $\|H_{wz}(s)\|_\infty = \|\mathfrak{W}_{wz}\|_\infty$; e.g., see [70]. This fact motivates the the notation $\|\mathfrak{W}_{wz}\|_\infty$ in (3.1.7).

Lemma 3.1.2 *(Strict Bounded Real Lemma) The following statements are equivalent:*

(i) A is stable and $\|C(sI-A)^{-1}B\|_\infty < 1$;

(ii) There exists a matrix $\tilde{P} > 0$ such that

$$A'\tilde{P} + \tilde{P}A + \tilde{P}BB'\tilde{P} + C'C < 0;$$

(iii) The Riccati equation

$$A'P + PA + PBB'P + C'C = 0 \tag{3.1.8}$$

has a stabilizing solution $P \geq 0$.

Furthermore, if these statements hold then $P < \tilde{P}$.

Proof: We first establish the equivalence of the statements (i) - (iii).

(i)⇒(ii). It follows from condition (i) that there exists an $\varepsilon \geq 0$ such that

$$C(j\omega - A)^{-1}BB'(-j\omega I - A')^{-1}C' \leq (1-\varepsilon)I \tag{3.1.9}$$

for all $\omega \geq 0$. Let $\mu := \|C(sI-A)^{-1}\|_\infty$. Hence,

$$\frac{\varepsilon}{2\mu^2}C(j\omega - A)^{-1}(-j\omega I - A')^{-1}C' \leq \frac{\varepsilon}{2}I \tag{3.1.10}$$

for all $\omega \geq 0$. Adding equations (3.1.9) and (3.1.10), it follows that given any $\omega \geq 0$,

$$C(j\omega - A)^{-1}\tilde{B}\tilde{B}'(-j\omega I - A')^{-1}C' \leq (1-\frac{\varepsilon}{2})I \tag{3.1.11}$$

where $\tilde{B}$ is a non-singular matrix defined by

$$\tilde{B}\tilde{B}' = BB' + \frac{\varepsilon}{2\mu^2}I.$$

Furthermore, (3.1.11) implies

$$\tilde{B}'(-j\omega - A')^{-1}C'C(j\omega I - A)^{-1}\tilde{B} \leq (1-\frac{\varepsilon}{2})I \tag{3.1.12}$$

for all $\omega \geq 0$. Now let $\eta := \|(sI-A)^{-1}\tilde{B}\|_\infty$. Hence

$$\frac{\varepsilon}{2\eta^2}\tilde{B}'(-j\omega - A')^{-1}(j\omega I - A)^{-1}\tilde{B} \leq \frac{\varepsilon}{2}I. \tag{3.1.13}$$

Adding equations (3.1.12) and (3.1.13), it follows that given any $\omega \geq 0$,

$$\tilde{B}'(-j\omega - A')^{-1}\tilde{C}'\tilde{C}(j\omega I - A)^{-1}\tilde{B} \leq I \tag{3.1.14}$$

where $\tilde{C}$ is a non-singular matrix defined so that

$$\tilde{C}'\tilde{C} = C'C + \frac{\varepsilon}{2\eta^2}I.$$

Thus, $\|\tilde{C}(sI - A)^{-1}\tilde{B}\|_\infty \leq 1$. Furthermore, since $\tilde{B}$ and $\tilde{C}$ are non-singular, the triple $(A, \tilde{B}, \tilde{C})$ is minimal. Hence, using the standard Bounded Real Lemma given in [4], it follows that there exist a matrix $\tilde{P}$ such that

$$A'\tilde{P} + \tilde{P}A + \tilde{P}\tilde{B}\tilde{B}'\tilde{P} + \tilde{C}'\tilde{C} = 0.$$

That is,

$$A'\tilde{P} + \tilde{P}A + \tilde{P}BB'\tilde{P} + C'C + \frac{\varepsilon}{2\mu^2}\tilde{P}^2 + \frac{\varepsilon}{2\eta^2}I = 0.$$

Hence, condition (ii) holds.

(ii)⇒(iii). It follows from (ii) that there exist matrices $\tilde{P} > 0$ and $\tilde{R} > 0$ such that

$$A'\tilde{P} + \tilde{P}A + \tilde{P}BB'\tilde{P} + C'C + \tilde{R} = 0. \tag{3.1.15}$$

Hence, using a standard Lyapunov stability result, it follows that A is stable; see Lemma 12.2 of [251]. Furthermore if we compare Riccati equations (3.1.15) and (3.1.8), it follows from Corollary 3.1.1 that Riccati equation (3.1.8) will have a unique strong solution $P \leq \tilde{P}$. Moreover, using the fact that A is stable, it follows from a standard property of Lyapunov equations that $P \geq 0$; see Lemma 12.1 of [251].

We now show that P is in fact the stabilizing solution to (3.1.8). First let $S := \tilde{P} - P \geq 0$ and $\bar{A} := A + BB'P$. It follows from (3.1.8) and (3.1.15) that

$$\bar{A}'S + S\bar{A} + SBB'S + \tilde{R} = 0. \tag{3.1.16}$$

This equation will be used to show that $A+BB'P$ has no eigenvalues in the closed right half-plane. Indeed, suppose $\bar{A}$ has an eigenvalue $\alpha + j\omega$ in the closed right half-plane ($\alpha \geq 0$) with a corresponding eigenvector x. That is, $\bar{A}x = (\alpha + j\omega)x$. It follows from (3.1.16) that

$$(\alpha - j\omega)x^*Sx + (\alpha + j\omega)x^*Sx + x^*SBB'Sx + x^*\tilde{R}x = 0$$

and hence $x^*\tilde{R}x \leq 0$. However, this contradicts the fact that $\tilde{R} > 0$. Thus, we can conclude that $A + BB'P$ is stable and therefore, $P \geq 0$ is the stabilizing solution to (3.1.8). Hence, condition (iii) holds.

(iii)⇒(i). Suppose condition (iii) holds and let $\tilde{C} := B'P$. It follows that the matrix

$$A + B\tilde{C} = A + BB'P$$

is stable and hence the pair $(A, \tilde{C})$ is detectable. Furthermore, it follows from (3.1.8) that

$$A'P + PA + \tilde{C}'\tilde{C} \leq 0.$$

Hence, using a standard Lyapunov stability result, we can conclude that A is stable; see Lemma 12.2 of [251].

In order to show that $\|C(sI - A)^{-1}B\|_\infty < 1$, we first observe that equation (3.1.8) implies

$$\begin{aligned} &B'(-j\omega I - A')^{-1}C'C(j\omega I - A)^{-1}B \\ &\quad = I - \left[I - B'P(-j\omega I - A)^{-1}B\right]' \left[I - B'P(j\omega I - A)^{-1}B\right] \end{aligned} \tag{3.1.17}$$

for all $\omega \geq 0$. It follows that $\|C(sI - A)^{-1}B\|_\infty \leq 1$. Furthermore, note that

$$C(j\omega I - A)^{-1}B \to 0$$

as $\omega \to \infty$. Now suppose that there exists an $\bar{\omega} \geq 0$ such that

$$\|C(j\bar{\omega}I - A)^{-1}B\| = 1.$$

It follows from (3.1.17) that there exists a non-zero vector z such that

$$[I - B'P(j\bar{\omega}I - A)^{-1}B]z = 0.$$

Hence,

$$\det[I - B'P(j\bar{\omega}I - A)^{-1}B] = 0.$$

However, using a standard result on determinants, it follows that

$$\det[j\bar{\omega}I - A - BB'P] = \det[j\bar{\omega}I - A]\det[I - B'P(j\bar{\omega} - A)^{-1}B];$$

see for example Section A.11 of [94]. Thus,

$$\det[j\bar{\omega}I - A - BB'P] = 0.$$

However, this contradicts the fact that P is the stabilizing solution to (3.1.8). Hence, it follows that $\|C(sI - A)^{-1}B\|_\infty < 1$. Thus, we have now established condition (i).

In order to complete the proof of the lemma, we now suppose that statements (i) - (iii) hold. We must show that $P < \tilde{P}$. However, we have already proved that $P \leq \tilde{P}$. Now suppose that there exists a non-zero vector z such that $z'\tilde{P}z = z'Pz$. That is, $Sz = 0$ where $S = \tilde{P} - P \geq 0$ is defined as above. Applying this fact to equation (3.1.16), it follows that $z'\tilde{R}z = 0$. However, this contradicts the fact that $\tilde{R} > 0$. Thus, we must have $P < \tilde{P}$. This completes the proof of the lemma. ■

3.2 H^∞ control

The H^∞ control problem was originally introduced by Zames in [268] and has subsequently played a major role in the area of robust control theory. Consider a linear time invariant system

$$\begin{aligned} \dot{x} &= Ax + B_1 u + B_2 w; \\ z &= C_1 x + D_1 u; \\ y &= C_2 x + D_2 w \end{aligned} \tag{3.2.1}$$

where $x \in \mathbf{R}^n$ is the state, $u \in \mathbf{R}^m$ is the control input, $w \in \mathbf{R}^p$ is the disturbance input, $z \in \mathbf{R}^q$ is the controlled output, and $y \in \mathbf{R}^l$ is the measured output. Here $A, B_1, B_2, C_1, D_1, C_2, D_2$ are real constant matrices of appropriate dimensions. Suppose that the disturbance input satisfies $w(\cdot) \in \mathbf{L}_2[0,\infty)$. In the H^∞ state-feedback control problem, one seeks to find necessary and sufficient conditions for the existence of a causal linear state-feedback controller

$$u(t) = \mathcal{K}(t, x(\cdot)|_0^t), \quad t \geq 0, \tag{3.2.2}$$

such that the resulting closed loop system satisfies a certain internal stability condition as well as an H^∞ norm bound defined as follows

$$\|\mathfrak{W}_{wz}^K\|_\infty < \gamma. \tag{3.2.3}$$

Here $\mathfrak{W}_{wz}^K$ denotes the linear operator which maps $w(\cdot)$ to $z(\cdot)$ for the closed loop system defined by (3.2.1) and (3.2.2). Furthermore, $\|\mathfrak{W}_{wz}^K\|_\infty$ denotes the induced norm of this linear operator; see (3.1.7). In the particular case when $u(t) = Kx(t)$ and the matrix $A + B_1K$ is stable, the norm $\|\mathfrak{W}_{wz}^K\|_\infty$ can be re-written as

$$\|\mathfrak{W}_{wz}^K\|_\infty = \sup_{\omega \in \mathbf{R}} \|(C_1 + D_1K)(j\omega I - A - B_1K)^{-1}B_2\|.$$

The stability of the closed loop system matrix $A + B_1K$ is equivalent to the internal closed loop stability requirement mentioned above.

The output-feedback H^∞ control problem is similar to the state-feedback H^∞ control problem described above. However in this case, the form of the controller is different. That is, a causal output-feedback controller of the form

$$u(t) = \mathcal{K}(t, y(\cdot)|_0^t), \quad t \geq 0 \tag{3.2.4}$$

is required so that condition (3.2.3) is satisfied and the closed loop system is stable. In this case, the output $z(\cdot)$ in condition (3.2.3) is the controlled output of the closed loop system (3.2.1), (3.2.4) corresponding to the given disturbance input $w(\cdot)$ and zero initial conditions.

The early developments in H^∞ control theory were based on frequency domain and operator theoretic methods. A major limitation of this approach was that most of the results obtained were concerned with linear time-invariant systems; e.g., see [60]. Also, the frequency domain approach was unable to take into account non-zero initial conditions. Hence, the issue of closed loop transient response could not easily be addressed using these methods.

A significant breakthrough in H^∞ control theory was the introduction of state space methods; e.g., see [141, 52, 101, 102, 147, 222, 69, 271, 70]. The state space approach made it possible to remove a number of the limitations of the earlier methods of H^∞ control such as mentioned above. The reader who is interested in this subject may find it useful to read the paper [52] and the book [9]. The latter gives a systematic exposition of the topic from a game theoretic point of view and contains a rather complete literature survey on this subject. Note that it was the state space approach that allowed solutions to be obtained for time-varying H^∞ control problems as well as for finite-horizon versions of this problem; see [114, 222, 223, 90]. Also, the state space approach allows for non-zero initial conditions; see [100]. Among other generalizations of H^∞ control theory, the extension to infinite-dimensional systems (e.g., see [107, 223]) and nonlinear systems (e.g., see [78]) are also worth mentioning.

Throughout this book, we will repeatedly use state space solutions to various H^∞ control problems. For ease of reference, we collect the relevant H^∞ control results in this section. Also, to establish results on minimax optimal control for systems with multiplicative noise, a corresponding stochastic H^∞ control theory will also be needed. This section provides the necessary pre-requisites for the derivation of this stochastic H^∞ control theory in Section 8.2.

3.2.1 The standard H^∞ control problem

The game theoretic formulation of the standard H^∞ control problem is based on the observation that under an assumption of closed loop internal stability, condition (3.2.3) holds if and only if there exists an $\varepsilon > 0$ such that

$$\begin{aligned} J_\gamma(u(\cdot), w(\cdot)) &:= \int_0^\infty \left(\|z(t)\|^2 - \gamma^2\|w(t)\|^2\right) dt \\ &\leq -\varepsilon \int_0^\infty \|w(t)\|^2 dt \end{aligned} \tag{3.2.5}$$

for all $w(\cdot) \in \mathbf{L}_2[0,\infty)$. Here, we assume the initial condition for the system (3.2.1) is zero. In Section 3.2.2, the case of non-zero initial conditions will also be discussed.

The equivalence between condition (3.2.3) and condition (3.2.5) serves as the launching point for the game-theoretic approach to H^∞ control. This approach to the synthesis of feedback controllers solving a H^∞ control problem is based on the interpretation of the original H^∞ control problem as a dynamic game. In this dynamic game, the underlying system is a system of the form (3.2.1). The designer uses the functional $J_\gamma(\cdot)$ as a measure of the "cost" associated with a particular control strategy. The objective of this player is to design a controller of the form (3.2.2) or (3.2.4) which minimizes the performance index $J_\gamma(\cdot)$ under the worst possible disturbance input. This leads to a guaranteed performance given by the upper value of the dynamic game

$$\inf_K \sup_{w \in \mathbf{L}_2[0,\infty)} J_\gamma(K, w(\cdot)). \tag{3.2.6}$$

Here, the infimum over $K(\cdot)$ denotes the infimum over controllers of the form (3.2.2) in the state-feedback problem, or the infimum over controllers of the form (3.2.4) in the output-feedback case.

Let

$$\gamma^\infty := \inf\{\gamma\colon \text{ The upper value (3.2.6) is finite.}\}$$

An important fact in H^∞ control theory is that the constant γ^∞ is the minimum value of $\|\mathfrak{W}_{wz}^K\|_\infty$ that can be achieved by means of a controller $K(\cdot)$ of the form of (3.2.2) in the state-feedback case. Similarly, γ^∞ is the minimum value of $\|\mathfrak{W}_{wz}^K\|_\infty$ that can be achieved by means of a controller $K(\cdot)$ of the form (3.2.4) in the output-feedback case. Thus,

$$\gamma^\infty = \inf_{K(\cdot)} \|\mathfrak{W}_{wz}^K\|_\infty. \tag{3.2.7}$$

A controller which attains the infimum in (3.2.7) is referred to as *an H^∞ optimal* controller. This controller also attains the upper value of the game (3.2.6) corresponding to $\gamma = \gamma^\infty$. However in many cases, it is sufficient to have a design procedure for the synthesis of a *suboptimal* controller $K(\cdot)$ which guarantees that $\|\mathfrak{W}_{wz}^K\|_\infty < \gamma$ with $\gamma > \gamma^\infty$. This is because one can approach as close as desired to the optimal value $\inf_{K(\cdot)} \|\mathfrak{W}_{wz}^K\|_\infty$ by iterating on γ. A suboptimal controller attains the upper value of the game corresponding to $\gamma > \gamma^\infty$. If the system has zero initial condition then a suboptimal controller will lead to the satisfaction of condition (3.2.5). However, note that for the dynamic game (3.2.6), the initial condition of the system is not necessarily zero.

The above game theoretic interpretation of the H^∞ control problem has proved to be extremely useful as it has allowed for the extension of the H^∞ control methodology beyond the initial linear time-invariant infinite-horizon setting. In particular, it leads to a connection between H^∞ control and stochastic risk-sensitive optimal control; e.g., see [69, 170, 9] and also Section 3.3.

We now present some standard H^∞ control theory results which will be used throughout this book.

Infinite-horizon state-feedback H^∞ control

The solution to the infinite-horizon state-feedback H^∞ control problem for time-invariant systems is now presented. This result is one of the simplest results to establish in H^∞ control theory. It also illustrates the relationship between state space H^∞ control results and results obtained using the frequency domain approach.

A rigorous definition of the dynamic game associated with the infinite-horizon state-feedback H^∞ control problem requires us to describe sets of admissible policies for the minimizing and maximizing players. This is achieved by specifying the information pattern for each player. That is, we specify the nature of the dependence of the control or disturbance input variable on the state or measured output. The dynamic games arising in H^∞ control for time-invariant systems use

Closed Loop Perfect State (CLPS) and *Closed-Loop Imperfect State* (CLIS) information patterns; see [9]. The Closed Loop Perfect State information structure requires an admissible control input to be of the form (3.2.2), with the function $K(\cdot)$ being Borel measurable and Lipschitz continuous in the second argument. These requirements ensure that the differential equation describing the system governed by such a controller and driven by a square integrable disturbance input, admits a unique solution. Similarly, the Closed Loop Imperfect State information structure allows the controller to be of the form (3.2.4) with the function $K(\cdot)$ being Borel measurable and Lipschitz continuous in the second argument and such that the differential equation corresponding to the closed loop system admits a unique solution.

Consider the following algebraic Riccati equation

$$\begin{aligned}&(A - B_1G^{-1}D_1'C_1)'X + X(A - B_1G^{-1}D_1'C_1) + C_1'(I - D_1G^{-1}D_1')C_1 \\ &+X(B_1G^{-1}B_1' - \frac{1}{\gamma^2}B_2B_2')X = 0 \qquad (3.2.8)\end{aligned}$$

where $G := D_1'D_1 > 0$. Also, consider the corresponding Riccati differential equation

$$\begin{aligned}&\dot{X} + (A - B_1G^{-1}D_1'C_1)'X + X(A - B_1G^{-1}D_1'C_1) \\ &\quad +C_1'(I - D_1G^{-1}D_1')C_1 + X(B_1G^{-1}B_1' - \frac{1}{\gamma^2}B_2B_2')X = 0; \quad (3.2.9) \\ &X(T) = M\end{aligned}$$

where M is a given nonnegative-definite symmetric matrix.

The main result of [9] concerning the dynamic game associated with the infinite-horizon state-feedback H^∞ control problem is as follows:

Lemma 3.2.1 *Consider the differential game (3.2.6) with the CLPS information pattern and the initial condition $x(0) = x_0$. Assume that the pair*

$$\left(A - B_1G^{-1}D_1'C_1, (I - D_1G^{-1}D_1')C_1\right) \qquad (3.2.10)$$

is detectable. Then the following claims hold:

(i) Let $X_T(t)$ be the solution to equation (3.2.9) corresponding to $M = 0$. Then, for each fixed t, $X_T(t)$ is nondecreasing in T.

(ii) If there exists a nonnegative-definite solution to Riccati equation (3.2.8), then there is a minimal such solution denoted X^+. This matrix has the property

$$X^+ \geq X_T(t)$$

for all $T \geq 0$, where $X_T(t)$ is the solution to equation (3.2.9) defined in (i). Also, if the pair (3.2.10) is observable, then every nonnegative-definite solution to equation (3.2.8) is positive-definite.

(iii) *The differential game (3.2.6) with the CLPS information pattern has equal upper and lower values if and only if Riccati equation (3.2.8) admits a nonnegative-definite solution. In this case, the common value is*

$$\inf_{K(\cdot)} \sup_{w(\cdot)\in\mathbf{L}_2[0,\infty)} J_\gamma(K(\cdot),w(\cdot)) = x_0'X^+x_0. \tag{3.2.11}$$

In (3.2.11), the infimum is taken over CLPS controllers of the form (3.2.2).

(iv) *If the upper value (3.2.11) is finite for some $\gamma = \hat{\gamma} > 0$, then it is bounded and equal to the lower value for all $\gamma > \hat{\gamma}$.*

(v) *If the upper value (3.2.11) is finite for some $\gamma = \hat{\gamma} > 0$, then for all $\gamma > \hat{\gamma}$, the two closed loop matrices*

$$\bar{A} := A - B_1G^{-1}(B_1'X^+ + D_1'C_1); \tag{3.2.12}$$

$$A_\gamma := A - B_1G^{-1}(B_1'X^+ + D_1'C_1) + \frac{1}{\gamma^2}B_2B_2'X^+ \tag{3.2.13}$$

are stable.

(vi) *For $\gamma > \hat{\gamma}$, X^+ is the unique nonnegative-definite stabilizing solution to Riccati equation (3.2.8);* i.e., *X^+ is the unique solution to Riccati equation (3.2.8) in the class of nonnegative-definite matrices such that the matrix (3.2.13) is stable.*

(vii) *If $X^+ \geq 0$ defined in (ii) exists, then the feedback controller given by*

$$u(t) = K^*x(t); \quad K^* := -G^{-1}(B_1'X^+ + D_1'C_1) \tag{3.2.14}$$

attains the finite upper value of the differential game in the sense that

$$\sup_{w(\cdot)\in\mathbf{L}_2[0,\infty)} J_\gamma(K^*,w(\cdot)) = x_0'X^+x_0. \tag{3.2.15}$$

(viii) *The upper value of the differential game is bounded for some $\gamma > 0$ if and only if the pair (A, B_1) is stabilizable.*

(ix) *If $\gamma < \gamma^\infty$ and the upper value of the differential game is infinite, then Riccati equation (3.2.8) has no real solution which also satisfies the condition that the matrix A_γ defined above, is stable.*

Proof: By substituting

$$u = v - G^{-1}D_1'C_1x \tag{3.2.16}$$

into equation (3.2.1), the system under consideration becomes a system of the form

$$\begin{aligned} \dot{x} &= \tilde{A}x + B_1v + B_2w; \\ z &= \tilde{C}_1x + D_1v. \end{aligned} \tag{3.2.17}$$

In equation (3.2.17), the matrices $\tilde{A}$ and $\tilde{C}_1$ are defined by the equations

$$\tilde{A} = A - B_1 G^{-1} D_1' C_1; \quad \tilde{C}_1 = (I - D_1 G^{-1} D_1') C_1. \tag{3.2.18}$$

Note that a matrix K solves the differential game problem (3.2.11) if and only if the matrix $\tilde{K} = K + G^{-1} D_1' C_1$ solves a corresponding game problem associated with the system (3.2.17).

The pair (A, B_1) is stabilizable if and only if the pair $(\tilde{A}, B_1)$ is stabilizable; e.g., see Lemma 4.5.3 of [109]. Also, condition (3.2.10) implies that the pair $(\tilde{A}, \tilde{C}_1)$ is detectable. Hence, the pair $(\tilde{A}, (I - D_1 G^{-1} D_1')^{1/2} C_1)$ is detectable. Thus, the conditions of Theorems 4.8 and 9.7 of [9] are satisfied. The lemma now follows from these results of [9].

■

Remark 3.2.1 The above result was proved in [9] under simplifying assumptions that $C_1' D_1 = 0$ and $G = D_1' D_1 = I$. Note that such assumptions are standard simplifying assumptions used in H^∞ control theory. These assumptions are not critical to the solution of an H^∞ control problem. However, they do simplify the development of results. Note that many H^∞ control problems can be transformed into an equivalent form such that the transformed problem satisfies these simplifying conditions. This transformation is illustrated in the above proof; see also Section 4.5 of [9]. We will also use a similar procedure in Section 8.2.

The solution to the infinite-horizon state-feedback H^∞ control problem can be readily obtained from Lemma 3.2.1. We first present a result concerning the infinite-horizon disturbance attenuation problem with $x_0 = 0$; see also Theorem 4.11 of the reference [9]. Consider the scalar quantity

$$\hat{\gamma}_\infty^{\mathrm{CL}} := \inf\{\gamma > 0 : \text{ equation (3.2.8) has a nonnegative-definite solution}\}. \tag{3.2.19}$$

Note that it follows from Lemma 3.2.1 that the set whose infimum determines $\hat{\gamma}_\infty^{\mathrm{CL}}$ is nonempty if the pair (A, B_1) is stabilizable and the pair (3.2.10) is detectable. Hence, in this case $\hat{\gamma}_\infty^{\mathrm{CL}} < \infty$.

Theorem 3.2.1 *Assume that the pair (3.2.10) is detectable and the pair (A, B_1) is stabilizable. Then,*

$$\gamma_\infty^* := \inf_K \|\mathfrak{W}_{zw}^K\|_\infty = \hat{\gamma}_\infty^{\mathrm{CL}} \tag{3.2.20}$$

where the infimum is taken over the set of CLPS controllers of the form (3.2.2). Moreover, given any $\varepsilon > 0$, we have

$$\|\mathfrak{W}_{zw}^{K_{\gamma_\epsilon}}\|_\infty \le \gamma_\epsilon := \hat{\gamma}_\infty^{\mathrm{CL}} + \varepsilon$$

where

$$K_\gamma = -G^{-1}(B_1 X_\gamma^+ + D_1' C_1); \quad \gamma > \hat{\gamma}_\infty^{\mathrm{CL}}$$

with X_γ^+ being the unique minimal nonnegative-definite solution of (3.2.8). Furthermore, for any $\varepsilon > 0$, the controller K_{γ_ϵ} leads to a Bounded-Input Bounded-State stable system.

Proof: This result follows from Lemma 3.2.1; see also Theorem 4.11 of [9]. ■

Theorem 3.2.2 *Assume that the pair (3.2.10) is detectable. Then the following conditions are equivalent:*

(i) There exists a controller $K(\cdot)$ of the form (3.2.2) that achieves $\|\mathfrak{W}_{zw}^K\|_\infty < \gamma$ and exponentially stabilizes the system (3.2.1) with $w(\cdot) \equiv 0$.

(ii) The Riccati equation (3.2.8) has a unique nonnegative definite solution such that the matrices (3.2.12) and (3.2.13) are stable.

If condition (ii) holds, then the solution to the infinite horizon state feedback H^∞ control problem is given by a controller of the form (3.2.14).

Proof: (ii)⇒(i). This part of the theorem follows from Theorem 3.2.1. Indeed, condition (ii) implies that $\gamma > \hat{\gamma}_\infty^{\mathrm{CL}}$. Also, note that the pair (A, B_1) is stabilizable since the matrix (3.2.12) is stable. Hence, it follows from Theorem 3.2.1 that for a sufficiently small $\varepsilon > 0$ such that $\hat{\gamma}_\infty^{\mathrm{CL}} + \varepsilon < \gamma$, the controller K_{γ_ϵ} satisfies condition (i) of the theorem.

(i)⇒(ii). Since $\|\mathfrak{W}_{zw}^K\|_\infty < \gamma < \infty$, then the corresponding differential game has a finite upper value and the pair (A, B_1) is stabilizable; see claim (viii) of Lemma 3.2.1. Then, equation (3.2.20) of Theorem 3.2.1 implies that

$$\hat{\gamma}_\infty^{\mathrm{CL}} \le \|\mathfrak{W}_{zw}^K\|_\infty < \gamma.$$

Hence, claim (ii) follows from the definition of the constant $\hat{\gamma}_\infty^{\mathrm{CL}}$ and from Lemma 3.2.1.

We will now prove the second part of the theorem. To this end, we use the arguments of [222, 90]. We first note that since the matrix $\bar{A}$ is stable, then the closed loop system corresponding to the system (3.2.1) and the controller (3.2.14):

$$\begin{aligned} \dot{x} &= \bar{A}x + B_2 w; \\ z &= (C_1 - D_1 G^{-1}(B_1' X^+ + D_1' C_1))x \end{aligned} \tag{3.2.21}$$

is internally stable. Also by completing the square in equation (3.2.8), we obtain

$$\begin{aligned} &x(t)' X^+ x(t) + \int_0^t \{\|z(s)\|^2 - \gamma^2 \|w(s)\|^2\} ds \\ &= -\gamma^2 \int_0^t \|(w(s) - \frac{1}{\gamma^2} B_2' X^+ x(s)\|^2 ds, \end{aligned} \tag{3.2.22}$$

where $x(\cdot)$ is the solution to equation (3.2.21) corresponding to the initial condition $x(0) = 0$. Note, that the substitution of $w = \tilde{w} + \gamma^{-2} B_2' X^+ x$ into equation

(3.2.21) leads to the following system

$$\begin{aligned} \dot{x} &= A_\gamma x + B_2 \tilde{w}; \\ z &= (C_1 - D_1 G^{-1}(B_1' X^+ + D_1' C_1))x. \end{aligned} \tag{3.2.23}$$

Since the matrix A_γ is stable, then the solution to equation (3.2.23) corresponding to the initial condition $x(0) = 0$ satisfies the condition

$$\int_0^\infty \|x(t)\|^2 dt \leq c_0 \int_0^\infty \|\tilde{w}(t)\|^2 dt$$

where $c_0 > 0$ is a constant independent of $\tilde{w}$. That is, the mapping $\tilde{w}(\cdot) \to x(\cdot)$ generated by equation (3.2.23) is a bounded mapping $\mathbf{L}_2[0;\infty) \to \mathbf{L}_2[0;\infty)$. Therefore, the mapping $\tilde{w}(\cdot) \to w(\cdot) = \tilde{w}(\cdot) + \gamma^{-2} B_2' X x(\cdot)$ is also a bounded mapping $\mathbf{L}_2[0;\infty) \to \mathbf{L}_2[0;\infty)$. Thus, there exists a constant $c > 0$ such that

$$\int_0^\infty \|w(t)\|^2 dt \leq c \int_0^\infty \|\tilde{w}(t)\|^2 dt$$

for all $\tilde{w}(\cdot) \in \mathbf{L}_2[0;\infty)$. Then, equation (3.2.22) together with the above inequality implies

$$\begin{aligned} \int_0^\infty \{\|z(t)\|^2 - \gamma^2 \|w(t)\|^2\} dt &\leq -\gamma^2 \int_0^\infty \|\tilde{w}(t)\|^2 dt \\ &\leq -\frac{\gamma^2}{c} \int_0^\infty \|w(t)\|^2 dt. \end{aligned}$$

That is, condition (3.2.5) is satisfied.

■

In some of the proofs to be presented in this book, it will be convenient for us to use an H^∞ control result from [147]. This result is given below in Lemma 3.2.3. Note that reference [147] establishes this result under the assumption that the associated matrix pencil

$$\begin{bmatrix} A - j\omega I & B_1 \\ C_1 & D_1 \end{bmatrix} \tag{3.2.24}$$

has full column rank for all $\omega \geq 0$ and also $D_1' D_1 > 0$; see Theorem 3.4 of [147]. This assumption is a weaker assumption than that of detectability for the matrix pair (3.2.10) which we require in Lemma 3.2.3. However, we use the stronger assumption of detectability as this assumption is required in a number of the results presented in this book. We first prove the following lemma.

Lemma 3.2.2 *Consider a system (3.2.1) in which $G = D_1' D_1 > 0$. Then the matrix pair (3.2.10) is detectable if and only if the matrix*

$$\begin{bmatrix} A - sI & B_1 \\ C_1 & D_1 \end{bmatrix} \tag{3.2.25}$$

has full column rank for all s such that $\mathrm{Re}(s) \geq 0$. *Also, the matrix pair (3.2.10) is observable if and only if the matrix (3.2.25) has full column rank for all $s \in \mathbb{C}$. Furthermore, the matrix pair (3.2.10) has no unobservable modes on the imaginary axis if and only if the matrix (3.2.25) has full column rank for all s such that $s = j\omega$, $\omega \geq 0$.*

Proof: In the case where the matrix pair (3.2.10) has no unobservable mode on $j\omega$-axis, the lemma is proved in [271]; see Lemma 13.9 of this reference. In the case where the matrix pair (3.2.10) is detectable or observable, the proof follows along the same lines as the proof of Lemma 13.9 of [271] except that $j\omega$ must be replaced by s with $\mathrm{Re} s \geq 0$ or by $s \in \mathbb{C}$ respectively.

■

Lemma 3.2.3 *Consider a system (3.2.1) in which $G = D_1' D_1 > 0$ and the matrix pair (3.2.10) is detectable. Suppose there exists a state-feedback control $u = Kx$ such that the matrix $A + B_1 K$ is stable and*

$$\|(C_1 + D_1 K)(sI - A - B_1 K)^{-1} B_2\|_\infty < 1. \tag{3.2.26}$$

Also suppose that the matrix $X > 0$ satisfies the inequality[1]

$$\begin{aligned}&(A + B_1 K)' X + X (A + B_1 K)\\&+\frac{1}{\gamma^2} X B_2 B_2' X + (C_1 + D_1 K)' (C_1 + D_1 K) < 0.\end{aligned} \tag{3.2.27}$$

Then the Riccati equation (3.2.8) has a stabilizing solution $X^+ \geq 0$ such that $X^+ < X$.

Proof: This result follows from Theorem 3.4 of [147]. In order to invoke this theorem, we must ensure that the conditions of the lemma imply that the matrix pencil (3.2.24) has full column rank for all $\omega \geq 0$. This fact follows from Lemma 3.2.2.

■

Infinite-horizon output-feedback H^∞ control

Our discussion of the infinite-horizon output-feedback H^∞ control problem begins with one of the preliminary results presented in [147]. This result will be used in Section 7.2. The result shows that if there exists a strictly proper controller of the form

$$\begin{aligned}\dot{x}_c &= A_c x_c + B_c y,\\ u &= C_c x_c,\end{aligned} \tag{3.2.28}$$

[1] The existence of such a matrix X follows from the Strict Bounded Real Lemma; see Lemma 3.1.2. In fact, conditions (3.2.27) and (3.2.26) are equivalent.

solving the output-feedback H^∞ control problem for the system (3.2.1), then there exists a state-feedback control law and an output injection, each solving a related H^∞ control problem.

Lemma 3.2.4 *Consider the system (3.2.1) and suppose there exists a controller of the form (3.2.28) such that the matrix*

$$\begin{bmatrix} A & B_1C_c \\ B_cC_2 & A_c \end{bmatrix}$$

is stable and

$$\left\| \begin{bmatrix} C_1 & D_1C_c \end{bmatrix} \times \left\{ sI - \begin{bmatrix} A & B_1C_c \\ B_cC_2 & A_c \end{bmatrix} \right\}^{-1} \begin{bmatrix} B_2 \\ B_cD_2 \end{bmatrix} \right\|_\infty < 1.$$

Then the following conditions hold:

(i) There exists a state-feedback matrix K and a matrix $X > 0$ such that

$$\begin{aligned} &(A + B_1K)'X + X(A + B_1K) \\ &\quad + XB_2B_2'X + (C_1 + D_1K)'(C_1 + D_1K) < 0. \end{aligned} \tag{3.2.29}$$

Hence the matrix $A + B_1K$ is stable and

$$\|(C_1 + D_1K)(sI - A - B_1K)^{-1}B_2\|_\infty < 1. \tag{3.2.30}$$

Note that $(C_1 + D_1K)(sI - A - B_1K)^{-1}B_2$ is the transfer function from w to z which results from the state-feedback control law $u = Kx$.

(ii) There exists an output-injection matrix L and a matrix $Y > 0$ such that

$$\begin{aligned} &(A + LC_2)'X + X(A + LC_2) \\ &\quad + XC_1'C_1X + (B_1 + LD_2)(B_1 + LD_2)' < 0. \end{aligned} \tag{3.2.31}$$

Hence the matrix $A + LC_1$ is stable and

$$\|C_1(sI - A - LC_2)^{-1}(B_2 + LD_2)\|_\infty < 1. \tag{3.2.32}$$

Note that $C_1(sI - A - LC_2)^{-1}(B_2 + LD_2)$ is the transfer function from w to z which results from the output injection $\dot{x} = Ax + B_2w + Ly$.

(iii) Furthermore, the matrices X and Y satisfy $\rho(XY) < 1$ where $\rho(\cdot)$ denotes the spectral radius of a matrix.

Proof: See Theorem 2 of [172] and Theorem 3.3 of [147].

■

One would expect from the above lemma that the output-feedback counterpart to Theorem 3.2.2 will involve two Riccati equations. Indeed, one of these equations is the Riccati equation (3.2.8). The other equation is a "filter" type Riccati equation which resembles the Kalman Filter Riccati equation arising in LQG control theory:

$$\begin{aligned} &(A - B_2 D_2' \Gamma^{-1} C_2) Y + Y (A - B_2 D_2' \Gamma^{-1} C_2)' \\ &\quad + B_2 (I - D_2' \Gamma^{-1} D_2) B_2' - Y (C_2' \Gamma^{-1} C_2 - \frac{1}{\gamma^2} C_1' C_1) Y = 0 \quad (3.2.33) \end{aligned}$$

where $\Gamma := D_2 D_2' > 0$.

Lemma 3.2.5 *Assume that the matrix pairs (3.2.10) and (A, C_2) are detectable and the matrix pairs (A, B_1) and*

$$\left(A - B_2 D_2' \Gamma^{-1} C_2, B_2 (I - D_2' \Gamma^{-1} D_2)\right) \quad (3.2.34)$$

are stabilizable. Then the following conditions hold:

(i) *If the algebraic Riccati equations (3.2.8) and (3.2.33) both admit minimal nonnegative-definite stabilizing solutions X^+ and Y^+ respectively which satisfy the condition*

$$\rho(Y^+ X^+) < \gamma^2, \quad (3.2.35)$$

then the dynamic game defined by the system (3.2.1) and cost functional $J_\gamma(\cdot)$ has a finite upper value

$$\inf_{K(\cdot)} \sup_{w \in \mathbf{L}_2[0,\infty)} J_\gamma(K, w(\cdot)) \quad (3.2.36)$$

where the infimum is over CLIS controllers of the form (3.2.4).

(ii) *The optimal minimizing controller is defined by the state equations*

$$\begin{aligned} \dot{\check{x}} &= A\check{x} + B_1 u + \frac{1}{\gamma^2} Y^+ C_1' \check{z} \\ &\qquad + (Y^+ C_2' + B_2 D_2') \Gamma^{-1} (y - C_2 \check{x}); \quad (3.2.37) \\ \check{z} &= C_1 \check{x} + D_1 u; \\ u &= -G^{-1} (B_1' X^+ + D_1' C_1)(I - \frac{1}{\gamma^2} Y^+ X^+)^{-1} \check{x} \quad (3.2.38) \end{aligned}$$

or equivalently, by the state equations

$$\begin{aligned} \dot{\hat{x}} &= A\hat{x} + B_1 u + B_2 \hat{w} \\ &\quad + (I - \frac{1}{\gamma^2} Y^+ X^+)^{-1} (Y^+ C_2' + B_2 D_2') \Gamma^{-1} (y - \hat{y}); \quad (3.2.39) \\ u &= -G^{-1} (B_1' X^+ + D_1' C_1) \hat{x} \quad (3.2.40) \end{aligned}$$

where we have introduced the notation

$$\begin{aligned} \hat{y} &= C_2 \hat{x} + D_2 \hat{w}; \\ \hat{w} &= \frac{1}{\gamma^2} B_2' X^+ \hat{x}. \end{aligned}$$

(iii) *Under the controller (3.2.38), the $2n$-dimensional closed loop system (3.2.1), (3.2.37), (3.2.38) is Bounded-Input Bounded-State stable, as well as asymptotically stable under the worst case disturbances. That is, in addition to (3.2.13) and (3.2.12), the two matrices*

$$A - (Y^+ C_2 + B_2' D_2)\Gamma^{-1} C_2 + \frac{1}{\gamma^2} Y^+ C_1' C_1; \tag{3.2.41}$$

$$A - (Y^+ C_2 + B_2' D_2)\Gamma^{-1} C_2 \tag{3.2.42}$$

are stable[2].

(iv) *If any one of the conditions in (i) fails, then the upper bound for the dynamic game is infinite.*

Proof: See Theorems 5.5 and 5.6 of [9].

■

We are now in a position to present a necessary and sufficient condition for the existence of a solution to the infinite-horizon output-feedback H^∞ control problem. This necessary and sufficient condition follows from the above game theoretic result in a straightforward manner.

Theorem 3.2.3 *Assume that the pairs (3.2.10) and (A, C_2) are detectable and the pairs (A, B_1) and (3.2.34) are stabilizable. Then the following statements are equivalent:*

(i) *There exists a controller $K(\cdot)$ of the form (3.2.4) which exponentially stabilizes the system (3.2.1) with $w(\cdot) \equiv 0$ and achieves the required bound on the H^∞ norm of the closed loop operator: $\|\mathfrak{W}_{zw}^K\|_\infty < \gamma$.*

(ii) *The Riccati equations (3.2.8) and (3.2.33) have unique nonnegative-definite solutions satisfying the spectral radius condition (3.2.35) and such that the matrices (3.2.13), (3.2.12) and (3.2.41), (3.2.42) are stable*[3].

(iii) *There exists a linear controller of the form*

$$\begin{aligned} \dot{x}_c &= A_c x_c + B_c y \\ u &= C_c x_c + D_c y \end{aligned} \tag{3.2.43}$$

such that the resulting closed loop system is such that

$$\bar{A} := \begin{bmatrix} A + B_1 D_c C_2 & B_1 C_c \\ B_c C_2 & A_c \end{bmatrix}$$

---

[2] Thus, according to Definition 3.1.1, Y^+ is a stabilizing solution to Riccati equation (3.2.33).

[3] In other words, the Riccati equations (3.2.8) and (3.2.33) have stabilizing solutions $X^+ \geq 0$ and $Y^+ \geq 0$ such that $\rho(X^+ Y^+) < 1$.

is stable and

$$\left\| \left[C_1 + D_1 D_c C_2 \;\; D_1 C_c \right] \times \left(sI - \bar{A}\right)^{-1} \begin{bmatrix} B_2 + B_1 D_C D_2 \\ B_c D_2 \end{bmatrix} + D_1 D_c D_1 \right\|_\infty < 1.$$

That is, the controller (3.2.43) exponentially stabilizes the system (3.2.1) with $w(\cdot) \equiv 0$ and achieves the required bound on the H^∞ norm of the closed loop operator: $\|\mathfrak{W}_{zw}^K\|_\infty < \gamma$.

If condition (ii) holds, then a linear controller (3.2.43) solving the infinite-horizon output-feedback H^∞ control problem can be constructed according to equations (3.2.37) and (3.2.38), or equivalently, according to equations (3.2.39) and (3.2.40).

Remark 3.2.2 A version of the above result has also been established in [147] under slightly weaker assumptions; see Theorem 3.1 of [147]. However, the proof given in [147] is also applicable here. We need only to note that the detectability of the pair (3.2.10) together with the condition $G > 0$, implies that the matrix pencil (3.2.24) has full column rank. Also, the stabilizability of the pair (3.2.34) together with the condition $\Gamma > 0$, implies that the matrix pencil

$$\begin{bmatrix} A - j\omega I & B_2 \\ C_2 & D_2 \end{bmatrix}$$

has full row rank.

Remark 3.2.3 Note that in Theorem 3.2.3, the condition that the pair (A, C_2) is detectable and the pair (A, B_1) is stabilizable is only needed to ensure that the corresponding infinite-horizon disturbance attenuation problem admits a finite optimum attenuation level γ^∞; see Theorem 5.5 of [9]. In some problems considered in this book, it will be assumed that the H^∞ control problem under consideration has a solution (possibly nonlinear) and hence, the corresponding differential game has a finite upper value. In this case, the assumption of detectability of (A, C_2) and stabilizability of (A, B_1) will not be needed.

3.2.2 H^∞ control with transients

The control problem addressed in this section is that of designing a controller that stabilizes the nominal closed loop system and minimizes the induced norm from the disturbance input $w(\cdot)$ and the *initial condition* x_0 to the controlled output $z(\cdot)$. This problem is referred to as an H^∞ control problem with transients. This problem will be considered in both the finite-horizon and infinite-horizon cases.

The results presented in this section are based on results obtained in reference [100]. Reference [100] considers a time-invariant system of the form (3.2.1)[4]. However, the class of controllers considered in reference [100] are *time-varying* linear output-feedback controllers $K(\cdot)$ of the form

$$\begin{aligned} \dot{x}_c(t) &= A_c(t)x_c(t) + B_c(t)y(t); \qquad x_c(0) = 0; \\ u(t) &= C_c(t)x_c(t) + D_c(t)y(t) \end{aligned} \tag{3.2.44}$$

where $A_c(\cdot), B_c(\cdot), C_c(\cdot)$ and $D_c(\cdot)$ are bounded piecewise continuous matrix functions. Note, that the dimension of the controller state vector x_c may be arbitrary.

In the problem of H^∞ control with transients, the performance of the closed loop system consisting of the underlying system (3.2.1) and the controller (3.2.44), is measured with a *worst case closed loop performance measure* defined as follows. For a fixed time $T > 0$, a positive-definite symmetric matrix P_0 and a nonnegative-definite symmetric matrix X_T, the worst case closed loop performance measure is defined by

$$\gamma(K, X_T, P_0, T) := \sup \left\{ \frac{x(T)'X_T x(T) + \int_0^T \|z(t)\|^2 dt}{x(0)'P_0 x(0) + \int_0^T \|w(t)\|^2 dt} \right\} \tag{3.2.45}$$

where the supremum is taken over all $x(0) \in \mathbf{R}^n$, $w(\cdot) \in \mathbf{L}_2[0, T]$ such that $x(0)'P_0 x(0) + \int_0^T \|w(t)\|^2 dt > 0$. From this definition, the performance measure $\gamma(K, X_T, P_0, T)$ can be regarded as the induced norm of the linear operator which maps the pair $(x_0, w(\cdot))$ to the pair $(x(T), z(\cdot))$ for the closed loop system; see [100]. In this definition, T is allowed to be ∞ in which case $X_T := 0$ and the operator mentioned above is an operator mapping the pair $(x(0), w(\cdot))$ to $z(\cdot)$. Another special case arises when $x(0) = 0$. In this case, the supremum on the right hand side of (3.2.45) is taken over all $w(\cdot) \in \mathbf{L}_2[0, \infty)$ and the performance measure reduces to the standard H^∞ norm as defined in equation (3.1.7).

The H^∞ control problem with transients is now defined as follows. Let the constant $\gamma > 0$ be given.

Finite-Horizon Problem. Does there exist a controller of the form (3.2.44) such that

$$\gamma(K, X_T, P_0, T) < \gamma^2? \tag{3.2.46}$$

Infinite-Horizon Problem. Does there exist a controller of the form (3.2.44) such that the resulting closed loop system is exponentially stable and

$$\gamma(K, 0, P_0, \infty) < \gamma^2? \tag{3.2.47}$$

---

[4]The assumption that the underlying system is time-invariant is not a critical assumption in [100]. The extension of the proofs presented in [100] to the time-varying setting involves only a minor modification.

The results of reference [100] require that the coefficients of the system (3.2.1) satisfy a number of technical assumptions.

Assumption 3.2.1

(i) The matrices C_1 and D_1 satisfy the conditions

$$C_1'D_1 = 0, \ G := D_1'D_1 > 0.$$

(ii) The matrices B_2 and D_2 satisfy the conditions

$$D_2B_2' = 0, \ \Gamma := D_2D_2' > 0.$$

(iii) The pair (A, B_2) is stabilizable and the pair (A, C_1) is detectable.

(iv) The pair (A, B_1) is stabilizable and the pair (A, C_2) is detectable.

We noted in Remark 3.2.1 that the simplifying assumptions $C_1'D_1 = 0$, $D_2B_2' = 0$ are not critical to the solution of an H^∞ control problem. Indeed, the results of [100] can be easily generalized to remove these assumptions. Also note that in [100], slightly different versions of conditions (i) and (ii) were assumed. That is, it was assumed that $G = I, \Gamma = I$. However, the design problems considered in this book do not satisfy these assumptions in some cases. This fact required us to modify the assumptions of [100] and replace them with the more general requirement that $G > 0$ and $\Gamma > 0$. This relaxation of these assumptions has required us to correspondingly modify the Riccati equations which arise in the solution to the H^∞ control problem with transients.

The following results present necessary and sufficient conditions for the solvability of a corresponding H^∞ control problem with transients. These necessary and sufficient conditions are stated in terms of certain algebraic or differential Riccati equations.

Finite-horizon state-feedback H^∞ control with transients

Theorem 3.2.4 *Consider the system (3.2.1) for the case in which the full state is available for measurement;* i.e., $y = x$. *Suppose that conditions (i) and (ii) of Assumption 3.2.1 are satisfied and let $X_T \geq 0$ and $P_0 > 0$ be given matrices. Then the following statements are equivalent.*

(i) There exists a controller $K(\cdot)$ of the form (3.2.44) satisfying condition (3.2.46).

(ii) There exists a unique symmetric matrix $X(t)$, $t \in [0, T]$ such that

$$\begin{aligned} -\dot{X}(t) &= A'X(t) + X(t)A \\ &\quad -X(t)\left[B_1G^{-1}B_1' - \frac{1}{\gamma^2}B_2B_2'\right]X(t) + C_1'C_1; \quad (3.2.48) \\ X(T) &= X_T \end{aligned}$$

and $X(0) < \gamma^2 P_0$.

If condition (ii) holds, then the control law

$$u(t) = K(t)x(t); \quad K(t) = -G^{-1}B_1'X(t) \tag{3.2.49}$$

achieves the bound (3.2.46).

Proof: See Theorem 2.1 of [100].

■

Observation 3.2.1 Note that in reference [100], the above result was stated for the case in which the class of controllers under consideration includes only linear time-varying controllers of the form (3.2.44). However, it is straightforward to verify that the same proof can also be used to establish the result for the case in which nonlinear controllers are allowed. We will use this fact in Chapter 9 where the robustness properties of nonlinear controllers will be compared to those of linear controllers; see Section 9.4.

Infinite-horizon state-feedback H^∞ control with transients

The result given below is a natural extension of Theorem 3.2.2 to the case where the initial state of the system (3.2.1) is unknown.

Theorem 3.2.5 *Consider the system (3.2.1) for the case in which the full state is available for measurement;* i.e., $y = x$. *Suppose that Assumption 3.2.1 is satisfied and let $P_0 > 0$ be a given matrix. Then the following statements are equivalent.*

(i) There exists a controller $K(\cdot)$ of the form (3.2.44) satisfying condition (3.2.47) and such that the closed loop system corresponding to this controller is exponentially stable.

(ii) There exists a unique symmetric matrix X satisfying the Riccati equation

$$A'X + XA - X\left[B_1G^{-1}B_1' - \frac{1}{\gamma^2}B_2B_2'\right]X + C_1'C_1 = 0 \tag{3.2.50}$$

and such that the matrix $A + (\frac{1}{\gamma^2}B_2B_2' - B_1G^{-1}B_1')X$ is stable and $0 \leq X \leq \gamma^2 P_0$.

If condition (ii) holds, then the control law

$$u(t) = Kx(t); \quad K = -G^{-1}B_1'X \tag{3.2.51}$$

achieves the bound (3.2.47).

Proof: See Theorem 2.2 of [100].

■

Finite-horizon output-feedback H^∞ control with transients

We now turn to the case of output-feedback controllers.

Theorem 3.2.6 *Consider the system (3.2.1) and suppose that conditions (i) and (ii) of Assumption 3.2.1 are satisfied. Let $X_T \geq 0$ and $P_0 > 0$ be given matrices. Then there exists an output-feedback controller $K(\cdot)$ of the form (3.2.44) satisfying condition (3.2.46) if only if the following three conditions are satisfied.*

(i) There exists a unique symmetric matrix function $X(t)$ satisfying (3.2.48) and $X(0) < \gamma^2 P_0$.

(ii) There exists a symmetric matrix function $Y(t)$ defined for $t \in [0,T]$ such that

$$\begin{aligned} \dot{Y}(t) &= AY(t) + Y(t)A' \\ &\quad -Y(t)\left[C_2'\Gamma^{-1}C_2 - \frac{1}{\gamma^2}C_1C_1'\right]Y(t) + B_2'B_2; \quad (3.2.52) \\ Y(0) &= P_0^{-1}. \end{aligned}$$

(iii) $\rho(X(t)Y(t)) < \gamma^2$ for all $t \in [0,T]$.

If conditions (i)–(iii) are satisfied, a controller that achieves the bound (3.2.46) is given by equation (3.2.44) with

$$\begin{aligned} A_c(t) &= A + B_1C_c - B_c(t)C_2 + \frac{1}{\gamma^2}B_2B_2'X(t); \\ B_c(t) &= (I - \frac{1}{\gamma^2}Y(t)X(t))^{-1}Y(t)C_2'\Gamma^{-1}; \\ C_c(t) &= -G^{-1}B_1'X(t); \\ D_c(t) &\equiv 0. \end{aligned} \quad (3.2.53)$$

Proof: See Theorem 2.3 of [100].

■

Infinite-horizon output-feedback H^∞ control with transients

Theorem 3.2.7 *Consider the system (3.2.1) and suppose that Assumption 3.2.1 is satisfied. Let $P_0 > 0$ be a given matrix. Then there exists an output-feedback controller $K(\cdot)$ of the form (3.2.44) satisfying condition (3.2.47) if only if the following three conditions are satisfied.*

(i) There exists a unique nonnegative-definite symmetric matrix X satisfying equation (3.2.50) and such that the matrix

$$A + (1/\gamma^2)B_2B_2'X - B_1G^{-1}B_1'X$$

is stable and $0 \leq X \leq \gamma^2 P_0$.

(ii) There exists a symmetric bounded matrix function $Y(t) > 0$ defined for $t \in [0, \infty)$ satisfying (3.2.52) with initial condition $Y(0) = P_0^{-1}$ and such that the time-varying linear system

$$\dot{p}(t) = [A - Y(t)(C_2'\Gamma^{-1}C_2 - \frac{1}{\gamma^2}C_1'C_1)]p(t) \tag{3.2.54}$$

is exponentially stable.

(iii) The matrix function $(1 - \rho(\gamma^{-2}Y(t)X))^{-1}$ is positive-definite and bounded for $t \in [0, \infty)$.

Moreover, if a matrix $Y(t)$ with the above properties exists for all $t \geq 0$, then $Y_\infty = \lim_{t\to\infty} Y(t)$ exists and is the unique symmetric solution to the Riccati equation

$$AY_\infty + Y_\infty A' + Y_\infty(\frac{1}{\gamma^2}C_1'C_1 - C_2'\Gamma^{-1}C_2)Y_\infty + B_2B_2' = 0 \tag{3.2.55}$$

such that the matrix

$$A - Y_\infty\left[C_2'\Gamma^{-1}C_2 - (1/\gamma^2)C_1'C_1\right]$$

is stable and $Y_\infty \geq 0$.

If the conditions (i)–(iii) are met, then one controller that achieves closed loop stability and the bound (3.2.47) is given by equation (3.2.44) with

$$A_c(t) = A + B_1C_c - B_c(t)C_2 + \frac{1}{\gamma^2}B_2B_2'X;$$

$$B_c(t) = (I - \frac{1}{\gamma^2}Y(t)X)^{-1}Y(t)C_2'\Gamma^{-1};$$

Proof: See

3.2.3 H

Note that t
tems in a s
varying H^∞
More recen
required in
present tw

We focus on the H^∞ control problem for time-varying systems of the form

$$\begin{aligned} \dot{x} &= A(t)x + B_1(t)u + B_2(t)w; \\ z &= C_1(t)x + D_1(t)u; \\ y &= C_2(t)x + D_2(t)w. \end{aligned} \tag{3.2.57}$$

Consider a set of causal linear time-varying controllers $K(\cdot)$ of the form (3.2.2) or (3.2.4) such that the corresponding closed loop system gives rise to a linear operator $\mathfrak{W}_{wz}^K$ mapping $w(\cdot)$ into $z(\cdot)$ (under zero initial conditions). We seek to find a controller which achieves the following bound on the induced norm of this operator:

$$\|\mathfrak{W}_{wz}^K\|_\infty < \gamma,$$

for a given $\gamma > 0$. The norm $\|\mathfrak{W}_{wz}^K\|_\infty$ is defined in the standard manner:

$$\|\mathfrak{W}_{wz}^K\|_\infty := \sup_{w(\cdot)\in \mathbf{L}_2[0,T]} \frac{\left[\int_0^T \|z(t)\|^2 dt\right]^{1/2}}{\left[\int_0^T \|w(t)\|^2 dt\right]^{1/2}}$$

in the finite-horizon case and

$$\|\mathfrak{W}_{wz}^K\|_\infty := \sup_{w(\cdot)\in \mathbf{L}_2[0,\infty]} \frac{\|z(\cdot)\|_2}{\|w(\cdot)\|_2}$$

in the infinite-horizon case.

Finite-horizon state-feedback control of time-varying systems

Theorem 3.2.8 *Let $\gamma > 0$ be given and consider a time-varying system of the form (3.2.57) defined for $t \in [0,T]$ and such that the full state is available for measurement;* i.e., *$y = x$. Then the following statements are equivalent:*

(i) There exists a causal time-varying state-feedback controller $K(\cdot)$ of the form (3.2.2) such that $\|\mathfrak{W}_{wz}^K\| < \gamma$.

(ii) The Riccati differential equation

$$\begin{aligned} -\dot{X} &= (A - B_1 G^{-1} D_1' C_1)' X + X(A - B_1 G^{-1} D_1' C_1) \\ &\quad + C_1'(I - D_1 G^{-1} D_1')C_1 - X(B_1 G^{-1} B_1' - \frac{1}{\gamma^2} B_2 B_2')X; \\ X(T) &= 0 \end{aligned} \tag{3.2.58}$$

has a solution on $[0,T]$.

If condition (ii) holds, then the controller

$$\begin{aligned} u^*(t) &= K(t)x(t); \\ K(t) &:= -G^{-1}(t)(B_1(t)'X(t) + D_1(t)'C_1(t)) \end{aligned} \tag{3.2.59}$$

guarantees the specified bound $\|\mathfrak{W}_{wz}^K\|_\infty < \gamma$. *Furthermore, if the initial condition in (3.2.57) is* $x(0) = x_0$, *then*

$$\sup_{w(\cdot)\in \mathbf{L}_2[0,T]} J_\gamma(u^*(\cdot), w(\cdot)) = x_0' X(0) x_0.$$

Proof: See Theorems 2.1 and 2.4 of [114] or Theorem 4.1 of [90].

■

Infinite-horizon state-feedback control of time-varying systems

In Section 8.2, we will use the following infinite-horizon time-varying state-feedback H^∞ result due to Ichikawa [90].

Theorem 3.2.9 *Let* $\gamma > 0$ *be given and consider the time-varying linear system (3.2.57). Assume that the pair* (A, C_1) *is detectable and the pair* (A, B_1) *is stabilizable. Also, assume that* $D_1' C_1 = 0$[5]. *Then there exists a stabilizing feedback law* $u = Kx$ *such that* $\|\mathfrak{W}_{wz}^K\|_\infty < \gamma$ *if and only if there exists a nonnegative-definite bounded stabilizing solution to the Riccati differential equation*

$$-\dot{X} = A'X + XA + C_1'C_1 - X(B_1 G^{-1} B_1' - \frac{1}{\gamma^2} B_2 B_2')X. \quad (3.2.60)$$

If the above condition holds, then the controller

$$u(t) = K(t)x(t); \quad K(t) := -G^{-1} B_1(t)' X(t) \quad (3.2.61)$$

leads to a stable closed loop system and the satisfaction of the bound

$$\|\mathfrak{W}_{wz}^K\|_\infty < \gamma.$$

Proof: See Theorems 4.2 and 4.3 of [90].

■

3.3 Risk-sensitive control

The development of the minimax LQG control theory presented in Chapter 8 relies on the interpretation of the minimax LQG control problem as an output-feedback stochastic dynamic game. It turns out that under certain conditions, the upper value of this game can be obtained via a logarithmic transformation on the optimal value of an exponential-of-integral cost functional. Such cost functionals arise in risk-sensitive control theory. We defer a rigorous discussion of the equivalence between output-feedback stochastic dynamic games and risk-sensitive control to Chapter 8. In this section, we collect some preliminary results concerning risk-sensitive control which will be exploited in Chapter 8.

[5] As noted previously, this is a simplifying assumption which is straightforward to remove.

The linear exponential quadratic regulator (LEQR) problem introduced by Jacobson [92] is commonly viewed as a stochastic control counterpart to the deterministic state-feedback H^∞ control problem. One of the most important results of [92] was the fact that the LEQR problem is related to a differential game which is similar to that arising in H^∞ control. In particular, in the small noise limit, the optimal controller in the LEQR problem is identical to the central controller solving an associated deterministic H^∞ control problem and corresponding differential game. This observation has led to extensive research concerning the relationship between stochastic optimal control problems with exponential-of-integral cost, stochastic and deterministic differential games, and H^∞ control. The study of risk-sensitive control was initiated in [92] where the solution of the risk-sensitive control problem with full state measurement was derived. References [244, 12] addressed an output-feedback linear exponential quadratic Gaussian (LEQG) optimal control problem for stochastic systems using a certainty equivalence principle and the information state approach respectively; see also the monographs [245, 11]. The information state approach has proved to be useful in many more general problems. In particular, extensions of this approach to the nonlinear case were obtained in [93, 36].

The above results concern finite-horizon risk-sensitive control problems. The infinite-horizon risk-sensitive control problem turns out to be more involved than the finite-horizon problem. A number of interesting results have been obtained which establish a link between the infinite-horizon LEQR problem and H^∞ control via a minimum entropy principle [69, 68]. A stochastic ergodic game type characterization of the infinite-horizon risk-sensitive control problem was recently given in [170] using Donsker-Varadhan large deviations ideas; see also, [53, 39, 59]. This result was given for both the state-feedback and output-feedback cases. The approach of [170] allows one to deal with a very general class of nonlinear problems. In Chapter 8, we will focus on a linear quadratic version of the problem considered in [170]. In the work of Pan and Başar [139], a direct derivation of the solution to both the finite-horizon and infinite-horizon LEQG control problems was proposed. These results, which will be used in Chapter 8, are summarized in Sections 3.3.2 and 3.3.3.

3.3.1 *Exponential-of-integral cost analysis*

In this subsection, we review some results of [170] concerning the characterization of an exponential-of-integral cost functional. These results are closely related to results on the duality between free energy and relative entropy arising in the theory of large deviations; e.g., see [42, 220, 53] and also Appendix A. Note that the paper [170] considers a more general formulation of this problem than is required here. We will be interested in the particular case of linear-exponential-of-quadratic risk-sensitive control problems.

Consider the Ito system

$$dx(t) = Ax(t)dt + BdW(t); \qquad x(0) = x_0 \tag{3.3.1}$$

where $x \in \mathbf{R}^n$ and $W(t)$ is a standard Wiener process defined on a probability space $(\Omega, \mathcal{F}, P)$. In this subsection, we assume that the matrix A is stable and the pair (A, B) is controllable. We will be interested in the cost functional

$$\Im := \lim_{T \to \infty} \frac{2\tau}{T} \log \mathbf{E}_{0,x} \exp \left\{ \frac{1}{2\tau} \int_0^T x'(t) R x(t) dt \right\} \tag{3.3.2}$$

defined on solutions to the equation (3.3.1). Here R is a nonnegative-definite symmetric matrix, $\tau \neq 0$ is a scalar parameter and $\mathbf{E}_{0,x}$ denotes the expectation conditioned on $x(0) = x$.

Equation (3.3.1) defines a Markov diffusion process on $\mathbf{R}^n$. Let $P(t, x, \Lambda)$ be the transition probability function for this Markov process and let $\mathcal{M}(\mathbf{R}^n)$ denote the set of probability measures on $\mathbf{R}^n$. The Markov process $x(t), t \geq 0$ defined by equation (3.3.1) has a unique invariant probability measure $\mu \in \mathcal{M}(\mathbf{R}^n)$ since the matrix A is stable and the pair (A, B) is controllable; e.g., see [267]. That is, there exists a probability measure μ on $\mathbf{R}^n$ such that

$$\int_{\mathbf{R}^n} P(t, x, \Lambda) \mu(dx) = \mu(\Lambda)$$

for any Borel set $\Lambda \subseteq \mathbf{R}^n$. Furthermore, it is shown in [267] that the invariant probability measure μ is unique. This probability measure is given by

$$\mu(dx) = \frac{1}{(2\pi)^{(n/2)} (\det \Pi)^{1/2}} \exp \left(-\frac{1}{2} x' \Pi^{-1} x \right) dx$$

where the matrix Π is the positive-definite solution to the Lyapunov equation

$$A\Pi + \Pi A' + BB' = 0.$$

That is, in this particular case, the invariant probability measure μ is a Gaussian probability measure. It is shown in [170] that the Markov process defined by equation (3.3.1) and the invariant probability measure μ satisfy the basic assumptions required to apply the results of [170].

We now consider an infinite-horizon version of the free energy function (*c.f.* Definition A.1 of Appendix A):

$$\mathbb{E}(\psi) := \lim_{T \to \infty} \frac{2\tau}{T} \log \mathbf{E}_{0,x} \exp \left\{ \int_0^T \psi(x(t)) dt \right\}.$$

The paper [170] introduces the Legendre transform $h(\cdot)$ of the free energy function $\mathbb{E}(\cdot)$:

$$h(\nu) := \sup_{\psi \in C_b(\mathbf{R}^n)} \left(\int \psi \nu(dx) - \mathbb{E}(\psi) \right). \tag{3.3.3}$$

Thus, the function $h(\cdot)$ is an analog of the relative entropy; *c.f.* equation (A.2). Indeed, it is shown in [170] that the function $h(\cdot)$ is identical to the Donsker-Varadhan relative entropy between the probability measure ν and the reference probability measure μ; see Definition A.2 in Appendix A. Furthermore, the function $h(\cdot)$ and the exponential-of-integral cost $\Im$ satisfy the following large deviations type identity:

$$\Im = \sup_{\nu \in \mathcal{M}(\mathbf{R}^n)} \left(\int x' Rx \mu(dx) - h(\nu) \right). \tag{3.3.4}$$

This result, given in Theorem 2.1 of [170], allows for the interpretation of the exponential-of-integral cost $\Im$ as the optimal value in an optimal control problem with ergodic quadratic cost; *c.f.* [250]. To define this control problem, consider a controlled version of the system (3.3.1):

$$dx(t) = (Ax(t) + B\phi(x(t)))dt + BdW(t), \tag{3.3.5}$$

where the function $\phi(x)$ is the controller. The set $\mathfrak{D}$ of admissible controllers includes all functions $\phi(x)$ such that the stochastic differential equation (3.3.5) has a stationary solution $x_\phi(\cdot)$. Let $\nu^\phi(dx)$ denote the probability distribution of $x_\phi(t)$ for each t. Theorem 2.2 of [170] shows that under certain additional assumptions, the supremum with respect to $\nu \in \mathcal{M}(\mathbf{R}^n)$ in equation (3.3.4) can be replaced by the supremum with respect to ν^ϕ, $\phi \in \mathfrak{D}$:

$$\Im = \sup_{\phi(\cdot) \in \mathfrak{D}} \int \left(x' Rx - \tau \|\phi(x)\|^2 \right) \nu^\phi(dx). \tag{3.3.6}$$

One particular case where that additional assumption is satisfied and hence, the identity (3.3.6) is true, is of particular importance for our subsequent investigation. This result, originally given in Example 2.2 of [170], is presented in the following lemma.

Lemma 3.3.1 *Suppose the assumptions of this subsection are satisfied and that the Riccati equation*

$$XA + A'X + R + \frac{1}{\tau} XBB'X = 0 \tag{3.3.7}$$

admits a nonnegative-definite stabilizing solution. Then, the identity (3.3.6) holds. Furthermore, the supremum on the right hand side of equation (3.3.6) is attained by the linear function $\phi(x) = \frac{1}{\tau} B'Xx$. This leads to the following optimal cost:

$$\Im = \operatorname{tr} B_2 B_2' X. \tag{3.3.8}$$

That is, the quantity $\Im$ is independent of the initial condition of the system (3.3.1). Hence in (3.3.2), the conditional expectation $\mathbf{E}_{0,x}$ can be replaced by the unconditional expectation $\mathbf{E}$.

Remark 3.3.1 In [170], the pair (A, R) is required to be observable in order to guarantee that the Riccati equation (3.3.7) has a positive-definite stabilizing solution X. In this case, the control problem on the right hand side of equation (3.3.6) admits a solution as described in Lemma 3.3.1. However, such a control problem does not require that $X > 0$. For example, it is shown in [250] that this control problem with ergodic cost has a solution if $X \geq 0$ exists. Therefore, we do not require that the pair (A, R) be observable in Lemma 3.3.1.

A result extending Lemma 3.3.1 to the case of controlled diffusions also holds. In this case, the cost $\Im$ becomes a functional $\Im(u(\cdot))$ of the control input $u(\cdot)$. Then, the problem of risk-sensitive cost analysis turns into a control problem involving minimization of the risk-sensitive functional $\Im(u(\cdot))$ defined on solutions to a controlled version of the system (3.3.1). Also, the maximization problem on the right hand side of equation (3.3.6) is replaced by a stochastic dynamic game. Instead of the Riccati equation (3.3.7), the solution of this stochastic dynamic game involves a game type Riccati equation arising in H^∞ control. Furthermore, the saddle point of this game is expressed in a state-feedback form. From this result of [170], an interesting conclusion can be drawn. In stochastic differential games equivalent to LEQR/LEQG control problems, the worst case strategy of the maximizing player will be a linear feedback strategy. This observation has a well known counterpart in H^∞ control where the worst case disturbance allows for a representation in state-feedback form.

3.3.2 Finite-horizon risk-sensitive control

On the probability space defined in Section 2.4.2, we consider a system and measurement dynamics driven by a noise input $W(\cdot)$ and a control input $u(\cdot)$. These dynamics are described by the following stochastic differential equation of the form (2.4.6):

$$\begin{aligned} dx(t) &= (A(t)x(t) + B_1(t)u(t))dt + B_2(t)dW(t); \\ &\qquad\qquad\qquad\qquad\qquad\qquad\qquad x(0) = x_0; \\ dy(t) &= C_2(t)x(t)dt + D_2(t)dW(t). \quad y(0) = y_0. \end{aligned} \tag{3.3.9}$$

The system (3.3.9) is considered on a finite time interval $[0, T]$. In the above system, $x(t) \in \mathbf{R}^n$ is the state, $u(t) \in \mathbf{R}^m$ is the control input and $y(t) \in \mathbf{R}^l$ is the measured output. Also, note that the probability measure on $x_0 \times y_0$ will be as defined in (2.4.4).

In this section, we will assume that the system (3.3.9) satisfies all of the assumptions of Section 2.4.2. In particular, we will consider causal output-feedback controllers of the form (2.4.7) adapted to the filtration $\{\mathcal{Y}_t, t \geq 0\}$ generated by the observation process y. Also, we assume that $\Gamma(t) := D_2(t)D_2(t)' > 0$ for all $t \in [0; T]$.

Associated with the system (3.3.9), we consider a risk-sensitive cost functional

$$\begin{aligned}&\Im_T(u(\cdot))\\ &:= 2\tau \log \mathbf{E}\left\{ \exp\left(\frac{1}{2\tau}\left[x'(T)Mx(T) + \int_0^T F(x(t),u(t))dt \right]\right)\right\}\end{aligned} \tag{3.3.10}$$

where

$$F(x,u) := x'R(t)x + 2x'\Upsilon(t)u + u'G(t)u, \tag{3.3.11}$$

M is a nonnegative-definite symmetric matrix, $R(\cdot)$, $\Upsilon(\cdot)$ and $G(\cdot)$ are matrix valued functions, $R(\cdot)$ and $G(\cdot)$ are symmetric and $R(t) \geq 0$, $G(t) > 0$ for all t. Also, $\tau \neq 0$ is a scalar parameter.

In the finite-horizon risk-sensitive optimal control problem, an output-feedback controller of the form (2.4.7) is sought to minimize the cost functional (3.3.10):

$$\inf_{u(\cdot)} \Im_T(u(\cdot)). \tag{3.3.12}$$

As mentioned above, this problem is also known as the LEQG control problem. A mathematically rigorous definition of the class of admissible controllers in this control problem will be given below. The standard LQG control problem is recovered when $\tau \to \infty$. This case is referred to as the *risk-neutral* case. Also, the cases where $\tau > 0$ and $\tau < 0$ are known as *risk-averse* and *risk-seeking* control problems, respectively; see [244, 245]. The scalar $1/\tau$ is called the *risk-sensitivity* parameter. In the sequel, we will be mainly concerned with the case where $\tau > 0$ (risk aversion).

Note that the solution to the risk-sensitive optimal control problem (3.3.12) was originally given in reference [12]; see also [11]. Also, the paper [139] presents a more direct derivation of the solution to the control problem (3.3.12) which allows for a more general cost functional. The solution to the risk-sensitive optimal control problem given in [139] also allows for correlation between the system and measurement noises. Thus, in this book we apply the results of [139] to the problem (3.3.12). It is important to note that the optimal LEQG controller derived in [139, 12] is expressed in the form of a linear state-feedback control law combined with a state estimator.

In order to apply the results of [139], we require the following assumptions to be satisfied.

Assumption 3.3.1 *The following conditions hold:*

(i) The Riccati differential equation

$$\begin{aligned}\dot{Y} &= (A - B_2D_2'\Gamma^{-1}C_2)Y + Y(A - B_2D_2'\Gamma^{-1}C_2)'\\ &\quad -Y(C_2'\Gamma^{-1}C_2 - \frac{1}{\tau}R)Y + B_2(I - D_2'\Gamma^{-1}D_2)B_2';\\ Y(0) &= Y_0\end{aligned} \tag{3.3.13}$$

has a symmetric solution $Y : [0,T] \to \mathbf{R}^{n\times n}$ *such that* $Y(t) \geq c_0 I$ *for some* $c_0 > 0$ *and for all* $t \in [0,T]$.

(ii) $R - \Upsilon G^{-1}\Upsilon' \geq 0$ *and furthermore the Riccati differential equation*

$$\begin{aligned} \dot{X} &+ X(A - B_1 G^{-1}\Upsilon') + (A - B_1 G^{-1}\Upsilon')'X \\ &+(R - \Upsilon G^{-1}\Upsilon') - X(B_1 G^{-1} B_1' - \frac{1}{\tau} B_2 B_2')X = 0; \\ X(T) &= M \end{aligned} \quad (3.3.14)$$

has a nonnegative-definite symmetric solution $X : [0,T] \to \mathbf{R}^{n\times n}$.

(iii) For each $t \in [0,T]$, *the matrix* $I - \frac{1}{\tau}Y(t)X(t)$ *has only positive eigenvalues. That is,*

$$\rho(Y(t)X(t)) < \tau \quad (3.3.15)$$

for all $t \in [0,T]$.

We now present the definition of the class of admissible controllers considered in [139]. Consider the estimator state equations

$$\begin{aligned} d\check{x}(t) &= (A + \frac{1}{\tau}YR)\check{x}(t)dt + (B + \frac{1}{\tau}Y\Upsilon)u(t)dt \\ &\quad +(YC_2' + B_2 D_2')\Gamma^{-1}(dy(t) - C_2\check{x}(t)dt); \\ \check{x}(0) &= \check{x}_0. \end{aligned} \quad (3.3.16)$$

Also, let

$$\hat{x} := (I - \frac{1}{\tau}YX)^{-1}\check{x}; \quad \tilde{u} = u + G^{-1}(B_1'X + \Upsilon')\hat{x}.$$

Then as in [139],

$$\begin{aligned} d\hat{x}(t) &= (A - B_1 G^{-1}\Upsilon' - (B_1 G^{-1} B_1' - \frac{1}{\tau}B_2 B_2')X)\hat{x}(t)dt \\ &\quad +(I - \frac{1}{\tau}YX)^{-1}(B_1 + \frac{1}{\tau}Y\Upsilon)\tilde{u}(t)dt \\ &\quad +(I - \frac{1}{\tau}YX)^{-1}(YC_2' + B_2 D_2')\Gamma^{-1} \\ &\quad \times(dy(t) - (C_2 + \frac{1}{\tau}D_2 B_2' X)\hat{x}(t)dt). \end{aligned} \quad (3.3.17)$$

The class of admissible controllers considered in [139] are controllers of the form (2.4.7) such that the function $\mathcal{K}(t,y)$ is piecewise continuous in t and Lipschitz continuous in y. These controllers are also required to satisfy the following causality condition: The process

$$\zeta(t) = \exp\left\{\int_0^t \alpha'(s)dW(s) - \frac{1}{2}\int_0^t \|\alpha(s)\|^2 dt\right\} \quad (3.3.18)$$

with

$$\begin{aligned}\alpha(t) &:= B_2'Y\epsilon(t) - D_2'\Gamma^{-1}(C_2Y + D_2B_2')Y^{-1}e(t);\\ \epsilon(t) &:= x(t) - \check{x}(t); \quad e(t) := x(t) - \hat{x}(t)\end{aligned}$$

is a martingale on $[0, T]$. As observed in [139], this condition is satisfied for any linear controller. Hence, the class of admissible controllers includes the class of linear controllers. Thus, if the controller which attains the infimum of the functional (3.3.10) in the risk-sensitive optimal control problem (3.3.12) is linear, then this controller will necessarily be admissible.

As in [139], the solutions to Riccati differential equations (3.3.13), (3.3.14) will define the optimal controller in the risk-sensitive optimal control problem. Indeed, consider the controller state equations

$$\begin{aligned}d\check{x}(t) &= \left[A + \frac{1}{\tau}YR - (YC_2' + B_2D_2')\Gamma^{-1}C_2\right]\check{x}dt\\ &\quad -(B_1 + \frac{1}{\tau}Y\Upsilon)G^{-1}(B_1'X + \Upsilon')(I - \frac{1}{\tau}YX)^{-1}\check{x}dt\\ &\quad +(YC_2' + B_2D_2')\Gamma^{-1}dy(t);\\ \check{x}(0) &= \check{x}_0 \end{aligned} \tag{3.3.19}$$

obtained from (3.3.16) with the substitution of the feedback control law

$$u^*(t) = -G^{-1}(t)(B_1'(t)X(t) + \Upsilon'(t))[I - \frac{1}{\tau}Y(t)X(t)]^{-1}\check{x}(t). \tag{3.3.20}$$

Theorem 3.3.1 *Suppose that Assumption 3.3.1 is satisfied. The optimal controller in the risk-sensitive optimal control problem (3.3.12) is given by equation (3.3.20) where $\check{x}$ is generated by equation (3.3.19). Equivalently, the optimal controller is given by the equation*

$$u^*(t) = -G^{-1}(t)(B_1'(t)X(t) + \Upsilon'(t))\hat{x}(t) \tag{3.3.21}$$

where $\hat{x}$ is generated by the equation

$$\begin{aligned}d\hat{x}(t) &= (A - B_1G^{-1}\Upsilon' - (B_1G^{-1}B_1' - \frac{1}{\tau}B_2B_2')X)\hat{x}(t)dt\\ &\quad +(I - \frac{1}{\tau}YX)^{-1}(YC_2' + B_2D_2')\Gamma^{-1}\\ &\quad \times(dy(t) - (C_2 + \frac{1}{\tau}D_2B_2'X)\hat{x}(t)dt).\end{aligned} \tag{3.3.22}$$

The optimal cost can be written as

$$\begin{aligned}
&\inf_{u(\cdot)} \Im_T(u(\cdot)) \\
&= \check{x}_0' X(0)(I - \frac{1}{\tau} Y_0 X(0))^{-1} \check{x}_0 - \tau \log\det(I - \frac{1}{\tau} M Y(T)) \\
&+ \int_0^T \operatorname{tr} \begin{bmatrix} Y(t)R + (Y(t)C_2'(t) + B_2(t)D_2'(t))\Gamma^{-1}(t) \\ \times (C_2(t)Y(t) + D_2(t)B_2'(t))X(t)(I - \frac{1}{\tau}Y(t)X(t))^{-1} \end{bmatrix} dt.
\end{aligned} \tag{3.3.23}$$

Proof: See Theorem 2 of reference [139].

■

3.3.3 Infinite-horizon risk-sensitive control

In this subsection, a solution to the infinite-horizon version of the risk-sensitive optimal control problem (3.3.12) will be presented. We now consider a time-invariant version of the system (3.3.9) (*c.f.* equation (2.4.25)):

$$\begin{aligned}
dx(t) &= (Ax(t) + B_1 u(t))dt + B_2 dW(t); \quad x(0) = x_0; \\
dy(t) &= C_2 x(t)dt + D_2 dW(t); \quad y(0) = 0
\end{aligned} \tag{3.3.24}$$

where all coefficients are assumed to be constant matrices of corresponding dimensions. Also, $\Gamma := D_2 D_2' > 0$.

An infinite-horizon version of the functional (3.3.10) is the risk-sensitive cost functional

$$\Im(u(\cdot)) := \lim_{T\to\infty} \frac{2\tau}{T} \log \mathbf{E} \left\{ \exp\left(\frac{1}{2\tau} \int_0^T F(x(t), u(t))dt \right) \right\} \tag{3.3.25}$$

where the quadratic form $F(\cdot)$ is a time-invariant version of the corresponding quadratic form (3.3.11) defined by matrices R, Υ and G with $R = R' \geq 0$ and $G = G' > 0$:

$$F(x, u) := x'Rx + 2x'\Upsilon u + u'Gu. \tag{3.3.26}$$

The infinite-horizon risk-sensitive optimal control problem involves finding an output-feedback controller of the form (2.4.7) which minimizes the cost functional (3.3.25):

$$\inf_{u(\cdot)} \Im(u(\cdot)). \tag{3.3.27}$$

The derivation of the solution to the risk-sensitive optimal control problem (3.3.27) makes use of the following time-invariant versions of Riccati equations

(3.3.13) and (3.3.14):

$$\begin{aligned}(A - B_2D_2'\Gamma^{-1}C_2)Y_\infty + Y_\infty(A - B_2D_2'\Gamma^{-1}C_2)' \\ -Y_\infty(C_2'\Gamma^{-1}C_2 - \frac{1}{\tau}R)Y_\infty + B_2(I - D_2'\Gamma^{-1}D_2)B_2' = 0;\end{aligned} \tag{3.3.28}$$

$$\begin{aligned}X_\infty(A - B_1G^{-1}\Upsilon') + (A - B_1G^{-1}\Upsilon')'X_\infty \\ +(R - \Upsilon G^{-1}\Upsilon') - X_\infty(B_1G^{-1}B_1' - \frac{1}{\tau}B_2B_2')X_\infty = 0;\end{aligned} \tag{3.3.29}$$

c.f. equations (3.2.33) and (3.2.8).

Assumption 3.3.2 *The following conditions are satisfied:*

(i) The matrix $R - \Upsilon G^{-1}\Upsilon'$ is nonnegative-definite and the pair

$$(A - B_1G^{-1}\Upsilon', R - \Upsilon G^{-1}\Upsilon')$$

is detectable. Also, the pair (3.2.34) is stabilizable.

(ii) The algebraic Riccati equation (3.3.28) admits a minimal positive-definite solution Y_∞.

(iii) The algebraic Riccati equation (3.3.29) admits a minimal nonnegative-definite solution X_∞.

(iv) The matrix $I - \frac{1}{\tau}Y_\infty X_\infty$ has only positive eigenvalues. That is the spectral radius of the matrix $Y_\infty X_\infty$ satisfies the condition

$$\rho(Y_\infty X_\infty) < \tau. \tag{3.3.30}$$

The following lemma shows that the matrices X_∞ and Y_∞ can be expressed as limits of the solutions to the corresponding finite-horizon Riccati differential equations (3.3.14) and (3.3.13).

Lemma 3.3.2 *Suppose that Assumption (3.3.2) holds. Then,*

(i) The Riccati differential equation (3.3.14) with $M = 0$ has a nonnegative-definite symmetric solution $X_T(t)$ such that $X_T(t) \to X_\infty$ as $T \to \infty$.

(ii) If $Y_\infty \geq Y_0 > 0$, then the Riccati differential equation (3.3.13) has a positive-definite symmetric solution $Y(t; Y_0)$ such that $Y(t; Y_0) \to Y_\infty$ as $t \to \infty$.

(iii) For each $T > 0$ and $t \in [0, T]$, the matrices $Y(t; Y_0)$ and $X_T(t; M)$ satisfy condition (3.3.15).

Proof: Part (i) of the lemma follows from Lemma 3.2.1.

In order to prove part (ii) of the lemma, we consider a system

$$\dot{\eta} = A'\eta + C_2'\nu + R^{1/2}\omega; \quad \eta(0) = \eta_0; \qquad (3.3.31)$$
$$\zeta = B_2'\eta + D_2'\nu$$

where $\eta \in R^n$ is the state, $\nu \in \mathbf{R}^l$ is the control input, $\omega \in \mathbf{R}^q$ is the disturbance input, $\zeta \in \mathbf{R}^p$ is the controlled output. Also, for a given $T > 0$, we consider the cost functionals

$$\bar{J}_\tau^{T,Y_0}(\nu(\cdot),\omega(\cdot)) := \eta(T)'Y_0\eta(T) + \int_0^T \left(\|\zeta(t)\|^2 - \tau\|\omega(t)\|^2\right) dt;$$
$$\bar{J}_\tau(\nu(\cdot),\omega(\cdot)) := \int_0^\infty \left(\|\zeta(t)\|^2 - \tau\|\omega(t)\|^2\right) dt \qquad (3.3.32)$$

where $\omega(\cdot) \in \mathbf{L}_2[0,T]$. The proof of claim (ii) of the lemma makes use of some results concerning the differential games associated with the system (3.3.31) and the cost functionals (3.3.32).

The first differential game to be considered in relation to equation (3.3.28) is the infinite-horizon differential game with the CLPS information pattern:

$$\inf_{\nu(\cdot)} \sup_{\omega(\cdot)\in\mathbf{L}_2[0,\infty)} \bar{J}_\tau(\nu(\cdot),\omega(\cdot)).$$

Since the Riccati equation (3.3.28) has a minimal positive-definite solution Y_∞, it follows from Lemma 3.2.1 that

$$\inf_{\nu(\cdot)} \sup_{\omega(\cdot)\in\mathbf{L}_2[0,\infty)} \bar{J}_\tau(\nu(\cdot),\omega(\cdot)) = \eta_0'Y_\infty\eta_0.$$

Also, the optimal controller solving the above differential game is given by

$$\nu^* = -\Gamma^{-1}(D_2B_2' + C_2Y_\infty)\eta.$$

Hence,

$$\sup_{\omega(\cdot)\in\mathbf{L}_2[0,\infty)} \bar{J}_\tau(\nu^*(\cdot),\omega(\cdot)) = \eta_0'Y_\infty\eta_0. \qquad (3.3.33)$$

For this controller and any $\omega(\cdot) \in \mathbf{L}_2[0,T]$, $Y_\infty \geq Y_0$ implies

$$\bar{J}_\tau^{T,Y_0}(\nu^*(\cdot),\omega(\cdot)) \leq \bar{J}_\tau^{T,Y_\infty}(\nu^*(\cdot),\omega(\cdot)).$$

Hence,

$$\begin{aligned}\inf_{\nu\in\mathbf{L}_2[0,T]} \sup_{\omega(\cdot)\in\mathbf{L}_2[0,T]} \bar{J}_\tau^{T,Y_0}(\nu(\cdot),\omega(\cdot)) &\leq \sup_{\omega(\cdot)\in\mathbf{L}_2[0,T]} \bar{J}_\tau^{T,Y_0}(\nu^*(\cdot),\omega(\cdot)) \\ &\leq \sup_{\omega(\cdot)\in\mathbf{L}_2[0,T]} \bar{J}_\tau^{T,Y_\infty}(\nu^*(\cdot),\omega(\cdot)) \\ &= \eta_0'Y_\infty\eta_0. \qquad (3.3.34)\end{aligned}$$

Here we have used (3.3.33) and the principle of optimality. Furthermore using (3.3.34), Theorem 4.1 of [9] implies that the Riccati differential equation

$$\begin{aligned}\dot{\bar{Y}} &+ (A - B_2D_2'\Gamma^{-1}C_2)\bar{Y} + \bar{Y}(A - B_2D_2'\Gamma^{-1}C_2)' \\ &- \bar{Y}(C_2'\Gamma^{-1}C_2 - \frac{1}{\tau}R)\bar{Y} + B_2(I - D_2'\Gamma^{-1}D_2)B_2' = 0; \\ \bar{Y}(T) &= Y_0 \end{aligned} \tag{3.3.35}$$

does not have a conjugate point in the interval $[0, T]$. The solution to equation (3.3.35) is denoted by $\bar{Y}_T(t; Y_0)$. This implies that the Riccati differential equation (3.3.13) has a symmetric solution on the interval $[0, T]$, defined by $Y(t; Y_0) := \bar{Y}_T(T - t; Y_0)$. Also, it follows from Theorem 4.1 of [9] that

$$\inf_{\nu \in \mathbf{L}_2[0,T]} \sup_{\omega(\cdot) \in \mathbf{L}_2[0,T]} \bar{J}_\tau^{T,Y_0}(\nu(\cdot), \omega(\cdot)) = \eta_0' \bar{Y}_T(0; Y_0)\eta_0 = \eta_0' Y(T; Y_0)\eta_0.$$

This fact together with (3.3.34) implies that

$$Y(T; Y_0) \leq Y_\infty \tag{3.3.36}$$

for any $Y_0 > 0$ such that $Y_0 \leq Y_\infty$.

Now $Y_0 > 0$ implies

$$\bar{J}_\tau^{T,0}(\nu(\cdot), \omega(\cdot)) \leq \bar{J}_\tau^{T,Y_0}(\nu(\cdot), \omega(\cdot))$$

for all $\nu(\cdot) \in \mathbf{L}_2[0, T]$ and $\omega(\cdot) \in \mathbf{L}_2[0, T]$. Furthermore, since $T > 0$ was arbitrary, we have,

$$\begin{aligned}\eta_0' Y(T; 0)\eta_0 &= \inf_{\nu(\cdot) \in \mathbf{L}_2[0,T]} \sup_{\omega(\cdot) \in \mathbf{L}_2[0,T]} \bar{J}_\tau^{T,0}(\nu(\cdot), \omega(\cdot)) \\ &\leq \inf_{\nu(\cdot) \in \mathbf{L}_2[0,T]} \sup_{\omega(\cdot) \in \mathbf{L}_2[0,T]} \bar{J}_\tau^{T,Y_0}(\nu(\cdot), \omega(\cdot)) \\ &= \eta_0' Y(T; Y_0)\eta_0 \end{aligned} \tag{3.3.37}$$

for all $T > 0$. Combining this fact with (3.3.36), it follows that

$$Y(T; 0) \leq Y(T; Y_0) \leq Y_\infty \tag{3.3.38}$$

for all $T > 0$.

Note that Lemma 3.2.1 guarantees that the matrix function $Y(T; 0)$ is nondecreasing as $T \to \infty$. Hence,

$$0 < Y_0 \leq Y(T; 0) \leq Y(T; Y_0)$$

for all $T > 0$. Furthermore, since the function $Y(T; 0)$ is nondecreasing and bounded by Y_∞, it follows that the limit $\lim_{T\to\infty} Y(T; 0)$ exists and is bounded by Y_∞. Moreover, since Y_∞ is the minimal nonnegative-definite solution to equation (3.3.28), then

$$\lim_{T\to\infty} Y(T; 0) = Y_\infty.$$

Thus, it follows from inequality (3.3.38) that $\lim_{T\to\infty} Y(T;Y_0) = Y_\infty$. This concludes the proof of part (ii) of the lemma.

Part (iii) of the lemma is proved by contradiction. Suppose that there exists a $T > 0$ and a $t \in [0,T]$ such that $Y(t;Y_0)X_T(t) \geq \tau I$. Then, using the fact that $X_T(t) \leq X_\infty$ and $Y(t;Y_0) \leq Y_\infty$, we obtain

$$X_\infty \geq X_T(t) \geq \tau Y^{-1}(t;Y_0) \geq \tau Y_\infty^{-1}.$$

This implies $Y_\infty X_\infty \geq \tau I$. This leads to a contradiction with Assumption 3.3.2. Thus, part (iii) of the lemma follows.

■

It follows from Lemma 3.3.2 that the satisfaction of Assumption 3.3.2 also guarantees the satisfaction of Assumption 3.3.1 for the corresponding time-invariant system (3.3.24). These two assumptions are needed in order to define the class of admissible controllers in the infinite-horizon risk-sensitive optimal control problem considered in this subsection.

Associated with Riccati equations (3.3.28) and (3.3.29), we consider time-invariant versions of the estimator state equations (3.3.16) and (3.3.17):

$$\begin{aligned} d\check{x} &= (A + \frac{1}{\tau}Y_\infty R)\check{x}dt + (B_1 + \frac{1}{\tau}Y_\infty \Upsilon)udt \\ &\quad +(Y_\infty C_2' + B_2D_2')\Gamma^{-1}(dy(t) - C_2\check{x}dt); \\ \check{x}(0) &= \check{x}_0; \end{aligned} \tag{3.3.39}$$

$$\begin{aligned} d\hat{x} &= (A - B_1G^{-1}\Upsilon' - (B_1G^{-1}B_1' - \frac{1}{\tau}B_2B_2')X_\infty)\hat{x}dt \\ &\quad +(I - \frac{1}{\tau}Y_\infty X_\infty)^{-1}(B_1 + \frac{1}{\tau}Y_\infty\Upsilon)\tilde{u}dt \\ &\quad +(I - \frac{1}{\tau}Y_\infty X_\infty)^{-1}(Y_\infty C_2' + B_2D_2')\Gamma^{-1} \\ &\quad \times \left(dy - (C_2 + \frac{1}{\tau}D_2B_2'X_\infty)\hat{x}dt\right) \end{aligned} \tag{3.3.40}$$

where

$$\begin{aligned} \hat{x} &:= (I - \frac{1}{\tau}Y_\infty X_\infty)^{-1}\check{x}; \\ \tilde{u} &:= u + G^{-1}(B_1'X_\infty + \Upsilon')\hat{x}. \end{aligned}$$

The paper [139] defines the class of admissible infinite-horizon risk-sensitive controllers as those controllers of the form (2.4.7) which are admissible for any finite-horizon risk sensitive optimal control problem (3.3.12) with underlying system (3.3.24) and a time-invariant version of the cost functional (3.3.10) corresponding to $M = 0$. That is, admissible infinite-horizon risk-sensitive controllers are required to satisfy the following causality condition: The processes

$$\begin{aligned}\zeta(t) &= \exp\left\{\int_0^t \alpha'(s)dW(s) - \frac{1}{2}\int_0^t \|\alpha(s)\|^2 dt\right\};\\ \zeta_T(t) &= \exp\left\{\int_0^t \alpha_T'(s)dW(s) - \frac{1}{2}\int_0^t \|\alpha_T(s)\|^2 dt\right\}\end{aligned}$$

with

$$\begin{aligned}\alpha(t) &:= B_2'Y_\infty\epsilon(t) - D_2'\Gamma^{-1}(C_2Y_\infty + D_2B_2')Y_\infty^{-1}e(t);\\ \alpha_T(t) &:= B_2'Y(t;Y_0)\epsilon_T(t)\\ &\quad -D_2'\Gamma^{-1}(C_2Y(t;Y_0) + D_2B_2')Y^{-1}(t;Y_0)e_T(t);\\ \epsilon(t) &:= x(t) - \check{x}(t); \quad e(t) := x(t) - \hat{x}(t);\\ \epsilon_T(t) &:= x(t) - \check{x}_T(t); \quad e_T(t) := x(t) - \hat{x}_T(t)\end{aligned}$$

are martingales on $[0,T]$. In the above equations, the subscript T refers to quantities defined in the corresponding finite-horizon risk-sensitive optimal control problem with $M = 0$.

Theorem 3.3.2 *Consider the risk-sensitive control problem (3.3.27) with underlying system (3.3.24). Suppose that Assumption 3.3.2 is satisfied. If $Y_\infty \geq Y_0$, then the optimal controller solving the risk-sensitive control problem (3.3.27) is given by*

$$\begin{aligned}u^* &= -G^{-1}(B_1'X_\infty + \Upsilon')[I - \frac{1}{\tau}Y_\infty X_\infty]^{-1}\check{x}\\ &= -G^{-1}(B_1'X_\infty + \Upsilon')\hat{x} \end{aligned} \tag{3.3.41}$$

where $\check{x}$ is generated by the filter (3.3.39), or equivalently, $\hat{x}$ is generated by the filter (3.3.40). The corresponding optimal value of the risk-sensitive cost is given by

$$\begin{aligned}&\inf_u \Im(u(\cdot))\\ &= \operatorname{tr}\left[\begin{matrix}Y_\infty R+\\ (Y_\infty C_2' + B_2D_2')\Gamma^{-1}(C_2Y_\infty + D_2B_2')X_\infty(I - \frac{1}{\tau}Y_\infty X_\infty)^{-1}\end{matrix}\right].\end{aligned} \tag{3.3.42}$$

Proof: See Theorem 3 of [139].

■

3.4 Quadratic stability

In this section, we consider the relation between the quadratic stability of uncertain systems with norm-bounded uncertainty and H^∞ control theory. The uncertain systems under consideration are those described by the state equations

$$\dot{x}(t) = [A + B_2\Delta(t)C_1]x(t); \quad x(t_0) = x_0; \tag{3.4.1}$$

where $x(t) \in \mathbf{R}^n$ is the *state*, x_0 is the *initial condition*, $\Delta(t)$ is a time varying *matrix of uncertain parameters* satisfying the bound $\Delta(t)'\Delta(t) \leq I$.

For this class of uncertain systems, we will consider a notion of robust stability known as quadratic stability; e.g., see [102].

Definition 3.4.1 *The uncertain system (3.4.1) is said to be quadratically stable if there exists a positive-definite matrix P such that*

$$2x'P[A + B_2\Delta C_1]x < 0 \tag{3.4.2}$$

for all non-zero $x \in \mathbf{R}^n$ and all matrices $\Delta : \Delta'\Delta \leq I$.

The following lemma is a version of the small gain theorem which relates the robust stability of an uncertain system to an H^∞ norm bound condition.

Lemma 3.4.1 *The uncertain system (3.4.1) is quadratically stable if and only if the following conditions are satisfied:*

(i) The matrix A is stable;

(ii) $\|C_1(sI - A)^{-1}B_2\|_\infty < 1$.

Proof: See [102].

■

Related to the property of quadratic stability is that of quadratic stabilizability. In this case, the class of uncertain systems under consideration is described by the state equations

$$\begin{aligned} \dot{x}(t) &= [A + B_2\Delta(t)C_1]x(t) + [B_1 + B_2\Delta(t)D_1]u(t); \quad x(t_0) = x_0; \\ y(t) &= C_2x(t); \end{aligned} \tag{3.4.3}$$

where, $x(t)$, x_0 and $\Delta(t)$ are as defined above, $y(t) \in \mathbf{R}^l$ is the *measured output* and $u(t)$ is the *control input*.

Definition 3.4.2 *The system (3.4.3) is said to be* quadratically stabilizable via linear output-feedback control *if there exists a linear time-invariant controller $K(s)$ such that with $u = -K(s)y$, the resulting closed loop system is quadratically stable for all uncertainties $\Delta(t)$ such that $\Delta'(t)\Delta(t) \leq I$.*

A necessary and sufficient condition for quadratic stabilizability can be found in [102].

3.5 A connection between H^∞ control and the absolute stabilizability of uncertain systems

In this section, we consider a robust control system design problem. As mentioned in the introduction, a central theme behind the robust control system design methodologies presented in this book is the connection between robust stabilization problems and H^∞ control. The results presented in this section give a simple illustration of this idea.

3.5.1 Definitions

Consider the following uncertain system of the form of (2.3.3):

$$\begin{aligned} \dot{x}(t) &= Ax(t) + B_1 u(t) + B_2 \xi(t); \\ z(t) &= C_1 x(t) + D_1 u(t); \\ y(t) &= C_2 x(t) + D_2 \xi(t) \end{aligned} \tag{3.5.1}$$

where $x(t) \in \mathbf{R}^n$ is the *state*, $\xi(t) \in \mathbf{R}^p$ is the *uncertainty input*, $u(t) \in \mathbf{R}^m$ is the *control input*, $z(t) \in \mathbf{R}^q$ is the *uncertainty output* and $y(t) \in \mathbf{R}^l$ is the *measured output*. The uncertainty in the above system is described by the following equation of the form (2.3.4):

$$\xi(t) = \phi(t, x(\cdot), u(\cdot)). \tag{3.5.2}$$

In this section, we consider the case of unstructured uncertainty. That is, we assume that the uncertainty inputs entering into the system are described by a single equation of the form (3.5.2). Furthermore, we deal with a single uncertainty constraint.

Definition 3.5.1 *An uncertainty (3.5.2) is an admissible uncertainty for the system (3.5.1) if the following conditions hold: Given any locally square integrable control input $u(\cdot)$ and any corresponding solution to equations (3.5.1), (3.5.2) defined on an existence interval $(0, t_*)$, then there exists a sequence $\{t_i\}_{i=1}^{\infty}$ and a constant $d \geq 0$ such that $t_i \to t_*$, $t_i \geq 0$ and*

$$\int_0^{t_i} (\|z(t)\|^2 - \|\xi(t)\|^2) dt \geq -d \quad \forall i. \tag{3.5.3}$$

Note that t_* and t_i may be equal to infinity.

Remark 3.5.1 As mentioned in Observation 2.3.1, the uncertain system (3.5.1), (3.5.3) allows for uncertainty satisfying a standard norm bound condition. In this case, the uncertain system would be described by the following state equations of the form (2.2.1):

$$\dot{x}(t) = [A + B_2 \Delta(t) C_1] x(t) + [B_1 + B_2 \Delta(t) D_1] u(t)$$

where $\Delta(t)$ is an uncertainty matrix such that $\|\Delta(t)\| \leq 1$; e.g., see [143, 102, 168]. Here, $\|\cdot\|$ denotes the standard induced matrix norm.

We now introduce notions of absolute stability and absolute stabilizability for the uncertain system (3.5.1)–(3.5.3).

Definition 3.5.2 *Consider the uncertain system (3.5.1), (3.5.3) with control input* $u(t) \equiv 0$. *This uncertain system is said to be* absolutely stable *if there exists a constant* $c > 0$ *such that the following conditions hold:*

(i) For any initial condition $x(0) = x_0$ *and uncertainty input* $\xi(\cdot) \in \mathbf{L}_2[0, \infty)$, *the system (3.5.1) has a unique solution which is defined on* $[0, \infty)$.

(ii) Given an admissible uncertainty for the uncertain system, then any corresponding solution to the equations (3.5.1), (3.5.2) satisfies $[x(\cdot), \xi(\cdot)] \in \mathbf{L}_2[0, \infty)$ *(hence,* $t_* = \infty$*) and*

$$\int_0^\infty (\|x(t)\|^2 + \|\xi(t)\|^2)dt \leq c[\|x_0\|^2 + d]. \tag{3.5.4}$$

Remark 3.5.2 Condition (3.5.4) requires that the $\mathbf{L}_2$ norm of any solution to (3.5.1) and (3.5.2) be bounded in terms of the norm of the initial condition and the 'measure of mismatch' d for this solution. In particular, since the uncertainty description given above allows for the uncertainty input $\xi(t)$ to depend dynamically on $x(t)$ and $u(t)$, in this case one could interpret the 'measure of mismatch' d as being due to a non-zero initial condition on the uncertainty dynamics.

Remark 3.5.3 It follows from the above definition that for any admissible uncertainty, an absolutely stable uncertain system (3.5.1), (3.5.3) (with $u(t) \equiv 0$) has the property that $x(t) \to 0$ as $t \to \infty$. Indeed, since $[x(\cdot), \xi(\cdot)] \in \mathbf{L}_2[0, \infty)$, we can conclude from (3.5.1) that $\dot{x}(\cdot) \in \mathbf{L}_2[0, \infty)$ (since $u(t) \equiv 0$). However, using the fact that $x(\cdot) \in \mathbf{L}_2[0, \infty)$ and $\dot{x}(\cdot) \in \mathbf{L}_2[0, \infty)$, it now follows that $x(t) \to 0$ as $t \to \infty$. Also, the system (3.5.1) with $\xi(t) \equiv 0$ and $u(t) \equiv 0$ will be asymptotically stable.

Remark 3.5.4 It is well known that a frequency domain necessary and sufficient condition for the absolute stability of the uncertain system (3.5.1), (3.5.3) (with $u(t) \equiv 0$) can be obtained via the use of the Kalman Yakubovich Popov Lemma; e.g., see [256, 258]. This frequency domain condition is also a necessary and sufficient condition for the quadratic stability of the corresponding uncertain system with norm-bounded uncertainty; e.g., see [256].

We now consider the problem of absolute stabilization for the uncertain system (3.5.1), (3.5.3) via a causal nonlinear output-feedback controller of the form (3.2.4) or a linear output-feedback controller of the form (3.2.28).

Definition 3.5.3 *The uncertain system (3.5.1), (3.5.3) is said to be* absolutely stabilizable via nonlinear control *if there exists an output-feedback controller of the form (3.2.4) and a constant* $c > 0$ *such that the following conditions hold:*

(i) For any initial condition $x(0) = x_0$ *and uncertainty input* $\xi(\cdot) \in \mathbf{L}_2[0, \infty)$, *the closed loop system defined by (3.5.1) and (3.2.4) has a unique solution which is defined on* $[0, \infty)$.

(ii) *The closed loop system defined by (3.5.1) and (3.2.4) with* $\xi(t) \equiv 0$ *is exponentially stable.*

(iii) *Given an admissible uncertainty* $\xi(\cdot)$ *for the uncertain system, then any corresponding solution to the equations (3.5.1), (3.2.4) satisfies* $[x(\cdot), u(\cdot), \xi(\cdot)] \in \mathbf{L}_2[0, \infty)$ *(hence,* $t_* = \infty$*) and*

$$\int_0^\infty (\|x(t)\|^2 + \|u(t)\|^2 + \|\xi(t)\|^2)dt \le c[\|x_0\|^2 + d]. \tag{3.5.5}$$

The uncertain system (3.5.1), (3.5.3) is said to be absolutely stabilizable via linear control *if there exists a linear output-feedback controller of the form (3.2.28) and a constant* $c > 0$ *such that conditions (i) - (iii) above are satisfied with the controller (3.2.4) replaced by the linear controller (3.2.28).*

Observation 3.5.1 Using a similar argument to that used in Remark 3.5.3, it follows that if the uncertain system (3.5.1), (3.5.3) is absolutely stabilizable, then the corresponding closed loop uncertain system (3.5.1), (3.5.3), (3.2.4) (or (3.5.1), (3.5.3), (3.2.28)) will have the property that $x(t) \to 0$ as $t \to \infty$ for any admissible uncertainty input $\xi(\cdot)$.

3.5.2 *The equivalence between absolute stabilization and H^∞ control*

In this section, we consider the problem of absolute stabilization for the uncertain system (3.5.1), (3.5.3). We show that this problem is equivalent to a standard H^∞ control problem. In this H^∞ control problem, the underlying linear system is described by the state equations (3.5.1). However in this case, $\xi(t) \in \mathbf{R}^p$ is the *disturbance input* and $z(t) \in \mathbf{R}^q$ is the *controlled output*. The H^∞ norm bound requirement is

$$\sup_{\xi(\cdot)\in \mathbf{L}_2[0,\infty), x(0)=0} \frac{\|z(\cdot)\|_2^2}{\|\xi(\cdot)\|_2^2} < 1. \tag{3.5.6}$$

As mentioned in Section 3.2, the controller (3.2.4) is said to solve the H^∞ control problem defined by (3.5.1) and (3.5.6) if the controller exponentially stabilizes the system (3.2.1) with $\xi(\cdot) \equiv 0$ and achieves the required bound on the H^∞ norm of the closed loop operator (3.5.6).

Theorem 3.5.1 *Consider the uncertain system (3.5.1), (3.5.3) and suppose that* $D_1'D_1 = G > 0$, $D_2 D_2' = \Gamma > 0$, *the pair (3.2.10) is detectable and the pair (3.2.34) is stabilizable. Then the following statements are equivalent:*

(i) *The uncertain system (3.5.1), (3.5.3) is absolutely stabilizable via a nonlinear controller of the form (3.2.4).*

(ii) The controller (3.2.4) solves the H^∞ control problem (3.5.1), (3.5.6).

(iii) There exist solutions $X > 0$ and $Y > 0$ to Riccati equations (3.2.8) and (3.2.33) such that the spectral radius of their product satisfies the condition

$$\rho(XY) < 1.$$

(iv) The linear controller (3.2.28) with

$$\begin{aligned} A_c &= A + B_1 C_c - B_c C_2 + (B_2 - B_c D_2) B_2' X; \\ B_c &= (I - YX)^{-1}(YC_2' + B_2 D_2')\Gamma^{-1}; \\ C_c &= -G^{-1}(B_1' X + D_1' C_1) \end{aligned} \tag{3.5.7}$$

solves the H^∞ control problem (3.5.1), (3.5.6).

(v) The uncertain system (3.5.1), (3.5.3) is absolutely stabilizable via the linear controller (3.2.28), (3.5.7).

Proof: (i)⇒(ii). In order to prove this statement, we first establish the following proposition.

Proposition 1 *If the controller (3.2.4) is absolutely stabilizing for the uncertain system (3.5.1), (3.5.3), then there exists a constant $\delta > 0$ such that*

$$\|z(\cdot)\|_2^2 \leq (1 - \delta)\|\xi(\cdot)\|_2^2 \tag{3.5.8}$$

for all solutions to the closed loop system (3.5.1), (3.2.4) with $x(0) = 0$ and $\xi(\cdot) \in \mathbf{L}_2[0, \infty)$.

To establish this proposition, let

$$F_0(z(\cdot), \xi(\cdot)) := \|z(\cdot)\|_2^2 - \|\xi(\cdot)\|_2^2$$

and

$$d = \begin{cases} -F_0(z(\cdot), \xi(\cdot)) & \text{if } F_0(z(\cdot), \xi(\cdot)) < 0; \\ 0 & \text{if } F_0(z(\cdot), \xi(\cdot)) \geq 0. \end{cases} \tag{3.5.9}$$

With these definitions, it can be seen that for all $\xi(\cdot) \in \mathbf{L}_2[0, \infty)$, the corresponding solution to the closed loop system (3.5.1), (3.2.4) will satisfy condition (3.5.3) with $t_i = \infty$. Hence with $x(0) = 0$, condition (3.5.5) implies that there exists a constant $c > 0$ such that

$$\|x(\cdot)\|_2^2 + \|u(\cdot)\|_2^2 + \|\xi(\cdot)\|_2^2 \leq cd \tag{3.5.10}$$

for all solutions of the closed loop system (3.5.1), (3.2.4) with $x(0) = 0$ and $\xi(\cdot) \in \mathbf{L}_2[0, \infty)$. We now consider two cases.

Case 1. $F_0(z(\cdot), \xi(\cdot)) \geq 0$. In this case, it follows from (3.5.9) that $d = 0$ and hence (3.5.10) implies that $x(t) \equiv 0$, $u(t) \equiv 0$ and $\xi(t) \equiv 0$ a.e.. Thus inequality (3.5.8) will be satisfied.

Case 2. $F_0(z(\cdot), \xi(\cdot)) < 0$. In this case, it follows from (3.5.9) and (3.5.10) that

$$\|\xi(\cdot)\|_2^2 \leq -cF_0(z(\cdot), \xi(\cdot)).$$

Hence

$$\|z(\cdot)\|_2^2 - \|\xi(\cdot)\|_2^2 \leq -c^{-1}\|\xi(\cdot)\|_2^2.$$

Therefore in this case, condition (3.5.8) will be satisfied with $\delta = c^{-1}$. Thus, the proposition has been established in both cases.

Now since (3.2.4) is an absolutely stabilizing controller for the uncertain system (3.5.1), (3.5.3), conditions (i) and (ii) of Definition 3.5.3 together with inequality (3.5.8) imply that this controller solves the H^∞ control problem (3.5.1), (3.5.6).

(ii)⇒(iii). This result follows directly using Theorem 3.2.3; see also [9, Chapter 5].

(iii)⇒(iv). This is a standard result from H^∞ control theory; see the second part of Theorem 3.2.3 and also [52, 147, 9].

(iv)⇒(v). Suppose the linear controller (3.2.28), (3.5.7) solves the H^∞ problem (3.5.1), (3.5.6). We now prove that this controller also solves the absolute stabilization problem for the uncertain system (3.5.1), (3.5.3).

The closed loop uncertain system defined by (3.5.1), (3.5.3), (3.2.28) and (3.5.7) may be rewritten as

$$\dot{h}(t) = Ph(t) + Q\xi(t) \tag{3.5.11}$$

with the integral quadratic constraint

$$\int_0^{t_i} (\|\Sigma h(t)\|^2 - \|\xi(t)\|^2)dt \geq -d \ \forall i \tag{3.5.12}$$

for some $d \geq 0$ and sequence $\{t_i\}_i^\infty$ such that $t_i \to t_*$; *c.f.* Definition 3.5.1. Here,

$$h = \begin{bmatrix} x \\ x_c \end{bmatrix}, \quad P = \begin{bmatrix} A & B_1C_c \\ B_cC_2 & A_c \end{bmatrix}, Q = \begin{bmatrix} B_2 \\ B_cD_2 \end{bmatrix}, \quad \Sigma = \begin{bmatrix} C_1 & D_1C_c \end{bmatrix}.$$

Since the controller (3.2.28) solves the H^∞ control problem (3.5.1), (3.5.6), the matrix P is stable and

$$\|\Sigma(sI - P)^{-1}Q\|_\infty < 1.$$

Hence, there exists a constant $\delta > 0$ such that

$$\left\| \begin{bmatrix} \Sigma \\ \delta I \end{bmatrix} (sI - P)^{-1}Q + \delta I \right\|_\infty \leq 1.$$

This inequality now implies

$$\int_0^\infty (\|\Sigma h(t)\|^2 - \|\xi(t)\|^2)dt \le -\delta \int_0^\infty (\|h(t)\|^2 + \|\xi(t)\|^2)dt \qquad (3.5.13)$$

for all $[h(\cdot), \xi(\cdot)] \in \mathbf{L}_2[0, \infty)$ satisfying (3.5.11) with $h(0) = 0$. Using Theorem 1 of [259], inequality (3.5.13) and the stability of the matrix P now implies the absolute stability of the uncertain system (3.5.11), (3.5.12). Thus with the controller (3.2.28), (3.5.7), conditions (i) and (ii) of Definition 3.5.3 will be satisfied. Also, since $x_c(0) = 0$ in the controller (3.2.28), it follows from condition (ii) of Definition 3.5.2 that for any initial condition $x(0) = x_0$ and any admissible uncertainty $\xi(\cdot)$, then $[x(\cdot), x_c(\cdot), \xi(\cdot)] \in \mathbf{L}_2[0, \infty)$ and

$$\int_0^\infty (\|x(t)\|^2 + \|x_c(t)\|^2 + \|\xi(t)\|^2)dt \le c[\|x_0\|^2 + d].$$

From this, it is straightforward to verify that for any initial condition $x(0) = x_0$ and any admissible uncertainty $\xi(\cdot)$, then $[x(\cdot), u(\cdot), \xi(\cdot)] \in \mathbf{L}_2[0, \infty)$ and

$$\int_0^\infty (\|x(t)\|^2 + \|u(t)\|^2 + \|\xi(t)\|^2)dt \le (\|C_c\|^2 + 1)c[\|x_0\|^2 + d].$$

Hence, the uncertain system (3.5.1), (3.5.3) is absolutely stabilizable using the linear controller (3.2.28), (3.5.7).

(v)⇒(i). This statement follows immediately from the definitions.

■

The following corollary follows from the relation between uncertain systems with an integral quadratic constraint on the uncertainty and uncertain systems with norm-bounded uncertainty; see Observation 2.3.1 on page 25.

Corollary 3.5.1 *Consider the following uncertain system with norm-bounded uncertainty:*

$$\begin{aligned} \dot{x}(t) &= [A + B_2\Delta(t)C_1]x(t) + [B_1 + B_2\Delta(t)D_1]u(t); \\ y(t) &= [C_2 + D_2\Delta(t)C_1]x(t) + D_2\Delta(t)D_1u(t); \\ &\quad \|\Delta(t)\| \le 1. \end{aligned} \qquad (3.5.14)$$

This uncertain system will be absolutely stabilizable via the linear controller (3.2.28), (3.5.7) if condition (iii) of Theorem 3.5.1 is satisfied.

4.
The S-procedure

4.1 Introduction

In this chapter, we present a collection of results concerning the so called S-procedure. The term "S-procedure" was introduced by Aizerman and Gantmacher in the monograph [2] (see also [65]) to denote a method which had been frequently used in the area of nonlinear control; e.g., see [119, 112, 169]. More recently, similar results have been extensively used in the Western control theory literature and the name S-procedure has remained; e.g., see [127, 185, 190, 210, 26]. As mentioned in Section 1.2, a feature of the S-procedure approach is that it allows for non-conservative results to be obtained for control problems involving structured uncertainty. In fact, the S-procedure provides a method for converting robust control problems involving structured uncertainty into parameter dependent problems involving unstructured uncertainty. Furthermore, S-procedure methods find application to many robust control problems not included in this book; e.g., see [196, 131, 202]. A general and systematic description of the S-procedure can be found, for example, in the monograph [67].

Let real-valued functionals $\mathcal{G}_0(x), \mathcal{G}_1(x), \ldots, \mathcal{G}_k(x)$ be defined on an abstract space $\mathcal{X}$. Also, let $\tau_1, \ldots, \tau_k$ be a collection of real numbers that form a vector $\tau = [\tau_1 \ \ldots \ \tau_k]'$ and let

$$\mathcal{S}(\tau, x) := \mathcal{G}_0(x) - \sum_{j=1}^{k} \tau_j \mathcal{G}_j(x). \tag{4.1.1}$$

We consider the following conditions on the functionals $\mathcal{G}_0(x), \mathcal{G}_1(x), \ldots, \mathcal{G}_k(x)$:

(i) $\mathcal{G}_0(x) \geq 0$ for all x such that $\mathcal{G}_1(x) \geq 0, \ldots, \mathcal{G}_k(x) \geq 0$.

(ii) There exists a collection of constants $\tau_1 \geq 0, \ldots, \tau_k \geq 0$ such that $\mathcal{S}(\tau, x) \geq 0$ for all $x \in \mathcal{X}$.

Note that in general, condition (ii) implies condition (i). The term S-procedure refers to the procedure of replacing condition (i) by the stronger condition (ii). One can easily find examples where condition (ii) does not follow from condition (i). However, if one imposes certain additional restrictions on the functionals $\mathcal{G}_0(x), \mathcal{G}_1(x), \ldots, \mathcal{G}_k(x)$, then the implication (i) implies (ii) may be true. In this case, the S-procedure is said to be *lossless* for the condition $\mathcal{G}_0(x) \geq 0$ and the constraints $\mathcal{G}_1(x) \geq 0, \ldots, \mathcal{G}_k(x) \geq 0$. In a similar fashion to the above, the losslessness of the S-procedure can be defined for the condition $\mathcal{G}_0(x) > 0$ and the constraints $\mathcal{G}_1(x) > 0, \ldots, \mathcal{G}_k(x) > 0$. Other combinations are also possible.

Conditions of the form (i) often arise in problems involving constructing Lyapunov functions. The absolute stability problem for Lur'e systems is one such example. In this problem, which involves a Lyapunov function of the Lur'e-Postnikov form

$$V(x) = x'Px + 2\theta \int_0^z \phi(\zeta)d\zeta,$$

a positive definite matrix P is sought to satisfy the condition:

$$\mathcal{G}_0(x, \xi) := -\left[2x'P(Ax + B\xi) + 2\theta x'PC(Ax + B\xi)\right] > 0$$

for all x, ξ such that

$$\mathcal{G}_1(x, \xi) := x'C\xi - \frac{1}{k}\xi^2 \geq 0.$$

In a typical application of the S-procedure, the functionals $\mathcal{G}_0(x), \mathcal{G}_1(x), \ldots, \mathcal{G}_k(x)$ depend on physical parameters. For instance in the above Lur'e system example, $\mathcal{G}_1(x, \xi)$ depends on the parameter k which characterizes the size of the uncertainty sector. One often seeks to find regions in the space of parameters where condition (i) is satisfied. The presence of multiple constraints $\mathcal{G}_1(x) \geq 0, \ldots, \mathcal{G}_k(x) \geq 0$ usually brings additional difficulty to the problem. However, if the S-procedure is applied, the problem can be reduced to one which does not involve multiple constraints.

Let $\mathcal{A}$ and $\mathcal{B}$ be domains in the set of physical parameters defined by conditions (i) and (ii), respectively. Since condition (ii) implies condition (i), then it immediately follows that $\mathcal{B} \subseteq \mathcal{A}$. That is the application of the S-procedure leads to a restriction on the set of admissible physical parameters. However, if the S-procedure is lossless, then $\mathcal{A} \equiv \mathcal{B}$ and no such restriction on the set of admissible physical parameters occurs.

In the next section, we present some well known theorems on the S-procedure. Also, some more recent results by Yakubovich, Megretski and Treil, and other authors will be presented in subsequent sections.

4.2 An S-procedure result for a quadratic functional and one quadratic constraint

Although the first applications of the S-procedure in control theory began in the 1950s, the corresponding mathematical ideas can be traced back as far as Hausdorff [77] and Töeplitz [224]. An interest in establishing the equivalence between conditions (i) and (ii) above for the case of two quadratic forms arose in connection to a problem of classifying definite matrix pencils. In the Western literature, the main result on this problem is commonly referred to as Finsler's Theorem; e.g., see [239]. This paper gives a comprehensive survey of this problem including an extensive Western bibliography and historical notes.

The following proof of the losslessness of the S-procedure for the case of two quadratic forms follows a proof given by Yakubovich in [256].

Theorem 4.2.1 *Let $\mathcal{X}$ be a real linear vector space and $\mathcal{G}_0(x)$, $\mathcal{G}_1(x)$ be quadratic functionals on $\mathcal{X}$. That is, $\mathcal{G}_0(x)$ and $\mathcal{G}_1(x)$ are functionals of the form*

$$\begin{aligned}\mathcal{G}_0(x) &= G_0(x,x) + g_0(x) + \gamma_0,\\ \mathcal{G}_1(x) &= G_1(x,x) + g_1(x) + \gamma_1,\end{aligned} \tag{4.2.1}$$

where $G_0(x_1,x_2)$ and $G_1(x_1,x_2)$ are bilinear forms on $\mathcal{X} \times \mathcal{X}$, $g_0(x)$, $g_1(x)$ are linear functionals on $\mathcal{X}$, and γ_0, γ_1 are constants. Assume that there exists a vector x_0 such that $\mathcal{G}_1(x_0) > 0$. Then, the following conditions are equivalent:

(i) $\mathcal{G}_0(x) \geq 0$ for all x such that $\mathcal{G}_1(x) \geq 0$;

(ii) There exists a constant $\tau \geq 0$ such that

$$\mathcal{G}_0(x) - \tau\mathcal{G}_1(x) \geq 0 \tag{4.2.2}$$

for all $x \in \mathcal{X}$.

Remark 4.2.1 Note that the condition $\mathcal{G}_1(x_0) > 0$ for some x_0 is required for the above theorem to hold. Indeed, let $\mathcal{X} = \mathcal{X}_1 \times \mathcal{X}_2$, $x = [x_1',\ x_2']'$, and

$$\mathcal{G}_0(x) = x_1'G_0x_2, \quad \mathcal{G}_1(x) = -x_1'G_1x_1,$$

where G_1 is a positive-definite matrix and $G_0 \neq 0$. It is easy to see that $\mathcal{G}_1(x) \geq 0$ only if $x_1 = 0$. Hence $\mathcal{G}_0(x) \geq 0$ for all x such that $\mathcal{G}_1(x) \geq 0$. However, the quadratic form $\mathcal{G}_0(x) - \tau\mathcal{G}_1(x)$ is not positive semi-definite for any $\tau \geq 0$.

Theorem 4.2.1 is based on the following lemma which will be proved in the next subsection.

Lemma 4.2.1 *Let $\mathcal{X}$ be a real linear vector space and $\mathcal{G}_0(x)$, $\mathcal{G}_1(x)$ be quadratic forms on $\mathcal{X}$ satisfying the following condition: $\mathcal{G}_0(x) \geq 0$ for all $x \in \mathcal{X}$ such that $\mathcal{G}_1(x) \geq 0$. Then there exist constants $\tau_0 \geq 0$, $\tau_1 \geq 0$, such that $\tau_0 + \tau_1 > 0$ and*

$$\tau_0\mathcal{G}_0(x) - \tau_1\mathcal{G}_1(x) \geq 0 \tag{4.2.3}$$

for all $x \in \mathcal{X}$.

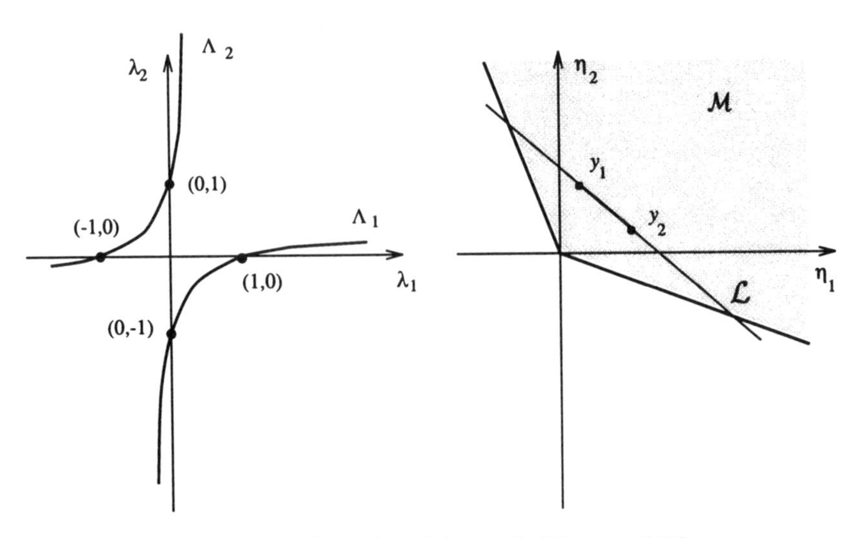

Figure 4.2.1. Illustration of the proof of Theorem 4.2.2

4.2.1 *Proof of Theorem 4.2.1*

In order to prove Theorem 4.2.1, we will need the following result due to Dines [43]. Let $\mathcal{X}$ be a real linear vector space and let $\mathcal{G}_0(x)$ and $\mathcal{G}_1(x)$ be two quadratic forms on $\mathcal{X}$. Consider the mapping $\mathcal{X} \to \mathbf{R}^2$ defined by

$$(\eta_1, \eta_2) = (\mathcal{G}_0(x), \mathcal{G}_1(x)). \tag{4.2.4}$$

Also we define a set $\mathcal{M}$ as the range of this mapping:

$$\mathcal{M} = \{(\eta_1, \eta_2) : (\eta_1, \eta_2) = (\mathcal{G}_0(x), \mathcal{G}_1(x))\}. \tag{4.2.5}$$

Theorem 4.2.2 (Dines) *The mapping (4.2.4) maps the space $\mathcal{X}$ onto a convex cone $\mathcal{M}$.*

Proof: It is obvious that the set $\mathcal{M}$ is a cone. It remains to show that the set $\mathcal{M}$ is a convex set. Consider two points in $\mathcal{M}$

$$y_1 = (\mathcal{G}_0(x_1), \mathcal{G}_1(x_1)), \quad y_2 = (\mathcal{G}_0(x_2), \mathcal{G}_1(x_2)). \tag{4.2.6}$$

We wish to prove that for every point $y_0 \in \{\lambda y_1 + (1-\lambda)y_2 : \lambda \in [0,1]\}$, there exists a vector $x_0 \in \mathcal{X}$ such that $y_0 = (\mathcal{G}_0(x_0), \mathcal{G}_1(x_0))$.

Let $\mathcal{L}$ be the line passing through the points y_1 and y_2. Also suppose this line is represented by the equation

$$a\eta_1 + b\eta_2 = c;$$

see Figure 4.2.1(b). Now consider the function

$$\varphi(\lambda_1, \lambda_2) := a\mathcal{G}_0(\lambda_1 x_1 + \lambda_2 x_2) + b\mathcal{G}_1(\lambda_1 x_1 + \lambda_2 x_2) \tag{4.2.7}$$

and the mapping

$$(\lambda_1, \lambda_2) \to y = (\mathcal{G}_0(\lambda_1 x_1 + \lambda_2 x_2), \mathcal{G}_1(\lambda_1 x_1 + \lambda_2 x_2)). \tag{4.2.8}$$

The mapping (4.2.8) maps the points $(1,0)$, $(0,1)$ into the points y_1, y_2 respectively. Also, the points $(-1,0)$, $(0,-1)$ are mapped into the points y_1, y_2 respectively since $\mathcal{G}_0(-x) = \mathcal{G}_0(x)$ and $\mathcal{G}_1(-x) = \mathcal{G}_1(x)$.

We now consider the set

$$\Lambda = \{(\lambda_1, \lambda_2) : \varphi(\lambda_1, \lambda_2) = c\}.$$

Obviously, the set Λ is not empty since it contains the points $(\pm 1, 0)$,$(0, \pm 1)$. From equation (4.2.7) and the fact that $\mathcal{G}_0(x)$ and $\mathcal{G}_1(x)$ are quadratic forms, it follows that

$$\varphi(\lambda_1, \lambda_2) = \alpha\lambda_1^2 + 2\beta\lambda_1\lambda_2 + \gamma\lambda_2^2$$

where α, β and γ are real constants. Depending on the values of these constants one of the following two cases will occur.

Case 1. $\alpha = \beta = \gamma = 0$. In this case, $\varphi(\lambda_1, \lambda_2) = 0$ for any pair $(\lambda_1, \lambda_2) \in \mathbf{R}^2$, Hence $\Lambda = \mathbf{R}^2$ and $c = 0$. That is, the line $\mathcal{L}$ passes through the origin and

$$a\mathcal{G}_0(\lambda_1 x_1 + \lambda_2 x_2) + b\mathcal{G}_1(\lambda_1 x_1 + \lambda_2 x_2) = 0$$

for all $(\lambda_1, \lambda_2) \in \mathbf{R}^2$. In particular, this condition is satisfied for any continuous curve $(\lambda_1(t), \lambda_2(t))$, $0 \le t \le 1$ in $\mathbf{R}^2$ which connects the points $(1,0)$ and $(0,1)$; *i.e.*,

$$a\mathcal{G}_0(\lambda_1(t)x_1 + \lambda_2(t)x_2) + b\mathcal{G}_1(\lambda_1(t)x_1 + \lambda_2(t)x_2) = 0$$

for all $t \in [0,1]$. Thus, the corresponding continuous curve

$$y(t) = (\mathcal{G}_0(\lambda_1(t)x_1 + \lambda_2(t)x_2), \mathcal{G}_1(\lambda_1(t)x_1 + \lambda_2(t)x_2)), \quad t \in [0,1],$$

which connects points y_1 and y_2, lies entirely in $\mathcal{L}$.

Case 2. At least one of the constants α, β, γ is not equal to zero. In this case, let $\delta = \alpha\gamma - \beta^2$. The set Λ which is defined by the equation $\varphi(\lambda_1, \lambda_2) = c$ is then a second order curve whose form depends on the value of δ. If $\delta > 0$, Λ will be an ellipse. If $\delta < 0$, Λ will be a hyperbola and if $\delta = 0$, Λ will be a pair of parallel lines. Also, since $\mathcal{G}_0(-x) = \mathcal{G}_0(x)$ and $\mathcal{G}_1(-x) = \mathcal{G}_1(x)$ for all $x \in \mathbf{R}^n$, the set Λ is symmetric with respect to the origin. Thus, the set Λ is either a simply connected set or else consists of two simply connected sets Λ_1 and Λ_2 as illustrated in Figure 4.2.1(a). There are two possible alternatives:

(a) The points $(1,0)$ and $(0,1)$ belong to the same branch of the set Λ; say Λ_1. In this case, there exists a continuous curve $(\lambda_1(t), \lambda_2(t))$, $0 \le t \le 1$, which belongs to this branch of the set Λ and connects the points $(1,0)$ and $(0,1)$. The corresponding continuous curve

$$y(t) = (\mathcal{G}_0(\lambda_1(t)x_1 + \lambda_2(t)x_2), \mathcal{G}_1(\lambda_1(t)x_1 + \lambda_2(t)x_2)), \quad t \in [0,1],$$

is contained in the line $\mathcal{L}$ and connects the points y_1 and y_2.

(b) The points $(1,0)$ and $(0,1)$ belong to different branches of the set Λ, say $(1,0) \in \Lambda_1$ and $(0,1) \in \Lambda_2$. Then, it follows from the symmetry of the set Λ that $(0,-1) \in \Lambda_1$ and $(-1,0) \in \Lambda_2$; see Figure 4.2.1(a). In this case, we can connect the points $(1,0)$ and $(0,-1)$ by a continuous curve which entirely lies in Λ_1. As in the previous case, the corresponding continuous curve

$$y(t) = (\mathcal{G}_0(\lambda_1(t)x_1 + \lambda_2(t)x_2), \mathcal{G}_1(\lambda_1(t)x_1 + \lambda_2(t)x_2)), \quad t \in [0,1]$$

is contained in the line $\mathcal{L}$ and connects the points y_1 and y_2.

Thus, in the both of the above cases, one can find a continuous curve $(\lambda_1(t), \lambda_2(t))$, $0 \leq t \leq 1$, such that the continuous curve

$$y(t) = (\mathcal{G}_0(\lambda_1(t)x_1 + \lambda_2(t)x_2), \mathcal{G}_1(\lambda_1(t)x_1 + \lambda_2(t)x_2)), \quad t \in [0,1]$$

is contained in the line $\mathcal{L}$ and connects the points y_1 and y_2. Hence, given any point y_0 on the line between y_1 and y_2, there exists a $t_0 \in [0,1]$ such that $y_0 = y(t_0)$. The corresponding vector $x_0 = \lambda_1(t_0)x_1 + \lambda_2(t_0)x_2$ then satisfies

$$\begin{aligned}
&(\mathcal{G}_0(x_0), \mathcal{G}_1(x_0)) \\
&\quad = (\mathcal{G}_0(\lambda_1(t_0)x_1 + \lambda_2(t_0)x_2), \mathcal{G}_1(\lambda_1(t_0)x_1 + \lambda_2(t_0)x_2)) \\
&\quad = y(t_0) \\
&\quad = y_0.
\end{aligned}$$

Thus, we have established the required convexity of the cone $\mathcal{M}$.

■

Proof of Lemma 4.2.1: In Theorem 4.2.2, we have shown that the set $\mathcal{M}$, which is the image of $\mathcal{X}$ under the mapping (4.2.4), is a convex cone. Let

$$\mathcal{Q} = \{(\eta_1, \eta_2) : \eta_1 < 0, \ \eta_2 \geq 0\}.$$

From the assumption that $\mathcal{G}_0(x) \geq 0$ for all x such that $\mathcal{G}_1(x) \geq 0$, it follows that $\mathcal{M} \cap \mathcal{Q} = \varnothing$. This fact can be established by contradiction. Indeed, suppose that there exists a point

$$y^0 = (\eta_1^0, \eta_2^0) \in \mathcal{M} \cap \mathcal{Q}.$$

Then, there exists a vector $x^0 \in \mathbf{R}^n$ such that

$$\eta_1^0 = \mathcal{G}_0(x^0), \eta_2^0 = \mathcal{G}_1(x^0).$$

Furthermore, since $y^0 = (\eta_1^0, \eta_2^0) \in \mathcal{Q}$, then $\eta_2^0 = \mathcal{G}_1(x^0) \geq 0$. Hence, the above assumption implies $\eta_1^0 = \mathcal{G}_0(x^0) \geq 0$. However, this fact contradicts the condition $y^0 = (\eta_1^0, \eta_2^0) \in \mathcal{Q}$ which requires that $\eta_1^0 < 0$. Thus, we must have $\mathcal{M} \cap \mathcal{Q} = \varnothing$.

Since $\mathcal{M}$ and $\mathcal{Q}$ are convex cones and $\mathcal{M} \cap \mathcal{Q} = \varnothing$, then there exists a line that separates these convex sets. This conclusion follows from the Separating Hyperplane Theorem; e.g., see Theorem 11.3 of [167]. That is, there exist constants τ_0,

τ_1, $|\tau_0| + |\tau_1| > 0$, such that

$$\tau_0\eta_1 - \tau_1\eta_2 \leq 0 \quad \text{for all } (\eta_1, \eta_2) \in \mathcal{Q}, \tag{4.2.9}$$

$$\tau_0\eta_1 - \tau_1\eta_2 \geq 0 \quad \text{for all } (\eta_1, \eta_2) \in \mathcal{M}. \tag{4.2.10}$$

Note that since $(-1, 0) \in \mathcal{Q}$, then condition (4.2.9) implies that $\tau_0 \geq 0$. Furthermore, since $(-\varepsilon, 1) \in \mathcal{Q}$ for all $\varepsilon > 0$, then it follows from (4.2.9) that $\tau_1 \geq -\tau_0\varepsilon$ for all $\varepsilon > 0$. Hence, $\tau_1 \geq 0$. Using the definition of the set $\mathcal{M}$, the lemma now follows from (4.2.10).

■

Proof of Theorem 4.2.1: (ii)⇒(i). This part of the proof follows immediately from the definitions.

(i)⇒(ii). We first prove this fact for the special case in which $\mathcal{G}_0(\cdot)$ and $\mathcal{G}_1(\cdot)$ are quadratic forms. In this case, condition (i) of Theorem 4.2.1 coincides with the assumption of Lemma 4.2.1. Hence, it follows from this lemma that there exist constants $\tau_0 \geq 0$, $\tau_1 \geq 0$, $\tau_0 + \tau_1 > 0$ such that condition (4.2.3) is satisfied.

The conditions of Theorem 4.2.1 also imply that $\tau_0 > 0$. Indeed, suppose that $\tau_0 = 0$. Then (4.2.3) implies that $\mathcal{G}_1(x_0) \leq 0$ since $\tau_1 > 0$. However, this contradicts the assumption that $\mathcal{G}_1(x_0) > 0$. Therefore, $\tau_0 > 0$ and hence condition (4.2.2) is satisfied with $\tau = \tau_1/\tau_0$.

Now consider the general case in which the quadratic functionals $\mathcal{G}_0(\cdot)$ and $\mathcal{G}_1(\cdot)$ are of the form (4.2.1). Without loss of generality we can suppose that $\mathcal{G}_1(0) > 0$; *i.e.*, $x_0 = 0$. Note that we can always transform the case $x_0 \neq 0$ to the case $x_0 = 0$ by replacing the original quadratic functionals (4.2.1) by the functionals $\tilde{\mathcal{G}_0}(\tilde{x}) := \mathcal{G}_0(\tilde{x} + x_0)$, $\tilde{\mathcal{G}_1}(\tilde{x}) := \mathcal{G}_1(\tilde{x} + x_0)$, where $\tilde{x} := x - x_0$.

Define

$$\begin{aligned} \mathcal{G}_0^0(x, \zeta) &:= G_0(x, x) + \zeta g_0(x) + \zeta^2\gamma_0; \\ \mathcal{G}_1^0(x, \zeta) &:= G_1(x, x) + \zeta g_1(x) + \zeta^2\gamma_1 \end{aligned} \tag{4.2.11}$$

where $\zeta \in \mathbf{R}$. Note that $\mathcal{G}_0^0(x, \zeta)$ and $\mathcal{G}_1^0(x, \zeta)$ are quadratic forms on the linear vector space $\mathcal{X} \times \mathbf{R}$. Also, note that $\mathcal{G}_1^0(0, 1) = \mathcal{G}_1(0) > 0$. We will now show that the quadratic forms (4.2.11) satisfy the following condition:

$$\mathcal{G}_0^0(x, \zeta) \geq 0 \quad \text{for all } (x, \zeta) \text{ such that} \quad \mathcal{G}_1^0(x, \zeta) \geq 0. \tag{4.2.12}$$

If $\zeta \neq 0$, condition (4.2.12) follows from condition (i) of the theorem since

$$\mathcal{G}_0^0(x, \zeta) = \zeta^2\mathcal{G}_0(\zeta^{-1}x), \; \mathcal{G}_1^0(x, \zeta) = \zeta^2\mathcal{G}_1(\zeta^{-1}x).$$

It remains to prove that condition (4.2.12) holds if $\zeta = 0$.

Indeed, consider $x \in \mathcal{X}$ such that $\mathcal{G}_1^0(x, 0) \geq 0$. Hence, $G_1(x, x) \geq 0$. Now two alternative cases are possible:

Case 1. $G_1(x, x) = 0$. In this case, $\mathcal{G}_1(\eta x) = \eta g_1(x) + \gamma_1$. Hence, $\mathcal{G}_1(\eta x) \geq 0$ for all η such that $|\eta|$ is sufficiently large and $\eta g_1(x) > 0$.

Case 2. $G_1(x, x) > 0$. In this case, $\mathcal{G}_1(\eta x) = \eta^2 G_1(x, x) + \eta g_1(x) + \gamma_1$. Hence, $\mathcal{G}_1(\eta x) \geq 0$ for all η such that $|\eta|$ is sufficiently large.

Summarizing the above facts and using equation (4.2.11) and condition (i) of the theorem, it follows that $\mathcal{G}_0(\eta x) \geq 0$ for all η such that $|\eta|$ is sufficiently large and $\eta g_1(x) > 0$. Therefore,

$$\mathcal{G}_0^0(x,0) = G_0(x,x) = \lim_{|\eta|\to\infty, \eta g_1(x)>0} \eta^{-2}\mathcal{G}_0(\eta x) \geq 0.$$

Thus, we have established that $\mathcal{G}_1^0(x,\zeta) \geq 0$ implies $\mathcal{G}_0^0(x,\zeta) \geq 0$. From the corresponding result for quadratic forms which has been established above, it follows that there exists a constant $\tau \geq 0$ such that $\mathcal{G}_0^0(x,\zeta) - \tau\mathcal{G}_1^0(x,\zeta) \geq 0$ for all $x \in \mathcal{X}$, $\zeta \in \mathbf{R}$. Setting $\zeta = 1$, it follows that condition (ii) of the Theorem is satisfied.

■

4.3 An S-procedure result for a quadratic functional and k quadratic constraints

The lossless S-procedure based on Theorem 4.2.1 can be applied to absolute stability problems in order to establish the existence of a Lyapunov function of the Lur'e-Postnikov form. However, non-conservative results are only obtained when the system contains a single scalar sector-bounded uncertainty. That is, the number of uncertainty constraints does not exceed one. The application of the S-procedure in the derivation of non-conservative results concerning robust stability of control systems with structured uncertainty requires the lossless S-procedure to be extended to the case of k quadratic constraints. However, the question as to whether a lossless version of the S-procedure exists in this case has long remained unsolved. In 1990, Megretski and Treil suggested a new approach to proving the losslessness of the S-procedure in the case of multiple constraints; see the technical report [126] and the journal version of this report [127]. The novelty of this approach concerned the use of a certain sequence of shift operators in proving the convexity of the closure of the image of a mapping $\mathcal{G}(\cdot) = [\mathcal{G}_0, \mathcal{G}_1, \ldots, \mathcal{G}_k]$ defined by given quadratic forms $\mathcal{G}_0, \mathcal{G}_1, \ldots, \mathcal{G}_k$ on a Hilbert space. In [261], Yakubovich extended the methodology of Megretski and Treil to the case of quadratic functionals on a Hilbert space. Since the result of [261] is more general than that of [127], we will present the result of [261].

Let $\mathcal{X}$ be a real Hilbert space. The inner product on this space is denoted by $\langle\cdot,\cdot\rangle$. Let

$$\mathcal{G}_j(x) = G_j^0(x) + \langle g_j, x\rangle + \gamma_j, \; j = 0, 1, \ldots, k \tag{4.3.1}$$

be continuous quadratic functionals. Here $g_j \in \mathcal{X}$ and $G_j^0(x) = \langle x, G_j x\rangle$ where the operators G_j are bounded self-adjoint operators, $G_j = G_j^*$. Consider an affine manifold $\mathfrak{L} = \mathfrak{M} + x_0$, where $x_0 \in \mathcal{X}$ is a vector, and $\mathfrak{M}$ is a linear subspace of the space $\mathcal{X}$.

Definition 4.3.1 *The quadratic functionals $\mathcal{G}_0, \ldots \mathcal{G}_k$ form an S-system if there exist bounded linear operators $\mathbf{T}_i : \mathcal{X} \to \mathcal{X}$, $i = 1, 2, \ldots$, such that*

(i) $\langle \mathbf{T}_i x_1, x_2 \rangle \to 0$ *as* $i \to \infty$ *for all* $x_1, x_2 \in \mathcal{X}$*;*

(ii) If $x \in \mathfrak{M}$*, then* $\mathbf{T}_i x \in \mathfrak{M}$ *for all* $i = 1, 2, \ldots$*;*

(iii) $G_j^0(\mathbf{T}_i x) \to G_j^0(x)$ *as* $i \to \infty$ *for all* $x \in \mathfrak{M}$ *and* $j = 0, 1, \ldots, k$*.*

Lemma 4.3.1 *Suppose the quadratic functionals $\mathcal{G}_0, \ldots \mathcal{G}_k$ form an S-system and define a mapping*

$$\mathcal{G}(x) := [\mathcal{G}_0(x), \ldots, \mathcal{G}_k(x)]' : \mathcal{X} \to \mathbf{R}^{k+1}.$$

Also, let the set $\mathcal{M} = \mathcal{G}(\mathfrak{L})$ be the image of $\mathfrak{L}$ under the mapping $\mathcal{G}$. Then $\mathrm{cl}\,\mathcal{M}$*, the closure of the set $\mathcal{M}$, is a convex set.*

Proof: Without loss of generality, we can consider the case $x_0 = 0$, $\mathfrak{L} = \mathfrak{M}$. To establish the result for this case, we first prove the following proposition.

Proposition 2 *For any $y_1 = \mathcal{G}(x_1)$, $y_2 = \mathcal{G}(x_2) \in \mathcal{M}$ there exists a sequence $\{x_i\}_{i=1}^{\infty} \subset \mathfrak{L}$ such that $y_i = \mathcal{G}(x_i) \to \frac{1}{2}(y_1 + y_2)$ as $i \to \infty$.*

For the given $y_1 = \mathcal{G}(x_1), y_2 = \mathcal{G}(x_2) \in \mathcal{M}, x_1, x_2 \in \mathfrak{M}$, we define:

$$\begin{aligned} x_1^{(i)} &= \mathbf{T}_i x_1, \quad x_2^{(i)} = \mathbf{T}_i x_2 \\ z^{(i)} &= \frac{1}{2}\left(x_1 + x_1^{(i)} + x_2 - x_2^{(i)}\right), \end{aligned}$$

where $\mathbf{T}_i$ is a sequence of linear bounded operators defined as in Definition 4.3.1. Then from condition (ii) of Definition 4.3.1, it follows that $z^{(i)} \in \mathfrak{M}$. Furthermore, from conditions (i) and (iii) of Definition 4.3.1, it follows that

$$\lim_{i \to \infty} \langle z^{(i)}, G_j z^{(i)} \rangle = \frac{1}{2}\left(\langle x_1, G_j x_1 \rangle + \langle x_2, G_j x_2 \rangle\right).$$

Furthermore, it follows from condition (i) of Definition 4.3.1 that for any $g \in \mathcal{X}$,

$$\lim_{i \to \infty} \langle g, z^{(i)} \rangle = \frac{1}{2}\left(\langle g, x_1 \rangle + \langle g, x_2 \rangle\right).$$

Therefore, $\mathcal{G}(z^{(i)}) \to \frac{1}{2}(y_1 + y_2)$. This completes the proof of the proposition.

We are now in a position to prove the lemma. Let $\bar{y}_1, \bar{y}_2 \in \mathrm{cl}\,\mathcal{M}$ be given and let

$$\bar{y}_1 = \lim_{n \to \infty} y_1^{(n)}, \; \bar{y}_2 = \lim_{n \to \infty} y_2^{(n)}.$$

Also, let $\{x_i^{(n)}\}_{i=1}^{\infty} \subset \mathfrak{L}$ be a sequence such that

$$\lim_{i\to\infty} \mathcal{G}(x_i^{(n)}) = \frac{1}{2}(y_1^{(n)} + y_2^{(n)})$$

for all n. The existence of such a sequence follows from Proposition 2. Then, there exists a subsequence $\{x_{i_n}^{(n)}\}_{n=1}^{\infty}$ such that $\lim_{n\to\infty} \mathcal{G}(x_{i_n}^{(n)}) = \frac{1}{2}(\bar{y}_1 + \bar{y}_2)$. Thus, if $\bar{y}_1, \bar{y}_2 \in \text{cl}\,\mathcal{M}$, then $\frac{1}{2}(\bar{y}_1 + \bar{y}_2) \in \text{cl}\,\mathcal{M}$. Hence, the set cl $\mathcal{M}$ is convex. ■

Theorem 4.3.1 *Suppose the quadratic functionals $\mathcal{G}_0, \ldots \mathcal{G}_k$ form an S-system and*

$$\mathcal{G}_0(x) \geq 0 \textit{ for all } x \in \mathfrak{L} \textit{ such that } \mathcal{G}_1(x) \geq 0, \ldots, \mathcal{G}_k(x) \geq 0. \quad (4.3.2)$$

Then, there exist constants $\tau_0 \geq 0$, $\tau_1 \geq 0, \ldots, \tau_k \geq 0$ such that $\sum_{j=0}^{k} \tau_j > 0$ and

$$\tau_0 \mathcal{G}_0(x) \geq \sum_{j=1}^{k} \tau_j \mathcal{G}_j(x) \quad \textit{for all } x \in \mathfrak{L}. \quad (4.3.3)$$

Furthermore, if there exists an $x_0 \in \mathcal{L}$ such that $\mathcal{G}_1(x_0) > 0, \ldots, \mathcal{G}_k(x_0) > 0$, then one may take $\tau_0 = 1$ in (4.3.3). That is, there exist constants $\tau_1 \geq 0, \ldots, \tau_k \geq 0$ such that

$$\mathcal{G}_0(x) \geq \sum_{j=1}^{k} \tau_j \mathcal{G}_j(x) \quad \textit{for all } x \in \mathfrak{L}. \quad (4.3.4)$$

Proof: The proof of this theorem follows the same separation theorem arguments as those used in the proof of Theorem 4.2.1. To establish the existence of the constants $\tau_1, \ldots, \tau_k$ satisfying the conditions of the theorem, a hyperplane is constructed to separate the closed convex set cl $\mathcal{M}$ and the open cone

$$\mathcal{Q} = \{\eta = (\eta_0, \ldots, \eta_k) : \eta_0 < 0,\ \eta_1 > 0, \ldots \eta_k > 0\}.$$

For details, we refer the reader to [261]. ■

Remark 4.3.1 It should be noted that condition (4.3.2) follows directly from (4.3.4). Thus, Theorem 4.3.1 shows that if the functionals $\mathcal{G}_0(\cdot), \ldots, \mathcal{G}_k(\cdot)$ form an S-system, then in fact conditions (4.3.2) and (4.3.4) are equivalent. Thus, the S-procedure involving these functionals is lossless. Hence, this version of the S-procedure will not introduce conservatism when applied to problems of absolute stability and absolute stabilizability.

Remark 4.3.2 For the case in which the quadratic functionals $\mathcal{G}_0(\cdot), \ldots, \mathcal{G}_k(\cdot)$ are quadratic forms, Theorem 4.3.1 reduces to the result of Megretski and Treil [127].

We now present an application of the above theorem which will be used in Chapters 5 and 6. This result is concerned with the solutions to a linear system of the following form:

$$\begin{aligned}\dot{\eta}(t) &= \Phi\eta(t) + \Lambda\mu(t); \\ \sigma(t) &= \Pi\eta(t)\end{aligned} \tag{4.3.5}$$

where Φ is a stability matrix. For this system with a given initial condition $\eta(0) = \eta_0$, a corresponding set $\mathfrak{L} \subset \mathbf{L}_2[0,\infty)$ is defined as follows:

$$\mathfrak{L} := \left\{ \lambda(\cdot) = \begin{bmatrix} \sigma(\cdot) \\ \mu(\cdot) \end{bmatrix} : \begin{array}{l} \sigma(\cdot), \mu(\cdot) \text{ are related by (4.3.5),} \\ \mu(\cdot) \in \mathbf{L}_2[0,\infty) \text{ and } \eta(0) = \eta_0 \end{array} \right\}. \tag{4.3.6}$$

Also, we consider the following set of integral functionals mapping from $\mathfrak{L}$ into $\mathbf{R}$:

$$\begin{aligned}\mathcal{G}_0(\lambda(\cdot)) &:= \int_0^\infty \lambda(t)' M_0 \lambda(t) dt + \gamma_0; \\ \mathcal{G}_1(\lambda(\cdot)) &:= \int_0^\infty \lambda(t)' M_1 \lambda(t) dt + \gamma_1; \\ &\vdots \\ \mathcal{G}_k(\lambda(\cdot)) &:= \int_0^\infty \lambda(t)' M_k \lambda(t) dt + \gamma_k\end{aligned} \tag{4.3.7}$$

where $M_0, M_1, \ldots, M_k$ are given matrices and $\gamma_0, \gamma_1, \ldots, \gamma_k$ are given constants.

Theorem 4.3.2 *Consider a system of the form (4.3.5), a set $\mathfrak{L}$ of the form (4.3.6) and a set of functionals of the form (4.3.7). Suppose that these functionals have the following properties:*

(i) $\mathcal{G}_0(\lambda(\cdot)) \leq 0$ for all $\lambda(\cdot) \in \mathfrak{L}$ such that

$$\mathcal{G}_1(\lambda(\cdot)) \geq 0, \; \mathcal{G}_2(\lambda(\cdot)) \geq 0, \; \ldots, \; \mathcal{G}_k(\lambda(\cdot)) \geq 0; \tag{4.3.8}$$

(ii) There exists a $\lambda(\cdot) \in \mathfrak{L}$ such that

$$\mathcal{G}_1(\lambda(\cdot)) > 0, \; \mathcal{G}_2(\lambda(\cdot)) > 0, \; \ldots, \; \mathcal{G}_k(\lambda(\cdot)) > 0. \tag{4.3.9}$$

Then there will exist constants $\tau_1 \geq 0, \tau_2 \geq 0, \ldots, \tau_k \geq 0$ such that

$$\mathcal{G}_0(\lambda(\cdot)) + \sum_{j=1}^{k} \tau_j \mathcal{G}_j(\lambda(\cdot)) \leq 0$$

for all $\lambda(\cdot) \in \mathfrak{L}$.

Proof: This theorem will be established using Theorem 4.3.1. In order to apply this result, we will establish that the functionals $-\mathcal{G}_0(\cdot)$, $\mathcal{G}_1(\cdot), \ldots, \mathcal{G}_k(\cdot)$ form an S-system on $\mathfrak{L}$. We choose the operators $\mathbf{T}_i$ in Definition 4.3.1 to be shift operators defined as follows:

$$\mathbf{T}_i \lambda(t) = \begin{cases} 0 & \text{if } 0 \leq t \leq i; \\ \lambda(t-i) & \text{if } t > i. \end{cases}$$

Now given any two functions $\lambda_1(\cdot), \lambda_2(\cdot) \in \mathbf{L}_2[0,\infty)$, the Cauchy Schwartz inequality gives

$$\begin{aligned} \left| \int_0^\infty \langle \mathbf{T}_i \lambda_1(t), \lambda_2(t) \rangle dt \right|^2 &= \left| \int_i^\infty \langle \lambda_1(t-i), \lambda_2(t) \rangle dt \right|^2 \\ &\leq \int_0^\infty \|\lambda_1(t)\|^2 dt \int_i^\infty \|\lambda_2(t)\|^2 dt. \end{aligned}$$

However, the term $\int_i^\infty \|\lambda_2\|^2 dt$ in the above inequality tends to zero as $i \to \infty$. Thus, condition (i) of Definition 4.3.1 is satisfied.

Now observe that $\mathcal{G}_j(\mathbf{T}_i \lambda(\cdot)) = \mathcal{G}_j(\lambda(\cdot))$ for all integers i, all $\lambda(\cdot) \in \mathbf{L}_2[0,\infty)$ and $j = 0, 1, \ldots, k$. Thus, condition (iii) of Definition 4.3.1 is satisfied.

To establish condition (ii) of Definition 4.3.1, we note that the subspace $\mathfrak{M}$ corresponding to the affine space $\mathfrak{L}$ is defined by

$$\mathfrak{M} := \left\{ \lambda(\cdot) = \begin{bmatrix} \sigma(\cdot) \\ \mu(\cdot) \end{bmatrix} : \begin{array}{c} \sigma(\cdot), \mu(\cdot) \text{ are related by (4.3.5),} \\ \mu(\cdot) \in \mathbf{L}_2[0,\infty) \text{ and } \eta(0) = 0 \end{array} \right\}.$$

This subspace is invariant under the operators $\mathbf{T}_i$. Hence, condition (ii) of Definition 4.3.1 is satisfied and therefore, the functionals $-\mathcal{G}_0(\cdot)$, $\mathcal{G}_1(\cdot), \ldots, \mathcal{G}_k(\cdot)$ form an S-system. Furthermore, it follows from condition (i) in the theorem that condition (4.3.2) of Theorem 4.3.1 is satisfied. Hence, using condition (ii) of the theorem, Theorem 4.3.1 implies that there exists constants $\tau_1 \geq 0$, $\tau_2 \geq 0, \ldots,$ $\tau_k \geq 0$ such that

$$\mathcal{G}_0(\lambda(\cdot)) + \sum_{j=1}^{k} \tau_j \mathcal{G}_j(\lambda(\cdot)) \leq 0$$

for all $\lambda(\cdot) \in \mathfrak{L}$; see (4.3.4). This completes the proof of the theorem.

■

4.4 An S-procedure result for nonlinear functionals

In this section, we present a result which extends the S-procedure theorem of Section 4.3. This extension involves a more general set of integral functionals defined over the space of solutions to a stable *nonlinear* time-invariant system. A slightly weaker version of this result appeared in the papers [185, 190].

Consider a nonlinear system

$$\dot{x}(t) = g(x(t), \xi(t)) \tag{4.4.1}$$

and associated set of functions

$$w_0(x(t), \xi(t)); \quad w_1(x(t), \xi(t)); \quad \ldots; \quad w_k(x(t), \xi(t)). \tag{4.4.2}$$

Also, consider the corresponding integral functionals

$$\begin{aligned}
\mathcal{G}_0(x(\cdot), \xi(\cdot)) &= \int_0^\infty w_0(x(t), \xi(t))dt; \\
\mathcal{G}_1(x(\cdot), \xi(\cdot)) &= \int_0^\infty w_1(x(t), \xi(t))dt; \\
&\vdots \\
\mathcal{G}_k(x(\cdot), \xi(\cdot)) &= \int_0^\infty w_k(x(t), \xi(t))dt.
\end{aligned} \tag{4.4.3}$$

The system (4.4.1), the set of functions (4.4.2) and the function $w_0(\cdot, \cdot)$ are assumed to satisfy the following assumptions:

Assumption 4.4.1 *The functions $g(\cdot, \cdot), w_0(\cdot, \cdot), \ldots, w_k(\cdot, \cdot)$ are Lipschitz continuous.*

Assumption 4.4.2 *For all $\xi(\cdot) \in \mathbf{L}_2[0, \infty)$ and all initial conditions $x(0) \in \mathbf{R}^n$, $x(\cdot)$ the corresponding solution to (4.4.1) belongs to $\mathbf{L}_2[0, \infty)$ and the corresponding quantities $\mathcal{G}_0(x(\cdot), \xi(\cdot)), \ldots, \mathcal{G}_k(x(\cdot), \xi(\cdot))$ are finite.*

Assumption 4.4.3 *Given any $\varepsilon > 0$, there exists a constant $\delta > 0$ such that the following condition holds: For any input function*

$$\xi_0(\cdot) \in \mathbf{L}_2[0, \infty) \text{ such that } \|\xi_0(\cdot)\|_2^2 \leq \delta$$

and any $x_0 \in \mathbf{R}^n$ such that $\|x_0\| \leq \delta$, let $x_0(t)$ denote the corresponding solution to (4.4.1) with initial condition $x_0(0) = x_0$. Then

$$|\mathcal{G}_j(x_0(\cdot), \xi_0(\cdot))| < \varepsilon$$

for $j = 0, 1, \ldots, k$.

Note that Assumption 4.4.3 is a stability type assumption on the system (4.4.1).

For the system (4.4.1) satisfying the above assumptions, we define a set $\Omega \subset \mathbf{L}_2[0, \infty) \times \mathbf{L}_2[0, \infty)$ as follows: Ω is the set of pairs $\{x(\cdot), \xi(\cdot)\}$ such that $\xi(\cdot) \in \mathbf{L}_2[0, \infty)$ and $x(\cdot)$ is the corresponding solution to (4.4.1) with initial condition $x(0) = 0$.

Theorem 4.4.1 *Consider the system (4.4.1) and associated functions (4.4.2). Suppose that Assumptions 4.4.1 – 4.4.3 are satisfied. If*

$$\mathcal{G}_0\left(x(\cdot),\xi(\cdot)\right) \geq 0$$

for all $\{x(\cdot),\xi(\cdot)\} \in \Omega$ *such that*

$$\mathcal{G}_1\left(x(\cdot),\xi(\cdot)\right) \geq 0, \ldots, \mathcal{G}_k\left(x(\cdot),\xi(\cdot)\right) \geq 0,$$

then there exist constants $\tau_0 \geq 0, \tau_1 \geq 0, \ldots, \tau_k \geq 0$ *such that* $\sum_{j=0}^{k} \tau_j > 0$ *and*

$$\tau_0\mathcal{G}_0\left(x(\cdot),\xi(\cdot)\right) \geq \tau_1\mathcal{G}_1\left(x(\cdot),\xi(\cdot)\right) + \cdots + \tau_k\mathcal{G}_k\left(x(\cdot),\xi(\cdot)\right) \tag{4.4.4}$$

for all $\{x(\cdot),\xi(\cdot)\} \in \Omega$.

In order to prove this theorem, we will use the following preliminary convex analysis result. However, we first introduce some notation; e.g., see [167]. Given sets $S \subset \mathbf{R}^n$ and $T \subset \mathbf{R}^n$ and a constant $\lambda \in \mathbf{R}$, then

$$\begin{aligned} S + T &:= \{x + y : x \in S, y \in T\}; \\ \lambda S &:= \{\lambda x : x \in S\}; \\ \operatorname{cone} S &:= \{\alpha x : x \in S, \alpha \geq 0\}. \end{aligned}$$

Here, $\operatorname{cone} S$ is referred to as the conic hull of the set S. Also, $\operatorname{cl} S$ denotes the closure of the set S.

Lemma 4.4.1 *Consider a set* $\mathcal{M} \subset \mathbf{R}^k$ *with the property that* $a + b \in \operatorname{cl}\mathcal{M}$ *for all* $a, b \in \mathcal{M}$. *If* $x_1 \geq 0$ *for all vectors* $\left[\, x_1 \; x_2 \; \cdots \; x_k \,\right]' \in \mathcal{M}$ *such that* $x_2 < 0, \ldots x_k < 0$, *then there exist constants* $\tau_1 \geq 0, \tau_2 \geq 0, \ldots, \tau_k \geq 0$ *such that* $\sum_{j=1}^{k} \tau_j = 1$ *and*

$$\tau_1 x_1 + \tau_2 x_2 + \cdots + \tau_k x_k \geq 0$$

for all $\left[\, x_1 \; x_2 \; \cdots \; x_k \,\right]' \in \mathcal{M}$.

Proof: In order to establish this result, we first establish the following propositions.

Proposition 3 *Given any two sets* $S \subset \mathbf{R}^n$ *and* $T \subset \mathbf{R}^n$ *and any scalars* $\alpha \in \mathbf{R}$ *and* $\beta \in \mathbf{R}$, *then*

$$\alpha \operatorname{cl} S + \beta \operatorname{cl} T \subseteq \operatorname{cl}(\alpha S + \beta T).$$

To establish this proposition, let $z \in \alpha \operatorname{cl} S + \beta \operatorname{cl} T$ be given. Hence, there exists $x \in \operatorname{cl} S$ and $y \in \operatorname{cl} T$ such that $z = \alpha x + \beta y$. Furthermore, there exist sequences $\{x_i\}_{i=1}^{\infty} \subset S$ and $\{y_i\}_{i=1}^{\infty} \subset T$ such that $x_i \to x$ and $y_i \to y$. Hence, $\alpha x_i + \beta y_i \to z$. However, $\alpha x_i + \beta y_i \in \alpha S + \beta T$ for all i. Thus, we must have $z \in \operatorname{cl}(\alpha S + \beta T)$. This completes the proof of the proposition.

Proposition 4 *The set $\mathcal{M}$ defined in Lemma 4.4.1 has the property that*

$$\operatorname{cl}\mathcal{M} + \operatorname{cl}\mathcal{M} \subseteq \operatorname{cl}\mathcal{M}.$$

To establish this proposition, first note that it follows from Proposition 3 that

$$\operatorname{cl}\mathcal{M} + \operatorname{cl}\mathcal{M} \subseteq \operatorname{cl}(\mathcal{M} + \mathcal{M}).$$

However, since $a + b \in \operatorname{cl}\mathcal{M}$ for all $a, b \in \mathcal{M}$, we must have $\mathcal{M} + \mathcal{M} \subseteq \operatorname{cl}\mathcal{M}$ and hence $\operatorname{cl}(\mathcal{M} + \mathcal{M}) \subseteq \operatorname{cl}\mathcal{M}$. Combining these two facts, it follows that $\operatorname{cl}\mathcal{M} + \operatorname{cl}\mathcal{M} \subseteq \operatorname{cl}\mathcal{M}$.

Proposition 5 *Given any set $S \subseteq \mathbf{R}^n$ then*

$$\operatorname{cone}(\operatorname{cl} S) \subseteq \operatorname{cl}(\operatorname{cone} S).$$

To establish this proposition, let $x \in \operatorname{cone}(\operatorname{cl} S)$ be given. We can write $x = \alpha y$ where $\alpha \geq 0$ and $y \in \operatorname{cl} S$. Hence, there exists a sequence $\{y_i\}_{i=1}^{\infty} \subset S$ such that $y_i \to y$. Therefore $\alpha y_i \to \alpha y = x$. However, $\alpha y_i \in \operatorname{cone} S$ for all i. Thus, $x \in \operatorname{cl}(\operatorname{cone} S)$. This completes the proof of the proposition.

Proposition 6 *The set* $\operatorname{cl}(\operatorname{cone}\mathcal{M})$ *is convex. Here, $\mathcal{M}$ is defined as in Lemma 4.4.1.*

To establish this proposition, let $\lambda \in [0, 1]$ be given and consider two points $x_1, x_2 \in \operatorname{cone}\mathcal{M}$. We can write $x_1 = \alpha_1 z_1$ and $x_2 = \alpha_2 z_2$ where $\alpha_1 \geq 0$, $\alpha_2 \geq 0$, $z_1 \in \mathcal{M}$, and $z_2 \in \mathcal{M}$. Hence $z_1 \in \operatorname{cl}\mathcal{M}$ and $z_2 \in \operatorname{cl}\mathcal{M}$. It follows from Proposition 4 that

$$nz_1 + mz_2 \in \operatorname{cl}\mathcal{M}$$

for all positive integers n and m. Now consider a sequence of rational numbers $\left\{\frac{n_i}{m_i}\right\}_{i=1}^{\infty}$ such that

$$\frac{n_i}{m_i} \to \frac{\lambda\alpha_1}{(1-\lambda)\alpha_2}.$$

Since, n_i and m_i are positive integers for all i, it follows that

$$n_i z_1 + m_i z_2 \in \operatorname{cl}\mathcal{M}$$

for all i. Therefore,

$$(1-\lambda)\alpha_2\left(\frac{n_i}{m_i}z_1 + z_2\right) = \frac{(1-\lambda)\alpha_2}{m_i}(n_i z_1 + m_i z_2) \in \operatorname{cone}(\operatorname{cl}\mathcal{M})$$

for all i. Hence, using Proposition 5, it follows that

$$(1-\lambda)\alpha_2\left(\frac{n_i}{m_i}z_1 + z_2\right) \in \operatorname{cl}(\operatorname{cone}\mathcal{M})$$

for all i. However,

$$
\begin{aligned}
(1-\lambda)\alpha_2\left(\frac{n_i}{m_i}z_1+z_2\right) &\to (1-\lambda)\alpha_2\left(\frac{\lambda\alpha_1}{(1-\lambda)\alpha_2}z_1+z_2\right)\\
&= \lambda\alpha_1 z_1+(1-\lambda)\alpha_2 z_2\\
&= \lambda x_1+(1-\lambda)x_2.
\end{aligned}
$$

Thus, since $\mathrm{cl}(\mathrm{cone}\,\mathcal{M})$ is a closed set, it follows that

$$\lambda x_1+(1-\lambda)x_2 \in \mathrm{cl}(\mathrm{cone}\,\mathcal{M}).$$

Moreover, $x_1 \in \mathrm{cone}\,\mathcal{M}$ and $x_2 \in \mathrm{cone}\,\mathcal{M}$ were chosen arbitrarily and hence, it follows that

$$\lambda\,\mathrm{cone}\,\mathcal{M}+(1-\lambda)\,\mathrm{cone}\,\mathcal{M}\subseteq \mathrm{cl}(\mathrm{cone}\,\mathcal{M}).$$

Therefore,

$$\mathrm{cl}(\lambda\,\mathrm{cone}\,\mathcal{M}+(1-\lambda)\,\mathrm{cone}\,\mathcal{M})\subseteq \mathrm{cl}(\mathrm{cone}\,\mathcal{M}).$$

Now using Proposition 3, we conclude

$$\lambda\,\mathrm{cl}(\mathrm{cone}\,\mathcal{M})+(1-\lambda)\,\mathrm{cl}(\mathrm{cone}\,\mathcal{M})\subseteq \mathrm{cl}(\mathrm{cone}\,\mathcal{M}).$$

Thus, $\mathrm{cl}(\mathrm{cone}\,\mathcal{M})$ is a convex set.

Using the above proposition, we are now in a position to complete the proof of the lemma. Indeed suppose $x_1 \geq 0$ for all vectors $\begin{bmatrix} x_1 & x_2 & \cdots & x_k \end{bmatrix}' \in \mathcal{M}$ such that $x_2 < 0, \ldots, x_k < 0$. Also, let the convex cone $\mathcal{Q} \subset \mathbf{R}^k$ be defined by

$$\mathcal{Q} := \{r=[r_1, r_2, \ldots, r_k]' : r_1<0, r_2<0, \ldots, r_k<0\}.$$

It follows that the intersection of the sets $\mathcal{Q}$ and $\mathcal{M}$ is empty. Also, since $\mathcal{Q}$ is a cone, it is straightforward to verify that $\mathcal{Q}\cap \mathrm{cone}\,\mathcal{M}=\varnothing$. Furthermore, the fact that $\mathcal{Q}$ is an open set implies $\mathcal{Q}\cap \mathrm{cl}(\mathrm{cone}\,\mathcal{M})=\varnothing$. Thus, using the Separating Hyperplane Theorem (Theorem 11.3 of [167]), it follows that there exists a hyperplane separating the convex cones $\mathcal{Q}$ and $\mathrm{cl}(\mathrm{cone}\,\mathcal{M})$ properly. Moreover using Theorem 11.7 of [167], it follows that the hyperplane can be chosen to pass through the origin. Thus there exists a non-zero vector $\tau \in \mathbf{R}^k$ such that

$$x'\tau \geq 0 \tag{4.4.5}$$

for all $x \in \mathrm{cl}(\mathrm{cone}\,\mathcal{M})$ and

$$x'\tau \leq 0 \tag{4.4.6}$$

for all $x \in \mathcal{Q}$. Now write $\tau = \begin{bmatrix} \tau_1 & \tau_2 & \ldots & \tau_k \end{bmatrix}'$. It follows from (4.4.6) that $\tau_i \geq 0$ for all i. Also, since $\tau \neq 0$, we can take coefficients $\tau_1, \tau_2, \ldots, \tau_k$ such that $\sum_{j=1}^{k}\tau_j = 1$. Moreover, it follows from (4.4.5) that

$$\tau_1 x_1+\tau_2 x_2+\cdots+\tau_k x_k \geq 0$$

for all $\begin{bmatrix} x_1 & x_2 & \cdots & x_k \end{bmatrix}' \in \mathcal{M}$. This completes the proof of the lemma.

■

Proof of Theorem 4.4.1: Consider the system (4.4.1) with an uncertainty input $\xi_0(\cdot) \in \mathbf{L}_2[0,\infty)$. Let $x_1(t)$ denote the corresponding solution to equation (4.4.1) with initial condition $x_1(0) = 0$. Also, for any $x_0 \in \{x_0 \in \mathbf{R}^n : \|x_0\| \leq \delta_0\}$, $\delta_0 > 0$, let $x_2(t)$ denote the corresponding solution to (4.4.1) with initial condition $x_2(0) = x_0$. In order to prove Theorem 4.4.1, we establish the following proposition.

Proposition 7 *Given any $\epsilon_0 > 0$ and any input $\xi_0(\cdot) \in \mathbf{L}_2[0,\infty)$, there exists a constant $\delta_0 > 0$ such that the following condition holds: For any $x_0 \in \{x_0 \in \mathbf{R}^n : \|x_0\| \leq \delta_0\}$,*

$$|\mathcal{G}_j\left(x_1(\cdot),\xi_0(\cdot)\right) - \mathcal{G}_j\left(x_2(\cdot),\xi_0(\cdot)\right)| < \epsilon_0$$

for $j = 0, 1, \ldots, k$.

To establish this proposition, let $\epsilon_0 > 0$ be a given constant and let δ be the constant defined as in Assumption 4.4.3 corresponding to $\varepsilon = \epsilon_0/4$. According to Assumption 4.4.2, $[x_1(\cdot), \xi_0(\cdot)] \in \mathbf{L}_2[0,\infty)$. Therefore, there exists a $T > 0$ such that $\|x_1(T)\| \leq \delta/2$ and $\int_T^\infty \|\xi_0(t)\|^2 dt \leq \delta$. Also, Assumption 4.4.1 implies that there exists a constant $\delta_0 > 0$ such that for all $\|x_0\| < \delta_0$, the solution $x_2(\cdot)$ to the system (4.4.1) with input $\xi_0(\cdot)$ and initial condition $x_2(0) = x_0$ satisfies the condition

$$\int_0^T |w_j(x_1(t),\xi_0(t)) - w_j(x_2(t),\xi_0(t))|dt < \frac{\epsilon_0}{2} \tag{4.4.7}$$

and

$$\|x_2(T) - x_1(T)\| < \frac{\delta}{2}. \tag{4.4.8}$$

Since $\|x_1(T)\| \leq \frac{\delta}{2}$, we have from (4.4.8) that $\|x_2(T)\| \leq \delta$. Furthermore, Assumption 4.4.3 and the time invariance of the system (4.4.1) imply that

$$\begin{aligned}&\int_T^\infty |w_j(x_1(t),\xi_0(t)) - w_j(x_2(t),\xi_0(t))|dt \\ &\leq \int_T^\infty (|w_j(x_1(t),\xi_0(t))| + |w_j(x_2(t),\xi_0(t))|)dt \leq \frac{\epsilon_0}{4} + \frac{\epsilon_0}{4} = \frac{\epsilon_0}{2}.\end{aligned}$$

From this and inequality (4.4.7), the proposition follows immediately.

Now suppose $\mathcal{G}_0\left(x(\cdot),\xi(\cdot)\right) \geq 0$ for all $\{x(\cdot),\xi(\cdot)\} \in \Omega$ such that

$$\mathcal{G}_1\left(x(\cdot),\xi(\cdot)\right) \geq 0, \ldots, \mathcal{G}_k\left(x(\cdot),\xi(\cdot)\right) \geq 0$$

and let

$$\mathcal{M} := \left\{ \left[\, \mathcal{G}_0\left(x(\cdot),\xi(\cdot)\right) \ \ldots \ \mathcal{G}_k\left(x(\cdot),\xi(\cdot)\right) \right]' \in \mathbf{R}^{k+1} : \{x(\cdot),\xi(\cdot)\} \in \Omega \right\}.$$

It follows that $\eta_0 \geq 0$ for all vectors $\left[\eta_0 \cdots \eta_k\right]' \in \mathcal{M}$ such that $\eta_1 \geq 0$, $\dots, \eta_k \geq 0$. Now let $[x_a(\cdot), \xi_a(\cdot)] \in \Omega$ and $[x_b(\cdot), \xi_b(\cdot)] \in \Omega$ be given. Since $x_a(\cdot) \in \mathbf{L}_2[0,\infty)$, there exists a sequence $\{T_i\}_{i=1}^{\infty}$ such that $T_i > 0$ for all i, $T_i \to \infty$ and $x_a(T_i) \to 0$ as $i \to \infty$. Now consider the corresponding sequence $\{[x_i(\cdot), \xi_i(\cdot)]\}_{i=1}^{\infty} \subset \Omega$, where

$$\xi_i(t) = \begin{cases} \xi_a(t) & \text{for } t \in [0, T_i); \\ \xi_b(t - T_i) & \text{for } t \geq T_i. \end{cases}$$

We will establish that

$$\mathcal{G}_j(x_i(\cdot), \xi_i(\cdot)) \to \mathcal{G}_j(x_a(\cdot), \xi_a(\cdot)) + \mathcal{G}_j(x_b(\cdot), \xi_b(\cdot))$$

as $i \to \infty$ for $j = 0, 1, \dots, k$. Indeed, let $j \in \{0, 1, \dots, k\}$ be given and fix i. Now suppose $\tilde{x}_b^i(\cdot)$ is the solution to equation (4.4.1) with input $\xi(\cdot) = \xi_b(\cdot)$ and initial condition $\tilde{x}_b^i(0) = x_a(T_i)$. It follows from the time invariance of the system (4.4.1) that $x_i(t) \equiv \tilde{x}_b^i(t - T_i)$. Hence,

$$\begin{aligned} \mathcal{G}_j(x_i(\cdot), \xi_i(\cdot)) &= \int_0^{\infty} w_j(x_i(t), \xi_i(t))dt \\ &= \int_0^{T_i} w_j(x_a(t), \xi_a(t))dt + \int_{T_i}^{\infty} w_j(x_i(t), \xi_b(t - T_i))dt \\ &= \int_0^{T_i} w_j(x_a(t), \xi_a(t))dt + \mathcal{G}_j(\tilde{x}_b^i(\cdot), \xi_b(\cdot)). \end{aligned}$$

Using the fact that $x_a(T_i) \to 0$, Proposition 7 implies

$$\mathcal{G}_j(\tilde{x}_b^i(t), \xi_b(t)) \to \mathcal{G}_j(x_b(\cdot), \xi_b(\cdot))$$

as $i \to \infty$. Also,

$$\int_0^{T_i} w_j(x_a(t), \xi_a(t))dt \to \mathcal{G}_j(x_a(\cdot), \xi_a(\cdot)).$$

Hence,

$$\mathcal{G}_j(x_i(\cdot), \xi_i(\cdot)) \to \mathcal{G}_j(x_a(\cdot), \xi_a(\cdot)) + \mathcal{G}_j(x_b(\cdot), \xi_b(\cdot)).$$

It follows from the above reasoning that the set $\mathcal{M}$ has the property that $a + b \in \operatorname{cl} \mathcal{M}$ for all a, $b \in \mathcal{M}$. Hence, Lemma 4.4.1 implies that there exist constants $\tau_0 \geq 0, \dots, \tau_1 \geq 0$ such that $\sum_{j=0}^{k} \tau_j > 0$ and

$$\tau_0 x_0 \geq \tau_1 x_1 + \cdots + \tau_k x_k$$

for all $\left[x_0 \cdots x_k\right]' \in \mathcal{M}$. That is, condition (4.4.4) is satisfied.

■

4.5 An S-procedure result for averaged sequences

In this section, we present another S-procedure result which can be applied to finite-horizon control problems with structured uncertainties. Note that the results of Sections 4.3 and 4.4 cannot be applied to these problems.

The following S-procedure result first appeared in the papers [191, 196, 198].

Theorem 4.5.1 *Consider a set* $\mathcal{P} = \{\nu = [\nu_0\ \nu_1\ \cdots\ \nu_k]'\} \subset \mathbf{R}^{k+1}$ *with the following property: Given any finite sequence* $\{\nu^1, \nu^2, \ldots, \nu^q\} \subset \mathcal{P}$ *such that* $\sum_{i=1}^q \nu_1^i \geq 0, \ldots, \sum_{i=1}^q \nu_k^i \geq 0$*, then* $\sum_{i=1}^q \nu_0^i \geq 0$*. Then there exist constants* $\tau_0 \geq 0, \tau_1 \geq 0, \ldots, \tau_k \geq 0$ *such that* $\sum_{j=0}^k \tau_j > 0$ *and*

$$\tau_0 \nu_0 \geq \tau_1 \nu_1 + \cdots + \tau_k \nu_k \tag{4.5.1}$$

for all $\nu = [\nu_0\ \nu_1\ \cdots\ \nu_k]' \in \mathcal{P}$*.*

Proof: Let the set $\mathcal{M} \subset \mathbf{R}^{k+1}$ be formed by taking finite sums of elements in the set $\mathcal{P}$. That is,

$$\mathcal{M} := \{\nu = \sum_{i=1}^q \nu^i : \{\nu^i\}_{i=1}^q \text{is any finite sequence contained in } \mathcal{P}\ \}.$$

The set $\mathcal{M}$ has the following property: Given any $a \in \mathcal{M}$ and $b \in \mathcal{M}$, then $a+b \in \mathcal{M}$. Indeed, if $a \in \mathcal{M}$ and $b \in \mathcal{M}$, then we can write $a = \sum_{i=1}^r \nu^i$, $b = \sum_{i=1}^h y^i$ where $\{\nu^i\}_{i=1}^r \subset \mathcal{P}$, $\{y^i\}_{i=1}^h \subset \mathcal{P}$. We now construct a sequence $\{z^i\}_{i=1}^{r+h}$ as follows: $z^i = \nu^i$ for $1 \leq i \leq r$ and $z^i = y^{i-r}$ for $r+1 \leq i \leq r+h$. Then, by definition $\{z^i\}_{i=1}^{r+h} \subset \mathcal{P}$. Also, $\sum_{i=1}^{r+h} z^i \in \mathcal{M}$. However, $\sum_{i=1}^{r+h} z^i = a+b$. Thus, we must have $a+b \in \mathcal{M}$. Also, it follows from the assumed property of the set $\mathcal{P}$ that the set $\mathcal{M}$ will have the property that $\nu_0 \geq 0$ for all vectors $[\nu_0\ \nu_1\ \cdots\ \nu_k]' \in \mathcal{M}$ such that $\nu_1 \geq 0,\ \nu_2 \geq 0,\ \ldots,\ \nu_k \geq 0$. Hence, using Lemma 4.4.1, there exist constants $\tau_0 \geq 0, \tau_1 \geq 0, \ldots, \tau_k \geq 0$ such that $\sum_{j=0}^k \tau_j > 0$ and inequality (4.5.1) holds for all $\nu = [\nu_0\ \nu_1\ \cdots\ \nu_k]' \in \mathcal{M}$. However, any point $\nu \in \mathcal{P}$ will also be contained in the set $\mathcal{M}$. Hence, inequality (4.5.1) holds for all $\nu \in \mathcal{P}$. This completes the proof.

■

Observation 4.5.1 If the assumptions of the above theorem are satisfied and there exists an element $\nu = [\nu_0\ \nu_1\ \cdots\ \nu_k]' \in \mathcal{P}$ such that $\nu_1 > 0, \ldots, \nu_k > 0$, then the constant τ_0 in (4.5.1) may be taken as $\tau_0 = 1$.

4.6 An S-procedure result for probability measures with constrained relative entropies

In this section, we present an S-procedure result that will be used in Section 8. Strictly speaking, this result is a standard result on Lagrange multipliers which is well known in the theory of convex optimization; e.g., see Theorem 1 on page 217 in the book [118]. We include this result in this chapter for the sake of completeness and for ease of reference.

Let $\mathcal{X}$ be a linear vector space and let $\mathcal{P}$ be a convex subset of $\mathcal{X}$. Also, let real valued convex functionals $\mathcal{G}_0(\cdot), \mathcal{G}_1(\cdot), \ldots, \mathcal{G}_k(\cdot)$ be defined on $\mathcal{P}$. It is assumed that the functional $\mathcal{G}_0(\cdot)$ is convex and the functionals $\mathcal{G}_i(\cdot)$, $i = 1, \ldots, k$, are concave mappings from $\mathcal{P}$ to $\mathbf{R}$.

Lemma 4.6.1 *Suppose that the functionals $\mathcal{G}_0(\cdot)$, $\mathcal{G}_i(\cdot)$, $i = 1, \ldots, k$, satisfy the following condition:*

$$\mathcal{G}_0(\lambda) \geq 0 \text{ for any } \lambda \in \mathcal{P} \text{ such that } \mathcal{G}_1(\lambda) \geq 0, \ldots \mathcal{G}_k(\lambda) \geq 0.$$

Then there exist constants $\tau_i \geq 0$, $i = 0, \ldots, k$, such that for any $\lambda \in \mathcal{P}$,

$$\tau_0 \mathcal{G}_0(\lambda) \geq \sum_{i=1}^{k} \tau_i \mathcal{G}_i(\lambda).$$

Furthermore, if there exists a point $\lambda_0 \in \mathcal{P}$ such that

$$\mathcal{G}_1(\lambda_0) > 0, \ldots, \mathcal{G}_k(\lambda_0) > 0$$

then we can choose $\tau_0 = 1$, and

$$\mathcal{G}_0(\lambda) \geq \sum_{i=1}^{k} \tau_i \mathcal{G}_i(\lambda).$$

A special case of Lemma 4.6.1 will be used in Section 8.4. In this special case, the space $\mathcal{X}$ is a set of Lebesgue measures defined on a measurable space $(\Omega, \mathcal{F})$. Also, the set $\mathcal{P}$ is a convex set of probability measures Q on $(\Omega, \mathcal{F})$ as defined in Section 2.4.2. We now present the corresponding version of Lemma 4.6.1.

For a given $T > 0$, let $(\Omega, \mathcal{F}, P)$ be a probability space and $\mathcal{P}$ be a convex set of probability measures Q characterized by the relative entropy condition $h(Q\|P) < \infty$; see Section 2.4.2.

Consider the following real valued measurable functionals defined on the space $C([0,T], \mathbf{R}^n) \times \mathbf{R}^n \times \Omega$:

$$g_0(x(\cdot), x_0, \omega), \ g_1(x(\cdot), x_0, \omega). \tag{4.6.1}$$

Associated with these functionals and the system (2.4.6), we define the following functionals on the set $\mathcal{P}$

$$\begin{aligned} \mathcal{G}_0(Q) &= \mathbf{E}^Q \left[g_0 \left(x(\cdot)|_0^T, x_0, \omega \right) \right], \\ \mathcal{G}_1(Q) &= \mathbf{E}^Q \left[g_1 \left(x(\cdot)|_0^T, x_0, \omega \right) \right] - h(Q\|P), \end{aligned} \tag{4.6.2}$$

where $x(\cdot)$ is the solution to (2.4.6) corresponding to the initial condition $x(0) = x_0$. Here x_0 is the given $\mathcal{F}_0$-measurable random variable and $Q \in \mathcal{P}$.

Theorem 4.6.1 *Suppose the system (2.4.6) and functionals (4.6.1) satisfy the following condition:*

$$\mathcal{G}_0(Q) \geq 0 \text{ for any } Q \in \mathcal{P} \text{ such that } \mathcal{G}_1(Q) \geq 0.$$

Then there exist constants $\tau_0 \geq 0$, $\tau_1 \geq 0$, $\tau_0 + \tau_1 > 0$, such that for any probability measure $Q \in \mathcal{P}$,

$$\tau_0 \mathcal{G}_0(Q) \geq \tau_1 \mathcal{G}_1(Q).$$

Furthermore, if there exists a probability measure $Q \in \mathcal{P}$ such that

$$\mathcal{G}_1(Q) > 0$$

then we can choose $\tau_0 = 1$, and

$$\mathcal{G}_0(Q) \geq \tau_1 \mathcal{G}_1(Q).$$

Proof: Observe that the functional $\mathcal{G}_0(\cdot)$ is a linear functional of Q and the functional $\mathcal{G}_1(\cdot)$ is a concave functional of Q defined on the convex set $\mathcal{P}$. This follows from the convexity of the relative entropy functional; e.g., see [53]. Consequently, the set $\mathcal{M}$ defined by

$$\mathcal{M} = \left\{ \begin{array}{c} [a_0 \ a_1]' : a_0 \geq \mathcal{G}_0(Q), a_1 \geq -\mathcal{G}_1(Q) \\ \text{for some } Q \in \mathcal{P} \end{array} \right\},$$

is a convex subset in $\mathbf{R}^2$. The remainder of the proof uses standard separating hyperplane arguments based on the convexity of the set $\mathcal{M}$. See the proof of Theorem 1 of [118], page 217, for details.

■

5.
Guaranteed cost control of time-invariant uncertain systems

5.1 Introduction

As mentioned in Chapter 1, one approach to the robust linear quadratic regulator problem takes as its launching point the Riccati equation approach to the quadratic stabilization of uncertain systems and its connection to H^∞ control; see [150, 143, 102, 252, 253, 41]. In these papers, the controller is obtained by solving a certain game type Riccati equation arising in an associated H^∞ control problem. The solution to the Riccati equation also defines a fixed quadratic Lyapunov function for the closed loop uncertain system. Hence, the controller obtained using this approach leads to a quadratically stable closed loop system. This approach can also be used to obtain a controller which not only guarantees robust stability but also guarantees a certain level of robust performance. In this case, the quadratic Lyapunov function is used to give an upper bound on the closed loop value of a quadratic cost functional; e.g., see [35, 13, 14]. Such a controller is referred to as a quadratic guaranteed cost controller. We consider the problem of quadratic guaranteed cost control in Section 5.2. The uncertainties under consideration in Section 5.2 are norm-bounded uncertainties. The main result to be presented on this problem is a Riccati equation approach to the construction of an optimal state-feedback quadratic guaranteed cost control. This result was originally obtained by Petersen and McFarlane in references [155, 154, 156]. Note that the result of Petersen and McFarlane goes beyond the earlier results of [35, 13, 14] in that it leads to an "optimal" quadratic guaranteed cost controller whereas the earlier results were sub-optimal in this sense. That is, for the case of norm-bounded uncertainty, the Riccati equation approach presented in Section 5.2

yields an upper bound on the closed loop value of the cost functional which is at least as good as any that can be obtained via the use of a fixed quadratic Lyapunov function.

In Section 5.3, the connection between the Riccati equation approach to H^∞ control and the robust linear quadratic regulator problem is given further consideration. We present a Riccati equation approach to robust LQR control which is closely related to the method developed in Section 5.2. It is shown that for the class of uncertain systems considered in Section 5.3, this controller actually solves a minimax optimal control problem. This section also extends the results presented in Section 5.2 in that structured uncertainty is considered. That is, we allow for multiple uncertainties in the underlying system as shown in Figure 1.2.2. In the case of a single uncertainty, the computation required in this result involves solving a Riccati equation dependent on a single parameter. In the case of multiple uncertainties, the computation required involves solving a Riccati equation dependent on multiple parameters. The results presented in Section 5.3 were originally published in the paper [184].

The results presented in Section 5.3 are limited in their application since it is assumed that full state measurements are available. Section 5.4 extends these results to the output-feedback case. We present a necessary and sufficient condition for the existence of an output-feedback guaranteed cost controller leading to a specified cost bound. This condition is given in terms of the existence of suitable solutions to an algebraic Riccati equation and a Riccati differential equation. The resulting guaranteed cost controller constructed using this approach is in general a linear *time-varying* controller. The emergence of a linear time-varying controller is related to the fact that a deterministic model for noise and uncertainty is being used in the case of an output-feedback guaranteed cost control problem. In Chapter 8, a similar output-feedback guaranteed cost control problem will be considered using a stochastic description of noise and uncertainty. In this case, a linear time-invariant controller is obtained. Also, it should be noted that while the results presented in Section 5.3 apply to uncertain systems with multiple uncertainty blocks (structured uncertainty), the results of Section 5.4 apply only to uncertain systems with a single uncertainty block (unstructured uncertainty).

Note that both the quadratic stability and absolute stability approaches to the guaranteed cost control problem use a fixed quadratic Lyapunov function to give an upper bound on the quadratic cost functional. However, the question remains as to whether a better controller could be obtained if one goes beyond the class of quadratic Lyapunov functions. In Section 5.5, we present a result which gives a positive answer to this question. It is shown that for the class of uncertainties under consideration, a guaranteed cost controller can be constructed by solving a modified Riccati equation which depends on parameters additional to those considered in the parameter dependent Riccati equation approach of Section 5.3. This modified Riccati equation arises from a minimax problem with an "uncertainty dependent" non-quadratic cost functional. Furthermore, the controller synthesized on the basis of this modified Riccati equation leads to a closed loop system which is absolutely stable with an uncertainty dependent Lyapunov function of the Lur'e-

Postnikov form. Also, we present an example which shows that this method of non-quadratic guaranteed cost control can lead to improved performance in terms of the bound on the quadratic cost functional.

5.2 Optimal guaranteed cost control for uncertain linear systems with norm-bounded uncertainty

5.2.1 *Quadratic guaranteed cost control*

In this section, the class of uncertain systems under consideration is described by the following state equations of the form (2.2.1):

$$\dot{x}(t) = [A + B_2\Delta(t)C_1]\,x(t) + [B_1 + B_2\Delta(t)D_1]\,u(t); \quad x(0) = x_0 \qquad (5.2.1)$$

where $x(t) \in \mathbf{R}^n$ is the *state*, $u(t) \in \mathbf{R}^m$ is the *control input*, and $\Delta(t)$ is a time-varying *matrix of uncertain parameters* satisfying the bound (2.2.2); that is,

$$\Delta(t)'\Delta(t) \leq I.$$

Associated with this system is the cost functional

$$J = \int_0^\infty [x(t)'Rx(t) + u(t)'Gu(t)]\,dt \qquad (5.2.2)$$

where $R > 0$ and $G > 0$; e.g., see [108].

Definition 5.2.1 *A control law* $u(t) = Kx(t)$ *is said to define a* quadratic guaranteed cost control *with associated* cost matrix $X > 0$ *for the system (5.2.1) and cost functional (5.2.2) if*

$$x'\,[R + K'GK]\,x + 2x'X\,[A + B_2\Delta(C_1 + D_1K) + B_1K]\,x < 0 \qquad (5.2.3)$$

for all non-zero $x \in \mathbf{R}^n$ *and all matrices* $\Delta : \Delta'\Delta \leq I$.

The connection between this definition and the definition of quadratic stabilizability enables us to extend the methods of quadratic stabilizability (e.g., see [150, 143, 102]) to allow for a guaranteed level of performance; see Section 3.4 for a definition of quadratic stability and quadratic stabilizability.

Theorem 5.2.1 *Consider the system (5.2.1) with cost functional (5.2.2) and suppose the control law* $u(t) = Kx(t)$ *is a quadratic guaranteed cost control with cost matrix* $X > 0$. *Then the closed loop uncertain system*

$$\dot{x}(t) = [A + B_2\Delta(t)(C_1 + D_1K) + B_1K]\,x(t); \quad \Delta(t)'\Delta(t) \leq I \qquad (5.2.4)$$

is quadratically stable[1]*. Furthermore, the corresponding value of the cost function (5.2.2) satisfies the bound*

$$J \leq x_0' X x_0$$

for all admissible uncertainties $\Delta(t)$.

Conversely, if there exists a control law $u(t) = Kx(t)$ *such that the resulting closed loop system (5.2.4) is quadratically stable, then for all positive-definite matrices* R *and* G*, this control law is a quadratic guaranteed cost control with some cost matrix* $\tilde{X} > 0$.

Proof: If the control law $u(t) = Kx(t)$ is a quadratic guaranteed cost control with cost matrix $X > 0$, it follows from (5.2.3) that

$$x' \begin{pmatrix} [A + B_2\Delta(C_1 + D_1K) + B_1K]' X \\ +X [A + B_2\Delta(C_1 + D_1K) + B_1K] \end{pmatrix} x < 0$$

for all $x \neq 0$ and $\Delta : \Delta'\Delta \leq I$. Hence, the closed loop uncertain system (5.2.4) is quadratically stable. Now define $V(x) := x'Xx$. For the closed loop system (5.2.4), it follows from (5.2.3) that

$$x(t)'Rx(t) + u(t)'Gu(t) \leq -\frac{d}{dt}V(x(t)).$$

Integrating this inequality from zero to infinity, implies

$$J = \int_0^\infty [x(t)'Rx(t) + u(t)'Gu(t)]\, dt \leq V(x_0) - V(x(\infty)).$$

However, we have established that the closed loop system (5.2.4) is quadratically stable and hence, $x(\infty) = 0$. Thus,

$$J \leq V(x_0) = x_0' X x_0$$

as required.

Conversely, if there exists a control law $u(t) = Kx(t)$ such that the resulting closed loop system (5.2.4) is quadratically stable, then there exists a matrix $X > 0$ and a constant $\varepsilon > 0$ such that

$$\varepsilon x'(R + K'GK)x + 2x'X [A + B_2\Delta(C_1 + D_1K) + B_1K]\, x < 0$$

for all non-zero $x \in \mathbf{R}^n$ and all $\Delta : \Delta'\Delta \leq I$. Thus, this control law is a quadratic guaranteed cost control with cost matrix $\tilde{X} := X/\varepsilon > 0$.

■

---

[1] See Definition 3.4.1 on page 95.

Observation 5.2.1 Note that the bound obtained in Theorem 5.2.1 depends on the initial condition x_0. To remove this dependence on x_0, we will assume x_0 is a zero mean random variable satisfying $\mathbf{E}\{x_0 x_0'\} = I$. In this case, we consider the expected value of the cost functional:

$$\overline{J} = \mathbf{E}\left\{\int_0^\infty [x(t)'Rx(t) + u(t)'Gu(t)]\, dt\right\}.$$

The cost bound obtained in Theorem 5.2.1 then yields

$$\overline{J} = \mathbf{E}\{J\} \leq E\{x_0' X x_0\} = \mathrm{tr}\{X\}.$$

5.2.2 *Optimal controller design*

We now present a Riccati equation approach to constructing an optimal quadratic guaranteed cost controller. Note that this approach uses the notion of stabilizing solution for an algebraic Riccati equation; see Definition 3.1.1 on page 55.

Theorem 5.2.2 *Suppose there exists a constant $\varepsilon > 0$ such that the Riccati equation*

$$\begin{aligned}
&\left(A - B_1 (\varepsilon G + D_1'D_1)^{-1} D_1'C_1\right)' X \\
&+X \left(A - B_1 (\varepsilon G + D_1'D_1)^{-1} D_1'C_1\right) \\
&+ \varepsilon X B_2 B_2' X - \varepsilon X B_1 (\varepsilon G + D_1'D_1)^{-1} B_1' X \\
&+\frac{1}{\varepsilon} C_1' \left(I - D_1 (\varepsilon G + D_1'D_1)^{-1} D_1'\right) C_1 + R = 0
\end{aligned} \qquad (5.2.5)$$

has a solution $X > 0$ and consider the control law

$$u(t) = -\left(\varepsilon G + D_1'D_1\right)^{-1} (\varepsilon B_1' X + D_1'C_1)x(t). \qquad (5.2.6)$$

Then given any $\delta > 0$, there exists a matrix $\tilde{X} > 0$ such that

$$X < \tilde{X} < X + \delta I$$

and (5.2.6) is a quadratic guaranteed cost control for the system (5.2.1) with cost matrix $\tilde{X}$.

Conversely, given any quadratic guaranteed cost control with cost matrix $\tilde{X} > 0$, there exists a constant $\varepsilon > 0$ such that Riccati equation (5.2.5) has a stabilizing solution $X^+ > 0$ where $X^+ < \tilde{X}$.

Proof: To establish the first part of the theorem, let the control law $u(t) = Kx(t)$ be defined as in (5.2.5) and (5.2.6). If we define $\bar{A} := A + B_1 K$ and $\tilde{G} := \varepsilon G + D_1'D_1$, then it follows from (5.2.5) that

$$\begin{aligned}
&\bar{A}'X + X\bar{A} + \varepsilon X B_1 \tilde{G}^{-1} B_1' X + \varepsilon X B_2 B_2' X \\
&+\frac{1}{\varepsilon} C_1' \left(I - D_1 \tilde{G}^{-1} D_1'\right) C_1 + R = 0.
\end{aligned}$$

It is straightforward but tedious to verify that this equation is equivalent to the Riccati equation

$$\bar{A}'X + X\bar{A} + \varepsilon X B_2 B_2{}' X + \frac{1}{\varepsilon}\bar{C}'\bar{C} + \bar{R} = 0$$

where $\bar{C} := C_1 + D_1 K$ and $\bar{R} := R + K'GK$. Now let $\hat{X} := \varepsilon X$. It follows that $\hat{X}$ satisfies the Riccati equation

$$\bar{A}'\hat{X} + \hat{X}\bar{A} + \hat{X} B_2 B_2{}' \hat{X} + \bar{C}'\bar{C} + \varepsilon\bar{R} = 0.$$

Hence, given any constant $\varepsilon_1 \in (0, \varepsilon)$, then

$$\bar{A}'\hat{X} + \hat{X}\bar{A} + \hat{X} B_2 B_2{}' \hat{X} + \bar{C}'\bar{C} + \varepsilon_1\bar{R} < 0. \tag{5.2.7}$$

Now given any $\delta > 0$, let the constant $\varepsilon_1 \in (0, \varepsilon)$ be chosen so that the matrix

$$\tilde{X} := \frac{\varepsilon}{\varepsilon_1} X = \frac{1}{\varepsilon_1}\hat{X} > X$$

is such that $\tilde{X} < X + \delta I$. It follows from (5.2.7) that $\tilde{X}$ satisfies the inequality

$$\bar{A}'\tilde{X} + \tilde{X}\bar{A} + \varepsilon_1 \tilde{X} B_2 B_2{}' \tilde{X} + \frac{1}{\varepsilon_1}\bar{C}'\bar{C} + \bar{R} < 0. \tag{5.2.8}$$

We will use this quadratic matrix inequality to establish (5.2.3). Indeed, given any admissible uncertainty matrix $\Delta : \Delta'\Delta \leq I$, it follows from (5.2.8) and a standard inequality (e.g., see [143]) that

$$\begin{aligned}
&\left(\bar{A} + B_2\Delta\bar{C}\right)' \tilde{X} + \tilde{X}\left(\bar{A} + B_2\Delta\bar{C}\right) + \bar{R} \\
&\quad \leq \bar{A}'\tilde{X} + \tilde{X}\bar{A} + \varepsilon_1 \tilde{X} B_2 B_2{}' \tilde{X} + \frac{1}{\varepsilon_1}\bar{C}'\bar{C} + \bar{R} \\
&\quad < 0.
\end{aligned}$$

This completes the proof of the first part of the theorem.

To establish the second part of the theorem, we suppose $u(t) = Kx(t)$ is a quadratic guaranteed cost control for the system (5.2.1) with cost matrix $\tilde{X}$. If we define $\bar{A} := A + B_1 K$, $\bar{C} := C_1 + D_1 K$ and $\bar{R} := R + K'GK$, it follows from inequality (5.2.3) that

$$x'\left[\bar{R} + \bar{A}'\tilde{X} + \tilde{X}\bar{A}\right]x + 2x'\tilde{X}B_2\Delta\bar{C}x < 0$$

for all non-zero $x \in \mathbf{R}^n$ and all $\Delta : \Delta'\Delta \leq I$. Now as in Lemma 3.1 and Observation 3.1 of [143], this implies that

$$x'\left[\bar{R} + \bar{A}'\tilde{X} + \tilde{X}\bar{A}\right]x + 2\|B_2{}'\tilde{X}x\| \cdot \|\bar{C}x\| < 0$$

for all non-zero $x \in \mathbf{R}^n$. Thus,

$$\bar{R} + \bar{A}'\tilde{X} + \tilde{X}\bar{A} < 0$$

and

$$\left\{x'\left[\bar{R} + \bar{A}'\tilde{X} + \tilde{X}\bar{A}\right]x\right\}^2 > 4x'\tilde{X}B_2B_2{}'\tilde{X}x \cdot x'\bar{C}'\bar{C}x$$

for all $x \neq 0$. Hence, using Theorem 4.7 of [150], it follows that there exists a constant $\varepsilon > 0$ such that

$$\varepsilon^2\tilde{X}B_2B_2{}'\tilde{X} + \varepsilon\left[\bar{A}'\tilde{X} + \tilde{X}\bar{A} + \bar{R}\right] + \bar{C}'\bar{C} < 0.$$

From this, it follows that the matrix $\tilde{X} > 0$ satisfies the inequality

$$\begin{aligned}[A + \sqrt{\varepsilon}B_1\tilde{K}]'\tilde{X} + \tilde{X}[A + \sqrt{\varepsilon}B_1\tilde{K}] + \varepsilon\tilde{X}B_2B_2{}'\tilde{X} + R \\ +\varepsilon\tilde{K}'G\tilde{K} + (\frac{1}{\sqrt{\varepsilon}}C_1 + D_1\tilde{K})'(\frac{1}{\sqrt{\varepsilon}}C_1 + D_1\tilde{K}) < 0\end{aligned} \quad (5.2.9)$$

where $\tilde{K} := \frac{1}{\sqrt{\varepsilon}}K$. Associated with this inequality is a state-feedback H^∞ problem defined by the system

$$\begin{aligned}\dot{x} &= Ax + \sqrt{\varepsilon}B_2w + \sqrt{\varepsilon}B_1u; \\ z &= \begin{bmatrix} \frac{1}{\sqrt{\varepsilon}}C_1 \\ 0 \\ R^{\frac{1}{2}} \end{bmatrix} x + \begin{bmatrix} D_1 \\ \sqrt{\varepsilon}G^{\frac{1}{2}} \\ 0 \end{bmatrix} u.\end{aligned} \quad (5.2.10)$$

We wish to apply Lemma 3.2.3 on page 70 to this system. Indeed, since $R > 0$ and $G > 0$, it is straightforward to verify using Lemma 3.2.2 that this system satisfies the assumptions required in order to apply Lemma 3.2.3. Now suppose that the state feedback control $u = \tilde{K}x$ is applied to this system. Using Lemma 3.2.3, it follows from (5.2.9) that the Riccati equation (5.2.5) has a stabilizing solution $X^+ \geq 0$ such that $X^+ < \tilde{X}$. Furthermore since $X^+ \geq 0$ satisfies (5.2.5), $R > 0$ implies $X^+ > 0$. This completes the proof of the theorem.

■

Observation 5.2.2 Using the above theorem and Theorem 5.2.1, it is straightforward to verify that if Riccati equation (5.2.5) has a solution $X > 0$ then the resulting closed loop system is quadratically stable and the closed loop value of the cost functional (5.2.2) satisfies the bound

$$J \leq x_0'Xx_0$$

for all $x_0 \in \mathbf{R}^n$ and all admissible uncertainties $\Delta(t)$.

We now present a number of results concerning the construction of an optimal value for ε in (5.2.5).

Theorem 5.2.3 *Suppose that for $\varepsilon = \tilde{\varepsilon} > 0$, the Riccati equation (5.2.5) has a solution $X > 0$. Then for any $\varepsilon \in (0, \tilde{\varepsilon})$, the Riccati equation (5.2.5) will have a stabilizing solution $X^+ > 0$.*

Proof: Suppose the Riccati equation (5.2.5) has a solution $X > 0$ with $\varepsilon = \tilde{\varepsilon}$ and let $\tilde{\Pi} := \frac{1}{\tilde{\varepsilon}} X^{-1} > 0$. It follows that $\tilde{\Pi}$ satisfies the Riccati equation

$$
\begin{aligned}
-(A - B_1\hat{G}^{-1}D_1'C_1)\tilde{\Pi} \; - \; & \tilde{\Pi}(A - B_1\hat{G}^{-1}D_1'C_1)' + B_1\hat{G}^{-1}B_1' - B_2B_2' \\
- \; & \tilde{\Pi}\{C_1{}'(I - D_1\hat{G}^{-1}D_1')C_1 + \tilde{\varepsilon}R\}\tilde{\Pi} = 0
\end{aligned}
\tag{5.2.11}
$$

where $\hat{G} := \tilde{\varepsilon}G + D_1'D_1$. Now let $\varepsilon \in (0, \tilde{\varepsilon})$ be given and consider the Riccati equation

$$
\begin{aligned}
-(A - B_1\tilde{G}^{-1}D_1'C_1)\Pi \; - \; & \Pi(A - B_1\tilde{G}^{-1}D_1'C_1)' + B_1\tilde{G}^{-1}B_1' - B_2B_2' \\
- \; & \Pi\{C_1{}'(I - D_1\tilde{G}^{-1}D_1')C_1 + \varepsilon R\}\Pi = 0
\end{aligned}
\tag{5.2.12}
$$

where $\tilde{G} := \varepsilon G + D_1'D_1$. We wish to use Lemma 3.1.1 to compare these two Riccati equations. Thus, we consider the matrix

$$
\tilde{M} := \begin{bmatrix} B_1\hat{G}^{-1}B_1' - B_2B_2' & -A + B_1\hat{G}^{-1}D_1'C_1 \\ -A' + C_1'D_1\hat{G}^{-1}B_1' & -\tilde{\varepsilon}R - C_1'(I - D_1\hat{G}^{-1}D_1')C_1 \end{bmatrix}
$$

associated with the Riccati equation (5.2.11) and the matrix

$$
M := \begin{bmatrix} B_1\tilde{G}^{-1}B_1' - B_2B_2' & -A + B_1\tilde{G}^{-1}D_1'C_1 \\ -A' + C_1'D_1\tilde{G}^{-1}B_1' & -\varepsilon R - C_1'(I - D_1\tilde{G}^{-1}D_1')C_1 \end{bmatrix}
$$

associated with Riccati equation (5.2.12). It follows by a straightforward algebraic manipulation that

$$
M - \tilde{M} = \begin{bmatrix} B_1 \\ C_1'D_1 \end{bmatrix} (\tilde{G}^{-1} - \hat{G}^{-1}) \begin{bmatrix} B_1' & D_1'C_1 \end{bmatrix} + \begin{bmatrix} 0 & 0 \\ 0 & (\tilde{\varepsilon} - \varepsilon)R \end{bmatrix}. \tag{5.2.13}
$$

Using the fact that $\varepsilon < \tilde{\varepsilon}$, it is now straightforward to verify that $M - \tilde{M} \geq 0$. Thus, using Lemma 3.1.1, it follows that Riccati equation (5.2.12) has a strong solution $\Pi^+ : \Pi^+ \geq \tilde{\Pi} > 0$.

To establish that Π^+ is in fact a stabilizing solution to (5.2.12), we first observe that Riccati equation (5.2.11) can be re-written in the form

$$
-\begin{bmatrix} I & \tilde{\Pi} \end{bmatrix} \tilde{M} \begin{bmatrix} I \\ \tilde{\Pi} \end{bmatrix} = 0. \tag{5.2.14}
$$

Similarly, Riccati equation (5.2.12) can be re-written in the form

$$
-\begin{bmatrix} I & \Pi \end{bmatrix} M \begin{bmatrix} I \\ \Pi \end{bmatrix} = 0.
$$

Thus using (5.2.13) and (5.2.14), it follows that $\tilde{\Pi} > 0$ satisfies the inequality

$$-[I\ \tilde{\Pi}]\,M\begin{bmatrix} I \\ \tilde{\Pi}\end{bmatrix} = -[I\ \tilde{\Pi}]\,\tilde{M}\begin{bmatrix} I \\ \tilde{\Pi}\end{bmatrix} - [I\ \tilde{\Pi}]\,(M-\tilde{M})\begin{bmatrix} I \\ \tilde{\Pi}\end{bmatrix} < 0.$$

That is, there exists a matrix $V > 0$ such that

$$\begin{aligned}(A - B_1\tilde{G}^{-1}D_1'C_1)\tilde{\Pi} + \tilde{\Pi}(A - B_1\tilde{G}^{-1}D_1'C_1)' - B_1\tilde{G}^{-1}B_1' + B_2B_2' \\ + \tilde{\Pi}\{C_1'(I - D_1\tilde{G}^{-1}D_1')C_1 + \varepsilon R\}\tilde{\Pi} + V = 0.\end{aligned} \tag{5.2.15}$$

Now subtracting Riccati equation (5.2.15) from (5.2.12), it follows that the matrix $Z := \Pi^+ - \tilde{\Pi} \geq 0$ satisfies the equation

$$\overline{A}Z + Z\overline{A}' + \tilde{V} = 0$$

where

$$\overline{A} := -A + B_1\tilde{G}^{-1}D_1'C_1 - \Pi^+\left[\varepsilon R + C_1'(I - D_1\tilde{G}^{-1}D_1')C_1\right]$$

and

$$\tilde{V} := Z\{C_1'(I - D_1\tilde{G}^{-1}D_1')C_1 + \varepsilon R\}Z + V > 0.$$

A standard Lyapunov argument now implies that the matrix $\overline{A}$ is stable; see Lemma 12.2 in [251]. Hence, Π^+ is a stabilizing solution to (5.2.12).

We now let $X^+ := \frac{1}{\varepsilon}(\Pi^+)^{-1} > 0$. It is straightforward to verify using Riccati equation (5.2.12) that X^+ is a stabilizing solution to Riccati equation (5.2.5). This completes the proof of the theorem.

■

Theorem 5.2.2 implies that a quadratic guaranteed cost control which minimizes the cost bound obtained in Observation 5.2.1 can be obtained by choosing $\varepsilon > 0$ to minimize $\mathrm{tr}(X^+)$ where X^+ is the stabilizing solution to Riccati equation (5.2.5). We now show that $\mathrm{tr}(X^+)$ is a convex function of ε.

Theorem 5.2.4 *Suppose Riccati equation (5.2.5) has a positive-definite stabilizing solution $X^+(\varepsilon)$ for each ε in the interval $(0,\overline{\varepsilon})$. Then $\mathrm{tr}(X^+(\varepsilon))$ is a convex function of ε over $(0,\overline{\varepsilon})$.*

Proof: If $X^+ > 0$ is a stabilizing solution to (5.2.5), it is straightforward to verify that $\Pi^+ := (X^+)^{-1} > 0$ is the stabilizing solution to the Riccati equation

$$\begin{aligned}&-(A - B_1\left(\varepsilon G + D_1'D_1\right)^{-1}D_1'C_1)\Pi \\ &-\Pi(A - B_1\left(\varepsilon G + D_1'D_1\right)^{-1}D_1'C_1)' \\ &+\varepsilon B_1\left(\varepsilon G + D_1'D_1\right)^{-1}B_1' - \varepsilon B_2B_2' \\ &-\Pi\{\frac{1}{\varepsilon}C_1'(I - D_1\left(\varepsilon G + D_1'D_1\right)^{-1}D_1')C_1 + R\}\Pi = 0.\end{aligned} \tag{5.2.16}$$

Now let $\dot{\Pi}^+ := \frac{d}{d\varepsilon}\Pi^+$, and $\ddot{\Pi}^+ := \frac{d^2}{d\varepsilon^2}\Pi^+$. Differentiating Riccati equation (5.2.16) twice, leads to the following equation (after a some straightforward but tedious algebraic manipulations):

$$\begin{array}{l}
-\overline{A}\ddot{\Pi}^+ - \ddot{\Pi}^+\overline{A}' \\
+2[(\varepsilon I + D_1G^{-1}D_1')^{-1}(D_1G^{-1}B_1' - C_1\Pi^+) + C_1\dot{\Pi}^+]'(\varepsilon I + D_1G^{-1}D_1')^{-1} \\
\times[(\varepsilon I + D_1G^{-1}D_1')^{-1}(D_1G^{-1}B_1' - C_1\Pi^+) + C_1\dot{\Pi}^+] = 0
\end{array}$$

where

$$\begin{array}{rl}
\overline{A} := & -A + B_1\left(G + D_1'D_1\right)^{-1}D_1'C_1 \\
& -\Pi^+\left[R + \frac{1}{\varepsilon}C_1'(I - D_1\left(\varepsilon G + D_1'D_1\right)^{-1}D_1')C_1\right].
\end{array}$$

However, since the matrix $\overline{A}$ is stable, it follows from a standard result on the Lyapunov equation that $\ddot{\Pi}^+ \leq 0$; e.g., see Lemma 12.1 of [251].

Now let $\dot{X}^+ := \frac{d}{d\varepsilon}X^+$, and $\ddot{X}^+ := \frac{d^2}{d\varepsilon^2}X^+$. Then $\dot{X}^+ = -X^+\dot{\Pi}^+X^+$ and $\ddot{X}^+ = 2X^+\dot{\Pi}^+X^+\dot{\Pi}^+X^+ - X^+\ddot{\Pi}^+X^+$. Thus since $\ddot{\Pi}^+ \leq 0$, we must have $\ddot{P}^+ \geq 0$. Hence, $tr(X^+)$ will be a convex function of ε.

■

Observation 5.2.3 We now show that in the case where there is no uncertainty in the input matrix 'B' ($D_1 = 0$) and the uncertainty matrices Δ are constant, the controller obtained in Theorem 5.2.2 will have the desirable 60° phase margin and 6 dB gain margin properties of the standard Linear Quadratic Regulator; e.g., see [3]. Indeed, suppose $D_1 = 0$ and there exists an $\varepsilon > 0$ such that Riccati equation (5.2.5) has a solution $X > 0$. Now let Δ be a given admissible uncertainty matrix and let $K = -G^{-1}B_1'X$. It follows that X satisfies inequality (5.2.3). Now substitute $K = -G^{-1}B_1'X$ into this inequality. It follows that there exists a matrix $V \geq 0$ such that

$$(A + B_2\Delta C_1)'X + X(A + B_2\Delta C_1) - XB_1G^{-1}B_1'X + R + V = 0.$$

Thus, the control law $u(t) = Kx(t)$ is in fact the solution to the linear quadratic regulator problem corresponding to the system

$$\dot{x}(t) = [A + B_2\Delta C_1]x(t) + B_1u(t)$$

with cost functional

$$\int_0^\infty (x(t)'[R + V]x(t) + u(t)'G_2u(t))\,dt.$$

Hence, as in [3], this control law must lead to the required gain and phase margin properties of a linear quadratic regulator.

5.2.3 *Illustrative example*

In this section, we present an example to illustrate the theory developed above.

Consider the uncertain system described by the state equations

$$\dot{x}(t) = \begin{bmatrix} 1+\Delta(t) & -2+\Delta(t) \\ 16 & 8 \end{bmatrix} x(t) + \begin{bmatrix} 0 \\ 8 \end{bmatrix} u(t); \quad x(0) = x_0$$

where $\Delta(t)$ is a scalar uncertain parameter subject to the bound $|\Delta(t)| \leq 1$ and x_0 is a zero mean Gaussian random vector with identity covariance. We wish to construct an optimal quadratic guaranteed cost control for this system which minimizes the bound on the cost index

$$\mathbf{E}\left\{ \int_0^\infty x_1^2(t) + x_2^2(t) + u^2(t) dt \right\}.$$

This uncertain system and cost function are of the form (5.2.1), (5.2.2) where

$$\begin{aligned} A &= \begin{bmatrix} 1 & -2 \\ 16 & 8 \end{bmatrix}; \quad B_1 = \begin{bmatrix} 0 \\ 8 \end{bmatrix}; \quad B_2 = \begin{bmatrix} 1 \\ 0 \end{bmatrix}; \\ C_1 &= \begin{bmatrix} 1 & 1 \end{bmatrix}; \quad D_1 = 0; \\ R &= \begin{bmatrix} 1 & 0 \\ 0 & 1 \end{bmatrix}; \quad G = 1. \end{aligned}$$

In order to construct the required optimal quadratic guaranteed cost control, we must find the value of the parameter $\varepsilon > 0$ which minimizes the trace of the stabilizing solution to Riccati equation (5.2.5). This Riccati equation was found to have a positive-definite solution for ε in the range $(0, 0.89)$. A plot of $\mathrm{tr}(X^+)$ versus ε is shown in Figure 5.2.1. This plot illustrates the convexity property established in Theorem 5.2.4. Also, from this plot, we determine the optimum value of ε to be $\varepsilon^* = 0.33$. Corresponding to this value of ε, we obtain the following stabilizing solution to Riccati equation (5.2.5):

$$X^+ = \begin{bmatrix} 18.7935 & -1.3418 \\ -1.3418 & 0.5394 \end{bmatrix} > 0.$$

The corresponding optimal value of the cost bound is $\mathrm{tr}(X^+) = 19.333$. Also, equation (5.2.6) gives the corresponding optimal quadratic guaranteed cost control matrix

$$K = \begin{bmatrix} 10.7347 & -4.3154 \end{bmatrix}.$$

To illustrate the gain and phase margin properties established in Observation 5.2.3, we now consider the loop gain transfer function corresponding to this system and the optimal quadratic guaranteed control calculated above;

$$L(s) := -K\left(sI - A - B_2 \Delta C_1\right)^{-1} B_1.$$

Here Δ is a *constant* uncertain parameter satisfying the bound $|\Delta| \leq 1$. Nyqist

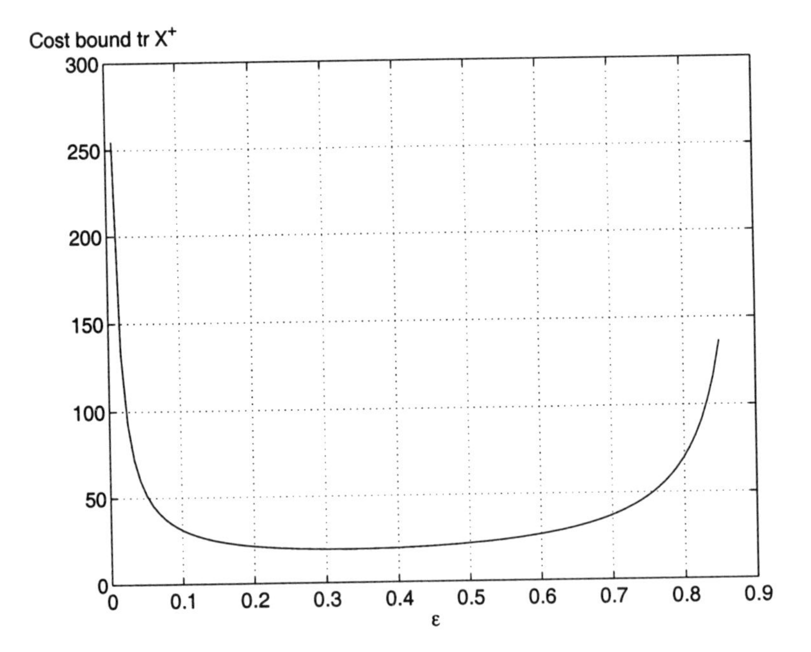

Figure 5.2.1. $\mathrm{tr}(X^+)$ versus ε.

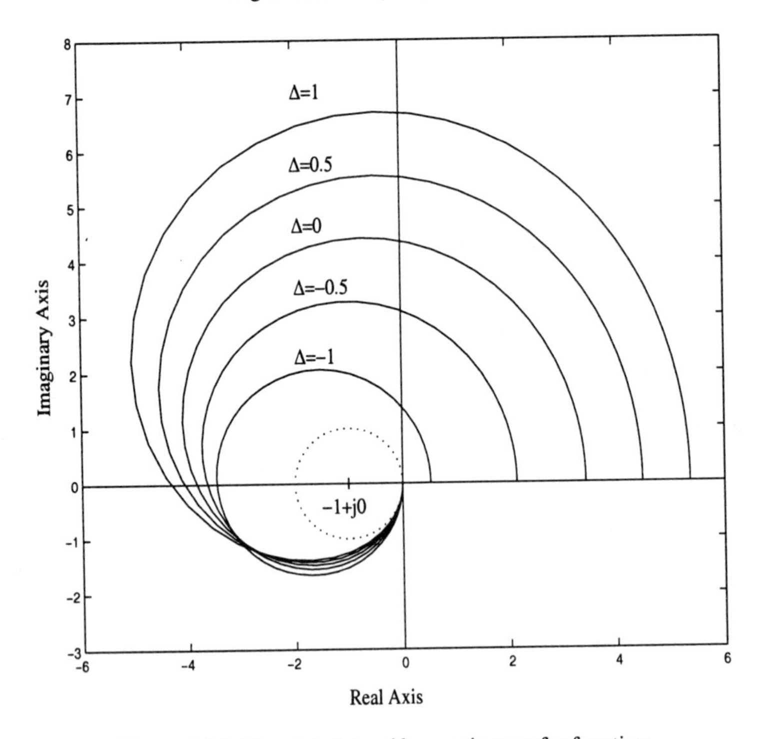

Figure 5.2.2. Nyquist plots of loop gain transfer function.

plots of this loop gain transfer function are shown in Figure 5.2.2 for various values of the uncertain parameter Δ. From this diagram, we can see that each of the Nyquist plots avoids the circle of center $-1 + j0$ and radius 1. From this the required gain and phase margin properties follow.

5.3 State-feedback minimax optimal control of uncertain systems with structured uncertainty

In the previous section, we developed an extension of the linear quadratic regulator to the case in which the controlled system is uncertain. This controller was obtained by solving a certain game type Riccati equation. An important feature of this approach is that it is computationally tractable. The calculations required involve solving a parametrized game type Riccati equation and optimizing over a single parameter. The solution to the Riccati equation also defined a fixed quadratic Lyapunov function for the closed loop uncertain system. Hence, the closed loop uncertain system was quadratically stable. The controller in the previous section not only guaranteed robust stability but also guaranteed an optimal level of robust performance. In this section, we present an extension of the Riccati equation results of previous section and show that the controller obtained is a minimax optimal controller for a class of uncertain systems with structured uncertainty.

The minimax optimal control approach presented in this section is a natural extension of the LQR approach to the case of uncertain systems. A minimax optimal controller minimizes the maximum, over all admissible uncertainties, of the value of the cost functional. Some results concerning such minimax control problems can be found in [19, 217]. However, these results have proved to be too computationally difficult to be widely useful. The main result presented in this section develops a Riccati equation approach to robust LQR control which is closely related to the method presented in Section 5.2 and the papers [155, 154, 156]. It is shown that for the class of uncertain systems with structured uncertainty introduced in Chapter 2, this controller actually solves a minimax optimal control problem. Note that in the case of multiple uncertainties, the computation required by the approach presented in this section involves solving a Riccati equation dependent on multiple parameters. The main results presented in this section were originally published in the paper [184].

The key feature of the approach used in this section, which enables minimax optimality to be established, is the class of uncertainties being considered. For the uncertain systems considered in this section, the uncertainties are required to satisfy certain integral quadratic constraints; see Definition 2.3.1. Another important technical feature of this section is the use of the S-procedure result of Megretski and Treil [127], and Yakubovich [261]; see Section 4.3. This enables the results of Section 5.2 to be extended to allow for multiple structured uncertainties.

One of the limitations of the results of this section is that the controller is designed for a known initial condition. Although this requirement may be too restrictive for some applications, it arises naturally in a number of practical control problems. In this section, we include an example of a robust tracking problem. In this example, it is natural to design the controller for a specific initial condition. For robust control problems in which the initial condition is unknown, the results of this section may also be useful in providing a benchmark against which the guaranteed cost results such as those presented in Section 5.2 and in the papers [155, 154, 156], can be compared.

5.3.1 Definitions

In this section we consider a class of uncertain linear systems described by the following state equations of the form (2.3.3):

$$\begin{aligned}\dot{x}(t) &= Ax(t) + B_1u(t) + \sum_{j=1}^{k} B_{2j}\xi_j(t);\\ z_1(t) &= C_{11}x(t) + D_{11}u(t);\\ z_2(t) &= C_{12}x(t) + D_{12}u(t);\\ &\vdots\\ z_k(t) &= C_{1k}x(t) + D_{1k}u(t)\end{aligned} \tag{5.3.1}$$

where $x(t) \in \mathbf{R}^n$ is the state, $u(t) \in \mathbf{R}^m$ is the control input, $z_1(t) \in \mathbf{R}^{q_1}$, $z_2(t) \in \mathbf{R}^{q_2}, \ldots, z_k(t) \in \mathbf{R}^{q_k}$ are the uncertainty outputs and $\xi_1(t) \in \mathbf{R}^{p_1}$, $\xi_2(t) \in \mathbf{R}^{p_2}, \ldots, \xi_k(t) \in \mathbf{R}^{p_k}$ are the uncertainty inputs.

Associated with the system (5.3.1), is the following cost functional of the form (5.2.2):

$$J = \int_0^\infty [x(t)'Rx(t) + u(t)'Gu(t)]dt \tag{5.3.2}$$

where $R = R' > 0$ and $G = G' > 0$ are given weighting matrices.

System Uncertainty

The uncertainty in the above system is described by equations of the form (2.3.4):

$$\begin{aligned}\xi_1(t) &= \phi_1(t, x(\cdot), u(\cdot))\\ \xi_2(t) &= \phi_2(t, x(\cdot), u(\cdot))\\ &\vdots\\ \xi_k(t) &= \phi_k(t, x(\cdot), u(\cdot))\end{aligned} \tag{5.3.3}$$

satisfying the following integral uncertainty constraint as described in Definition 2.3.1:

Definition 5.3.1 *Let $S_1 > 0$, $S_2 > 0$, ..., $S_k > 0$, be given positive-definite matrices. Then an uncertainty of the form (5.3.3) is an admissible uncertainty for the system (5.3.1) if the following conditions hold: Given any locally square integrable control input $u(\cdot)$ and any corresponding solution to equations (5.3.1), (5.3.3) with an interval of existence $(0, t_\star)$ (that is, $t_\star$ is the upper time limit for which the solution exists), then there exists a sequence $\{t_i\}_{i=1}^{\infty}$ such that $t_i \to t_\star$, $t_i \geq 0$ and*

$$\int_0^{t_i} (\|z_j(t)\|^2 - \|\xi_j(t)\|^2)dt \geq -x(0)'S_jx(0) \tag{5.3.4}$$

for all i and for $s = 1, 2, \ldots, k$. Also, note that $t_\star$ and t_i may be equal to infinity.

Remark 5.3.1 The above IQC (5.3.4) is a special case of the IQC (2.3.5) in Definition 2.3.1 where $d_1 = x_0'S_1x_0$, ..., $d_k = x_0'S_kx_0$. Also, as mentioned in Observation 2.3.1, the uncertain system (5.3.1), (5.3.4) allows for uncertainty satisfying a standard norm bound condition. In this case, the uncertain system would be described by the the following state equations of the form (2.2.1)

$$\dot{x}(t) = [A + \sum_{j=1}^{k} B_{2j}\Delta_j(t)C_{1j}]x(t) + [B_1 + \sum_{j=1}^{k} B_{2j}\Delta_j(t)D_{1j}]u(t)$$

where $\Delta_j(t)$ are uncertainty matrices such that $\|\Delta_j(t)\| \leq 1$. Here $\|\cdot\|$ denotes the standard induced matrix norm; e.g., see Chapter 2, Section 5.2 and the papers [143, 102, 168]. To verify that such uncertainty is admissible for the uncertain system (5.3.1), (5.3.3), let

$$\xi_j(t) = \Delta_j(t)[C_{1j}x(t) + D_{1j}u(t)]$$

where $\|\Delta(t)_j\| \leq 1$ for all $t \geq 0$. Then the uncertainty inputs $\xi_j(\cdot)$ satisfy condition (5.3.4) with any t_i and with any $S_j > 0$.

We will consider a problem of optimizing of the worst case of the cost functional (5.3.2) via a linear state-feedback controller of the form

$$\begin{aligned} \dot{x}_c(t) &= A_cx_c(t) + B_cx(t); \quad x_c(0) = 0; \\ u(t) &= C_cx_c(t) + D_cx(t) \end{aligned} \tag{5.3.5}$$

where A_c, B_c, C_c and D_c are given matrices. Note, the dimension of the state vector x_c may be arbitrary. When a controller of the form (5.3.5) is applied to the uncertain system (5.3.1), (5.3.4), the resulting closed loop uncertain system is described by the state equations:

$$\begin{aligned} \dot{h}(t) &= \hat{A}h(t) + \hat{B}_2\xi(t); \\ z(t) &= \hat{C}h(t); \\ u(t) &= \hat{K}h(t) \end{aligned} \tag{5.3.6}$$

where

$$h(t) = \begin{bmatrix} x(t) \\ x_c(t) \end{bmatrix}; \quad \xi(t) = \begin{bmatrix} \xi_1(t) \\ \xi_2(t) \\ \vdots \\ \xi_k(t) \end{bmatrix}; \; z(t) = \begin{bmatrix} z_1(t) \\ z_2(t) \\ \vdots \\ z_k(t) \end{bmatrix};$$

$$\hat{A} = \begin{bmatrix} A + B_1 D_c & B_1 C_c \\ B_c & A_c \end{bmatrix}; \quad \hat{B}_2 = \begin{bmatrix} B_{21} & \dots & B_{2k} \\ 0 & \dots & 0 \end{bmatrix};$$

$$\hat{C} = \begin{bmatrix} C_{11} + D_{11} D_c & D_{11} C_c \\ C_{12} + D_{12} D_c & D_{12} C_c \\ \vdots & \\ C_{1k} + D_{1k} D_c & D_{1k} C_c \end{bmatrix}; \quad \hat{K} = \begin{bmatrix} D_c & C_c \end{bmatrix}. \tag{5.3.7}$$

The uncertainties for this closed loop uncertain system will be described by equations of the form

$$\begin{aligned} \xi_1(t) &= \phi_1(t, h(\cdot)); \\ \xi_2(t) &= \phi_2(t, h(\cdot)); \\ &\vdots \\ \xi_k(t) &= \phi_k(t, h(\cdot)) \end{aligned} \tag{5.3.8}$$

where the integral quadratic constraint given above is satisfied with the substitution $u(t) = \hat{K} h(t)$.

Definition 5.3.2 *The controller (5.3.5) is said to be a* guaranteed cost controller *for the uncertain system (5.3.1), (5.3.4) with cost functional (5.3.2) and initial condition $x(0) = x_0$ if the following conditions hold:*

(i) The matrix $\hat{A}$ defined in (5.3.7) is stable.

(ii) There exists a constant $c_0 > 0$ such that the following conditions hold: For all admissible uncertainties, the solution to the closed loop system (5.3.6), (5.3.4) corresponding to the initial condition $h(0) = \begin{bmatrix} x_0' & 0 \end{bmatrix}'$ satisfies

$$[x(\cdot), u(\cdot), \xi_1(\cdot), \dots, \xi_k(\cdot)] \in \mathbf{L}_2[0, \infty)$$

and hence, $t_\star = \infty$. Also, the corresponding value of the cost functional (5.3.2) satisfies the bound $J \leq c_0$.

An uncertain system (5.3.1), (5.3.4) with the cost function (5.3.2) which admits a guaranteed cost controller (5.3.5) with initial condition $x(0) = x_0$ is said to be guaranteed cost stabilizable *with this initial condition.*

In the sequel, will show that if there exists a dynamic state-feedback guaranteed cost controller of the form (5.3.5), then there will also exist a static state-feedback guaranteed cost controller of the form

$$u(t) = K_c x(t). \tag{5.3.9}$$

The following definition is a modification to the standard definition of absolute stability for the closed loop uncertain system (5.3.6), (5.3.4).

Definition 5.3.3 *The closed loop uncertain system (5.3.6), (5.3.4) is said to be* absolutely stable *if there exists a constant $c > 0$ such that the following conditions hold:*

(i) For any initial condition $h(0) = h_0$ and any uncertainty inputs $\xi_j(\cdot) \in \mathbf{L}_2[0,\infty)$, the system (5.3.6) has a unique solution which is defined on $[0,\infty)$.

(ii) Given any admissible uncertainty for the uncertain system (5.3.6), (5.3.8), then all corresponding solutions to equations (5.3.6), (5.3.4) satisfy $[h(\cdot), \xi_1(\cdot), \ldots, \xi_k(\cdot)] \in \mathbf{L}_2[0,\infty)$ (hence, $t_\star = \infty$) and

$$\|h(\cdot)\|_2^2 + \sum_{j=1}^{k} \|\xi_j(\cdot)\|_2^2 \leq c\|h_0\|^2. \tag{5.3.10}$$

Remark 5.3.2 We previously dealt with a similar stability property in Section 3.5. There, we observed that for any admissible uncertainty, an absolutely stable uncertain system (5.3.6), (5.3.4) has the property that $h(t) \to 0$ as $t \to \infty$; see Remark 3.5.3 on page 97. Also, the system (5.3.6) with $\xi(t) \equiv 0$ will be asymptotically stable.

5.3.2 Construction of a guaranteed cost controller

The results of this section give a procedure for constructing a guaranteed cost controller for the uncertain system (5.3.1), (5.3.4) which minimizes the worst case value of the cost functional (5.3.2). The required controller, which will be of the form (5.3.9), is constructed by solving a parameter dependent algebraic Riccati equation. This equation is defined as follows: Let $\tau_1 > 0, \ldots, \tau_k > 0$ be given constants and consider the following Riccati equation

$$\begin{aligned}(A - B_1 G_\tau^{-1} D_\tau' C_\tau)' X_\tau + X_\tau (A - B_1 G_\tau^{-1} D_\tau' C_\tau) - X_\tau B_1 G_\tau^{-1} B_1' X_\tau \\ + X_\tau \tilde{B}_2 \tilde{B}_2' X_\tau + C_\tau' (I - D_\tau G_\tau^{-1} D_\tau') C_\tau = 0\end{aligned} \tag{5.3.11}$$

where

$$C_\tau := \begin{bmatrix} R^{1/2} \\ 0 \\ \sqrt{\tau_1}C_{11} \\ \vdots \\ \sqrt{\tau_k}C_{1k} \end{bmatrix}; \quad D_\tau := \begin{bmatrix} 0 \\ G^{1/2} \\ \sqrt{\tau_1}D_{11} \\ \vdots \\ \sqrt{\tau_k}D_{1k} \end{bmatrix};$$

$$G_\tau := D_\tau' D_\tau; \quad \tilde{B}_2 := \left[\frac{1}{\sqrt{\tau_1}}B_{21} \ \dots \ \frac{1}{\sqrt{\tau_k}}B_{2k} \right]. \tag{5.3.12}$$

The parameters $\tau_1 > 0, \dots, \tau_k > 0$ are to be chosen so that this Riccati equation has a positive-definite solution $X_\tau > 0$. Hence, we consider a set $\mathcal{T}_0 \subset \mathbf{R}^k$ defined as follows:

$$\mathcal{T}_0 := \left\{ \begin{array}{l} \tau = [\tau_1 \ \tau_2 \ \dots \ \tau_k] \in \mathbf{R}^k : \tau_1 > 0, \ \tau_2 > 0, \ \dots, \ \tau_k > 0 \text{ and} \\ \qquad\qquad\qquad\qquad\qquad\quad \text{Riccati equation (5.3.11)} \\ \qquad\qquad\qquad\qquad\qquad\quad \text{has a solution } X_\tau > 0 \end{array} \right\}. \tag{5.3.13}$$

If X_τ is the *minimal positive-definite* solution to Riccati equation (5.3.11), then the corresponding guaranteed cost controller to be considered is given by

$$u(t) = -G_\tau^{-1}[B_1' X_\tau + D_\tau' C_\tau]x(t). \tag{5.3.14}$$

Remark 5.3.3 For the case of a single uncertain parameter (*i.e.*, $k = 1$), the Riccati equation (5.3.11) is identical to the Riccati equation (5.2.5) which appears in Section 5.2 with the substitution $\varepsilon = \frac{1}{\tau}$. Thus in this case, the Riccati equation properties established in Section 5.2 carry over to Riccati equation (5.3.11). Also in this case, the control law (5.3.14) is identical to the control law given in Section 5.2. However, there are a number of differences between the results of Section 5.2 and the results given in this section. In this section, we assume that the initial condition x_0 is known whereas in Section 5.2, x_0 was assumed to be a random variable with unity covariance. Also, this section considers a different uncertainty class than was considered in Section 5.2.

We are now in a position to state our first result concerning the construction of a guaranteed cost controller for the uncertain system (5.3.1), (5.3.4).

Theorem 5.3.1 *Consider the uncertain system (5.3.1), (5.3.4) with cost function (5.3.2). Then for any $\tau = [\tau_1, \dots, \tau_k]' \in \mathcal{T}_0$, the corresponding controller (5.3.14) is a guaranteed cost controller for this uncertain system with any initial condition $x_0 \in \mathbf{R}^n$. Furthermore, the corresponding value of the cost function (5.3.2) satisfies the bound*

$$J \le x_0' X_\tau x_0 + \sum_{j=1}^{k} \tau_j x_0' S_j x_0. \tag{5.3.15}$$

for all admissible uncertainties and moreover, the closed loop uncertain system (5.3.1), (5.3.4), (5.3.14) is absolutely stable.

Proof: Suppose $\tau = [\tau_1, \dots, \tau_k]' \in \mathcal{T}_0$ is given and let $X_\tau > 0$ be the corresponding minimal positive-definite solution to Riccati equation (5.3.11). Also, let $x_0 \in \mathbf{R}^n$ be a given initial condition for the uncertain system (5.3.1), (5.3.4). In order to prove that the controller (5.3.14) is a guaranteed cost controller, we consider the closed loop uncertain system formed from the uncertain system (5.3.1), (5.3.4) and this controller:

$$\begin{aligned}\dot{x}(t) &= \left(A - B_1 G_\tau^{-1}[B_1' X_\tau + D_\tau' C_\tau]\right) x(t) + B_2 \xi(t);\\ z(t) &= \bar{C} x(t);\\ u(t) &= -G_\tau^{-1}[B_1' X_\tau + D_\tau' C_\tau] x(t)\end{aligned} \qquad (5.3.16)$$

where

$$B_2 = \begin{bmatrix} B_{21} & \dots & B_{2k} \end{bmatrix},$$
$$\bar{C} = \begin{bmatrix} C_{11} - D_{11} G_\tau^{-1}[B_1' X_\tau + D_\tau' C_\tau] \\ \vdots \\ C_{1k} - D_{1k} G_\tau^{-1}[B_1' X_\tau + D_\tau' C_\tau] \end{bmatrix}.$$

Furthermore, for the given initial condition $x(0) = x_0$, we define a functional $J_\tau(x(\cdot), u(\cdot), z(\cdot), \xi(\cdot))$ as follows:

$$J_\tau(x(\cdot), u(\cdot), z(\cdot), \xi(\cdot)) := J(x(\cdot), u(\cdot)) + \sum_{j=1}^{k} \tau_j (\|z_j(\cdot)\|_2^2 - \|\xi_j(\cdot)\|_2^2). \qquad (5.3.17)$$

Proposition 1 *Suppose $X_\tau > 0$ is the minimal positive-definite solution to Riccati equation (5.3.11) and consider the system (5.3.16) with initial condition $x(0) = x_0$. Then*

$$\sup_{\xi(\cdot) \in \mathbf{L}_2[0,\infty)} J_\tau(x(\cdot), u(\cdot), z(\cdot), \xi(\cdot)) = x_0' X_\tau x_0$$

and furthermore, the system (5.3.16) is asymptotically stable.

In order to establish this proposition, we consider a differential game (see [9]). In this differential game, the underlying system is described by the state equations

$$\dot{x}(t) = Ax(t) + B_1 u(t) + \tilde{B}_2 w(t), \quad x(0) = x_0 \qquad (5.3.18)$$

and the cost function is given by

$$L(u(\cdot), w(\cdot)) = \int_0^\infty (\|C_\tau x(t) + D_\tau u(t)\|^2 - \|w(t)\|^2) dt \qquad (5.3.19)$$

where $\tilde{B}_2, C_\tau$ and D_τ are defined as in (5.3.12). Here $u(t)$ is the *minimizing player input* and $w(t)$ is the *maximizing player input.* With these definitions, observe that the system (5.3.1) may be re-written as (5.3.18) with

$$w(\cdot) = [\sqrt{\tau_1}\xi_1(\cdot), \dots, \sqrt{\tau_k}\xi_k(\cdot)].$$

Similarly with this substitution, the functional $L(u(\cdot), w(\cdot))$ is equal to the functional $J_\tau(x(\cdot), u(\cdot), z(\cdot), \xi(\cdot))$ defined in (5.3.17).

We now apply Lemma 3.2.1 on page 65 (see also Theorem 4.8 and Section 4.5.1 of [9]) to the differential game defined above. This result requires that the pair (A, B_1) is stabilizable and the pair

$$(A - B_1 G_\tau^{-1} D_\tau' C_\tau, (I - D_\tau G_\tau^{-1} D_\tau') C_\tau) \tag{5.3.20}$$

is detectable. These requirements are met in our case. Indeed, let

$$\bar{A} := A - B_1 G_\tau^{-1} [B_1' X_\tau + D_\tau' C_\tau]. \tag{5.3.21}$$

It follows from Riccati equation (5.3.11) that the closed loop matrix (5.3.21) satisfies the equation

$$\begin{aligned} &\bar{A}' X_\tau + X_\tau \bar{A} + X_\tau B_1 G_\tau^{-1} B_1' X_\tau + X_\tau \tilde{B}_2 \tilde{B}_2' X_\tau \\ &+ C_\tau' (I - D_\tau G_\tau^{-1} D_\tau') C_\tau = 0. \end{aligned}$$

Furthermore, it follows from the form of C_τ defined in (5.3.12) that $C_\tau'(I - D_\tau G_\tau^{-1} D_\tau') C_\tau > 0$. Hence, using a standard Lyapunov argument, $X_\tau > 0$ implies the matrix $\bar{A}$ is stable and hence, the pair (A, B_1) is stabilizable. The detectability of the matrix pair (5.3.20) also follows from the fact that $C_\tau'(I - D_\tau G_\tau^{-1} D_\tau') C_\tau > 0$.

Now, it follows from Lemma 3.2.1 that

$$\sup_{w(\cdot) \in \mathbf{L}_2[0,\infty)} L(u^0(\cdot), w(\cdot)) = x_0' X_\tau x_0$$

where $u^0(\cdot)$ is defined by (5.3.14). Hence

$$\sup_{\xi(\cdot) \in \mathbf{L}_2[0,\infty)} J_\tau(x(\cdot), u(\cdot), z(\cdot), \xi(\cdot)) = x_0' X_\tau x_0.$$

Furthermore, as noted above, the matrix $\bar{A}$ is stable. That is, the system (5.3.16) is asymptotically stable. This completes the proof of the proposition.

Now consider the closed loop uncertain system described by equations (5.3.16) and (5.3.4). Given an admissible uncertainty for this system, let $\lambda^0(\cdot) = [x^0(\cdot), u^0(\cdot), z^0(\cdot), \xi^0(\cdot)]$ be a corresponding solution to the system with initial condition $x^0(0) = x_0$. Furthermore, let the sequence $\{t_i\}_{i=1}^{\infty}$ be as defined in Definition 5.3.1 and consider a corresponding sequence $\lambda^i(\cdot) =$

$[x^i(\cdot), u^i(\cdot), z^i(\cdot), \xi^i(\cdot)]$ of vector functions defined by the initial condition $x^i(0) = x_0$ and inputs $\xi^i(\cdot)$:

$$\xi^i(t) := \begin{cases} \xi^0(t) & \text{for } t \in (0, t_i); \\ 0 & \text{for } t \geq t_i. \end{cases}$$

It is clear that $\lambda^i(t) = \lambda^0(t)$ for $t \in (0, t_i)$. Also, since the matrix $\bar{A}$ is stable and $\xi^i(\cdot) \in \mathbf{L}_2[0, \infty)$, we must have $\lambda^i(\cdot) \in \mathbf{L}_2[0, \infty)$. Hence, using Proposition 1, it follows that

$$x_0' X_\tau x_0 \geq J(x^i(\cdot), u^i(\cdot)) + \sum_{j=1}^{k} \tau_j (\|z_j^i(\cdot)\|_2^2 - \|\xi_j^i(\cdot)\|_2^2).$$

This, combined with inequality (5.3.4) implies that

$$J(x^i(\cdot), u^i(\cdot)) \leq x_0' X_\tau x_0 + \sum_{j=1}^{k} \tau_j x_0' S_j x_0 \tag{5.3.22}$$

for all i. However, it follows from Definition 5.3.1 that $t_i \to t_\star$ where $t_\star$ is the upper limit of the existence interval of the solution $x^0(\cdot)$. Hence, using the fact that $R > 0$ in the functional J, it follows from (5.3.22) that $t_\star = \infty$ and $\lambda^0(\cdot) \in \mathbf{L}_2[0, \infty)$. Furthermore, (5.3.22) also implies that (5.3.15) will be satisfied.

We now prove that the closed loop uncertain system (5.3.16), (5.3.4) is absolutely stable. It was shown above that the matrix $\bar{A}$ is stable and hence condition (i) of Definition 5.3.3 is satisfied. We also proved above that for any admissible uncertainties, all solutions to the closed loop system are such that $[x(\cdot), u(\cdot), \xi(\cdot)] \in \mathbf{L}_2[0, \infty)$. Hence, it remains only to establish condition (5.3.10). However, we have already established inequality (5.3.15) and since $R > 0$, this implies that there exists a constant $c_1 > 0$ such that

$$\|x(\cdot)\|_2^2 \leq c_1 \|x_0\|^2$$

for all the solutions to the closed loop uncertain system. Furthermore, the constraint (5.3.4) and equation (5.3.14) imply that there exist constants $c_2 > 0$ and $c_3 > 0$ such that

$$\|\xi_j(\cdot)\|_2^2 \leq c_2 \|x(\cdot)\|_2^2 + c_3 \|x_0\|^2$$

for all the solutions to the closed loop uncertain system and all $j = 1, \ldots, k$. Therefore,

$$\begin{aligned} \|x(\cdot)\|_2^2 + \sum_{j=1}^{k} \|\xi_j(\cdot)\|_2^2 &\leq (kc_2 + 1)\|x(\cdot)\|_2^2 + kc_3 \|x_0\|^2 \\ &\leq \{(kc_2 + 1)c_1 + kc_3\} \|x_0\|^2. \end{aligned}$$

Thus, condition (ii) of Definition 5.3.3 is satisfied (with $x(t)$ replacing $h(t)$ in this case) and hence, the closed loop uncertain system is absolutely stable. This completes the proof of the Theorem.

■

The following result shows that the controller construction given in Theorem 5.3.1 can be used to construct a controller which approaches the minimax optimum.

Theorem 5.3.2 *Consider the uncertain system (5.3.1), (5.3.4) with cost functional (5.3.2) and suppose that* $B_{21} \neq 0, \ldots, B_{2k} \neq 0$. *Then:*

(i) Given a non-zero initial condition $x_0 \in \mathbf{R}^n$, *the uncertain system (5.3.1), (5.3.4) will be guaranteed cost stabilizable with initial condition* $x(0) = x_0$ *if and only if the set* $\mathcal{T}_0$ *defined in (5.3.13) is not empty.*

(ii) Suppose the set $\mathcal{T}_0$ *is not empty and let* Ξ *be the set of all admissible uncertainties for the uncertain system (5.3.1), (5.3.4) as defined in Definition 5.3.1. Also, for any initial condition* $x_0 \neq 0$, *let* Θ *denote the set of all guaranteed cost controllers of the form (5.3.5) for the uncertain system with this initial condition. Then*

$$\inf_{u(\cdot)\in\Theta} \sup_{\xi(\cdot)\in\Xi} J = \inf_{\tau\in\mathcal{T}_0} [x_0' X_\tau x_0 + \sum_{j=1}^{k} \tau_j x_0' S_j x_0]. \tag{5.3.23}$$

Before proving this theorem, we will first present a preliminary lemma relating to the existence of a guaranteed cost controller for the uncertain system (5.3.1), (5.3.3).

Lemma 5.3.1 *Let* $x_0 \in \mathbf{R}^n$ *be a given non-zero initial condition and let* $c_1 > 0$ *be a given constant. Then the following statements are equivalent:*

(i) The uncertain system (5.3.1), (5.3.3) with the cost function (5.3.2) and initial condition x_0 *has a guaranteed cost controller of the form (5.3.5) such that*

$$\sup_{\xi(\cdot)\in\Xi} J < c_1. \tag{5.3.24}$$

(ii) There exists a $\tau \in \mathcal{T}_0$ *such that*

$$x_0' X_\tau x_0 + \sum_{j=1}^{k} \tau_j x_0' S_j x_0 < c_1. \tag{5.3.25}$$

Proof: (i)⇒(ii). Let x_0 be a given non-zero initial condition and suppose the uncertain system (5.3.1), (5.3.4) is guaranteed cost stabilizable (with initial condition x_0) via the controller (5.3.5). We wish to apply Theorem 4.3.2 to the corresponding closed loop system (5.3.6). In this case, we define the set $\mathcal{M}_0$ to be the set of the vector functions

$$\lambda(\cdot) = [x(\cdot), u(\cdot), z(\cdot), \xi_1(\cdot), \ldots, \xi_k(\cdot)] = [x(\cdot), u(\cdot), z(\cdot), \xi(\cdot)] \in \mathbf{L}_2[0, \infty)$$

connected by (5.3.6) with initial condition $h(0) = [x_0' \;\; 0]'$. Also, the integral functionals $\mathcal{G}_1, \mathcal{G}_2, \dots, \mathcal{G}_k$ mapping from $\mathcal{M}_0$ into $\mathbf{R}$ are defined as follows: For each $s = 1, 2, \dots, k$,

$$\mathcal{G}_j(x(\cdot), u(\cdot), z(\cdot), \xi(\cdot)) := \|z_j(\cdot)\|_2^2 - \|\xi_j(\cdot)\|_2^2 + x_0' S_j x_0.$$

It follows from condition (5.3.24) that there exists a constant $\varepsilon > 0$ such that the cost function J given in (5.3.2) satisfies the bound

$$(1 + \varepsilon) J(x(\cdot), u(\cdot)) \le c_1 - \varepsilon \tag{5.3.26}$$

for all $\lambda(\cdot) = [x(\cdot), u(\cdot), z(\cdot), \xi(\cdot)] \in \mathcal{M}_0$ such that

$$\mathcal{G}_1(\lambda(\cdot)) \ge 0, \; \dots, \; \mathcal{G}_k(\lambda(\cdot)) \ge 0.$$

We now define the functional $\mathcal{G}_0$ as

$$\mathcal{G}_0(x(\cdot), u(\cdot), z(\cdot), \xi(\cdot)) := (1 + \varepsilon) J(x(\cdot), u(\cdot)) - c_1 + \varepsilon.$$

Using this definition and inequality (5.3.26), it follows that condition (i) of Theorem 4.3.2 is satisfied. Also observe that since $S_1 > 0, S_2 > 0, \dots, S_k > 0$, if we choose $\xi(t) \equiv 0$ then the corresponding vector function $\lambda(\cdot) \in \mathcal{M}_0$ is such that $\mathcal{G}_1(\lambda(\cdot)) > 0, \mathcal{G}_k(\lambda(\cdot)) > 0$. That is, condition (ii) of Theorem 4.3.2 is satisfied. Hence, using this result, it follows that there exist constants $\tau_1 \ge 0, \dots, \tau_k \ge 0$ such that

$$\mathcal{G}_0(\lambda(\cdot)) + \sum_{j=1}^{k} \tau_j \mathcal{G}_j(\lambda(\cdot)) \le 0 \tag{5.3.27}$$

for all $\lambda(\cdot) \in \mathcal{M}_0$.

We now define a functional J_τ as in the proof of Theorem 5.3.1; see (5.3.17):

$$J_\tau(x(\cdot), u(\cdot), z(\cdot), \xi(\cdot)) := J(x(\cdot), u(\cdot)) + \sum_{j=1}^{k} \tau_j (\|z_j(\cdot)\|_2^2 - \|\xi_j(\cdot)\|_2^2). \tag{5.3.28}$$

Moreover, if we define the functional

$$J_\tau^\varepsilon(x(\cdot), u(\cdot), z(\cdot), \xi(\cdot)) := \varepsilon J(x(\cdot), u(\cdot)) + J_\tau(x(\cdot), u(\cdot), z(\cdot), \xi(\cdot)), \tag{5.3.29}$$

it follows from (5.3.27) that

$$J_\tau^\varepsilon(\lambda(\cdot)) \le c_1 - \varepsilon - \sum_{j=1}^{k} \tau_j x_0' S_j x_0 \tag{5.3.30}$$

for all $\lambda(\cdot) \in \mathcal{M}_0$.

Now let $\mathcal{M}_{00}$ be the set of the vector functions

$$\lambda(\cdot) = [x(\cdot), u(\cdot), z(\cdot), \xi(\cdot)] \in \mathbf{L}_2[0, \infty)$$

related by (5.3.6) with the initial condition $h(0) = 0$.

Proposition 2 $J_\tau^\varepsilon(\lambda(\cdot)) \leq 0$ *for all* $\lambda(\cdot) \in \mathcal{M}_{00}$.

We establish this proposition by contradiction. Suppose the proposition does not hold. Then there exists a vector function $\lambda(\cdot) = [x(\cdot), u(\cdot), z(\cdot), \xi(\cdot)] \in \mathcal{M}_{00}$ such that $J_\tau^\varepsilon(\lambda(\cdot)) > 0$. Associated with this vector function, let $h(\cdot)$ be the corresponding solution to (5.3.6). Also, let $h_0(\cdot)$ be the solution to (5.3.6) with initial condition $h(0) = \begin{bmatrix} x_0' & 0 \end{bmatrix}'$ and input $\xi(\cdot) \equiv 0$. Associated with $h_0(\cdot)$, let $\lambda_0(\cdot) = [x_0(\cdot), u_0(\cdot), z_0(\cdot), 0]$ be the corresponding element of $\mathcal{M}_0$. It follows from the linearity of the system (5.3.6) that

$$a\lambda(\cdot) + \lambda_0(\cdot) \in \mathcal{M}_0$$

for all $a \in \mathbf{R}$. Moreover, since $J_\tau^\varepsilon(\cdot)$ is a quadratic functional, we can write

$$J_\tau^\varepsilon(a\lambda(\cdot) + \lambda_0(\cdot)) = a^2 J_\tau^\varepsilon(\lambda(\cdot)) + af(\lambda(\cdot), \lambda_0(\cdot)) + J_\tau^\varepsilon(\lambda_0(\cdot))$$

where $f(\cdot,\cdot)$ is a corresponding bilinear form. However, since $J_\tau^\varepsilon(\lambda(\cdot)) > 0$, it follows that

$$\lim_{a\to\infty} J_\tau^\varepsilon(a\lambda(\cdot) + \lambda_0(\cdot)) = \infty$$

which contradicts (5.3.30). This completes the proof of the proposition.

Proposition 3 $\tau_j > 0$ *for all* $j = 1, \ldots, k$.

To establish this proposition, we note that Proposition 2 implies

$$\begin{aligned} J_\tau^\varepsilon(x(\cdot), u(\cdot), z(\cdot), \xi(\cdot)) &= \int_0^\infty \Bigg[(1+\varepsilon)(x(t)'Rx(t) + u(t)'Gu(t)) \\ &\qquad + \sum_{j=1}^k \tau_j(\|z_j(t)\|^2 - \|\xi_j(t)\|^2)\Bigg] dt \\ &\leq 0 \end{aligned} \tag{5.3.31}$$

for all vector functions $\lambda(\cdot) = [x(\cdot), u(\cdot), z(\cdot), \xi(\cdot)] \in \mathbf{L}_2[0,\infty)$ connected by (5.3.6) with initial condition $h(0) = 0$.

Now suppose $\tau_j = 0$ for some j and consider an input function $\xi(\cdot)$ defined so that $\xi_j(\cdot) \not\equiv 0$ and $\xi_s(\cdot) \equiv 0$ for $s \neq j$. For such an input, it follows from (5.3.31) that we must have $J_\tau^\varepsilon(x(\cdot), u(\cdot), z(\cdot), \xi(\cdot)) = 0$. Furthermore, since $R > 0$ and $G > 0$, this implies that $x(\cdot) \equiv 0$ and $u(\cdot) \equiv 0$. However, since $B_{2j} \neq 0$, we can choose $\xi_j(\cdot)$ such that $B_{2j}\xi_j(\cdot) \not\equiv 0$. Thus, we have $x(\cdot) \equiv 0$ and $B_{2j}\xi_j(\cdot) \not\equiv 0$ which leads to a contradiction with state equations (5.3.6). Hence, we can conclude that $\tau_j > 0$ for all $j = 1, \ldots, k$. This completes the proof of the proposition.

Proposition 4 *The Riccati equation (5.3.11) has a minimal positive-definite solution* $X_\tau > 0$.

To establish this proposition, we consider the functional J_τ defined in (5.3.28). Since $J_\tau^\varepsilon = \varepsilon J + J_\tau$, Proposition 2 implies

$$J_\tau(\lambda(\cdot)) \leq -\varepsilon J(x(\cdot), u(\cdot)) \tag{5.3.32}$$

for all $\lambda(\cdot) \in \mathcal{M}_{00}$. Now consider the system

$$\begin{aligned} \dot{x}(t) &= Ax(t) + B_1 u(t) + \tilde{B}_2 w(t); \\ \tilde{z}(t) &= C_\tau x(t) + D_\tau u(t) \end{aligned} \tag{5.3.33}$$

where $\tilde{B}_2$, C_τ and D_τ are defined as in (5.3.12) and

$$w(\cdot) = [\sqrt{\tau_1}\xi_1(\cdot), \ldots, \sqrt{\tau_k}\xi_k(\cdot)]. \tag{5.3.34}$$

Since we established in Proposition 3 that $\tau_j > 0$ for all s, this system is well defined.

For this system with zero initial condition, we let $w(\cdot)$ be the disturbance input and let $z(\cdot)$ be the controlled output. Using these definitions, it follows that the functional J_τ defined in (5.3.28) may be re-written as

$$J_\tau(x(\cdot), u(\cdot), \xi(\cdot)) = \|\tilde{z}(\cdot)\|_2^2 - \|w(\cdot)\|_2^2.$$

However, the matrices R and G defining the cost functional (5.3.2) are positive-definite. Hence, condition (5.3.32) implies that

$$\sup_{x(0)=0, w(\cdot)\in \mathbf{L}_2[0,\infty)} \frac{\|z(\cdot)\|_2^2}{\|w(\cdot)\|_2^2} < 1. \tag{5.3.35}$$

That is, the controller (5.3.5) solves a standard H^∞ control problem defined by the system (5.3.33) and the H^∞ condition (5.3.35). Therefore, using a result from H^∞ control theory, it follows that the Riccati equation (5.3.11) has a minimal positive-definite solution $X_\tau > 0$; see Theorem 3.2.2. This completes the proof of the proposition.

Proposition 5 *Let $X_\tau > 0$ be the minimal positive-definite solution to Riccati equation (5.3.11). Then the functional J_τ defined in (5.3.28) satisfies the inequality*

$$\sup_{\lambda(\cdot)\in\mathcal{M}_0} J_\tau(\lambda(\cdot)) \geq x_0' X_\tau x_0. \tag{5.3.36}$$

This proposition is established in a similar fashion to Proposition 1 in the proof of Theorem 5.3.1; *i.e.*, we consider a differential game defined by the system (5.3.18) and cost functional (5.3.19). Now observe that the system (5.3.1) may be re-written as (5.3.18) with the disturbance input $w(\cdot)$ defined as in (5.3.34). Similarly, with the substitution (5.3.34), the functional $L(u(\cdot), w(\cdot))$ defined in (5.3.19) is equal to the functional $J_\tau(x(\cdot), u(\cdot), z(\cdot), \xi(\cdot))$ defined in (5.3.28).

Now, according to Lemma 3.2.1 (see also Theorem 4.8 and Section 4.5.1 of [9]),

$$\sup_{w(\cdot)\in \mathbf{L}_2[0,\infty)} L(u(\cdot), w(\cdot)) \geq x_0' X_\tau x_0$$

where $u(\cdot)$ is defined by (5.3.5). Hence

$$\sup_{\xi(\cdot)\in \mathbf{L}_2[0,\infty)} J_\tau(x(\cdot), u(\cdot), z(\cdot), \xi(\cdot)) \geq x_0' X_\tau x_0.$$

That is, inequality (5.3.36) is satisfied. This completes the proof of the proposition.

We are now in a position to complete the proof of the lemma. Indeed, Proposition 4 implies the existence of a solution to Riccati equation (5.3.11), $X_\tau > 0$. Hence, $[\tau_1, \tau_2, \ldots, \tau_k]' \in \mathcal{T}_0$. Also, Proposition 5 implies that

$$\sup_{\lambda(\cdot)\in\mathcal{M}_0} J_\tau(\lambda(\cdot)) \geq x_0' X_\tau x_0.$$

This and inequalities (5.3.29) and (5.3.30) imply

$$c_1 - \sum_{j=1}^{k} x_0' S_j x_0 - \varepsilon \geq \sup_{\lambda(\cdot)\in\mathcal{M}_0} J_\tau^\varepsilon(\lambda(\cdot)) \geq \sup_{\lambda(\cdot)\in\mathcal{M}_0} J_\tau(\lambda(\cdot)) \geq x_0' X_\tau x_0.$$

Therefore, we have

$$x_0' X_\tau x_0 + \sum_{j=1}^{k} \tau_j x_0' S_j x_0 \leq c_1 - \varepsilon$$

which implies the inequality (5.3.25).

(ii)⇒(i). This statement follows immediately from Theorem 5.3.1. This completes the proof of the lemma.

■

Proof of Theorem 5.3.2: Statements (i) and (ii) of the theorem follow directly from Lemma 5.3.1.

■

Remark 5.3.4 In the above theorem, there is no guarantee that a minimax optimal controller will exist. That is, the infimum on the right hand side of (5.3.23) may not be achieved. In fact, there may arise "singular" problems in which the minimax optimum cannot be achieved but rather is only approached in the limit as one considers controllers of higher and higher gain.

As mentioned in the introduction, the controller obtained by application of Theorems 5.3.2 and 5.3.1 will be the minimax optimal controller for a specific value of the initial condition $x(0) = x_0$. The tracking example given in the next subsection is an example in which the requirement of a known initial condition arises naturally. However, in many applications, the initial condition will be unknown. There are a number of approaches to this problem. In Section 5.2 as well as in references [155, 154, 156], the initial condition is assumed to be a random variable with unity covariance and the controller was designed to minimize the maximum value of the expectation of the cost function. Referring to equation (5.3.23) in Theorem 5.3.2, the optimum value of the parameter $\tau \in \mathcal{T}_0$ would be obtained by performing the optimization

$$\inf_{\tau \in \mathcal{T}_0} \left\{ \mathrm{tr}[X_\tau + \sum_{j=1}^{k} \tau_j S_j] \right\}.$$

It was shown in Section 5.2 that this leads to a controller which is an optimal "quadratic guaranteed cost" controller. However, there is no guarantee that this will lead to a minimax optimal controller.

It can be seen from the above that the results of this section cannot be directly applied in the case in which the initial condition is unknown. However, they do provide a benchmark against which the results of Section 5.2 can be compared.

5.3.3 *Illustrative example*

We now present an example to illustrate the type of problems for which the above results can be directly applied. The problem is a tracking problem in which a state-feedback controller is to be designed so that the system output will track a step input. A feature of this example is that it is natural to design the controller for a specific initial condition.

The system to be controlled was obtained from a proposed benchmark problem for robust control; e.g., see [246] and [61]. The system consists of two carts connected by a spring as shown in Figure 2.4.2 on page 35. In our example, we assume the masses are $m_1 = 1$ and $m_2 = 1$ and the spring constant k is treated as an uncertain parameter subject to the bound $0.5 \leq k \leq 2.0$. (This may reflect the nonlinear nature of the true spring.) From this, we obtain a corresponding uncertain system of the form (5.3.1), (5.3.4) described by the state equations

$$\begin{aligned}
\dot{x} &= \begin{bmatrix} 0 & 0 & 1 & 0 \\ 0 & 0 & 0 & 1 \\ -1.25 & 1.25 & 0 & 0 \\ 1.25 & -1.25 & 0 & 0 \end{bmatrix} x + \begin{bmatrix} 0 \\ 0 \\ 0 \\ 1 \end{bmatrix} u + \begin{bmatrix} 0 \\ 0 \\ -0.75 \\ 0.75 \end{bmatrix} \xi; \\
z &= \begin{bmatrix} 1 & -1 & 0 & 0 \end{bmatrix} x; \\
y &= \begin{bmatrix} 1 & 0 & 0 & 0 \end{bmatrix} x
\end{aligned} \qquad (5.3.37)$$

where the uncertainty is subject to the integral quadratic constraint:

$$\int_0^{t_i} (\|z(t)\|^2 - \|\xi(t)\|^2) dt \geq -x_0' S_1 x_0.$$

Here the times t_i are chosen arbitrarily so that $t_i \to \infty$. The matrix S_1 is defined by

$$S_1 = \begin{bmatrix} 0.1 & 0 & 0 & 0 \\ 0 & 0.1 & 0 & 0 \\ 0 & 0 & 0.1 & 0 \\ 0 & 0 & 0 & 0.1 \end{bmatrix} > 0.$$

In the system (5.3.37), the state variables are $x = [x_1\ x_2\ x_3\ x_4]'$ where $x_3 = \dot{x}_1$ and $x_4 = \dot{x}_2$. It is assumed that all of these state variables are available for measurement. The control problem to be solved involves finding a controller which absolutely stabilizes the system and also ensures that the output y tracks a reference step input. In order to apply our minimax control results to this problem, we apply a standard technique involving the Internal Model Principle; e.g., see Chapter 4 of [3].

First, it will be convenient to introduce a new set of state variables for the system (5.3.37): $\tilde{x} = [\tilde{x}_1\ \tilde{x}_2\ \tilde{x}_3\ \tilde{x}_4]'$ where $\tilde{x}_1 = (x_1 + x_2)/2$, $\tilde{x}_2 = \dot{\tilde{x}}_1$, $\tilde{x}_3 = (x_1 - x_2)/2$ and $\tilde{x}_4 = \dot{\tilde{x}}_3$. With this change of state variables, the system (5.3.37) becomes

$$\begin{aligned} \dot{\tilde{x}} &= \begin{bmatrix} 0 & 1 & 0 & 0 \\ 0 & 0 & 0 & 0 \\ 0 & 0 & 0 & 1 \\ 0 & 0 & -2.5 & 0 \end{bmatrix} \tilde{x} + \begin{bmatrix} 0 \\ 0.5 \\ 0 \\ -0.5 \end{bmatrix} u + \begin{bmatrix} 0 \\ 0 \\ 0 \\ -0.75 \end{bmatrix} \xi; \\ z &= \begin{bmatrix} 0 & 0 & 2 & 0 \end{bmatrix} \tilde{x}; \\ y &= \begin{bmatrix} 1 & 0 & 1 & 0 \end{bmatrix} \tilde{x}. \end{aligned} \tag{5.3.38}$$

Also, the reference input signal $\tilde{y}$ can be described by the state space model

$$\begin{aligned} \dot{\eta} &= 0; \\ \tilde{y} &= \eta. \end{aligned} \tag{5.3.39}$$

We now define a new state vector $\bar{x} = [\tilde{x}_1 - \eta\ \tilde{x}_1\ \tilde{x}_3\ \tilde{x}_4]'$ and combine state equations (5.3.38) and (5.3.39). This leads to the state equations

$$\begin{aligned} \dot{\bar{x}} &= \begin{bmatrix} 0 & 1 & 0 & 0 \\ 0 & 0 & 0 & 0 \\ 0 & 0 & 0 & 1 \\ 0 & 0 & -2.5 & 0 \end{bmatrix} \bar{x} + \begin{bmatrix} 0 \\ 0.5 \\ 0 \\ -0.5 \end{bmatrix} u + \begin{bmatrix} 0 \\ 0 \\ 0 \\ -0.75 \end{bmatrix} \xi; \\ z &= \begin{bmatrix} 0 & 0 & 2 & 0 \end{bmatrix} \bar{x}; \\ y - \tilde{y} &= \begin{bmatrix} 1 & 0 & 1 & 0 \end{bmatrix} \bar{x}. \end{aligned} \tag{5.3.40}$$

If the original system (5.3.37) has an initial condition of $x(0) = 0$ and the reference input is a unit step function, this corresponds to an initial condition of $\bar{x}(0) = [-1\ 0\ 0\ 0]'$ for the system (5.3.40). Thus, the minimax control approach

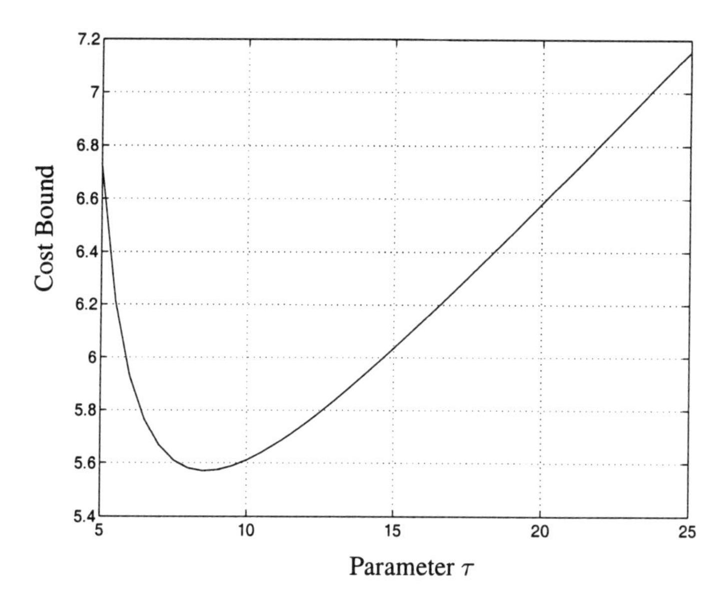

Figure 5.3.1. Cost bound $\bar{x}_0'[X_\tau + \tau S_1]\bar{x}_0$ versus the parameter τ

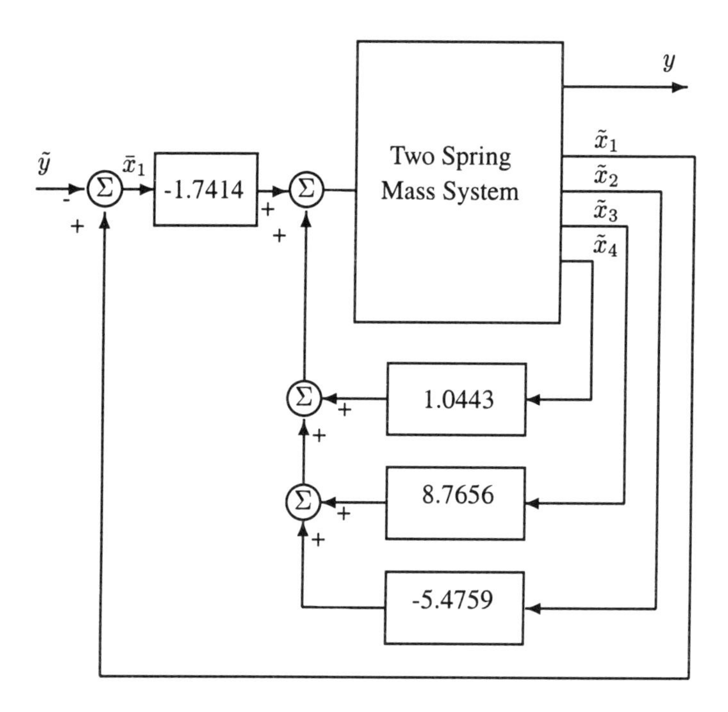

Figure 5.3.2. Control system block diagram

developed above can be applied to this system assuming a known initial condition of $\bar{x}(0) = \bar{x}_0 = [-1\ 0\ 0\ 0]'$.

In order to construct the required controller, we consider the following cost functional of the form (5.3.2):

$$\int_0^\infty [(y - \tilde{y})^2 + 0.1\|\bar{x}\|^2 + u^2]dt. \tag{5.3.41}$$

Hence, the matrices R and G are given by

$$R = \begin{bmatrix} 1.1 & 0 & 1 & 0 \\ 0 & 0.1 & 0 & 0 \\ 1 & 0 & 1.1 & 0 \\ 0 & 0 & 0 & 0.1 \end{bmatrix} > 0; \quad G = 1. \tag{5.3.42}$$

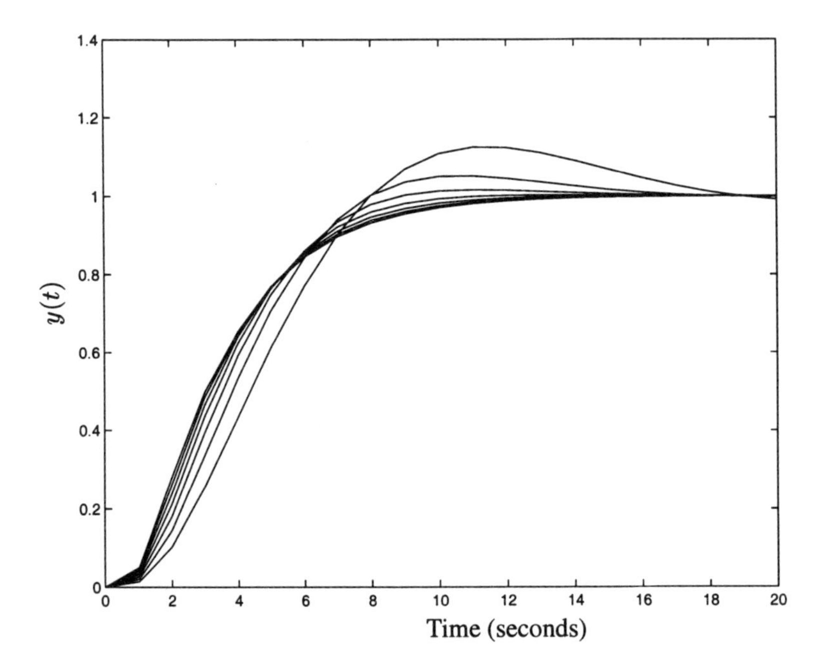

Figure 5.3.3. Control system step response for various spring constants

We now apply the results of the previous subsection to the uncertain system (5.3.40) with cost functional (5.3.41). This involves solving Riccati equation (5.3.11) for a series of values of the parameter τ. This Riccati equation was found to have a positive-definite solution for $\tau \in [3.5, \infty)$. A plot of $\bar{x}_0'[X_\tau + \tau S_1]\bar{x}_0$ versus τ is shown in Figure 5.3.1. From this figure, we can see that the optimal value of the parameter τ is $\tau = 8.59$. With this value of τ, we obtain the following minimal positive-definite solution to Riccati equation (5.3.11):

$$X_\tau = \begin{bmatrix} 4.7112 & 8.9200 & -7.3907 & 5.4358 \\ 8.9200 & 24.5254 & -25.0641 & 13.5709 \\ -7.3907 & -25.0641 & 56.0113 & -7.5326 \\ 5.4358 & 13.5709 & -7.5326 & 15.6574 \end{bmatrix}.$$

Using equation (5.3.14) this leads to the following minimax optimal controller

$$u = \begin{bmatrix} -1.7421 & -5.4772 & 8.7657 & 1.0432 \end{bmatrix} \bar{x}.$$

Referring back to the system (5.3.38), this equation leads to a controller of the form shown in Figure 5.3.2. To verify the robust tracking properties of this control system, Figure 5.3.3 shows the step response of the system for various values of the spring constant parameter k.

5.4 Output-feedback minimax optimal control of uncertain systems with unstructured uncertainty

5.4.1 *Definitions*

In this section, we consider an output-feedback version of the guaranteed cost control problem considered in the Section 5.3. The main result presented in this section gives a necessary and sufficient condition for the existence of a *time-varying* output-feedback guaranteed cost controller for an uncertain system with unstructured uncertainty and an unknown initial condition. The results presented in this section were originally published in the papers [158, 179, 200]. In Subsection 5.4.3, we consider the problem of constructing a guaranteed cost controller which minimizes the guaranteed cost bound. While it appears that the problem of constructing an *optimal* guaranteed cost controller is computationally intractable within the current framework, this section presents a computationally tractable scheme for constructing a sub-optimal guaranteed cost controller. Subsection 5.4.4 presents an example to illustrate the main results. A robust tracking problem is considered and a suitable time-varying guaranteed cost controller is constructed using the sub-optimal scheme of subsection 5.4.3.

Consider an uncertain system described by an equation of the form (2.3.3) in which the uncertainty is unstructured:

$$\begin{aligned} \dot{x}(t) &= Ax(t) + B_1u(t) + B_2\xi(t); \\ z(t) &= C_1x(t) + D_{12}u(t); \\ y(t) &= C_2x(t) + D_{21}\xi(t) \end{aligned} \tag{5.4.1}$$

where $x(t) \in \mathbf{R}^n$ is the *state*, $\xi(t) \in \mathbf{R}^p$ is the *uncertainty input*, $u(t) \in \mathbf{R}^m$ is the *control input*, $z(t) \in \mathbf{R}^q$ is the *uncertainty output* and $y(t) \in \mathbf{R}^l$ is the *measured output*.

The uncertainty in the above system is described by a single equation of the form (2.3.4):

$$\xi(t) = \phi(t, x(\cdot), u(\cdot)) \tag{5.4.2}$$

where the integral quadratic constraint (5.3.4) is satisfied. That is, in Definition 5.3.1, $k = 1$ and $S_1 = S$ where S is a given weighting matrix. The class of all such admissible uncertainty inputs is again denoted Ξ.

We now consider a problem of guaranteed cost control for the uncertain system (5.4.1), (5.3.4) via a finite dimensional *time-varying* linear output-feedback controller of the form

$$\begin{aligned} \dot{x}_c(t) &= A_c(t)x_c(t) + B_c(t)y(t); \quad x_c(0) = 0; \\ u(t) &= C_c(t)x_c(t) + D_c(t)y(t) \end{aligned} \tag{5.4.3}$$

where $A_c(\cdot), B_c(\cdot), C_c(\cdot)$ and $D_c(\cdot)$ are bounded piecewise continuous matrix functions defined on $[0, \infty)$. Note that the dimension of the controller state x_c may be arbitrary.

As in the guaranteed cost control problems considered in the previous sections, associated with the system (5.4.1), (5.3.4), is a cost functional J defined by equation (5.3.2) with given weighting matrices $R = R' > 0$ and $G = G' > 0$.

Definition 5.4.1 *Let $P = P' > 0$ be a given positive-definite matrix associated with the system (5.4.1), (5.3.4). The controller (5.4.3) is said to be a* guaranteed cost controller *for the uncertain system (5.4.1), (5.3.4) with the cost functional (5.3.2) and cost bound matrix P if the following conditions hold:*

(i) The linear closed loop system (5.4.1), (5.4.3) with $\xi(\cdot) \equiv 0$ is exponentially stable.

(ii) For all admissible uncertainty inputs $\xi(\cdot) \in \Xi$, the corresponding solution to the closed loop uncertain system (5.4.1), (5.3.4), (5.4.3) is such that $[x(\cdot), u(\cdot), \xi(\cdot)] \in \mathbf{L}_2[0, \infty)$ (hence in Definition 5.3.1, the constant $t_* = \infty$) and the corresponding value of the cost functional (5.3.2) satisfies the bound

$$\sup_{\xi(\cdot)\in\Xi,\ x(0)\neq 0} \left\{ \frac{J(x(\cdot), u(\cdot), \xi(\cdot))}{x(0)'Px(0)} \right\} < 1. \tag{5.4.4}$$

An uncertain system (5.4.1), (5.3.4) with the cost functional (5.3.2) which admits a guaranteed cost controller (5.4.3) with cost bound matrix P is said to be guaranteed cost stabilizable *with this cost bound matrix.*

Observation 5.4.1 It follows from the above definition that if the uncertain system (5.4.1), (5.3.4) is guaranteed cost stabilizable, then the corresponding closed loop uncertain system (5.4.1), (5.3.4), (5.4.3) will have the property that $x(t) \to 0$ as $t \to \infty$ for any admissible uncertainty $\xi(\cdot)$; see Remark 3.5.3 on page 97.

Also, it follows from inequality (5.4.4) that

$$J(x(\cdot), u(\cdot), \xi(\cdot)) < x(0)'Px(0) \tag{5.4.5}$$

for all admissible uncertainties and for all non-zero $x(0)$.

5.4.2 A necessary and sufficient condition for guaranteed cost stabilizability

We now present the main result of this section which gives a necessary and sufficient condition for guaranteed cost stabilizability of the uncertain system (5.4.1), (5.4.2) with the cost function (5.3.2). This result requires that the coefficients of the system (5.4.1) satisfy a number of technical assumptions; e.g., see Section 3.2.2 and reference [100]. These assumptions correspond to Assumption 3.2.1 of Section 3.2.2. We repeat them here for ease of reference.

Assumption 5.4.1

(i) The matrices C_1 and D_{12} satisfy the condition $C_1'D_{12} = 0$.

(ii) The matrix D_{21} satisfies the condition $D_{21}D_{21}' > 0$.

(iii) The pair (A, B_2) is stabilizable and $B_2 \neq 0$.

(iv) The pair (A, B_1) is stabilizable and the pair (A, C_2) is detectable.

(v) The matrices B_2 and D_{21} satisfy the condition $B_2D_{21}' = 0$.

Let $\tau > 0$ be a given constant. We will consider the following pair of algebraic Riccati equations of the form (3.2.50) and (3.2.55):

$$A'X + XA + X(\hat{B}_2\hat{B}_2' - B_1G_\tau^{-1}B_1')X + C_\tau'C_\tau = 0; \tag{5.4.6}$$

$$AY_\infty + Y_\infty A' + Y_\infty(C_\tau'C_\tau - C_2'\Gamma_\tau^{-1}C_2)Y_\infty + \hat{B}_2\hat{B}_2' = 0; \tag{5.4.7}$$

and the following Riccati differential equation of the form (3.2.52):

$$\dot{Y}(t) = AY(t) + Y(t)A' + Y(t)(C_\tau'C_\tau - C_2'\Gamma_\tau^{-1}C_2)Y(t) + \hat{B}_2\hat{B}_2' \tag{5.4.8}$$

where

$$\begin{aligned} C_\tau &:= \begin{bmatrix} R^{\frac{1}{2}} \\ 0 \\ \sqrt{\tau}C_1 \end{bmatrix}; \quad D_\tau := \begin{bmatrix} 0 \\ G^{\frac{1}{2}} \\ \sqrt{\tau}D_{12} \end{bmatrix}; \\ \hat{D}_{21} &:= \frac{1}{\sqrt{\tau}}D_{21}; \quad \hat{B}_2 := \frac{1}{\sqrt{\tau}}B_2; \\ G_\tau &:= D_\tau'D_\tau; \quad \Gamma_\tau := D_{21}D_{21}'. \end{aligned} \tag{5.4.9}$$

We are now in a position to present the main result of this section.

Theorem 5.4.1 *Consider the uncertain system (5.4.1), (5.4.2) with cost functional (5.3.2) and suppose that Assumption 5.4.1 is satisfied. Let $P = P' > 0$ be a given matrix. Then the uncertain system (5.4.1), (5.4.2) with cost functional (5.3.2) is guaranteed cost stabilizable with the cost bound matrix P if and only if there exists a constant $\tau > 0$ such that the following conditions hold:*

(i) There exists a unique solution X to Riccati equation (5.4.6) such that the matrix $A + (\hat{B}_2\hat{B}_2 - B_1 G_\tau^{-1} B_1')X$ is stable[2] and $P - \tau S > X \geq 0$.

(ii) $Y(\cdot)$, the solution to Riccati differential equation (5.4.8) with initial condition $Y(0) = (P - \tau S)^{-1}$, is positive-definite and bounded for all $t \geq 0$ and such that the linear time-varying system

$$\dot{h}(t) = \left[A + Y(t)(C_\tau' C_\tau - C_2' \Gamma_\tau^{-1} C_2)\right] h(t) \tag{5.4.10}$$

is exponentially stable.

(iii) There exists a constant $\varepsilon > 0$ such that

$$\rho(Y(t)X) < 1 - \varepsilon$$

for all $t \geq 0$. Here $\rho(Y(t)X)$ denotes the spectral radius of the matrix $Y(t)X$.

Moreover, if conditions (i)-(iii) hold then $\lim_{t\to\infty} Y(t) = Y_\infty$ where Y_∞ is the unique nonnegative-definite solution of Riccati equation (5.4.7) such that the matrix $A + Y_\infty(C_\tau' C_\tau - C_2' \Gamma_\tau^{-1} C_2)$ is stable.[3] Also, the controller (5.4.3) with coefficients

$$\begin{aligned} A_c(t) &= A + B_1 C_c - B_c(t) C_2 + \hat{B}_2 \hat{B}_2' X; \\ B_c(t) &= (I - Y(t)X)^{-1} Y(t) C_2' \Gamma_\tau^{-1}; \\ C_c &= -G_\tau^{-1} B_1' X; \\ D_c &= 0; \end{aligned} \tag{5.4.11}$$

is a guaranteed cost controller for the uncertain system (5.4.1), (5.4.2) with cost functional (5.3.2) and cost bound matrix P.

Proof: *Necessity.* Suppose that there exists a controller of the form (5.4.3) such that condition (5.4.4) holds. Inequality (5.4.4) for the cost functional (5.3.2) together with the positive-definiteness of the matrices R and G implies that there exists a constant $\delta > 0$ such that

$$J(x(\cdot), u(\cdot)) + \delta \int_0^\infty (\|x(t)\|^2 + \|u(t)\|^2) dt \leq x(0)' P x(0) \tag{5.4.12}$$

---

[2] That is, X is the stabilizing solution to Riccati equation (5.4.6).

[3] That is, Y_∞ is the stabilizing solution to Riccati equation (5.4.7).

for all solutions to the closed loop system (5.4.1), (5.4.2), (5.4.3). Let $\mathcal{M}$ be the Hilbert space of all vector valued functions

$$\lambda(\cdot) := [x(\cdot), u(\cdot), \xi(\cdot)] \in \mathbf{L}_2[0, \infty)$$

connected by equations (5.4.1) and (5.4.3). We also introduce continuous quadratic functionals $\mathcal{G}_0(\cdot)$ and $\mathcal{G}_1(\cdot)$ defined on $\mathcal{M}$ by the equations

$$\begin{aligned} \mathcal{G}_0(\lambda(\cdot)) &:= J(x(\cdot), u(\cdot)) + \delta \int_0^\infty (\|x(t)\|^2 + \|u(t)\|^2) dt - x(0)' P x(0); \\ \mathcal{G}_1(\lambda(\cdot)) &:= \int_0^\infty (\|z(t)\|^2 - \|\xi(t)\|^2) dt + x(0)' S x(0). \end{aligned} \quad (5.4.13)$$

Since condition (5.4.12) holds for all solutions to the system (5.4.1), (5.4.3) with inputs $\xi(\cdot)$ satisfying condition (5.3.4), then we have $\mathcal{G}_0(\lambda(\cdot)) \leq 0$ for all $\lambda(\cdot) \in \mathcal{M}$ such that $\mathcal{G}_1(\lambda(\cdot)) \geq 0$. Also observe that since $S > 0$, if we choose $\xi(t) \equiv 0$ and any $x(0) \neq 0$, then the corresponding vector function $\lambda(\cdot) \in \mathcal{M}$ is such that $\mathcal{G}_1(\lambda(\cdot)) > 0$. Hence, using the S-procedure Theorem 4.2.1, it follows that there exists a constant $\tau \geq 0$ such that the inequality

$$\mathcal{G}_0(\lambda(\cdot)) + \tau \mathcal{G}_1(\lambda(\cdot)) \leq 0 \quad (5.4.14)$$

is satisfied for all $\lambda(\cdot) \in \mathcal{M}$. Now inequality (5.4.14), with the functionals $\mathcal{G}_0(\cdot)$ and $\mathcal{G}_1(\cdot)$ defined as in (5.4.13), may be re-written as

$$\begin{aligned} & J(x(\cdot), u(\cdot)) + \int_0^\infty [\delta(\|x(t)\|^2 + \|u(t)\|^2)] + \tau(\|z(t)\|^2 - \|\xi(t)\|^2) dt \\ & \leq x(0)' [P - \tau S] x(0). \end{aligned} \quad (5.4.15)$$

This inequality, with $\xi(t) \equiv 0$, implies that $P - \tau S > 0$.

We now prove $\tau > 0$ by contradiction. Indeed, suppose $\tau = 0$. According to condition (iii) of Assumption 5.4.1, $B_2 \neq 0$. Hence, we can choose an input $\xi(\cdot) \in \mathbf{L}_2[0, \infty)$ such that $B_2 \xi(\cdot) \neq 0$. Now consider the solution to the closed loop system (5.4.1), (5.4.3) with this input and with initial condition $x(0) = 0$. Then inequality (5.4.15) with $\tau = 0$ implies that $x(t) \equiv 0$ and $u(t) \equiv 0$. However since $B_2 \xi(\cdot) \neq 0$, we have a contradiction with state equation (5.4.1). Hence, $\tau > 0$.

Now consider the system

$$\begin{aligned} \dot{x}(t) &= A x(t) + B_1 u(t) + \hat{B}_2 w(t); \\ \hat{z}(t) &= C_\tau x(t) + D_\tau u(t); \\ y(t) &= C_2 x(t) + \hat{D}_{21} w(t) \end{aligned} \quad (5.4.16)$$

where $\hat{D}_{21}$, $\hat{B}_2, C_\tau$ and D_τ are defined as in (5.4.9) and $w(\cdot) = \sqrt{\tau}\xi(\cdot)$. Since we have established that $\tau > 0$, this system is well defined.

For this system, let $w(\cdot)$ be the disturbance input and let $\hat{z}(\cdot)$ be the controlled output. Using these definitions, it follows that inequality (5.4.15) may be re-written as

$$\int_0^\infty (\|\hat{z}(t)\|^2 - \|w(t)\|^2)dt + \delta \int_0^\infty (\|x(t)\|^2 + \|u(t)\|^2)dt \leq x(0)'[P - \tau S]x(0). \quad (5.4.17)$$

This implies that the following condition holds:

$$\sup \left\{ \frac{\int_0^\infty \|\hat{z}(t)\|^2 dt}{x(0)'[P - \tau S]x(0) + \int_0^\infty \|w(t)\|^2 dt} \right\} < 1 \quad (5.4.18)$$

where the supremum is taken over all $x(0) \in \mathbf{R}^n$ and $w(\cdot) \in \mathbf{L}_2[0, \infty)$ such that

$$x(0)'[P - \tau S]x(0) + \int_0^\infty \|w(t)\|^2 dt > 0.$$

That is, the controller (5.4.3) solves the H^∞ control problem defined by the system (5.4.16) and H^∞ norm bound condition (5.4.18); see Section 3.2.2. Since $R > 0$ in the cost functional (5.3.2), it follows that the pair (A, C_τ) is detectable. This, together with Assumption 5.4.1, implies that Assumption 3.2.1 required by Theorem 3.2.7 is satisfied. Therefore, Theorem 3.2.7 implies that conditions (i)-(iii) are satisfied. This completes the proof of this part of the theorem.

Sufficiency. Suppose there exists a constant $\tau > 0$ such that conditions (i)-(iii) hold. Then Theorem 3.2.7 implies that $\lim_{t\to\infty} Y(t) = Y_\infty$ where $Y(t)$ is the solution to Riccati equation (5.4.8) with initial condition $Y(0) = (P - \tau S)^{-1}$ and $Y_\infty \geq 0$ is the stabilizing solution to Riccati equation (5.4.7). Also, this theorem implies that the controller (5.4.3) with coefficients (5.4.11) solves the H^∞ control problem (5.4.18) for the system (5.4.16) with coefficients (5.4.9). It follows from condition (5.4.18) and the positive-definiteness of the matrices R and G in the cost functional (5.3.2), that there exists a constant $\delta > 0$ such that inequality (5.4.17) holds for all $[x(\cdot), u(\cdot), w(\cdot)] \in \mathbf{L}_2[0, \infty)$ connected by (5.4.16), (5.4.3) and (5.4.11). This fact implies that inequality (5.4.15) is satisfied for the closed loop system (5.4.1), (5.4.3), (5.4.11) with any uncertainty input $\xi(\cdot) \in \mathbf{L}_2[0, \infty)$. Combining inequality (5.4.15) with condition (5.3.4) as in the proof of Theorem 5.3.1, we conclude that $[x(\cdot), u(\cdot), \xi(\cdot)] \in \mathbf{L}_2[0, \infty)$ for any admissible input $\xi(\cdot)$ and condition (5.4.4) holds. This completes the proof of the theorem.

■

5.4.3 *Optimizing the guaranteed cost bound*

In the previous subsection, we gave a procedure for constructing a controller which leads to a guaranteed cost bound of the form (5.4.5). However, it is desirable to construct a controller which minimizes this cost bound. In this subsection, we present a sub-optimal procedure for minimizing the guaranteed cost bound.

We first note that the bound (5.4.5) depends on the initial condition $x(0)$. To remove this dependence, we will assume $x(0)$ is contained in the unit ball; *i.e.*, $x(0) \in \mathcal{B}_1 := \{x \in \mathbf{R}^n : \|x\| \leq 1\}$. The cost bound (5.4.5) then yields

$$\max_{x(0)\in\mathcal{B}_1} \{J(x(\cdot), u(\cdot), \xi(\cdot))\} < \lambda_{\max}(P) \tag{5.4.19}$$

where $\lambda_{\max}(P)$ denotes the maximum eigenvalue of the matrix P. Thus, we wish to construct a guaranteed cost controller which minimizes $\lambda_{\max}(P)$. One approach to this problem would be to solve an optimization problem to minimize $\lambda_{\max}(P)$ where the unknowns are the matrix P and the constant $\tau > 0$ and the optimization is subject to the constraints defined by conditions (i) - (iii) of Theorem 5.4.1. However, this is a difficult optimization problem and would be computationally intractable. Thus, we present a sub-optimal scheme to minimize $\lambda_{\max}(P)$. This scheme leads to a tractable optimization procedure involving only a one parameter search. The basis of this scheme is the following result concerning the monotonicity of solutions to the Riccati differential equation (5.4.8). This result is closely related to the monotonicity results of [22] and [21].

Theorem 5.4.2 *Consider the Riccati equation (5.4.8) with initial condition $Y(0) = Y_0$. Suppose there exists a corresponding solution to this Riccati equation $Y(t)$ defined on the existence interval $[0, t_*)$. If*

$$AY_0 + Y_0A' + Y_0(C_\tau'C_\tau - C_2'\Gamma_\tau^{-1}C_2)Y_0 + \hat{B}_2\hat{B}_2' < 0, \tag{5.4.20}$$

then $Y(t)$ is monotonically decreasing and $t_* = \infty$.

In order to prove this theorem, we will use the following lemma taken from [21].

Lemma 5.4.1 *Consider the time-varying Lyapunov matrix differential equation*

$$\dot{Z}(t) = Z(t)M(t) + M(t)'Z(t) + W(t); \quad Z(0) = Z_0. \tag{5.4.21}$$

Also, let $\Phi(t, \tau)$ denote the state transition matrix associated with $M(t)$. Then the solution to (5.4.21) is given by

$$Z(t) = \Phi(t, 0)'Z_0\Phi(t, 0) + \int_0^t \Phi(t, \tau)W(\tau)\Phi(t, \tau)d\tau. \tag{5.4.22}$$

Proof of Theorem 5.4.2: Let $Y(t)$ be the solution to Riccati differential equation (5.4.8) with initial condition $Y(0) = Y_0$ and suppose $Y(t)$ is defined on the existence interval $[0, t_*)$. If Y_0 satisfies inequality (5.4.20), it follows from equation (5.4.8) that $\dot{Y}(0) < 0$.

We now observe that the following equation is obtained by differentiating (5.4.8):

$$\begin{aligned}\ddot{Y}(t) &= [A + Y(t)(C_\tau'C_\tau - C_2'\Gamma_\tau^{-1}C_2)]\dot{Y}(t) \\ &\quad + \dot{Y}(t)[A + Y(t)(C_\tau'C_\tau - C_2'\Gamma_\tau^{-1}C_2)]'.\end{aligned}$$

Hence, using Lemma 5.4.1 with $Z(t) = \dot{Y}(t)$ and

$$M(t) = [A + Y(t)(C_\tau' C_\tau - C_2' \Gamma_\tau^{-1} C_2)]',$$

we can write

$$\dot{Y}(t) = \Phi(t,0)' \dot{Y}(0) \Phi(t,0)$$

for all $t \in [0, t_*)$. However, we have established that $\dot{Y}(0) < 0$. Hence, $\dot{Y}(t) < 0$ for all $t \in [0, t_*)$. That is, $Y(t)$ is monotonically decreasing. Furthermore, since $Y(t)$ is monotonically decreasing, the Riccati equation (5.4.8) cannot have a finite escape time. Hence, $t_* = \infty$.

■

One approach to finding a matrix Y_0 such that condition (5.4.20) is satisfied is to choose Y_0 as the solution to the Riccati equation

$$AY_0 + Y_0 A' + Y_0 (C_\tau' C_\tau - C_2' \Gamma_\tau^{-1} C_2) Y_0 + \hat{B}_2 \hat{B}_2' + \sigma I = 0 \qquad (5.4.23)$$

where $\sigma > 0$. This motivates a sub-optimal procedure for constructing a guaranteed cost controller which minimizes the cost bound (5.4.19). This procedure involves a two parameter search over the parameters $\tau > 0$ and $\sigma > 0$. First let $\varepsilon > 0$ be a given sufficiently small constant. Then the parameters $\tau > 0$ and $\sigma > 0$ must satisfy the following conditions:

(a) The Riccati equation (5.4.6) has a nonnegative-definite stabilizing solution X.

(b) The Riccati equation (5.4.7) has a nonnegative-definite stabilizing solution Y_∞.

(c) The Riccati equation (5.4.23) has a positive-definite stabilizing solution Y_0.

(d) The matrix Y_0 satisfies the inequality

$$Y_0 \leq (X + \varepsilon I)^{-1}. \qquad (5.4.24)$$

If these conditions are satisfied, the theorem to follow shows that there exists a corresponding guaranteed cost controller with cost bound matrix $P = Y_0^{-1} + \tau S$.

Theorem 5.4.3 *Let $\varepsilon > 0$, $\tau > 0$ and $\sigma > 0$ be given such that conditions (a)-(d) above are satisfied. Then conditions (i)-(iii) of Theorem 5.4.1 are satisfied with the cost bound matrix $P = Y_0^{-1} + \tau S$.*

Proof: Condition (i) of Theorem 5.4.1 follows directly from condition (a) above. To establish condition (ii) of Theorem 5.4.1, we first note that Theorem 5.4.2 implies that with initial condition $Y(0) = Y_0$, Riccati equation (5.4.8) has a monotonically decreasing solution $Y(t)$ defined on $[0, \infty)$. Furthermore, since $Y(t)$ is monotonically decreasing, it must be bounded for all $t \geq 0$. Also, it follows from the differential game interpretation of Riccati equation (5.4.8) that $Y(t) \geq 0$ for all $t \geq 0$; e.g., see [9]. Hence, since $Y(t)$ is monotonically decreasing, there must exist a limiting solution $\bar{Y} = \lim_{t\to\infty} Y(t) \geq 0$ satisfying Riccati equation (5.4.7).

Proposition 1 $\bar{Y} = Y_\infty \geq 0$.

To establish this proposition, first suppose $\bar{Y}$ is such that the matrix $A + \bar{Y}(C_\tau' C_\tau - C_2'\Gamma_\tau^{-1}C_2)$ has at least one eigenvalue in the open right half of the complex plane. Also let $Z(t) := Y(t) - \bar{Y} > 0$. Then

$$\dot{Z}(t) = \hat{A}(t)Z(t) + Z(t)\hat{A}(t)'$$

where

$$\hat{A}(t) := A + \frac{1}{2}\left(Y(t) + \bar{Y}\right)(C_\tau' C_\tau - C_2'\Gamma_\tau^{-1}C_2).$$

Hence using Lemma 5.4.1, it follows that

$$Z(t) = \hat{\Phi}(t,0)Z(0)\hat{\Phi}(t,0)' \tag{5.4.25}$$

where $\hat{\Phi}(t,0)$ is the state transition matrix associated with $\hat{A}(t)$. Since $Z(0) > 0$ and $Z(t) \to 0$ as $t \to \infty$, equation (5.4.25) implies $\hat{\Phi}(t,0) \to 0$ as $t \to \infty$. Hence, the system

$$\dot{h}(t) = \hat{A}(t)h(t)$$

is asymptotically stable. However, $\lim_{t\to\infty} \hat{A}(t) = A + \bar{Y}(C_\tau' C_\tau - C_2'\Gamma_\tau^{-1}C_2)$ which leads to a contradiction with the assumption that the matrix $A + \bar{Y}(C_\tau' C_\tau - C_2'\Gamma_\tau^{-1}C_2)$ has at least one eigenvalue in the open right half of the complex plane. Thus, we can conclude that the matrix $A + \bar{Y}(C_\tau' C_\tau - C_2'\Gamma_\tau^{-1}C_2)$ has all of its eigenvalues in the closed left half of the complex plane. That is, $\bar{Y}$ is a strong solution to Riccati equation (5.4.7); see Definition 3.1.1.

If we now recall Riccati equation (5.4.23) and use Theorem 2.1 of [144], it follows that the Riccati equation (5.4.7) can have at most one strong solution. However, the stabilizing solution Y_∞ is a strong solution of (5.4.7). Thus, we must have $\bar{Y} = Y_\infty$. This completes the proof of the proposition.

Using the above proposition, we have $Y(t) > Y_\infty \geq 0$ for all $t > 0$. Furthermore, $\lim_{t\to\infty} Y(t) = Y_\infty$ and the system

$$\dot{h}(t) = \left[A + Y_\infty(C_\tau' C_\tau - C_2'\Gamma_\tau^{-1}C_2)\right] h(t)$$

is stable. From this it follows that $Y(t)$ is such that the system (5.4.10) is exponentially stable.

From the above, we have shown that $Y(t)$ is the solution to Riccati equation (5.4.8) required in condition (ii) of Theorem 5.4.1. Furthermore since $P =$

$Y_0^{-1} + \tau S$, it has the required initial condition $Y(0) = Y_0 = (P - \tau S)^{-1}$. Thus, condition (ii) of Theorem 5.4.1 is satisfied. Moreover, since $Y(t)$ is monotonically decreasing, it follows from inequality (5.4.24) that $Y(0) = Y_0 \le (X + \varepsilon I)^{-1}$ and hence $\rho(Y(t)X) \le \rho(Y(0)X) < 1 - \varepsilon$ for all $t \ge 0$. Therefore, condition (iii) of Theorem 5.4.1 is satisfied. This completes the proof of the theorem.

■

Construction of a Sub-optimal Controller

The above theorem motivates the following procedure for the construction of a sub-optimal time-varying guaranteed cost controller for the uncertain system (5.4.1). First choose $\varepsilon > 0$ to be a sufficiently small constant. Then choose the parameters $\tau > 0$ and $\sigma > 0$ to minimize $\lambda_{\max}(Y_0^{-1} + \tau S)$ subject to conditions (a) – (d) above. Once suitable values for $\tau > 0$ and $\sigma > 0$ have been determined, the corresponding time-varying guaranteed cost controller of the form (5.4.3) is constructed by solving the Riccati differential equation (5.4.8) with initial condition Y_0. The required controller coefficients are then defined as in (5.4.11).

Note that the stabilizing solution to Riccati equation (5.4.23) will be monotonically increasing with $\sigma > 0$. Hence for each value of $\tau > 0$, the optimal value of σ will the largest value such that inequality (5.4.24) is satisfied.

5.4.4 *Illustrative example*

In this example, we reconsider the tracking problem considered as an illustrative example in Section 5.3. However, we now wish to design an output-feedback controller so that a specified system output will track a step input.

The system to be controlled is shown in Figure 2.4.2. Recall that in the example in Section 5.3, we assumed that the masses of the carts are $m_1 = 1$ and $m_2 = 1$ and the spring constant k is treated as an uncertain parameter subject to the bound $0.5 \le k \le 2.0$. From this, we obtain a corresponding uncertain system of the form (5.4.1), (5.4.2) described by the state equations

$$\begin{aligned} \dot{x} &= \begin{bmatrix} 0 & 0 & 1 & 0 \\ 0 & 0 & 0 & 1 \\ -1.25 & 1.25 & 0 & 0 \\ 1.25 & -1.25 & 0 & 0 \end{bmatrix} x + \begin{bmatrix} 0 \\ 0 \\ 0 \\ 1 \end{bmatrix} u + \begin{bmatrix} 0 & 0 & 0 \\ 0 & 0 & 0 \\ -0.70 & 0 & 0 \\ 0.80 & 0 & 0 \end{bmatrix} \xi; \\ z &= \begin{bmatrix} 1 & -1 & 0 & 0 \end{bmatrix} x; \\ y &= \begin{bmatrix} 1 & 0 & 0 & 0 \\ 0 & 1 & 0 & 0 \end{bmatrix} x + \begin{bmatrix} 0 & 0.05 & 0 \\ 0 & 0 & 0.05 \end{bmatrix} \xi; \\ y_T &= \begin{bmatrix} 1 & 0 & 0 & 0 \end{bmatrix} x \end{aligned} \tag{5.4.26}$$

where the uncertainty is subject to the integral quadratic constraint:

$$\int_0^{t_i} \|\xi(t)\|^2 dt \le \int_0^{t_i} \|z(t)\|^2 dt + x_0' S x_0.$$

Note that we have modified this example slightly from the example considered in Section 5.3. This modification involves adding two extra components to the uncertainty input vector ξ and adding a corresponding small but non-zero matrix D_{21}. These extra uncertainty inputs correspond to sensor noise and uncertainty. This modification was carried out in order to satisfy condition (ii) of Assumption 5.4.1. The output y_T is the output which will be required to track a step input. The matrix S is the same matrix which was used in Subsection 5.3.3.

In the system (5.4.26), the state variables are $x = [x_1\ x_2\ x_3\ x_4]'$ where $x_3 = \dot{x}_1$ and $x_4 = \dot{x}_2$. The control problem to be solved involves finding a controller which absolutely stabilizes the system and also ensures that the output y_T tracks a reference step input as in Subsection 5.3.3. In order to apply the results presented in this section, we again apply the Internal Model Principle.

As in Subsection 5.3.3, we first introduce a new set of state variables for the system (5.4.26): $\tilde{x} = [\tilde{x}_1\ \tilde{x}_2\ \tilde{x}_3\ \tilde{x}_4]'$ where $\tilde{x}_1 = (x_1 + x_2)/2$, $\tilde{x}_2 = \dot{\tilde{x}}_1$, $\tilde{x}_3 = (x_1 - x_2)/2$ and $\tilde{x}_4 = \dot{\tilde{x}}_3$. With this change of state variables, the system (5.4.26) becomes

$$\begin{aligned}
\dot{\tilde{x}} &= \begin{bmatrix} 0 & 1 & 0 & 0 \\ 0 & 0 & 0 & 0 \\ 0 & 0 & 0 & 1 \\ 0 & 0 & -2.5 & 0 \end{bmatrix} \tilde{x} + \begin{bmatrix} 0 \\ 0.5 \\ 0 \\ -0.5 \end{bmatrix} u + \begin{bmatrix} 0 & 0 & 0 \\ 0.05 & 0 & 0 \\ 0 & 0 & 0 \\ -0.75 & 0 & 0 \end{bmatrix} \xi; \\
z &= \begin{bmatrix} 0 & 0 & 2 & 0 \end{bmatrix} \tilde{x}; \\
y &= \begin{bmatrix} 1 & 0 & 1 & 0 \\ 1 & 0 & -1 & 0 \end{bmatrix} \tilde{x} + \begin{bmatrix} 0 & 0.05 & 0 \\ 0 & 0 & 0.05 \end{bmatrix} \xi; \\
y_T &= \begin{bmatrix} 1 & 0 & 1 & 0 \end{bmatrix} \tilde{x}.
\end{aligned} \tag{5.4.27}$$

Also, the reference input signal $\tilde{y}_T$ is described by the state space model (5.3.39). Moreover, as in Subsection 5.3.3, we define a new state vector $\bar{x} = [\tilde{x}_1 - \eta\ \tilde{x}_2\ \tilde{x}_3\ \tilde{x}_4]'$ and combine state equations (5.4.27) and (5.3.39). This leads to the state equations

$$\begin{aligned}
\dot{\bar{x}} &= \begin{bmatrix} 0 & 1 & 0 & 0 \\ 0 & 0 & 0 & 0 \\ 0 & 0 & 0 & 1 \\ 0 & 0 & -2.5 & 0 \end{bmatrix} \bar{x} + \begin{bmatrix} 0 \\ 0.5 \\ 0 \\ -0.5 \end{bmatrix} u + \begin{bmatrix} 0 & 0 & 0 \\ 0.05 & 0 & 0 \\ 0 & 0 & 0 \\ -0.75 & 0 & 0 \end{bmatrix} \xi; \\
z &= \begin{bmatrix} 0 & 0 & 2 & 0 \end{bmatrix} \bar{x}; \\
\bar{y} &= \begin{bmatrix} 1 & 0 & 1 & 0 \\ 1 & 0 & -1 & 0 \end{bmatrix} \bar{x} + \begin{bmatrix} 0 & 0.05 & 0 \\ 0 & 0 & 0.05 \end{bmatrix} \xi; \\
y_T - \tilde{y}_T &= \begin{bmatrix} 1 & 0 & 1 & 0 \end{bmatrix} \bar{x}
\end{aligned} \tag{5.4.28}$$

where

$$\bar{y} := y - \begin{bmatrix} 1 \\ 1 \end{bmatrix} \eta.$$

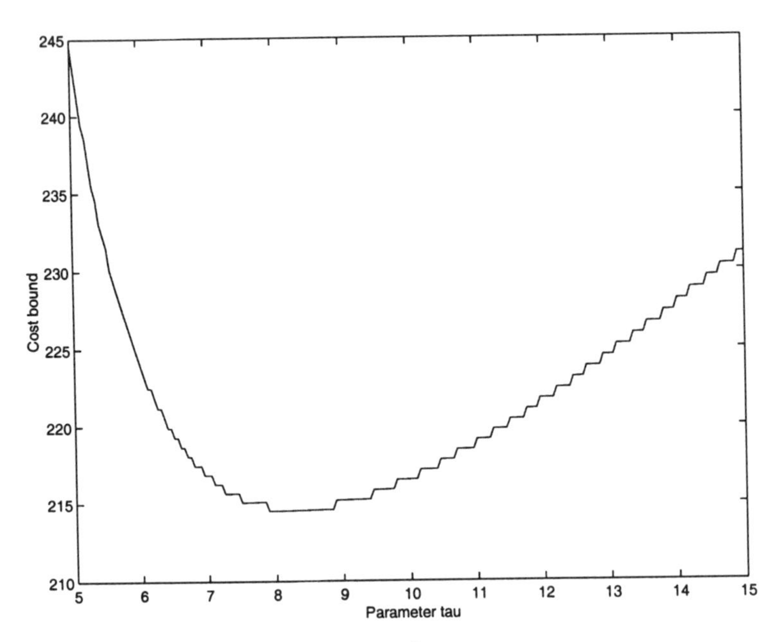

Figure 5.4.1. Cost bound $\lambda_{\max}[Y_0^{-1} + \tau S]$ versus the parameter τ

Also, as in Subsection 5.3.3, we consider the following cost functional:

$$\int_0^\infty [(y_T - \tilde{y}_T)^2 + 0.1\|\bar{x}\|^2 + u^2]dt. \tag{5.4.29}$$

Hence, the matrices R and G are given as in equation (5.3.42) on page 154.

We now apply the sub-optimal time-varying control scheme of the previous subsection to the uncertain system (5.4.28) with cost function (5.4.29). We first set the constant ε to $\varepsilon = 0.001$. Then, for each value of $\tau > 0$, the Riccati equation (5.4.6) is solved and then a Fibonacci search is carried out to find the maximum value of $\sigma > 0$ such that inequality (5.4.24) is satisfied. A plot of the resulting value of $\lambda_{\max}[Y_0^{-1} + \tau S]$ versus τ is shown in Figure 5.4.1. From this figure, we can see that the optimal value of the parameter τ is $\tau = 7.5$. The corresponding optimal value of σ is $\sigma = 0.0186$. With these values of τ and σ we obtain a cost bound of $\lambda_{\max}[Y_0^{-1} + \tau S] = 214.45$.

With the optimal values of τ and σ obtained above, the following positive-definite stabilizing solutions to Riccati equations (5.4.6) and (5.4.23) were obtained:

$$X = \begin{bmatrix} 4.3594 & 7.7806 & -6.3124 & 4.6821 \\ 7.7806 & 20.5282 & -20.9729 & 11.0126 \\ -6.3124 & -20.9729 & 49.7520 & -5.4993 \\ 4.6821 & 11.0126 & -5.4993 & 13.8674 \end{bmatrix};$$

$$Y_0 = \begin{bmatrix} 0.0059 & 0.0049 & -0.0001 & -0.0004 \\ 0.0049 & 0.0231 & -0.0004 & -0.0024 \\ -0.0001 & -0.0004 & 0.0068 & 0.0083 \\ -0.0004 & -0.0024 & 0.0083 & 0.0598 \end{bmatrix}.$$

Furthermore, with initial condition $Y(0) = Y_0$, Riccati differential equation (5.4.8) was then solved and a corresponding time-varying controller of the form (5.4.3) was constructed with coefficients defined as in (5.4.11). Then, referring back to system (5.4.27), the required tracking control system is constructed as shown in Figure 5.4.2. To verify the robust tracking properties of this control system, Figure 5.4.3 shows the step response of the system for various values of the spring constant parameter k.

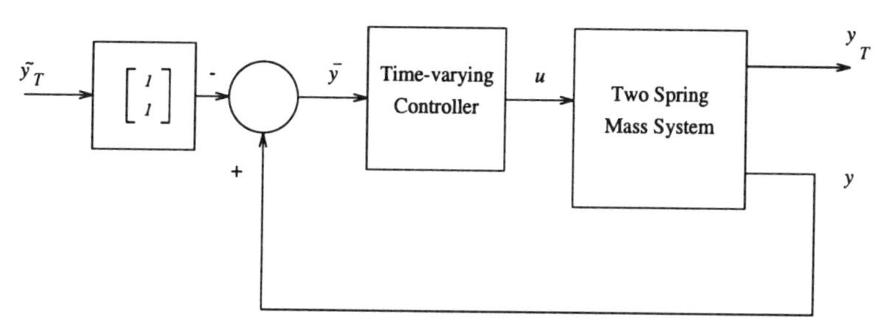

Figure 5.4.2. Control system block diagram

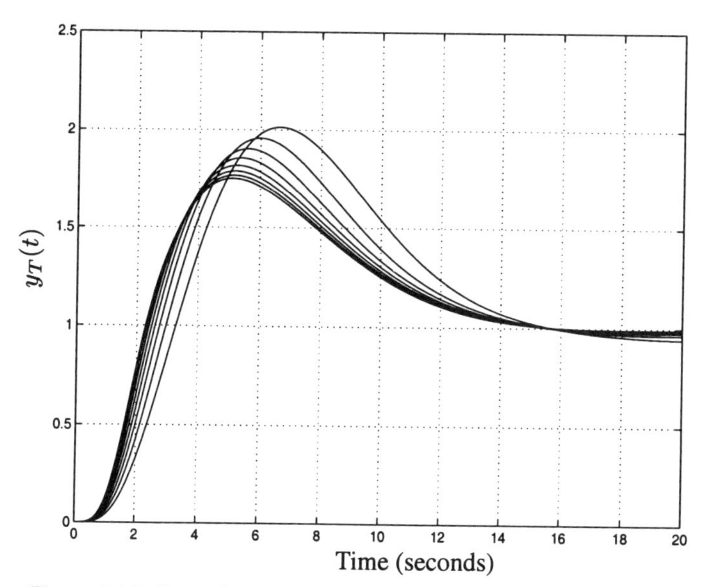

Figure 5.4.3. Control system step response for various spring constants

5.5 Guaranteed cost control via a Lyapunov function of the Lur'e-Postnikov form

5.5.1 *Problem formulation*

In this section, we again consider the problem of guaranteed cost stabilization. In previous sections, a fixed quadratic Lyapunov function was used to give an upper bound on a quadratic cost functional. In this section, we present a result which shows that if one goes beyond the class of fixed quadratic Lyapunov functions, a controller which guarantees an improved level of performance can be obtained. This guaranteed cost controller can be constructed by solving a modified Riccati equation which depends on parameters in addition to those considered in the parameter-dependent Riccati equation of Section 5.3. The results presented in this section were originally published in the paper [232].

Consider the following uncertain system of the form (2.3.3):

$$\begin{aligned} \dot{x}(t) &= Ax(t) + B_1 u(t) + B_2 \xi(t); \\ z(t) &= Cx(t) \end{aligned} \tag{5.5.1}$$

where $x(t) \in \mathbf{R}^n$ is the state, $u(t) \in \mathbf{R}^{m_1}$ is the control input, $z(t) \in \mathbf{R}^p$ is a vector assembling all uncertainty outputs and $\xi(t) \in \mathbf{R}^{m_2}$ is a vector assembling all uncertainty inputs.

We suppose that the uncertainty input satisfies the following "structured norm bound" condition. There exist vectors $r_i \in \mathbf{R}^p$ and $g_i \in \mathbf{R}^{m_2}$, $i = 1, \dots, k$, such that:

$$|g_i'\xi(t)| \leq |r_i'z(t)| \qquad \forall t \geq 0, \tag{5.5.2}$$

for each $i = 1, \dots, k$. An uncertainty satisfying this constraint is said to be *admissible*. The set of admissible uncertainties is denoted by Ξ_{NB}.

Consider the following additional constraint on the uncertainty: There exists an $i \in \{1, \dots, k\}$ and a locally integrable function $\zeta_i(\sigma)$, $|\zeta_i(\sigma)| \leq |\sigma|$, such that

$$\begin{aligned} \zeta_i(\sigma) &\leq \inf_{\{t:\ r_i'z(t)=\sigma,\ r_i'\dot{z}(t)\geq 0\}} g_i'\xi(t), \quad \text{and} \\ \zeta_i(\sigma) &\geq \sup_{\{t:\ r_i'z(t)=\sigma,\ r_i'\dot{z}(t)\leq 0\}} g_i'\xi(t). \end{aligned} \tag{5.5.3}$$

The set of admissible uncertainties satisfying the additional constraint (5.5.3) is denoted by Ξ^0. It is obvious that $\Xi^0 \subset \Xi_{NB}$.

The set of i's such that a function ζ_i in (5.5.3) exists is denoted by $\Im$. This set will play a special role in the sequel. We will not necessarily assume that the set $\Im$ is nonempty and (5.5.3) holds. However, if condition (5.5.3) is satisfied, then the corresponding function $\zeta_i(\cdot)$ will generate a Lyapunov function of the Lur'e-Postnikov form. Alternatively, if (5.5.3) does not hold and therefore, the set $\Im$ is empty, a fixed quadratic Lyapunov function may still exist.

Note that in the particular case where $k = p = m_2$ and

$$r_i = [\underbrace{0 \ \ldots \ 0}_{i-1} \ 1 \ 0 \ \ldots \ 0]';$$

$$g_i = [\underbrace{0 \ \ldots \ 0}_{i-1} \ 1 \ 0 \ \ldots \ 0]';$$

condition (5.5.2) amounts to the standard structured uncertainty bound. Indeed, if $z = [z_1 \ \ldots \ z_k]'$ and $\xi = [\xi_1 \ \ldots \ \xi_k]'$, then with the above choice of r_i and g_i, it follows from (5.5.2) that

$$|\xi_i(t)| \leq |z_i(t)|$$

for all $i = 1, \ldots, k$. Condition (5.5.2) however allows one to consider more general bounds on the uncertainty such as for example, the condition $|\sum_{i=1}^k \alpha_i \xi_i(t)| \leq |\sum_{i=1}^k \beta_i z_i(t)|$. Obviously, the uncertainty input $\xi(\cdot)$ may satisfy this condition even if the constraints $|\xi_i(t)| \leq |z_i(t)|$ are not satisfied.

It is easy to see that the admissible uncertainty inputs contained in Ξ_{NB} satisfy the integral quadratic constraints introduced in Section 2.3. Indeed, given any locally square integrable control input $u(\cdot)$ and any corresponding solution to equation (5.5.1) with an interval of existence $[0, t^*)$, then it follows from (5.5.2) that for any positive constants $d_1 > 0, \ldots, d_k > 0$ and a sequence $\{t_j\}_{j=1}^{\infty} \subset [0, t^*)$ such that $t_j \to t^*$,

$$\int_0^{t_j} \left(|r_i' z(t)|^2 - |g_i' \xi(t)|^2\right) dt \geq -d_i \tag{5.5.4}$$

for all j and for $i = 1, \ldots, k$. We denote this fact by writing $\Xi_{NB} \subset \Xi_{\mathrm{IQC}}$.

The constraint (5.5.3) allows us to take into account additional properties of the uncertainty. Such properties may relate to the way in which the uncertainty block reacts to an increase or decrease in the uncertainty output (which is the input to the uncertainty block). For example, an increasing uncertainty output may lead to a larger uncertainty input than that which occurs when the uncertainty output decreases. This situation is illustrated in Figure 5.5.1. A practical example in which such an uncertainty may arise is the presence of an uncertain backlash in a series of gears. It is straightforward to verify that the uncertainty in Figure 5.5.1 satisfies condition (5.5.3).

Consider the following quadratic cost functional of the form (5.2.2):

$$J(u(\cdot)) = \int_0^{\infty} (x'Rx + u'Gu)dt. \tag{5.5.5}$$

In the functional (5.5.5), $R = R' > 0$ and $G = G' > 0$ are given weighting matrices of corresponding dimensions. The following constrained optimization problem associated with the system (5.5.1), the constraints (5.5.2), (5.5.3) and the cost functional (5.5.5) is the focus of this section. We seek to minimize the worst case value of the cost functional (5.5.5)

$$\sup_{\xi(\cdot) \in \Xi^0} J(u(\cdot))$$

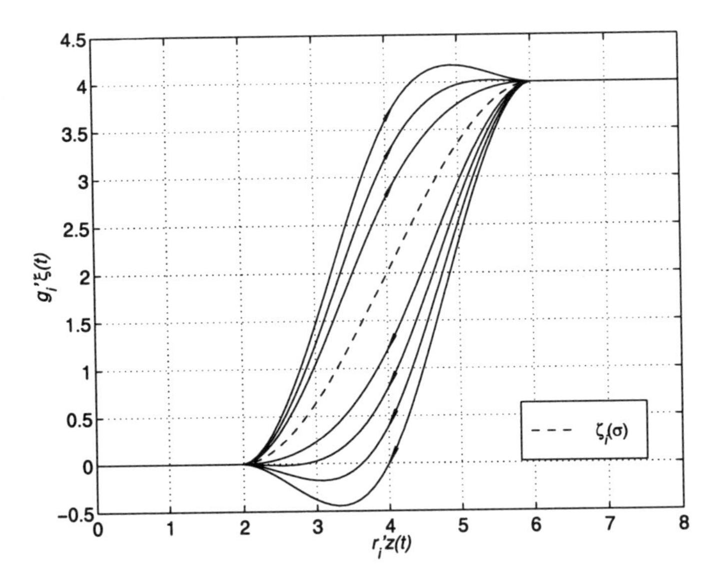

Figure 5.5.1. Plot of $g_i'\xi(t)$ versus $r_i'z(t)$ corresponding to an uncertainty input $\xi(\cdot)$ which satisfies condition (5.5.3). The arrows indicate the direction corresponding to increasing t.

via a state-feedback controller $u(\cdot) \in \mathcal{M}$. The class $\mathcal{M}$ of controllers under consideration are causal state-feedback controllers of the form

$$u(t) = \mathcal{K}(t, x(\cdot)|_0^t).$$

Here, $\mathcal{K}(\cdot)$ is a function such that the state equation (5.5.1) admits a unique solution; e.g., see [9]. Let V^0 denote the upper value of this minimax optimization problem:

$$V^0 := \inf_{u(\cdot)\in\mathcal{M}} \sup_{\xi(\cdot)\in\Xi^0} J(u(\cdot)). \tag{5.5.6}$$

Also, we consider the upper values of corresponding minimax optimization problems with norm-bounded and integral quadratic constraint uncertainty descriptions. We define

$$V_{NB} := \inf_{u(\cdot)} \sup_{\xi(\cdot)\in\Xi_{NB}} J(u(\cdot));$$

$$V_{\mathrm{IQC}} := \inf_{d_1>0,\dots,d_k>0} \left[\inf_{u(\cdot)} \sup_{\xi(\cdot)\in\Xi_{\mathrm{IQC}}} J(u(\cdot))\right].$$

It is easy to see that the condition

$$\Xi^0 \subset \Xi_{NB} \subset \Xi_{\mathrm{IQC}}$$

implies:

$$V^0 \le V_{NB};$$
$$V^0 \le V_{\mathrm{IQC}}.$$

The minimax problem $\inf_{u(\cdot)} \sup_{\xi(\cdot)\in\Xi_{IQC}} J(u(\cdot))$ was considered in Section 5.3. In that section, it was shown that for the class of uncertainties satisfying (5.5.4)

$$\inf_{u(\cdot)} \sup_{\xi(\cdot)\in\Xi_{IQC}} J(u(\cdot)) = \inf_{\tau\in\mathcal{T}_0} [x_0' X_\tau x_0 + \sum_{j=1}^{k} \tau_j d_j]$$

where X_τ satisfies (5.3.11) and $\mathcal{T}_0$ is defined in (5.3.13). In this section, the constants $d_1 > 0, \ldots, d_k > 0$ were chosen arbitrarily and hence, we consider the infimum over $d_1 > 0, \ldots, d_k > 0$ in the definition of V_{IQC}.

Now consider the following Riccati equation

$$A'X + XA + R + \sum_{i=1}^{k} \tau_i C' r_i r_i' C - X(B_1 G^{-1} B_1' - B_2 H^{-1} B_2')X = 0 \tag{5.5.7}$$

where $\tau_1 > 0, \ldots, \tau_k > 0$ are given constants, and

$$H = \sum_{i=1}^{k} \tau_i g_i g_i'.$$

Let $\mathcal{T}_0$ be the set of vectors $\tau \in \mathbf{R}^k$, $\tau_1 > 0, \ldots, \tau_k > 0$ such that the Riccati equation (5.5.7) has a solution $X_\tau > 0$. Adapting Theorem 5.3.2 to the current set up and notation, it follows from the inequality $V^0 \leq V_{IQC}$ that for a given initial condition $x(0)$,

$$V^0 \leq \inf_{\tau\in\mathcal{T}_0} x'(0) X_\tau x(0). \tag{5.5.8}$$

We now consider the quadratic stability results in Section 5.1 of [26] applied to the closed loop uncertain system with norm-bounded uncertainty defined by (5.5.1), (5.5.2) and the control law $u = -G^{-1}B_1' X_\tau x$. From this, it follows that

$$V^0 \leq V_{NB} \leq \inf_{\tau\in\mathcal{T}_0} x'(0) X_\tau x(0).$$

These observations imply that both the approach based on overbounding the uncertainty by an integral quadratic constraint and the quadratic stability approach lead to the same upper bound (5.5.8) on the value V^0.

In light of inequality (5.5.8), one might expect that there exists a controller stabilizing the uncertain system (5.5.1), (5.5.2), (5.5.3), and guaranteeing an improved performance compared with the performance obtained by directly converting the norm-bounded uncertainty into an integral quadratic constraint uncertainty or by ignoring condition (5.5.3). In the next subsection, we consider the construction of such a controller.

5.5.2 Controller synthesis via a Lyapunov function of the Lur'e-Postnikov form

Letting $\bar{R} = [r_1 \ \dots \ r_k]'$ and $\bar{G} = [g_1 \ \dots \ g_k]'$, consider the following set of real $k \times k$ diagonal matrices

$$\mathcal{T}_1 = \left\{ \Theta: \begin{array}{c} \Theta = \mathrm{diag}[\theta_1, \dots, \theta_k], \ \theta_i = 0 \text{ if } i \notin \Im, \ \theta_i \geq 0 \text{ if } i \in \Im, \\ B_1'C'\bar{R}'\Theta\bar{G} = 0, \ B_2'C'\bar{R}'\Theta\bar{G} + \bar{G}'\Theta\bar{R}CB_2 < H \end{array} \right\}. \quad (5.5.9)$$

Also, consider the modified Riccati equation

$$\begin{aligned} & A'(I + B_2E_\theta^{-1}\bar{G}'\Theta\bar{R}C)'X + X(I + B_2E_\theta^{-1}\bar{G}'\Theta\bar{R}C)A \\ & + R + \sum_{i=1}^{k} \tau_i C' r_i r_i' C + A'C'\bar{R}'\Theta\bar{G}E_\theta^{-1}\bar{G}'\Theta\bar{R}CA \\ & - X(B_1G^{-1}B_1' - B_2E_\theta^{-1}B_2')X = 0. \end{aligned} \quad (5.5.10)$$

where $E_\theta := H - B_2'C'\bar{R}'\Theta\bar{G} - \bar{G}'\Theta\bar{R}CB_2 > 0$.

Theorem 5.5.1 *If for some diagonal matrix $\Theta \in \mathcal{T}_1$ and constants $\tau_1 \geq 0$, $\dots$, $\tau_k \geq 0$, there exists a positive-definite solution $X_{\tau,\theta}$ to the Riccati equation (5.5.10) such that the matrix*

$$(I + B_2E_\theta^{-1}\bar{G}'\Theta\bar{R}C)A - B_1G^{-1}B_1'X_{\tau,\theta} + B_2E_\theta^{-1}B_2'X_{\tau,\theta}$$

stable matrix and $X_{\tau,\theta} > C'\bar{R}'\Theta\bar{R}C$, then the state-feedback controller

$$u(t) = K_{\tau,\theta}x(t); \qquad K_{\tau,\theta} = -G^{-1}B_1'X_{\tau,\theta} \quad (5.5.11)$$

is an absolutely stabilizing controller for the uncertain system (5.5.1), (5.5.2), (5.5.3). Furthermore, this controller guarantees the following upper bound on the value of the constrained minimax optimization problem (5.5.6):

$$\begin{aligned} V^0 & \leq \sup_{\xi \in \Xi^0} J(K_{\tau,\theta}x(\cdot)) \\ & \leq x'(0)(X_{\tau,\theta} + C'\bar{R}'\Theta\bar{R}C)x(0). \end{aligned} \quad (5.5.12)$$

Remark 5.5.1 A similar upper bound on the quadratic cost was obtained in [26] for an analysis problem involving Lur'e systems; see Section 8.3.1 of [26]. In this case, $B_1 = 0$ and hence $B_1'C'\bar{R}'\Theta\bar{G} = 0$. Also, the vectors r_i, g_i were defined as follows:

$$r_i = [\underbrace{0 \ \dots \ 0}_{i-1} \ 1 \ 0 \ \dots \ 0]';$$

$$g_i = [\underbrace{0 \ \dots \ 0}_{i-1} \ 1 \ 0 \ \dots \ 0]'.$$

In this case, the upper bound (5.5.12) can be obtained from a corresponding linear matrix inequality (LMI). However in the synthesis problem where $B_1 \neq 0$ and $u = Kx$, the method of [26] leads to the nonlinear matrix inequality

$$\begin{bmatrix} \begin{pmatrix} (A+B_1K)'X + X(A+B_1K) \\ +R+K'GK+C'\mathbf{T}C \end{pmatrix} & XB_2 + A'C'\Theta \\ B_2'X + \Theta CA & -E_\theta \end{bmatrix} \leq 0, \tag{5.5.13}$$

where $\mathbf{T} := \mathrm{diag}[\tau_1, \ldots \tau_k]$. This inequality relates the unknown Lyapunov function matrix X, controller gain matrix K and matrices of scaling parameters T and Θ.

Note that the standard change of variables $Q = X^{-1}, Y = KQ$ does not allow one to remove the nonlinearity in (5.5.13). That is, the method of matrix inequalities does not "linearize" the problem of searching for a suitable solution to the Riccati equation. Thus, for the guaranteed cost control problem under consideration, the method of [26] does not seem to have any advantages over the approach based on Theorem 5.5.1. Moreover, it is worth noting that condition (5.5.3) defines a larger class of uncertainties as compared to the uncertainty class considered in [26]. Therefore, Theorem 5.5.1 can be regarded as being less conservative than the results of [26].

Proof of Theorem 5.5.1: We first introduce the notation:

$$\begin{aligned} &\tilde{A} = (I + B_2E_\theta^{-1}\bar{G}'\Theta\bar{R}C)A; \qquad \tilde{B}_1 = B_1G^{-1/2}; \qquad \tilde{B}_2 = B_2E_\theta^{-1/2}; \\ &\tilde{C} = \left(R + \sum_{i=1}^{k} \tau_i C'r_ir_i'C + A'C'\bar{R}'\Theta\bar{G}E_\theta^{-1}\bar{G}'\Theta\bar{R}CA\right)^{1/2}; \\ &\tilde{K} = -\tilde{B}_1'X_{\tau,\theta}; \qquad \tilde{A}_K = \tilde{A} + \tilde{B}_1\tilde{K}. \end{aligned} \tag{5.5.14}$$

With this notation, the condition that $X_{\tau,\theta}$ satisfies (5.5.10) becomes

$$\tilde{A}_K'X_{\tau,\theta} + X_{\tau,\theta}\tilde{A}_K + \tilde{C}'\tilde{C} + \tilde{K}'\tilde{K} + X_{\tau,\theta}\tilde{B}_2\tilde{B}_2'X_{\tau,\theta} = 0. \tag{5.5.15}$$

Thus, it follows from the assumption in the statement of the theorem that the algebraic Riccati equation (5.5.15) has a positive-definite solution $X_{\tau,\theta}$ such that the matrix $\tilde{A}_K + \tilde{B}_2\tilde{B}_2'X_{\tau,\theta}$ is stable. That is, $X_{\tau,\theta}$ is a stabilizing solution to (5.5.15).

We now consider the unconstrained differential game

$$\inf_{u\in\mathcal{M}} \sup_{\xi\in\mathbf{L}_2} J_{\tau,\theta}(u(\cdot), \xi(\cdot)) \tag{5.5.16}$$

for the system (5.5.1) with cost functional

$$\begin{aligned} &J_{\tau,\theta}(u(\cdot), \xi(\cdot)) \\ &= \int_0^\infty \left[\begin{array}{l} x'(t)Rx(t) + u'(t)Gu(t) \ + \sum_{i=1}^k \tau_i(|r_i'z(t)|^2 - |g_i'\xi(t)|^2) \\ \qquad\qquad\qquad\qquad\qquad + 2\xi(t)'\bar{G}'\Theta\bar{R}\dot{z}(t) \end{array} \right] dt \end{aligned} \tag{5.5.17}$$

where the control input $u(\cdot)$ of the system (5.5.1) corresponds to the minimizing player, and the uncertainty input $\xi(\cdot)$ of the system (5.5.1) corresponds to the maximizing player. In this differential game, we seek to find a state-feedback controller $u(\cdot) \in \mathcal{M}$ minimizing the worst case of the cost functional (5.5.17).

Proposition 1 *The controller solving the differential game (5.5.1), (5.5.16) is defined by equation (5.5.11). Furthermore, the upper value of this differential game is equal to $x'(0)X_{\tau,\theta}x(0)$.*

To establish this proposition, we note that substituting $\tilde{u} = G^{1/2}u$, $w = E_\theta^{1/2}(\xi - E_\theta^{-1}\bar{G}'\Theta\bar{R}CAx)$ into (5.5.1) and using the notation (5.5.14) yields

$$\begin{aligned} \dot{x} &= \tilde{A}x + \tilde{B}_1\tilde{u} + \tilde{B}_2 w; \\ \tilde{z} &= \tilde{C}x. \end{aligned} \tag{5.5.18}$$

Also the functional (5.5.17) becomes a functional of the form

$$\tilde{J}(v(\cdot), w(\cdot)) = \int_0^\infty \left[\|\tilde{z}(t)\|^2 + \|\tilde{u}(t)\|^2 - \|w(t)\|^2\right] dt.$$

Thus, with the above substitutions, the differential game (5.5.1), (5.5.16) is equivalent to the H^∞ control problem of finding a controller satisfying the condition

$$\sup_{w(\cdot)\in\mathbf{L}_2} \frac{\int_0^\infty \left[\|\tilde{z}(t)\|^2 + \|\tilde{u}(t)\|^2\right] dt}{\int_0^\infty \|w(t)\|^2 dt} < 1.$$

It follows from the Strict Bounded Real Lemma (see Lemma 3.1.2 on page 58) that the existence of a stabilizing positive-definite solution to the Riccati equation (5.5.15) implies that the matrix $\tilde{A}_K$ is a stable matrix and the controller solving this H^∞ control problem has the form

$$\tilde{u} = \tilde{K}x, \tag{5.5.19}$$

where $\tilde{K}$ is defined in (5.5.14). Also, it follows from H^∞ control theory that the optimal value of the corresponding differential game is given by $x'(0)X_{\tau,\theta}x(0)$; e.g., see Section 3.2 and [9]. Note that since $R > 0$, then the pair $(\tilde{A}, \tilde{C})$ is observable, as required in order to apply the above result from H^∞ control theory. The substitution $u = G^{-1/2}\tilde{u}$ completes the proof of the proposition.

Proposition 2 *The controller (5.5.11) is an absolutely stabilizing controller for the uncertain system (5.5.1), (5.5.2) and (5.5.3).*

To establish this proposition, we first observe that it follows from the condition $\Theta \in \mathcal{T}_1$ and from Riccati equation (5.5.10) that for any $x \in \mathbf{R}^n$, $u \in \mathbf{R}^{m_1}$, and

$\xi \in \mathbf{R}^{m_2}$,

$$\begin{aligned}
&2(Ax + B_1 u + B_2\xi)'\tilde{X}x + C'\bar{R}'\Theta(\bar{R}Cx + \bar{G}\xi) \\
&+ \sum_{i=1}^{k} \tau_i(|r_i'Cx|^2 - |g_i'\xi|^2) + x'Rx + u'Gu \\
&= \|G^{1/2}(K_{\tau,\theta}x - u)\|^2 - \|Fx + M\xi\|^2,
\end{aligned} \tag{5.5.20}$$

where

$$\begin{aligned}
\tilde{X} &= X_{\tau,\theta} - C'\bar{R}'\Theta\bar{R}C; \\
F &= -E_\theta^{-1/2}(B_2'X_{\tau,\theta} + \bar{G}'\Theta\bar{R}CA); \\
M &= E_\theta^{1/2}.
\end{aligned}$$

Now consider the collection of functions $\zeta_i(\sigma)$, $i \in \Im$, satisfying condition (5.5.3). Also, define the following Lyapunov function of the Lur'e-Postnikov form

$$W(x) = x'\tilde{X}x + 2\sum_{i\in\Im}\theta_i \int_0^{r_i'z} (\sigma + \zeta_i(\sigma))d\sigma. \tag{5.5.21}$$

It follows from the assumption in the statement of the theorem that $\tilde{X} \geq 0$ and each function $\sigma + \zeta_i(\sigma)$ satisfies the sector bound condition

$$0 \leq (\sigma + \zeta_i(\sigma))\sigma \leq 2\sigma^2. \tag{5.5.22}$$

Hence,

$$0 \leq W(x) \leq x'(X_{\tau,\theta} + C'\bar{R}'\Theta\bar{R}C)x. \tag{5.5.23}$$

The function (5.5.21) will be used as a candidate Lyapunov function for the stabilization problem under consideration.

Let $x(t)$ be a solution to the closed loop system (5.5.1), (5.5.11) with initial condition $x(0) = x_0$. Note that for any $T > 0$, it follows from (5.5.3) that

$$\begin{aligned}
&\int_0^T (r_i'z(t) + g_i'\xi(t))r_i'\dot{z}(t)dt \\
&\geq \int_0^{r_i'z(T)} (\sigma + \zeta_i(\sigma))d\sigma - \int_0^{r_i'z(0)} (\sigma + \zeta_i(\sigma))d\sigma
\end{aligned} \tag{5.5.24}$$

for any $i \in \Im$. Hence, it follows from (5.5.20) that for any $T > 0$

$$\begin{aligned}
&W(x(t))\Big|_0^T + \int_0^T x'(t)(R + K_{\tau,\theta}'GK_{\tau,\theta})x(t)dt \\
&+ \int_0^T \sum_{i=1}^{k} \tau_i \left(|r_i'z(t)|^2 - |g_i'\xi(t)|^2\right) dt \\
&\leq -\int_0^T \|Fx(t) + M\xi(t)\|^2 dt \leq 0.
\end{aligned} \tag{5.5.25}$$

Since the uncertainty input $\xi(\cdot)$ satisfies the constraint (5.5.2), $W(x) \geq 0$, $R > 0$ and $G > 0$, then the inequality

$$\int_0^\infty x'(t)(R + K'_{\tau,\theta} G K_{\tau,\theta}) x(t) dt \leq W(x(0)) \tag{5.5.26}$$

follows from (5.5.25) and the fact that $T > 0$ has been chosen arbitrarily. This implies the absolute stability of the closed loop system (5.5.1), (5.5.11). Indeed, (5.5.26) and (5.5.2) imply that $x(\cdot) \in \mathbf{L}_2[0,\infty]$, $\xi(\cdot) \in \mathbf{L}_2[0,\infty]$, and hence $x(t) \to 0$ as $t \to \infty$. Also, it follows from (5.5.23) and (5.5.26) that the $\mathbf{L}_2$ norms of the state and uncertainty processes are uniformly bounded.

We are now in a position to complete the proof of the theorem. Indeed, it remains only to prove (5.5.12). It follows from (5.5.17) that for any admissible uncertainty input $\xi(\cdot)$

$$\begin{aligned} J_{\tau,\theta}(K_{\tau,\theta}x(\cdot), \xi(\cdot)) &= J(K_{\tau,\theta}x(\cdot)) + \sum_{i=1}^k \tau_i \int_0^\infty (|r_i' z(t)|^2 - |g_i' \xi(t)|^2) dt \\ &\quad + 2\int_0^\infty \xi(t)' \bar{G}' \Theta \bar{R} \dot{z}(t) dt. \end{aligned}$$

Since the controller $u = K_{\tau,\theta} x$ is an absolutely stabilizing controller and therefore, $x(t) \to 0$ as $t \to \infty$, it follows from (5.5.24) and (5.5.22) that

$$\begin{aligned} J_{\tau,\theta}(K_{\tau,\theta}x(\cdot), \xi(\cdot)) &\geq J(K_{\tau,\theta}x(\cdot)) + \sum_{i=1}^k \tau_i \int_0^\infty (|r_i' z(t)|^2 - |g_i' \xi(t)|^2) dt \\ &\quad - z'(0) \bar{R}' \Theta \bar{R} z(0). \end{aligned} \tag{5.5.27}$$

Indeed,

$$2\int_0^\infty \xi(t)' \bar{G}' \Theta \bar{R} \dot{z}(t) dt = 2 \sum_{i \in \Im} \theta_i \int_0^\infty [g_i' \xi(t)][r_i' \dot{z}(t)] dt.$$

Furthermore using (5.5.3), we obtain

$$\begin{aligned} 2\int_0^\infty \xi(t)' \bar{G}' \Theta \bar{R} \dot{z}(t) dt &\geq 2 \sum_{i \in \Im} \theta_i \int_0^\infty \zeta_i(r_i' z(t))[r_i' \dot{z}(t)] dt \\ &= 2 \sum_{i \in \Im} \theta_i \int_{r_i' z(0)}^0 \zeta_i(\sigma) d\sigma \\ &= -2 \sum_{i \in \Im} \theta_i \int_0^{r_i' z(0)} \zeta_i(\sigma) d\sigma. \end{aligned}$$

Also, it follows from (5.5.22) that

$$2\int_0^\sigma (\zeta_i(\sigma) + \sigma)\, d\sigma \leq 2 \int_0^\sigma 2\sigma d\sigma = 2\sigma^2.$$

Hence, $2\int_0^\sigma \zeta_i(\sigma)d\sigma \leq \sigma^2$ and therefore,

$$2\int_0^\infty \xi(t)'\bar{G}'\Theta\bar{R}\dot{z}(t)dt \geq -\sum_{i\in\Im}\theta_i|r_i'z(0)|^2$$

from which (5.5.27) follows.

Using (5.5.27), we obtain

$$\begin{aligned} J(K_{\tau,\theta}x(\cdot)) &\leq J_{\tau,\theta}(K_{\tau,\theta}x(\cdot),\xi(\cdot)) + x'(0)C'\bar{R}'\Theta\bar{R}Cx(0) \\ &\leq x'(0)(X_{\tau,\theta} + C'\bar{R}'\Theta\bar{R}C)x(0). \end{aligned} \quad (5.5.28)$$

Since the value on the right hand side of this inequality is independent of $\xi(\cdot)$, it follows that (5.5.12) holds.

■

Theorem 5.5.1 implies

$$V^0 \leq \inf x'(0)(X_{\tau,\theta} + C'\bar{R}'\Theta\bar{R}C)x(0), \quad (5.5.29)$$

where the infimum is with respect to the nonnegative quantities $\tau_1, \ldots, \tau_k$ and diagonal matrices Θ satisfying the conditions of Theorem 5.5.1.

It is important to note that in the case of $\Im = \varnothing$, the Lur'e-Postnikov Lyapunov function (5.5.21) reduces to a quadratic Lyapunov function $x'X_\tau x$. In this case, the upper bound (5.5.29) reduces to the upper bound (5.5.8).

Theorem 5.5.1 provides a sufficient condition[4] for the existence of an absolutely stabilizing controller (5.5.11) such that the closed loop system (5.5.1), (5.5.11), (5.5.2), (5.5.3) satisfies the bound (5.5.29). However, this controller is not a minimax optimal controller.

The upper bound on V^0 given by Theorem 5.5.1 differs from the bound (5.5.8) obtained using the results of Section 5.3. Hence, it is of interest to compare these two bounds. The example given in the next subsection illustrates the fact that (5.5.29) can give a better upper bound on the value of V^0 as compared to the bound (5.5.8). That is, the controller (5.5.11) corresponding to parameters $\tau_1, \ldots, \tau_k$ and Θ which attain the infimum in (5.5.29), guarantees better performance than the controller derived using Riccati equation (5.5.7).

5.5.3 *Illustrative Example*

We consider the tracking problem introduced in Section 5.3.3. In Section 5.3.3, this tracking problem was reduced to a control problem with an underlying uncertain system (5.3.40) and cost functional (5.3.41) with weighting matrices defined as in (5.3.42).

---

[4]For the case of unstructured scalar uncertainty of the form $\xi(t) = \phi(z(t))$, where $\phi(z)$ is a sector-bounded nonlinearity, necessary and sufficient conditions for absolute stabilizability with a Lyapunov function of the Lur'e-Postnikov form will be given in Section 7.2; see also [183, 237].

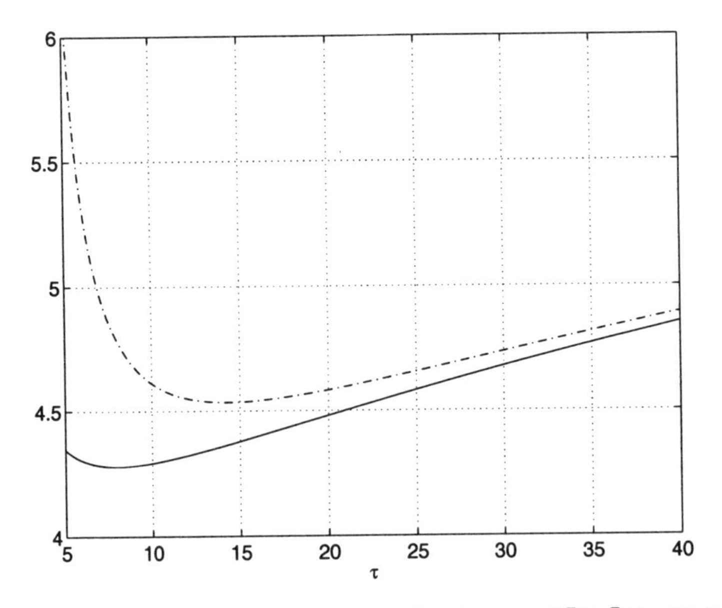

Figure 5.5.2. $x'(0)X_\tau x(0)$ (the dashed line) and $x'(0)(X_{\tau,\theta}+C'\bar{R}'\Theta\bar{R}C)x(0)$ (the solid line) versus the parameter τ.

In Section 5.3.3, the uncertainty in (5.3.40) was assumed to satisfy a certain integral quadratic constraint. In this section, we assume that the uncertainty satisfies constraints of the form (5.5.2) and (5.5.3) as follows:

$$|\xi(t)| \leq |z(t)| \tag{5.5.30}$$

and for all $\xi(\cdot)$, there exists a $\zeta(\cdot)$ such that

$$\inf_{\substack{\{t:\, z(t)=\sigma,\\ \dot{z}(t)\geq 0\}}} \xi(t) \geq \zeta(\sigma) \geq \sup_{\substack{\{t:\, z(t)=\sigma,\\ \dot{z}(t)\leq 0\}}} \xi(t). \tag{5.5.31}$$

Since we consider an uncertain system with a single uncertainty input and a single uncertainty output, the results of Section 5.2 can also be applied to this example. It follows from Theorems 5.2.1 and 5.2.2 (see also Observation 5.2.2), that $\inf_{\tau\in\mathcal{T}_0} x'(0)X_\tau x(0)$ is the optimal guaranteed cost bound which can be achieved by means of a quadratic Lyapunov function. Also, note that an uncertainty input satisfying the conditions (5.5.30), (5.5.31) will satisfy the uncertainty constraint considered in Section 5.3. Hence, a controller obtained using the approach of Section 5.3 can also be applied to stabilize the uncertain system (5.3.40), (5.5.30), (5.5.31). This controller is in fact the same controller as obtained using the approach of Section 5.2 since both are obtained from the Riccati equation (5.5.7). For the initial condition $x(0) = [-1\ 0\ 0\ 0]'$, and $\tau \in [5, 40]$, the guaranteed cost bound corresponding to this quadratically stabilizing controller is shown in Figure 5.5.2 (the dashed line). The best guaranteed cost bound which can be achieved by using this approach, is equal to 4.5343. That is, for $x(0) = [-1\ 0\ 0\ 0]'$, the approaches of Sections 5.2 and 5.3 lead to a controller which guarantees

$$V^0 \leq 4.5343.$$

The solid line in Figure 5.5.2 corresponds to the quantity

$$x'(0)(X_{\tau,\theta} + C'\bar{R}'\Theta\bar{R}C)x(0)$$

when we fix $\Theta = 1$. Thus, even when not performing the optimization with respect to Θ, the controller (5.5.11) guarantees a better upper bound on V^0 than the controller obtained using the approaches of Sections 5.2 and 5.3; *i.e.*,

$$V^0 \leq 4.2790.$$

The optimization with respect to the both parameters Θ and τ should give an even better guaranteed cost bound. This will lead to a controller guaranteeing a better level of performance.

5.6 Conclusions

The main conclusion to be reached in this chapter is the fact that the standard state-feedback linear quadratic regulator can be extended to a minimax optimal controller for a class of uncertain linear systems. A similar extension was also obtained in the output-feedback case. These extensions involve replacing the algebraic Riccati equations which occur in the LQR and LQG problems by parameter dependent Riccati equations of the game type. In this case, the corresponding control problem is a guaranteed cost control problem and the uncertain systems being considered are those in which the uncertainty satisfies an integral quadratic constraint. In such uncertain linear systems, an extremely rich class of uncertainties is allowed. Indeed, it is this richness of the uncertainty class which enables us to give a Riccati equation solution to the state-feedback guaranteed cost control problems.

An interesting consequence of the results presented in this chapter is the fact that in the output-feedback case, the resulting optimal guaranteed cost controller is time-varying even though the underlying uncertain system is time-invariant and an infinite-horizon problem is being considered.

The fact that in the output-feedback case, the optimal guaranteed cost controller is time-varying is of some theoretical interest. However, from a practical point of view, a time-invariant optimal guaranteed cost controller would be much more useful. It appears that to obtain such a time-invariant optimal guaranteed cost controller a further extension of the class of uncertainties is needed to allow for stochastic processes. This issue will be addressed in Sections 8.4 and 8.5.

6. Finite-horizon guaranteed cost control

6.1 Introduction

As mentioned in Chapter 1, the problem of directly extending the results of Chapter 5 to the case of a finite time horizon or to the case of time-varying uncertain systems, appears to be mathematically intractable. In this chapter, we overcome this difficulty by considering an alternative class of uncertain systems with structured uncertainty.

In Sections 6.2 and 6.3, we begin by considering uncertain systems referred to as uncertain systems with an averaged integral quadratic constraint. This class of uncertain systems was defined in Section 2.3.4. Within the framework of this class of uncertain systems, we are able present a mathematically tractable solution to the finite-horizon time-varying minimax optimal control problem. The state-feedback case is considered in Section 6.2 and the output-feedback case is considered in Section 6.3. The main results of Sections 6.2 and 6.3 are necessary and sufficient conditions for the existence of a guaranteed cost controller for an uncertain system with structured uncertainties satisfying an averaged integral quadratic constraint.

In section 6.4, we consider a robust control problem referred to as "robust control with a terminal state constraint." Given an uncertain linear system, it is desired to construct an output-feedback controller which steers a given initial state to the vicinity of the origin in a specified time. Also, it is required that the control energy required to achieve this is bounded. The main result of Section 6.4 gives a necessary and sufficient condition for an uncertain linear system to be robustly controllable with a given terminal state constraint.

Ideas similar to those used in Section 6.4 will allow us to solve a problem of robust control referred to as robust control with rejection of harmonic disturbances. This problem naturally arises as an extension of a disturbance rejection problem to the case in which the controlled plant is uncertain; e.g., see [1]. In the problem considered in Section 6.5, we will be interested in an optimal controller which minimizes the worst case of a quadratic function. The uncertainty in the system is a combination of standard uncertainty (subject to an integral quadratic constraint) and harmonic disturbances with known frequencies but unknown magnitudes and phases.

6.2 The uncertainty averaging approach to state-feedback minimax optimal control

6.2.1 *Problem statement*

Consider the following uncertain time-varying system of the form (2.3.13) defined on the finite time interval $[0, T]$:

$$\begin{aligned}\dot{x}(t) &= A(t)x(t) + B_1(t)u(t) + \sum_{j=1}^{k} B_{2j}(t)\xi_j(t)\\ z_1(t) &= C_{11}(t)x(t) + D_{11}(t)u(t),\\ &\vdots\\ z_k(t) &= C_{1k}(t)x(t) + D_{1k}(t)u(t),\end{aligned} \tag{6.2.1}$$

where $x(t) \in \mathbf{R}^n$ is the *state*, $u(t) \in \mathbf{R}^m$ is the *control input*, $z_1(t) \in \mathbf{R}^{q_1}$, $z_2(t) \in \mathbf{R}^{q_2}$, ..., $z_k(t) \in \mathbf{R}^{q_k}$ are the *uncertainty outputs*, $\xi_1(t) \in \mathbf{R}^{p_1}$, $\xi_2(t) \in \mathbf{R}^{p_2}$, ..., $\xi_k(t) \in \mathbf{R}^{p_k}$ are the *uncertainty inputs*, and $A(\cdot)$, $B_1(\cdot)$, $B_{21}(\cdot), \ldots, B_{2k}(\cdot)$, $C_{11}(\cdot), \ldots, C_{1k}(\cdot)$, $D_{11}(\cdot), \ldots, D_{1k}(\cdot)$ are bounded piecewise continuous matrix functions defined on $[0, T]$. Associated with the system (6.2.1) is the quadratic cost functional

$$J_0\left(x(\cdot), u(\cdot), \xi(\cdot)\right) := \int_0^T [x(t)'R(t)x(t) + u(t)'G(t)u(t)]dt \tag{6.2.2}$$

where $R(\cdot)$ and $G(\cdot)$ are given bounded piecewise continuous symmetric matrix weighting functions satisfying the condition: There exists a constant $\epsilon > 0$ such that $R(t) \geq \epsilon I$ and $G(t) \geq \epsilon I$ for all $t \in [0, T]$.

System Uncertainty

The uncertainty in the above system is described by the following set of equations of the form (2.3.4):

$$\xi_j(t) = \phi_j(t, x(\cdot)) \quad \text{for } j = 1, 2, \ldots, k. \tag{6.2.3}$$

Alternatively, the uncertainty inputs and uncertainty outputs may be collected together into two vectors as in Section 2.3.4:

$$\xi(t) := \begin{bmatrix} \xi_1(t) \\ \xi_2(t) \\ \vdots \\ \xi_k(t) \end{bmatrix}, \quad z(t) := \begin{bmatrix} z_1(t) \\ z_2(t) \\ \vdots \\ z_k(t) \end{bmatrix}.$$

Then, (6.2.3) can be re-written in the more compact form:

$$\xi(t) = \Phi(t, x(\cdot)). \tag{6.2.4}$$

In accordance with the uncertainty model described in Section 2.3.4, we consider finite sequences of uncertainty functions of the form (6.2.4) satisfying a certain averaged integral quadratic constraint; see Definition 2.3.4.

Definition 6.2.1 *Let $d_1 > 0, d_2 > 0, \ldots, d_k > 0$, be given positive constants associated with the system (6.2.1). We will consider sequences of uncertainty functions $\mathcal{S} = \{\Phi^1(\cdot), \Phi^2(\cdot), \ldots \Phi^q(\cdot)\}$ of arbitrary length q. A sequence of uncertainty functions $\mathcal{S}$ is an admissible* uncertainty sequence *for the system (6.2.1) if the following conditions hold: Given any $\Phi^i(\cdot) \in \mathcal{S}$, any control input $u^i(\cdot) \in \mathbf{L}_2[0,T]$, and any corresponding solution $\{x^i(\cdot), \xi^i(\cdot)\}$ to equations (6.2.1), (6.2.4) defined on $[0,T]$, then $\xi^i(\cdot) \in \mathbf{L}_2[0,T]$ and*

$$\frac{1}{q}\sum_{i=1}^{q} \int_0^T (\|\xi_j^i(t)\|^2 - \|z_j^i(t)\|^2)dt \leq d_j \tag{6.2.5}$$

for $s = 1, 2, \ldots, k$. The class of all such admissible uncertainty sequences is denoted Ξ.

Remark 6.2.1 Recall from Section 2.3.4 that the above uncertainty averaging description of structured uncertainty can be given a probabilistic interpretation; see Remark 2.3.2.

The class of controllers to be considered in this section are nonlinear full information controllers of the form

$$u(t) = U(t, x(\cdot)|_0^t, \xi(\cdot)|_0^t). \tag{6.2.6}$$

Remark 6.2.2 Note that full information controllers of the form (6.2.6) have access to both the state and uncertainty input. One setting in which full information controllers arise naturally occurs when the parameterization of all suboptimal H^∞ controllers is considered; e.g., see [52, 114]. However, it is known that the class of suboptimal controllers arising in the standard H^∞ control problem includes a controller for which the information provided by the uncertainty input is redundant. Such a controller is referred to as a central controller. This controller is identical to the linear state-feedback controller which solves the corresponding H^∞ control problem with the CLPS information pattern; see Section 3.2. This fact is used in the sequel.

In this section, the full information controllers of the form (6.2.6) are required to satisfy the following technical assumptions.

Assumption 6.2.1 *The controller $U(\cdot)$ is such that given any uncertainty input function $\xi(\cdot) \in \mathbf{L}_2[0,T]$ and any initial condition $x(0) \in \mathbf{R}^n$, there exists a solution to equations (6.2.1), (6.2.6) defined on $[0,T]$ and the corresponding control input $u(\cdot)$ satisfies $u(\cdot) \in \mathbf{L}_2[0,T]$.*

Assumption 6.2.2 *The controller $U(\cdot)$ is such that the following condition holds: Given any time $t^* > 0$, if $x(0) = 0$ and $\xi(t) = 0$ for $t \in [0,t^*]$, then the resulting control input $u(t) = U[t, x(\cdot)|_0^t, 0]$ satisfies $u(t) = 0$ for $t \in [0,t^*]$.*

The set of all controllers of the form (6.2.6) satisfying Assumptions 6.2.1 and 6.2.2 is denoted $\mathcal{U}$.

Definition 6.2.2 *Let $\mathcal{X}_0 \subset \mathbf{R}^n$ be a given set of possible initial conditions and let $\Psi(\cdot)$ be a given positive valued weighting function mapping from $\mathcal{X}_0$ to $\mathbf{R}^+$. A controller of the form (6.2.6) is said to be a* guaranteed cost controller *for the uncertain system (6.2.1), (6.2.5) with the cost function (6.2.2), initial condition set $\mathcal{X}_0$ and the weighting function $\Psi(\cdot)$ if the following condition holds: Given any $\mathcal{S} \in \Xi$, let q be the number of elements in this uncertainty sequence and let $\{x_0^i\}_{i=1}^q \subset \mathcal{X}_0$ be a corresponding sequence of initial conditions. Also, for each $\Phi^i(\cdot) \in \mathcal{S}$ and corresponding x_0^i, let $x^i(\cdot)$ be the corresponding solution to equations (6.2.1), (6.2.4), (6.2.6) with initial condition $x^i(0) = x_0^i$. Also, let $\xi^i(\cdot)$ be the corresponding uncertainty input defined by equation (6.2.4) and let $u^i(\cdot)$ be the control input defined by equation (6.2.6). Then,*

$$\sup_{\mathcal{S}\in\Xi} \sup_{\{x_0^i\}_{i=1}^q \subset \mathcal{X}_0} \frac{\sum_{i=1}^q J_0\left(x^i(\cdot), u^i(\cdot), \xi^i(\cdot)\right)}{\sum_{i=1}^q \Psi(x_0^i)} < 1. \tag{6.2.7}$$

Remark 6.2.3 According to the above definition, a guaranteed cost controller guarantees that the averaged value of the cost function is bounded in terms of the averaged initial condition for all admissible uncertainty sequences. An important special case of Definition 6.2.2 is the case in which the set $\mathcal{X}_0$ is a set of the form $\mathcal{X}_0 = \{x \in \mathbf{R}^n : \|x\| \leq 1\}$ and the function $\Psi(\cdot)$ takes on a constant value; *i.e.*, $\Psi(x) \equiv \gamma > 0$. In this case, condition (6.2.7) becomes

$$\sup_{\mathcal{S}\in\Xi} \sup_{\{x_0^i\}_{i=1}^q \subset \mathcal{X}_0} \frac{1}{q} \sum_{i=1}^q J_0\left(x^i(\cdot), u^i(\cdot), \xi^i(\cdot)\right) < \gamma. \tag{6.2.8}$$

6.2.2 *A necessary and sufficient condition for the existence of a state-feedback guaranteed cost controller*

In this section, we present a result which establishes a necessary and sufficient condition for the existence of a nonlinear state-feedback guaranteed cost controller (6.2.6) for the uncertain time-varying system (6.2.1), (6.2.5) with associated cost functional (6.2.2), initial condition set $\mathcal{X}_0$ and weighting function $\Psi(\cdot)$. This condition is given in terms of the existence of a suitable solution to a parameter dependent Riccati differential equation of the form (3.2.58). The Riccati equation under consideration is defined as follows: Let $\tau_1 > 0, \ldots, \tau_k > 0$ be given constants and consider the Riccati differential equation

$$\begin{aligned} -\dot{X} = {} & (A - B_1 G_\tau^{-1} D_\tau' C_\tau)' X + X(A - B_1 G_\tau^{-1} D_\tau' C_\tau) \\ & + X(\tilde{B}_2 \tilde{B}_2' - B_1 G_\tau^{-1} B_1') X + C_\tau'(I - D_\tau G_\tau^{-1} D_\tau') C_\tau \end{aligned} \tag{6.2.9}$$

where

$$C_\tau := \begin{bmatrix} R^{1/2} \\ 0 \\ \sqrt{\tau_1} C_{11} \\ \vdots \\ \sqrt{\tau_k} C_{1k} \end{bmatrix}; \quad D_\tau := \begin{bmatrix} 0 \\ G^{1/2} \\ \sqrt{\tau_1} D_{11} \\ \vdots \\ \sqrt{\tau_k} D_{1k} \end{bmatrix};$$

$$G_\tau := D_\tau' D_\tau; \quad \tilde{B}_2 := \left[\frac{1}{\sqrt{\tau_1}} B_{21} \; \cdots \; \frac{1}{\sqrt{\tau_k}} B_{2k} \right]. \tag{6.2.10}$$

Note that the Riccati differential equation (6.2.9) is a time-varying version of the corresponding algebraic Riccati equation (5.3.11). In a similar fashion to the infinite-horizon case considered in Section 5.3, if $X(t) > 0$ is the solution to Riccati equation (6.2.9), then the corresponding guaranteed cost controller will be a linear time-varying control law given by

$$u(t) = -G_\tau^{-1}(t)[B_1'(t) X(t) + D_\tau(t)' C_\tau(t)] x(t). \tag{6.2.11}$$

The main result of this section requires that the uncertain system (6.2.1), (6.2.5) with associated cost functional (6.2.2), initial condition set $\mathcal{X}_0$ and weighting function $\Psi(\cdot)$ satisfy the following assumption.

Assumption 6.2.3

(i) $\int_0^T \|B_{2j}(t)\| dt > 0$ *for* $j = 1, 2, \ldots, k$.

(ii) The set $\mathcal{X}_0$ *contains the origin;* i.e., $0 \in \mathcal{X}_0$.

(iii) The function $\Psi(\cdot)$ *is bounded on the set* $\mathcal{X}_0$.

Theorem 6.2.1 *Consider the uncertain system (6.2.1), (6.2.5) with associated cost functional (6.2.2), initial condition set* $\mathcal{X}_0$ *and weighting function* $\Psi(\cdot)$*. Also, suppose that Assumption 6.2.3 is satisfied. Then the following statements are equivalent:*

(i) There exists a nonlinear state-feedback guaranteed cost controller of the form (6.2.6) satisfying Assumptions 6.2.1 and 6.2.2.

(ii) There exist constants $\tau_1 > 0, \dots, \tau_k > 0$ such that the Riccati differential equation (6.2.9) has a solution $X(t)$ on the interval $[0, T]$ such that $X(T) = 0$ and

$$\sup_{x_0 \in \mathcal{X}_0} \frac{x_0' X(0) x_0 + \sum_{j=1}^{k} \tau_j d_j}{\Psi(x_0)} < 1. \tag{6.2.12}$$

If condition (ii) holds, then the uncertain system (6.2.1), (6.2.5) with associated cost function (6.2.2), initial condition set $\mathcal{X}_0$ and weighting function $\Psi(\cdot)$ admits a linear guaranteed cost controller of the form (6.2.11).

Proof: (i)⇒(ii). Suppose the uncertain system (6.2.1), (6.2.5) is such that there exists a guaranteed cost controller of the form (6.2.6) satisfying Assumptions 6.2.1 and 6.2.2. Then inequality (6.2.7) implies that there exists a constant $\delta > 0$ such that the following condition holds: Given any admissible uncertainty sequence $\mathcal{S} = \{\Phi^i(\cdot)\}_{i=1}^q \in \Xi$ and any corresponding sequence of initial conditions $\{x_0^i\}_{i=1}^q \subset \mathcal{X}_0$, let $[x^i(\cdot), u^i(\cdot), \xi^i(\cdot)]$ be the corresponding solution to equations (6.2.1), (6.2.6) and (6.2.4) with initial condition $x^i(0) = x_0^i$. Then

$$(1 + 2\delta) \sum_{i=1}^{q} J_0(x^i(\cdot), u^i(\cdot), \xi^i(\cdot)) \leq (1 - 2\delta) \sum_{i=1}^{q} \Psi(x_0^i). \tag{6.2.13}$$

For the closed loop system defined by (6.2.1) and (6.2.6), a set $\mathcal{M} \subset \mathbf{L}_2[0, T]$ is defined as follows:

$$\mathcal{M} := \left\{ \begin{array}{l} \lambda(\cdot) = [x(\cdot)\ u(\cdot)\ \xi(\cdot)] : x(\cdot), u(\cdot), \xi(\cdot) \text{ are related by (6.2.1),} \\ \qquad\qquad (6.2.6), \xi(\cdot) \in \mathbf{L}_2[0, T] \text{ and } x(0) \in \mathcal{X}_0 \end{array} \right\}. \tag{6.2.14}$$

Also, we consider the following set of integral functionals mapping from $\mathcal{M}$ into $\mathbf{R}$:

$$\begin{aligned} \mathcal{G}_0(x(\cdot), u(\cdot), \xi(\cdot)) &:= -(1 + 2\delta) J_0(x(\cdot), u(\cdot), \xi(\cdot)) + (1 - 2\delta)\Psi(x(0)); \\ \mathcal{G}_j(x(\cdot), u(\cdot), \xi(\cdot)) &:= \int_0^T (\|z_j(t)\|^2 - \|\xi_j(t)\|^2) dt + d_j; \end{aligned} \tag{6.2.15}$$

for $j = 1, 2, \dots, k$.

Now consider a sequence of vector functions $\{x^i(\cdot), u^i(\cdot), \xi^i(\cdot)\}_{i=1}^q \subset \mathcal{M}$ such that

$$\sum_{i=1}^q \mathcal{G}_1\left(x^i(\cdot), u^i(\cdot), \xi^i(\cdot)\right) \geq 0;$$
$$\sum_{i=1}^q \mathcal{G}_2\left(x^i(\cdot), u^i(\cdot), \xi^i(\cdot)\right) \geq 0;$$
$$\vdots$$
$$\sum_{i=1}^q \mathcal{G}_k\left(x^i(\cdot), u^i(\cdot), \xi^i(\cdot)\right) \geq 0.$$

It follows from Definition 6.2.1 that this sequence corresponds to an admissible uncertainty sequence $\mathcal{S} \in \Xi$. Hence, inequality (6.2.13) implies that

$$\sum_{i=1}^q \mathcal{G}_0\left(x^i(\cdot), u^i(\cdot), \xi^i(\cdot)\right) \geq 0. \tag{6.2.16}$$

We now apply the S-procedure Theorem 4.5.1 to the set $\mathcal{P} \subset \mathbf{R}^{k+1}$ defined as follows:

$$\mathcal{P} := \left\{ \nu = \begin{bmatrix} \mathcal{G}_0\left(x(\cdot), u(\cdot), \xi(\cdot)\right) \\ \vdots \\ \mathcal{G}_k\left(x(\cdot), u(\cdot), \xi(\cdot)\right) \end{bmatrix} : [x(\cdot), u(\cdot), \xi(\cdot)] \in \mathcal{M} \right\}.$$

From this definition and condition (6.2.16), it follows that the condition required by the S-procedure Theorem 4.5.1 is satisfied. Furthermore using the fact that

$$d_1 > 0, \ldots, d_k > 0,$$

it follows that if we choose a vector function $[x(\cdot), u(\cdot), 0] \in \mathcal{M}$ corresponding to the uncertainty input $\xi(t) \equiv 0$ and any initial condition $x(0) \in \mathcal{X}_0$, the corresponding values of the functionals $\mathcal{G}_1(\cdot), \ldots, \mathcal{G}_k(\cdot)$ will satisfy

$$\mathcal{G}_1\left(x(\cdot), u(\cdot), 0\right) > 0, \quad \ldots, \quad \mathcal{G}_k\left(x(\cdot), u(\cdot), 0\right) > 0.$$

Hence, there exists a point $\nu \in \mathcal{P}$ such that $\nu_1 > 0, \ldots \nu_k > 0$. Using Theorem 4.5.1 and Observation 4.5.1, it now follows that there exist constants $\tau_1 \geq 0, \ldots, \tau_k \geq 0$ such that

$$\nu_0 \geq \tau_1\nu_1 + \cdots + \tau_k\nu_k$$

for all $\nu = [\nu_0 \; \nu_1 \; \ldots \; \nu_k]' \in \mathcal{P}$. Hence,

$$\mathcal{G}_0\left(x(\cdot), u(\cdot), \xi(\cdot)\right) \geq \sum_{j=1}^k \tau_j \mathcal{G}_j\left(x(\cdot), u(\cdot), \xi(\cdot)\right) \tag{6.2.17}$$

for all $[x(\cdot), u(\cdot), \xi(\cdot)] \in \mathcal{M}$. Also, using this inequality, Assumption 6.2.3 and equations (6.2.15), we can conclude that the functional

$$\begin{aligned} &J_\delta^T(x(\cdot), u(\cdot), \xi(\cdot)) \\ &:= (1+\delta) J_0(x(\cdot), u(\cdot), \xi(\cdot)) + \sum_{j=1}^{k} \tau_j \int_0^T (\|z_j(t)\|^2 - \|\xi_j(t)\|^2) dt \end{aligned}$$

satisfies the following condition: There exists a constant $c > 0$ such that

$$J_\delta^T(x(\cdot), u(\cdot), \xi(\cdot)) \leq -\delta J_0(x(\cdot), u(\cdot), \xi(\cdot)) + c$$

for all $[x(\cdot), u(\cdot), \xi(\cdot)] \in \mathcal{M}$. Furthermore, using the fact that $R(t) \geq \epsilon I$ and $G(t) \geq \epsilon I$ for all $t \in [0, T]$, it now follows that there exists a constant $\varepsilon > 0$ such that

$$J_\delta^T(x(\cdot), u(\cdot), \xi(\cdot)) \leq -\varepsilon\delta \int_0^T \|u(t)\|^2 dt + c \qquad (6.2.18)$$

for all $[x(\cdot), u(\cdot), \xi(\cdot)] \in \mathcal{M}$. Also, we recall that Assumption 6.2.3 requires that the set $\mathcal{X}_0$ contains the origin. Hence, the following set is a subset of $\mathcal{M}$:

$$\mathcal{M}_0 := \{\lambda(\cdot) = [x(\cdot), u(\cdot), \xi(\cdot)] : [x(\cdot), u(\cdot), \xi(\cdot)] \in \mathcal{M},\ x(0) = 0\}.$$

Since $\mathcal{M}_0 \subset \mathcal{M}$, inequality (6.2.18) will also be satisfied on the set $\mathcal{M}_0$.

We now consider a new controller $u(\cdot) = U^*(\xi(\cdot))$ defined for the system (6.2.1) with initial condition $x(0) = 0$. This controller is defined as follows: Let $\xi_0(\cdot)$ be a given uncertainty input. If there exists a constant $\alpha > 0$ such that

$$J_\delta^T(x_\alpha(\cdot), u_\alpha(\cdot), \alpha\xi_0(\cdot)) \leq 0 \qquad (6.2.19)$$

where $[x_\alpha(\cdot), u_\alpha(\cdot)]$ is the solution to the closed loop system (6.2.1), (6.2.6) with initial condition $x_\alpha(0) = 0$ and the uncertainty input $\xi(\cdot) = \alpha\xi_0(\cdot)$, then we define the control input $u(\cdot)$ as

$$u(\cdot) = U^*(\xi_0(\cdot)) = \frac{1}{\alpha} u_\alpha(\cdot). \qquad (6.2.20)$$

If there does not exist such a constant $\alpha > 0$, then we define the control input $u(\cdot)$ as

$$u(\cdot) = U^*(\xi_0(\cdot)) \equiv 0. \qquad (6.2.21)$$

We now prove that

$$J_\delta^T(x(\cdot), u(\cdot), \xi_0(\cdot)) \leq 0 \qquad (6.2.22)$$

for all solutions to the closed loop system (6.2.1), (6.2.20), (6.2.21) with initial condition $x(0) = 0$. Indeed, if the controller is defined by (6.2.20), then using the fact that $J_\delta^T(\cdot)$ is a quadratic form, it follows from the linearity of the system (6.2.1) that

$$J_\delta^T(x(\cdot), u(\cdot), \xi_0(\cdot)) = \frac{1}{\alpha^2} J_\delta^T(x_\alpha(\cdot), u_\alpha(\cdot), \alpha\xi_0(\cdot)).$$

Hence, condition (6.2.22) follows from (6.2.19).

It remains to prove inequality (6.2.22) for the case in which the controller is defined by equation (6.2.21). Indeed, for any $\alpha > 0$, define $x_\alpha(\cdot)$ and $u_\alpha(\cdot)$ as the solutions to the closed loop system (6.2.1), (6.2.6) with initial condition $x_\alpha(0) = 0$ and uncertainty input $\xi(\cdot) = \alpha\xi_0(\cdot)$. Our definition of the controller $U^*(\cdot)$ implies that for this case,

$$J_\delta^T(x_\alpha(\cdot), u_\alpha(\cdot), \alpha\xi_0(\cdot)) > 0$$

for all $\alpha > 0$. This and inequality (6.2.18) together imply that

$$\int_0^T \|u_\alpha(t)\|^2 dt < \frac{c}{\epsilon\delta}$$

for all $\alpha > 0$. Now let $x_\alpha^*(\cdot)$ be the solution to the system (6.2.1) with the uncertainty input $\xi(\cdot) = \xi_0(\cdot)$, control input

$$u(\cdot) = u_\alpha^*(\cdot) = \frac{1}{\alpha} u_\alpha(\cdot)$$

and initial condition $x_\alpha^*(0) = 0$. Then, we have

$$\int_0^T \|u_\alpha^*(t)\|^2 dt < \frac{c}{\alpha^2\epsilon\delta} \tag{6.2.23}$$

Furthermore, using inequality (6.2.18), we have

$$J_\delta^T(x_\alpha^*(\cdot), u_\alpha^*(\cdot), \xi_0(\cdot)) = \frac{1}{\alpha^2} J_\delta^T(x_\alpha(\cdot), u_\alpha(\cdot), \alpha\xi_0(\cdot)) \le \frac{c}{\alpha^2}. \tag{6.2.24}$$

Moreover, inequality (6.2.23) implies that

$$\lim_{\alpha\to\infty} \int_0^T \|u_\alpha^*(t)\|^2 dt = 0.$$

Hence,

$$\lim_{\alpha\to\infty} J_\delta^T(x_\alpha^*(\cdot), u_\alpha^*(\cdot), \xi_0(\cdot)) = J_\delta^T(x^*(\cdot), 0, \xi_0(\cdot)) \tag{6.2.25}$$

where $x^*(\cdot)$ is the solution to the system (6.2.1) with control input $u(\cdot) \equiv 0$, uncertainty input $\xi(\cdot) = \xi_0(\cdot)$ and initial condition $x^*(0) = 0$. However, inequalities (6.2.24) and (6.2.25) together imply that

$$J_\delta^T(x^*(\cdot), 0, \xi_0(\cdot)) \le 0.$$

Thus, we have established the inequality (6.2.22) for this case.

We now prove that $\tau_j > 0$ for $j = 1, 2, \ldots, k$. This fact will be established by contradiction. Suppose $\tau_j = 0$ for some j and consider an input function $\xi(\cdot) \in \mathbf{L}_2[0,T]$ defined so that $\xi_j(\cdot) \not\equiv 0$ and $\xi_s(\cdot) \equiv 0$ for $s \neq j$. For such an input, it follows from the definition of $J_\delta^T(\cdot)$ and (6.2.22) that we must have

$$J_\delta^T((x(\cdot), u(\cdot), \xi(\cdot)) = 0$$

for all solutions to the closed loop system (6.2.1), (6.2.20), (6.2.21) with initial condition $x(0) = 0$. Furthermore using the fact that $R(t) \geq \epsilon I$ and $G(t) \geq \epsilon I$ for all $t \in [0,T]$, this implies that $x(\cdot) \equiv 0$ and $u(\cdot) \equiv 0$. However, it follows from Assumption 6.2.3 that we can choose $\xi_j(\cdot)$ such that $\int_0^T \|B_{2j}(t)\xi_j(t)\|dt \neq 0$. Thus, we have $x(\cdot) \equiv 0$, $u(\cdot) \equiv 0$ and $\int_0^T \|B_{2j}(t)\xi_j(t)\|dt \neq 0$ which leads to a contradiction with state equations (6.2.1). Hence, we can conclude that $\tau_j > 0$ for all $j = 1, \ldots, k$.

In order to establish the existence of a suitable solution to Riccati equation (6.2.9), we consider the system

$$\begin{aligned} \dot{x}(t) &= A(t)x(t) + B_1(t)u(t) + \tilde{B}_2(t)w(t); \\ \hat{z}(t) &= C_\tau(t)x(t) + D_\tau(t)u(t) \end{aligned} \tag{6.2.26}$$

where

$$\hat{z}(\cdot) = \begin{bmatrix} R^{\frac{1}{2}}x(\cdot) \\ G^{\frac{1}{2}}u(\cdot) \\ \sqrt{\tau_1}z_1(\cdot) \\ \vdots \\ \sqrt{\tau_k}z_k(\cdot) \end{bmatrix}; \quad w(\cdot) = \begin{bmatrix} \sqrt{\tau_1}\xi_1(\cdot) \\ \vdots \\ \sqrt{\tau_k}\xi_k(\cdot) \end{bmatrix}$$

and the matrices $\tilde{B}_2(\cdot), C_\tau(\cdot)$ and $D_\tau(\cdot)$ are defined as in (6.2.10). Since we have established above that $\tau_j > 0$ for all s, this system is well defined. It is straightforward to establish that with this notation, the state equations (6.2.26) are equivalent to the state equations (6.2.1). Furthermore, condition (6.2.22) may be re-written as

$$\int_0^T (\|\hat{z}(t)\|^2 - \|w(t)\|^2)dt \leq -\delta \int_0^T (x(t)'R(t)x(t) + u(t)'G(t)u(t))dt$$

for all solutions to the closed loop system (6.2.1), (6.2.20), (6.2.21) with initial condition $x(0) = 0$. Using the condition $R(t) \geq \epsilon I$ and $G(t) \geq \epsilon I$, this inequality implies that there exists a constant $\delta_1 > 0$ such that

$$\int_0^T (\|\hat{z}(t)\|^2 - \|w(t)\|^2)dt \leq -\delta_1 \int_0^T (\|x(t)\|^2 + \|u(t)\|^2)dt.$$

Furthermore, it follows from (6.2.26) that there exists a constant $c > 0$ such that

$$\|\hat{z}(t)\|^2 \leq c(\|x(t)\|^2 + \|u(t)\|^2).$$

Hence, there exists a constant $\delta_2 > 0$ such that

$$\int_0^T ((1+\delta_2)\|\hat{z}(t)\|^2 - \|w(t)\|^2)dt \leq 0$$

for all uncertainty inputs $w(\cdot) \in \mathbf{L}_2[0,T]$. From this, it follows that the controller $U^*(\cdot)$ solves a standard H^∞ control problem defined by the system (6.2.26) and the H^∞ norm bound condition:

$$\sup_{\substack{w(\cdot)\in\mathbf{L}_2[0,T],\\ x(0)=0}} \frac{\int_0^T \|\hat{z}(t)\|^2 dt}{\int_0^T \|w(t)\|^2 dt} < 1.$$

We now observe that it follows from the equations (6.2.20) and (6.2.21), that if the controller (6.2.6) satisfies Assumption 6.2.2, then the controller $U^*(\cdot)$ also satisfies Assumption 6.2.2. Indeed, given any uncertainty input $\xi_0(\cdot)$ and any $t^* > 0$, if $x(0) = 0$ and $\xi_0(t) = 0$ for $t \in [0,t^*]$, then $\alpha\xi_0(t) = 0$ for $t \in [0,t^*]$. Hence, using the fact that the controller (6.2.6) satisfies Assumption 6.2.2, it follows that $u_\alpha(t) = 0$ for $t \in [0,t^*]$. Furthermore, it now follows from (6.2.20) and (6.2.21) that the control $u(\cdot) = U^*(\xi_0(\cdot))$ is such that $u(t) = 0$ for $t \in [0,t^*]$. That is, the controller $U^*(\cdot)$ satisfies Assumption 6.2.2. Hence, Theorem 3.2.8 on page 80 implies the existence of a solution to Riccati equation (6.2.9) which satisfies the boundary condition $X(T) = 0$.

It remains only to show that inequality (6.2.12) is satisfied by this solution to Riccati equation (6.2.9). Indeed, it follows from inequality (6.2.17) that

$$\begin{aligned} &J_0\left(x(\cdot),u(\cdot),\xi(\cdot)\right) \\ &\quad+\sum_{j=1}^k \tau_j\Big(\int_0^T (\|z_j(t)\|^2 - \|\xi_j(t)\|^2)dt + d_j\Big) \leq (1-\delta)\Psi\left(x(0)\right) \end{aligned}$$

for all solutions of the closed loop system (6.2.1), (6.2.6) with an initial condition $x(0) \in \mathcal{X}_0$. That is,

$$\int_0^T (\|\hat{z}(t)\|^2 - \|w(t)\|^2)dt + \sum_{j=1}^k \tau_j d_j \leq (1-\delta)\Psi\left(x(0)\right) \tag{6.2.27}$$

for all the solutions of the closed loop system (6.2.26), (6.2.6) with an initial condition $x(0) \in \mathcal{X}_0$. However using a standard result from H^∞ control theory, it follows that the system (6.2.26) with initial condition $x(0) = x_0$ is such that

$$\inf_{u(\cdot)} \sup_{w(\cdot)\in\mathbf{L}_2[0,T]} \int_0^T (\|\hat{z}(t)\|^2 - \|w(t)\|^2)dt = x_0' X(0) x_0 \tag{6.2.28}$$

where $X(\cdot)$ is the solution to the Riccati equation (6.2.9) with boundary condition $X(T) = 0$; e.g., see [114]. Inequalities (6.2.27) and (6.2.28) together imply

$$x_0' X(0) x_0 + \sum_{j=1}^k \tau_j d_j \leq (1-\delta)\Psi(x_0).$$

From this, inequality (6.2.12) follows immediately. This completes the proof of this part of the theorem.

(ii)⇒(i). Suppose condition (ii) holds and consider the system defined by state equations (6.2.26). Using a standard result from H^∞ control theory, it follows that

$$\int_0^T (\|\hat{z}(t)\|^2 - \|w(t)\|^2)dt \leq x_0' X(0) x_0$$

for all solutions of the closed loop system (6.2.26), (6.2.11); see Theorem 3.2.8 and also [114]. Furthermore, using the notation given above, this inequality implies

$$J_0\left(x(\cdot), u(\cdot), \xi(\cdot)\right) + \sum_{j=1}^{k} \tau_j \int_0^T (\|z_j(t)\|^2 - \|\xi_j(t)\|^2)dt \leq x_0' X(0) x_0$$

for all solutions to the closed loop system (6.2.1), (6.2.11). Combining this inequality with inequalities (6.2.5) and (6.2.12) leads directly to the satisfaction of condition (6.2.7). Thus, the controller (6.2.11) is a guaranteed cost controller. This completes the proof of the theorem.

■

An important special case of the above theorem concerns the case in which the set $\mathcal{X}_0$ is the unit ball and the function $\Psi(\cdot)$ is constant. In this case, the guaranteed cost condition (6.2.7) reduces to condition (6.2.8) on the averaged value of the cost functional.

Notation. The set $\mathcal{T}_0 \subset \mathbf{R}^k$ is defined as follows:

$$\mathcal{T}_0 := \left\{ \begin{array}{l} \tau = [\tau_1\ \tau_2\ \ldots\ \tau_k] \in \mathbf{R}^k : \tau_1 > 0,\ \tau_2 > 0,\ \ldots,\ \tau_k > 0, \\ \text{Riccati equation (6.2.9) has a solution } X_\tau(t) \\ \text{on the interval } [0,T] \text{ and } X_\tau(T) = 0 \end{array} \right\}. \tag{6.2.29}$$

Corollary 6.2.1 *Consider the uncertain system (6.2.1), (6.2.5) with associated cost functional (6.2.2) and initial condition set*

$$\mathcal{X}_0 = \{x \in \mathbf{R}^n : \|x\| \leq 1\}$$

such that Assumption 6.2.3 is satisfied. Also, suppose the set $\mathcal{T}_0$ is non-empty. Then

$$\begin{aligned} &\inf_{U \in \mathcal{U}} \sup_{S \in \Xi} \sup_{\{x_0^i\}_{i=1}^q \subset \mathcal{X}_0} \frac{1}{q} \sum_{i=1}^{q} J_0\left(x^i(\cdot), u^i(\cdot), \xi^i(\cdot)\right) \\ &\qquad = \inf_{\tau \in \mathcal{T}_0} \left\{ \lambda_{\max}[X_\tau(0)] + \sum_{j=1}^{k} \tau_j d_j \right\}. \end{aligned}$$

Proof: The proof of this result follows directly from Theorem 6.2.1 for the case in which $\mathcal{X}_0 = \{x \in \mathbf{R}^n : \|x\| \leq 1\}$ and $\Psi(x_0) \equiv \gamma$ is a constant.

■

Remark 6.2.4 In order to apply the above results, a suitable value of the parameter vector τ must be found. In particular, in order to apply Corollary 6.2.1, a value of τ must be found which achieves the infimum in

$$\inf_{\tau \in \mathcal{T}_0} \left\{ \lambda_{\max}[X_\tau(0)] + \sum_{j=1}^{k} \tau_j d_j \right\}.$$

For the case of a single uncertainty block, the scalar τ can be found via a one parameter search. For the case of multiple uncertainty blocks, the optimization problem to find the vector τ becomes more involved. However, one possible approach to this problem would be to apply recent results from the theory of convex optimization; e.g., see [26].

6.3 The uncertainty averaging approach to output-feedback optimal guaranteed cost control

6.3.1 *Problem statement*

In this section, we address an output-feedback version of the optimal guaranteed cost control problem considered in Section 6.2. That is, an uncertain time-varying system of the form (6.2.1) is considered on a finite time interval $[0, T]$. However, we consider output-feedback control strategies. In such control strategies, the controller can depend only on the measured variable y defined by

$$y(t) = C_2(t)x(t) + \sum_{j=1}^{k} D_{2j}(t)\xi_j(t), \tag{6.3.1}$$

where $x(t) \in \mathbf{R}^n$ is the state of the system (6.2.1), $\xi_1(t) \in \mathbf{R}^{r_1}$, $\xi_2(t) \in \mathbf{R}^{r_2}, \ldots, \xi_k(t) \in \mathbf{R}^{r_k}$ are the uncertainty inputs for the system (6.2.1), and $C_2(\cdot)$, $D_{21}(\cdot)$, $D_{22}(\cdot)$, $\ldots$, $D_{2k}(\cdot)$, are bounded piecewise continuous matrix functions defined on $[0, T]$.

As in Section 6.2, the uncertainty in the system (6.2.1), (6.3.1) is described by the set of equations (6.2.3). Furthermore, the uncertainty is required to satisfy the averaged integral quadratic constraint (6.2.5) as defined in Definition 6.2.1; see also, Definition 2.3.4. The class of all admissible uncertainty sequences is also denoted Ξ. The reader is referred to Section 2.3.4 where this class of uncertainty is discussed in detail.

Associated with the uncertain system (6.2.1), (6.3.1), (6.2.5) is the quadratic cost functional (*c.f.* (6.2.2))

$$J_0\left(x(\cdot), u(\cdot), \xi(\cdot)\right) := x(T)'X_T x(T) + \int_0^T [x(t)'R(t)x(t) + u(t)'G(t)u(t)]dt \tag{6.3.2}$$

where $R(\cdot)$ and $G(\cdot)$ are given bounded piecewise continuous symmetric matrix weighting functions satisfying the condition: There exists a constant $\epsilon > 0$ such that $R(t) \geq \epsilon I$ and $G(t) \geq \epsilon I$ for all $t \in [0, T]$. Also, $X_T \geq 0$ is a given nonnegative-definite symmetric weighting matrix.

For a given uncertainty sequence $\mathcal{S} \in \Xi$, the corresponding average value of the cost functional is obtained by averaging the cost (6.3.2) over all uncertainties in the given uncertainty sequence. We then consider the worst case of this averaged cost over all admissible uncertainty sequences. Using the probabilistic interpretation of the averaged integral quadratic constraint uncertainty description (see Remark 2.3.2), the averaged cost would be the expected value of the cost function for a given uncertainty sequence. Thus, we would consider the worst case of the expected cost over all admissible uncertainty sequences.

The class of controllers to be considered in this section are time-varying linear output-feedback controllers of the form

$$\begin{aligned} \dot{x}_c &= A_c(t)x_c(t) + B_c(t)y(t); \qquad x_c(0) = 0; \\ u(t) &= C_c(t)x_c(t) + D_c(t)y(t) \end{aligned} \tag{6.3.3}$$

where $A_c(\cdot), B_c(\cdot), C_c(\cdot)$ and $D_c(\cdot)$ are bounded piecewise continuous matrix functions. Note, that the dimension of the controller state vector x_c may be arbitrary.

Let $P_0 > 0$ be a given positive-definite symmetric matrix and let $d_x > 0$ be a given constant. We will consider sequences of solutions of the system (6.2.1), (6.3.1) with initial conditions satisfying the following property: Given any $\mathcal{S} \in \Xi$, let q be the number of elements in this uncertainty sequence and let $\{x_0^i\}_{i=1}^q$ be a corresponding sequence of initial conditions, then

$$\frac{1}{q}\sum_{i=1}^{q} x^i(0)'P_0 x^i(0) \leq d_x. \tag{6.3.4}$$

The class of all such initial condition sequences is denoted $\mathcal{X}_0$.

Definition 6.3.1 *Let $\gamma > 0$ be a given positive constant. A controller of the form (6.3.3) is said to be a* guaranteed cost controller *for the uncertain system (6.2.1), (6.3.1), (6.2.5) with the cost functional (6.3.2), initial condition set $\mathcal{X}_0$ and cost bound γ if the following condition holds: Given any $\mathcal{S} \in \Xi$, let q be the number of elements in this uncertainty sequence and let $\{x_0^i\}_{i=1}^q \in \mathcal{X}_0$ be a corresponding sequence of initial conditions. Also, for each $\Phi^i(\cdot) \in \mathcal{S}$ and corresponding x_0^i,* *let* $[x^i(\cdot), u^i(\cdot), \xi^i(\cdot)]$ *be the corresponding solution to equations (6.2.1), (6.3.1), (6.2.4), (6.3.3) with initial condition* $x^i(0) = x_0^i$. *Then,*

$$\sup_{\mathcal{S}\in\Xi} \sup_{\{x_0^i\}_{i=1}^q \in \mathcal{X}_0} \left\{ \frac{1}{q} \sum_{i=1}^{q} J_0\left(x^i(\cdot), u^i(\cdot), \xi^i(\cdot)\right) \right\} < \gamma. \tag{6.3.5}$$

6.3.2 *A necessary and sufficient condition for the existence of a guaranteed cost controller*

We now present the main result of this section which establishes a necessary and sufficient condition for the existence of an output-feedback guaranteed cost controller for the time-varying uncertain system (6.2.1), (6.3.1), (6.2.5) with associated cost functional (6.3.2), initial condition set $\mathcal{X}_0$ and cost bound γ. This condition is given in terms of the existence of suitable solutions to a pair of parameter dependent Riccati differential equations. The Riccati equations under consideration are defined as follows: Let $\tau_1 > 0, \ldots, \tau_k > 0, \tau_{k+1} > 0$ be given constants and consider the Riccati equations:

$$\dot{X} = -A'X - XA + X(B_1 G_\tau^{-1} B_1' - \tilde{B}_2 \tilde{B}_2')X - C_\tau' C_\tau; \tag{6.3.6}$$

$$\begin{aligned} \dot{Y} &= Y(A - \tilde{B}_2 \tilde{D}_2' \Gamma_\tau^{-1} C_2)' + (A - \tilde{B}_2 \tilde{D}_2' \Gamma_\tau^{-1} C_2)Y \\ &\quad + Y(C_\tau C_\tau' - C_2' \Gamma_\tau^{-1} C_2)Y + \tilde{B}_2(I - \tilde{D}_2' \Gamma_\tau^{-1} \tilde{D}_2)\tilde{B}_2' \end{aligned} \tag{6.3.7}$$

where the matrices C_τ, D_τ, G_τ, $\tilde{B}_2$ are defined as in equation (6.2.10), and

$$\tilde{D}_2 := \left[\frac{1}{\sqrt{\tau_1}} D_{21} \ \cdots \ \frac{1}{\sqrt{\tau_k}} D_{2k} \right]; \ \Gamma_\tau := \tilde{D}_2 \tilde{D}_2'. \tag{6.3.8}$$

The main result of this section requires that the uncertain system (6.2.1), (6.3.1), (6.2.5) and associated cost functional (6.3.2) satisfy the following additional assumption.

Assumption 6.3.1

(i) $\int_0^T \|B_{2j}(t)\| dt > 0$ *for* $j = 1, 2, \ldots, k$.

(ii) $D_{1j}'(\cdot) C_{1j}(\cdot) \equiv 0$ *for* $j = 1, 2, \ldots, k$.

(iii) The matrix function $\Gamma(\cdot) := \sum_{j=1}^k D_{2j}(\cdot) D_{2j}'(\cdot)$ *is strictly positive;* i.e., *there exists a constant* $\epsilon > 0$ *such that* $\Gamma(t) \geq 0$ *for all* $t \in [0, T]$.

Theorem 6.3.1 *Consider the uncertain system (6.2.1), (6.3.1), (6.2.5) with associated cost functional (6.3.2) and initial condition set* $\mathcal{X}_0$. *Let* $\gamma > 0$ *be a given constant. Also, suppose that Assumption 6.3.1 is satisfied. Then there exists an output-feedback guaranteed cost controller of the form (6.3.3) with cost bound* γ *if and only if there exist constants* $\tau_1 > 0, \tau_2 > 0, \ldots, \tau_k > 0, \tau_{k+1} > 0$ *such that the following conditions hold:*

(i)

$$\sum_{j=1}^{k} \tau_j d_j + \tau_{k+1} d_x < \gamma; \tag{6.3.9}$$

(ii) The Riccati differential equation (6.3.7) with initial condition

$$Y(0) = \frac{1}{\tau_{k+1}} P_0^{-1}$$

has a solution $Y(t) > 0$ *on the interval* $[0,T]$*;*

(iii) The Riccati differential equation (6.3.6) with terminal condition

$$X(T) = X_T$$

has a solution $X(t)$ *on the interval* $[0,T]$ *such that* $X(0) < \tau_{k+1} P_0$*;*

(iv) The spectral radius of the product of the above solutions satisfies $\rho(Y(t)X(t)) < 1$ *for all* $t \in [0,T]$.

Furthermore, if there exist constants $\tau_1 > 0, \ldots, \tau_k > 0, \tau_{k+1} > 0$ *such that conditions (i)–(iv) hold, then the uncertain system (6.2.1), (6.3.1), (6.2.5) with associated cost functional (6.3.2) and initial condition set* $\mathcal{X}_0$ *admits a linear time-varying guaranteed cost controller of the form (6.3.3) with cost bound* γ *where*

$$\begin{aligned} A_c(t) &= A + B_1 C_c - B_c C_2 + (\tilde{B}_2 - B_c \tilde{D}_2)\tilde{B}_2' X; \\ B_c(t) &= (I - YX)^{-1}(YC_2' + \tilde{B}_2 \tilde{D}_2')\Gamma_\tau^{-1}; \\ C_c(t) &= -G_\tau^{-1} B_1' X; \quad D_c(t) = 0. \end{aligned} \tag{6.3.10}$$

Proof: Suppose the uncertain system (6.2.1), (6.3.1), (6.2.5) is such that there exists an output-feedback guaranteed cost controller of the form (6.3.3) with cost bound γ. Then inequality (6.3.5) implies that there exists a constant $\delta > 0$ such that the following condition holds: Given any admissible uncertainty sequence $\mathcal{S} = \{\Phi^i(\cdot)\}_{i=1}^q \in \Xi$ and any corresponding sequence of initial conditions $\{x_0^i\}_{i=1}^q \in \mathcal{X}_0$, let $[x^i(\cdot), u^i(\cdot), \xi^i(\cdot)]$ be the corresponding solution to equations (6.2.1), (6.3.1), (6.3.3) and (6.2.4) with initial condition $x^i(0) = x_0^i$. Then

$$\frac{(1+\delta)}{q} \sum_{i=1}^{q} J_0(x^i(\cdot), u^i(\cdot), \xi^i(\cdot)) \leq \gamma - \delta. \tag{6.3.11}$$

For the system closed loop system defined by (6.2.1), (6.3.1) and (6.3.3), a corresponding set $\mathcal{M} \subset \mathbf{L}_2[0,T]$ is defined as in equation (6.2.14). Also, we consider the following set of functionals mapping from $\mathcal{M}$ into $\mathbf{R}$:

$$\begin{aligned}
\mathcal{G}_0\left(x(\cdot),u(\cdot),\xi(\cdot)\right) &:= -(1+\delta)J_0(x(\cdot),u(\cdot),\xi^i(\cdot)) + \gamma - \delta; \\
\mathcal{G}_1\left(x(\cdot),u(\cdot),\xi(\cdot)\right) &:= \int_0^T (\|z_1(t)\|^2 - \|\xi_1(t)\|^2)dt + d_1; \\
&\vdots \\
\mathcal{G}_k\left(x(\cdot),u(\cdot),\xi(\cdot)\right) &:= \int_0^T (\|z_k(t)\|^2 - \|\xi_k(t)\|^2)dt + d_k; \\
\mathcal{G}_{k+1}\left(x(\cdot),u(\cdot),\xi(\cdot)\right) &:= -x(0)'P_0x(0) + d_x. \qquad (6.3.12)
\end{aligned}$$

Now consider a sequence of vector functions $\{x^i(\cdot),u^i(\cdot),\xi^i(\cdot)\}_{i=1}^q \subset \mathcal{M}$ such that

$$\sum_{i=1}^q \mathcal{G}_1\left(x^i(\cdot),u^i(\cdot),\xi^i(\cdot)\right) \geq 0, \ldots, \quad \sum_{i=1}^q \mathcal{G}_{k+1}\left(x^i(\cdot),u^i(\cdot),\xi^i(\cdot)\right) \geq 0.$$

If we recall Definition 6.2.1 and condition (6.3.4), it follows that this sequence corresponds to an admissible uncertainty sequence $\mathcal{S} \in \Xi$ and an initial condition sequence $\{x^i(0)\}_{i=1}^q \in \mathcal{X}_0$. Hence, inequality (6.3.11) implies that

$$\sum_{i=1}^q \mathcal{G}_0\left(x^i(\cdot),u^i(\cdot),\xi^i(\cdot)\right) \geq 0. \qquad (6.3.13)$$

We now apply Theorem 4.5.1 and Observation 4.5.1 (with k replaced by $k+1$) to the set $\mathcal{P} \subset \mathbf{R}^{k+2}$ defined as follows:

$$\mathcal{P} := \left\{ \nu = \begin{bmatrix} \mathcal{G}_0\left(x(\cdot),u(\cdot),\xi(\cdot)\right) \\ \vdots \\ \mathcal{G}_k\left(x(\cdot),u(\cdot),\xi(\cdot)\right) \\ \mathcal{G}_{k+1}\left(x(\cdot),u(\cdot),\xi(\cdot)\right) \end{bmatrix} : [x(\cdot),u(\cdot),\xi(\cdot)] \in \mathcal{M} \right\}.$$

In a similar fashion to the proof of Theorem 6.2.1, we can conclude that there exists constants $\tau_1 \geq 0, \ldots, \tau_{k+1} \geq 0$ such that

$$\nu_0 \geq \tau_1\nu_1 + \cdots + \tau_{k+1}\nu_{k+1} \qquad (6.3.14)$$

for all $\nu = [\nu_0\ \nu_1\ \cdots\ \nu_{k+1}]' \in \mathcal{P}$. Hence,

$$\mathcal{G}_0\left(x(\cdot),u(\cdot),\xi(\cdot)\right) \geq \sum_{j=1}^{k+1} \tau_j \mathcal{G}_j\left(x(\cdot),u(\cdot),\xi(\cdot)\right) \qquad (6.3.15)$$

for all $[x(\cdot),u(\cdot),\xi(\cdot)] \in \mathcal{M}$. Also, using this inequality and equations (6.3.12), we can conclude that the functional

$$\begin{aligned}
J_\delta^\tau\left(x(\cdot),u(\cdot),\xi(\cdot)\right) := &\ (1+\delta)J_0\left(x(\cdot),u(\cdot)\right) - \tau_{k+1}x(0)'P_0x(0) \\
&+ \sum_{j=1}^k \tau_j \int_0^T (\|z_j(t)\|^2 - \|\xi_j(t)\|^2)dt \qquad (6.3.16)
\end{aligned}$$

satisfies the following inequality

$$J_\delta^T(x(\cdot), u(\cdot), \xi(\cdot)) \leq \gamma - \delta - \sum_{j=1}^{k} \tau_j d_j - \tau_{k+1} d_x \tag{6.3.17}$$

for all $[x(\cdot), u(\cdot), \xi(\cdot)] \in \mathcal{M}$. Now consider the zero element $[0,0,0] \in \mathcal{M}$. Since $J_\delta^T(0,0,0) = 0$, inequality (6.3.17) implies that

$$\sum_{j=1}^{k} \tau_j d_j + \tau_{k+1} d_x \leq \gamma - \delta.$$

Hence, condition (6.3.9) holds.

We now prove that

$$J_\delta^T(x(\cdot), u(\cdot), \xi(\cdot)) \leq 0 \tag{6.3.18}$$

for all $[x(\cdot), u(\cdot), \xi(\cdot)] \in \mathcal{M}$. Indeed, if this is not true and

$$J_\delta^T(x^*(\cdot), u^*(\cdot), \xi^*(\cdot)) > 0 \tag{6.3.19}$$

for some $[x^*(\cdot), u^*(\cdot), \xi^*(\cdot)] \in \mathcal{M}$. Since the closed loop system (6.2.1), (6.3.1), (6.3.3) is linear, $[ax^*(\cdot), au^*(\cdot), a\xi^*(\cdot)] \in \mathcal{M}$ for any $a > 0$ and condition (6.3.19) implies that

$$\lim_{a\to\infty} J_\delta^T(ax^*(\cdot), au^*(\cdot), a\xi^*(\cdot)) = \infty.$$

However, this contradicts condition (6.3.17). Hence, we have proved condition (6.3.18).

We now prove that $\tau_j > 0$ for $j = 1, 2, \ldots, k, k+1$. For $j = 1, \ldots, k$, this fact is established exactly as in the proof of Theorem 6.2.1. Furthermore, if $\tau_{k+1} = 0$, we consider an input $\xi(\cdot) \equiv 0$ and a solution to the system (6.2.1), (6.3.1), (6.3.3) with any non-zero initial condition. Then from (6.3.16) and the fact that $R(t) \geq \epsilon I$ and $G(t) \geq \epsilon I$ for all $t \in [0, T]$, we must have $J_\delta^T(x(\cdot), u(\cdot), 0) > 0$ which contradicts condition (6.3.18). Thus, we have proved that $\tau_j > 0$ for all $j = 1, 2, \ldots, k+1$.

In order to establish the existence of suitable solutions to the Riccati equations (6.3.7) and (6.3.6), we consider the system

$$\begin{aligned} \dot{x}(t) &= A(t)x(t) + B_1(t)u(t) + \tilde{B}_2(t)\hat{\xi}(t); \\ y(t) &= C_2(t)x(t) + \tilde{D}_2(t)\hat{\xi}(t); \\ \hat{z}(t) &= C_\tau(t)x(t) + D_\tau(t)u(t) \end{aligned} \tag{6.3.20}$$

where

$$\begin{aligned} \hat{z}(\cdot) &:= \left[x(\cdot)' R^{\frac{1}{2}} \; u(\cdot)' G^{\frac{1}{2}} \; \sqrt{\tau_1} z_1(\cdot)' \; \ldots \; \sqrt{\tau_k} z_k(\cdot)' \right]'; \\ \hat{\xi}(\cdot) &:= \left[\sqrt{\tau_1} \xi_1(\cdot)' \; \ldots \; \sqrt{\tau_k} \xi_k(\cdot)' \right]' \end{aligned}$$

and the matrices $\tilde{B}_2(\cdot)$, $\tilde{D}_2(\cdot)$, $C_\tau(\cdot)$ and $D_\tau(\cdot)$ are defined as in (6.3.8), (6.2.10). Since we have established above that $\tau_j > 0$ for all s, this system is well defined. It is straightforward to establish that with this notation, the state equations (6.3.20) are equivalent to the state equations (6.2.1), (6.3.1). Furthermore, condition (6.3.18) for the functional (6.3.16) may be re-written as

$$x(T)'X_T x(T) + \int_0^T (\|\hat{z}(t)\|^2 - \|\hat{\xi}(t)\|^2)dt - \tau_{k+1}x(0)'P_0x(0) \\ \leq -\delta J_0(x(\cdot), u(\cdot)) \quad (6.3.21)$$

for all solutions to the closed loop system (6.3.20), (6.3.3). Moreover, condition (6.3.21) together with the fact that $R(t) \geq \epsilon I$ and $G(t) \geq \epsilon I$ for all $t \in [0, T]$ implies that the controller (6.3.3) solves the finite-horizon H^∞ control problem for the system (6.3.20) with the following H^∞ requirement

$$\sup \left\{ \frac{x(T)'X_T x(T) + \int_0^T \|\hat{z}(t)\|^2 dt}{\tau_{k+1}x(0)'P_0x(0) + \int_0^T \|\hat{\xi}(t)\|^2 dt} \right\} < 1 \quad (6.3.22)$$

where the supremum is taken over all $x(0) \in \mathbf{R}^n$, $\hat{\xi}(\cdot) \in \mathbf{L}_2[0, T]$ such that

$$\tau_{k+1}x(0)'P_0x(0) + \int_0^T \|\hat{\xi}(t)\|^2 dt > 0$$

(see Section 3.2.2 and reference [100]). Therefore, Theorem 3.2.6 implies that conditions (ii), (iii) and (iv) hold. Note that the assumptions of this theorem are satisfied. This fact follows from Assumption 6.3.1. Since condition (i) has been proved above, this completes the proof of this part of the theorem.

Now suppose that there exist constants $\tau_1 > 0, \ldots, \tau_k > 0, \tau_{k+1} > 0$ such that conditions (i)–(iv) hold and consider the system defined by state equations (6.3.20). We noted above that Assumption 6.3.1 implies the satisfaction of the conditions of Theorem 3.2.6. Using Theorem 3.2.6, it follows that condition (6.3.22) holds for all solutions to the closed loop system (6.3.20), (6.3.3), (6.3.10). Furthermore, using the notation given above, condition (6.3.22) implies that

$$J_0(x(\cdot), u(\cdot)) + \sum_{j=1}^{k} \tau_j \int_0^T (\|z_j(t)\|^2 - \|\xi_j(t)\|^2)dt - \tau_{k+1}x(0)'P_0x(0) \leq 0$$

for all solutions to the closed loop system (6.2.1), (6.3.1), (6.3.3), (6.3.10). Combining this inequality with inequalities (6.2.5), (6.3.4) and (6.3.9) leads directly to the satisfaction of condition (6.3.5). Thus, the controller (6.3.3), (6.3.10) is a guaranteed cost controller with cost bound γ. This completes the proof of the theorem.

■

6.4 Robust control with a terminal state constraint

6.4.1 Problem statement

This section considers a robust control problem referred to in Chapter 1 as robust control with a terminal state constraint. The problem of robust control with a terminal state constraint (see Definition 6.4.2) has a goal which is analogous to condition (6.3.22) in the guaranteed cost control problem of Section 6.3.

In the problem to be considered in this section, the underlying system is uncertain and only output-feedback is available. Hence, it is unreasonable to expect a controller to steer a given initial state exactly to the origin in a finite time. For example in [149], it was shown that even in the state-feedback case, an uncertain system must satisfy a very strong assumption for this to be possible. Thus, we consider the problem of steering a given initial state into an ellipsoidal bounding set. Furthermore, the size of this bounding set will depend on the size of the initial condition. The problem considered also imposes a penalty on the control energy used in steering the state into the ellipsoidal bounding set.

Consider an uncertain system of the form (2.3.13) defined on the finite time interval $[0,T]$ as follows:

$$\begin{aligned} \dot{x}(t) &= A(t)x(t) + B_1(t)u(t) + B_2(t)w(t); \\ z(t) &= C_1(t)x(t) + D_1(t)u(t); \\ y(t) &= C_2(t)x(t) + v(t) \end{aligned} \tag{6.4.1}$$

where $x(t) \in \mathbf{R}^n$ is the *state*, $w(t) \in \mathbf{R}^p$ and $v(t) \in \mathbf{R}^k$ are the *uncertainty inputs*, $u(t) \in \mathbf{R}^m$ is the *control input*, $z(t) \in \mathbf{R}^q$ is the *uncertainty output*, $y(t) \in \mathbf{R}^k$ is the *measured output*, $A(\cdot)$, $B_1(\cdot)$, $B_2(\cdot)$, $C_1(\cdot)$, $D_1(\cdot)$ and $C_2(\cdot)$ are bounded piecewise continuous matrix functions.

System Uncertainty

The uncertainty in the above system is represented by an equation

$$\left[\, w(t) \;\; v(t) \,\right] = \phi(t, x(\cdot)) \tag{6.4.2}$$

where the following integral quadratic constraint is satisfied (*c.f.* Definition 2.3.2):

Definition 6.4.1 *Let $S = S' > 0$ be a given positive-definite matrix, $Q(\cdot) = Q(\cdot)'$ and $N(\cdot) = N(\cdot)'$ be given bounded piecewise continuous weighting matrix functions such that for some $\delta > 0$, $Q(t) \geq \delta I$ and $N(t) \geq \delta I$ for all $t \in [0,T]$. An uncertainty of the form (6.4.2) is an admissible uncertainty for the system (6.4.1) if the following conditions hold: Given any control input $u(\cdot) \in \mathbf{L}_2[0,T]$, then any corresponding solution to equations (6.4.1), (6.4.2) is defined on $[0,T]$ and*

$$\int_0^T (w(t)'Q(t)w(t) + v(t)'N(t)v(t))dt \leq \int_0^T \|z(t)\|^2 dt + x(0)'Sx(0). \tag{6.4.3}$$

The class of all such admissible uncertainty inputs is denoted Ξ.

As noted in Section 2.3, the uncertainty description given above allows for $w(t)$ and $v(t)$ to depend dynamically on $z(t)$. It has been observed previously that an uncertainty description of this form allows for uncertainty satisfying a norm bound condition; see Observation 2.3.1. To see that this observation is true for the uncertain system (6.4.1), (6.4.2), one could consider an uncertain system described as follows:

$$\begin{aligned}
\dot{x}(t) &= [A(t) + B_2(t)\Delta(t)C_1(t)]x(t) + [B_1(t) + B_2(t)\Delta(t)D_1(t)]u(t); \\
y(t) &= C_2(t)x(t) + n(t); \\
&\quad \|\Delta(t)Q(t)^{\frac{1}{2}}\| \leq 1 \quad \forall t; \\
&\quad \int_0^T n(t)'N(t)n(t)dt \leq x(0)'Sx(0)
\end{aligned} \tag{6.4.4}$$

where $\Delta(t)$ is a time-varying uncertainty matrix and $n(t)$ is a measurement noise signal. To verify that such uncertainty is admissible for the uncertain system (6.4.1), (6.4.3), let $w(t) = \Delta(t)[C_1(t)x(t) + D_1(t)u(t)]$ and $v(t) = n(t)$. Then $w(\cdot)$ and $v(\cdot)$ satisfy condition (6.4.3).

Also note that when studying questions of stability and stabilizability for systems with a single uncertainty block satisfying an integral quadratic constraint, it has been shown in Section 3.5 that the question of robust stability is equivalent to a small gain condition. This is the same condition as obtained using the quadratic stability and frequency domain small gain approaches to robustness. Thus in this case, the integral quadratic constraint approach to modeling uncertainty is no more conservative than more commonly used approaches to modeling uncertainty. However, these approaches cannot easily be applied to the problem considered in this section. Also, it should be noted that it would be possible to extend the above class of uncertain systems to include multiple uncertainty blocks. However in this case, the S-procedure used in this section becomes conservative.

We now consider a problem of robust control with a terminal state constraint for the uncertain system (6.4.1), (6.4.3). The class of controllers to be considered are finite dimensional time-varying linear output-feedback controllers of the form (6.3.3). Note that the dimension of the controller state may be arbitrary.

Definition 6.4.2 *Let* $P_0 = P_0' > 0$ *and* $X_T = X_T' > 0$ *be given matrices and* $\gamma > 0$ *be a given constant. The controller (6.3.3) is said to be a* robust controller with respect to the terminal state constraint defined by P_0, X_T *and* γ, *if the closed loop uncertain system (6.4.1), (6.4.3), (6.3.3) satisfies the constraint*

$$\sup_{\substack{x(0)\in\mathbf{R}^n\setminus\{0\},\\ [w(\cdot),v(\cdot)]\in\Xi}} \frac{x(T)'X_Tx(T) + \gamma\int_0^T \|u(t)\|^2dt}{x(0)'P_0x(0)} < 1. \tag{6.4.5}$$

An uncertain system (6.4.1), (6.4.3) which admits a robust controller (6.3.3) with a terminal state constraint defined by P_0, X_T *and* γ *is said to be* robustly controllable *with respect to this terminal state constraint.*

Remark 6.4.1 Note that the constraint (6.4.5) includes a term weighting the control input $u(t)$. This corresponds to the situation in which one wishes to limit the size of the control input signal $u(t)$ as well as meet the terminal state constraint.

6.4.2 *A criterion for robust controllability with respect to a terminal state constraint*

The main result of this section requires that the system (6.4.1) satisfies the following additional assumption; *c.f.* Assumption 6.3.1.

Assumption 6.4.1

(i) The matrix functions $C_1(\cdot)$ and $D_1(\cdot)$ satisfy the condition $C_1(\cdot)'D_1(\cdot) \equiv 0$.

(ii) The matrix function $B_2(\cdot)$ satisfies the condition

$$\int_0^T \|B_2(t)\|dt > 0.$$

Let $\tau > 0$ be a given constant. We will consider the following pair of Riccati differential equations

$$\begin{aligned} -\dot{X}(t) &= A(t)'X(t) + X(t)A(t) + C_1(t)'C_1(t) \\ &\quad +X(t)\begin{pmatrix} B_2(t)Q^{-1}B_2(t)' \\ \quad -B_1(t)(D_1(t)'D_1(t) + \tau\gamma I)^{-1}B_1(t)' \end{pmatrix} X(t); \end{aligned} \tag{6.4.6}$$

$$\begin{aligned} \dot{Y}(t) &= A(t)Y(t) + Y(t)A(t)' + B_2(t)Q(t)^{-1}B_2(t)' \\ &\quad +Y(t)\left(C_1(t)'C_1(t) - C_2(t)'N(t)C_2(t)\right)Y(t). \end{aligned} \tag{6.4.7}$$

Theorem 6.4.1 *Consider the uncertain system (6.4.1), (6.4.3) and suppose that Assumption 6.4.1 is satisfied. Also, let $P_0 = P_0' > 0$ and $X_T = X_T' > 0$ be given matrices and $\gamma > 0$ be a given constant. Then the uncertain system (6.4.1), (6.4.3) is robustly controllable with respect to the terminal state constraint defined by P_0, X_T and $\gamma > 0$ if and only if there exists a constant $\tau > 0$ such that the following conditions hold:*

(i) There exists a solution $X(\cdot)$ to Riccati differential equation (6.4.6) defined on the interval $[0, T]$ with $X(T) = \tau X_T$. Furthermore, this solution satisfies $\tau P_0 - S > X(0) > 0$.

(ii) There exists a solution $Y(\cdot)$ to Riccati differential equation (6.4.7) defined on the interval $[0, T]$ with $Y(0) = (\tau P_0 - S)^{-1}$. Furthermore, this solution satisfies $Y(t) > 0$ for all $t \in [0, T]$.

(iii) The spectral radius $\rho(Y(t)X(t))$ satisfies

$$\rho(Y(t)X(t)) < 1$$

for all $t \in [0, T]$.

If conditions (i)-(iii) hold then a controller of the form (6.3.3) with coefficient matrices

$$\begin{aligned} A_c(t) &= A(t) + B_1(t)C_c(t) - B_c(t)C_2(t) + B_2(t)Q(t)^{-1}B_2(t)'X(t); \\ B_c(t) &= (I - Y(t)X(t))^{-1}Y(t)C_2(t)'N(t); \\ C_c(t) &= -(D_1(t)'D_1(t) + \tau\gamma I)^{-1}B_1(t)'X(t); \\ D_c(t) &= 0 \end{aligned} \tag{6.4.8}$$

is a robust controller with respect to the above terminal state constraint.

Proof: *Necessity.* Suppose that there exists a controller of the form (6.3.3) such that condition (6.4.5) holds. Since $P_0 > 0$, $X_T > 0$ and $\gamma > 0$, condition (6.4.5) implies that there exists an $\epsilon > 0$ such that

$$\begin{aligned} &(1+\epsilon)(x(T)'X_T x(T) + \gamma \int_0^T \|u(t)\|^2 dt) + \epsilon \int_0^T (\|x(t)\|^2 + \|u(t)\|^2)dt \\ &\leq x(0)'P_0 x(0) \end{aligned} \tag{6.4.9}$$

for all solutions to the closed loop uncertain system (6.4.1), (6.4.3), (6.3.3). Now let $\mathcal{M}$ be the Hilbert space of all vector valued functions

$$\lambda(\cdot) := [x(\cdot), u(\cdot), w(\cdot), v(\cdot)] \in \mathbf{L}_2[0, T]$$

connected by the equations (6.4.1), (6.3.3). We also introduce continuous quadratic functionals $\mathcal{G}_0(\cdot)$ and $\mathcal{G}_1(\cdot)$ defined on $\mathcal{M}$ by the equations

$$\begin{aligned} \mathcal{G}_0(\lambda(\cdot)) &:= (1+\epsilon)(x(T)'X_T x(T) + \gamma \int_0^T \|u(t)\|^2 dt \\ &\quad + \epsilon \int_0^T (\|x(t)\|^2 + \|u(t)\|^2)dt - x(0)'P_0 x(0); \\ \mathcal{G}_1(\lambda(\cdot)) &:= \int_0^T (\|z(t)\|^2 - w(t)'Q(t)w(t) - v(t)'N(t)v(t))dt \\ &\quad + x(0)'Sx(0). \end{aligned} \tag{6.4.10}$$

Since condition (6.4.9) holds for all solutions to the system (6.4.1), (6.3.3) with inputs $w(\cdot)$ and $v(\cdot)$ satisfying condition (6.4.3), then we have

$$\mathcal{G}_0(\lambda(\cdot)) \leq 0 \text{ for all } \lambda(\cdot) \in \mathcal{M} \text{ such that } \mathcal{G}_1(\lambda(\cdot)) \geq 0.$$

Since $S > 0$, if we let $w(t) \equiv 0$, $v(t) \equiv 0$ and $x(0) \neq 0$, the corresponding vector function $\lambda(\cdot) \in \mathcal{M}$ is such that $\mathcal{G}_1(\lambda(\cdot)) > 0$. Hence, using the S-procedure Theorem 4.3.2, it follows that there exists a constant $\mu \geq 0$ such that the inequality

$$\mathcal{G}_0(\lambda(\cdot)) + \mu\mathcal{G}_1(\lambda(\cdot)) \leq 0 \tag{6.4.11}$$

is satisfied for all $\lambda(\cdot) \in \mathcal{M}$. Also, Assumption 6.4.1 implies that we can choose an input $w(\cdot)$ such that $B_2(\cdot)w(\cdot) \neq 0$. Then, for the corresponding solution to the system (6.4.1), (6.4.3) with $x(0) = 0$, we have $\int_0^T (\|x(t)\|^2 + \|u(t)\|^2)dt > 0$. Therefore, the corresponding vector function $\lambda(\cdot) \in \mathcal{M}$ is such that $\mathcal{G}_0(\lambda(\cdot)) > 0$. Referring to inequality (6.4.11), this implies that $\mu \neq 0$. However since $\mu \geq 0$, we can now conclude that $\mu > 0$. Hence, if we define $\tau := 1/\mu > 0$, it follows from (6.4.11) that

$$\mathcal{G}_1(\lambda(\cdot)) + \tau\mathcal{G}_0(\lambda(\cdot)) \leq 0 \quad \forall\lambda(\cdot) \in \mathcal{M}. \tag{6.4.12}$$

Now with $\mathcal{G}_0(\cdot)$ and $\mathcal{G}_1(\cdot)$ defined as in (6.4.10), inequality (6.4.12) may be re-written as

$$\begin{aligned} &x(0)'[\tau P_0 - S]x(0) \\ &\geq \tau(1+\epsilon)(x(T)'X_T x(T) + \tau\gamma \int_0^T \|u(t)\|^2 dt \\ &\quad + \int_0^T \begin{bmatrix} \delta(\|x(t)\|^2 + \|u(t)\|^2) + \|z(t)\|^2 \\ -w(t)'Q(t)w(t) - v(t)'N(t)v(t) \end{bmatrix} dt \end{aligned} \tag{6.4.13}$$

where $\delta = \epsilon\tau$. If we let $w(t) \equiv 0$ and $v(t) \equiv 0$, it follows from this inequality that $\tau P_0 - S > 0$.

Now consider the system

$$\begin{aligned} \dot{x}(t) &= A(t)x(t) + \hat{B}_2(t)\hat{w}(t) + B_1(t)u(t); \\ \hat{z}(t) &= \begin{bmatrix} C_1(t)x(t) + D_1(t)u(t) \\ \sqrt{\tau\gamma}u(t) \end{bmatrix}; \\ y(t) &= C_2(t)x(t) + \hat{D}_2(t)\hat{w}(t) \end{aligned} \tag{6.4.14}$$

where

$$\begin{aligned} &\hat{w}(t) := \begin{bmatrix} Q(t)^{\frac{1}{2}}w(t) \\ N(t)^{\frac{1}{2}}v(t) \end{bmatrix}; \quad \hat{B}_2(t) := \begin{bmatrix} B_2(t)Q(t)^{-\frac{1}{2}} & 0 \end{bmatrix}; \\ &\hat{D}_2(t) := \begin{bmatrix} 0 & N(t)^{-\frac{1}{2}} \end{bmatrix}. \end{aligned}$$

For this system, we let $\hat{w}(\cdot)$ be the disturbance input and let $\hat{z}(\cdot)$ be the controlled output. Using these definitions, it follows that inequality (6.4.13) may be re-written as

$$\begin{aligned} &x(0)'[\tau P_0 - S]x(0) \\ &\geq \tau(1+\epsilon)x(T)'X_T x(T) \\ &\quad + \int_0^T \left[\delta(\|x(t)\|^2 + \|u(t)\|^2) + \|\hat{z}(t)\|^2 - \|\hat{w}(t)\|^2\right] dt. \end{aligned} \tag{6.4.15}$$

This implies that the following condition holds

$$\sup_{\substack{x(0)\in\mathbf{R}^n\setminus\{0\},\\ \hat{w}(\cdot)\in\mathbf{L}_2[0,T]}} \left\{ \frac{\tau x(T)'X_T x(T) + \int_0^T \|\hat{z}(t)\|^2 dt}{x(0)'[\tau P_0 - S]x(0) + \int_0^T \|\hat{w}(t)\|^2 dt} \right\} < 1. \tag{6.4.16}$$

That is, the controller (6.3.3) solves a H^∞ control problem defined by the system (6.4.14) and the H^∞ norm bound condition (6.4.16); see Section 3.2.2. Therefore, Theorem 3.2.6 of Section 3.2.2 implies that conditions (i)-(iii) are satisfied.

Sufficiency. Suppose there exists a constant $\tau > 0$ such that conditions (i)-(iii) hold. Then Theorem 3.2.6 of Section 3.2.2 implies that the controller (6.3.3) with the coefficients (6.4.8) solves the H^∞ control problem (6.4.16) for the uncertain system (6.4.14). It follows from condition (6.4.16) that there exists a constant $\delta > 0$ such that the inequality (6.4.15) holds for all $[x(\cdot), u(\cdot), \hat{w}(\cdot)] \in \mathbf{L}_2[0,T]$ connected by (6.4.14), (6.3.3), (6.4.8). This fact implies that inequality (6.4.13) is satisfied for the closed loop system (6.4.1), (6.3.3), (6.4.8) with any uncertainty inputs $[w(\cdot), v(\cdot)] \in \mathbf{L}_2[0,T]$. Combining inequality (6.4.13) with condition (6.4.3), we conclude that for any admissible uncertainty inputs $[w(\cdot), v(\cdot)] \in \Xi$, condition (6.4.5) holds.

■

6.4.3 *Illustrative example*

To illustrate the results of this section, we consider a problem of robust control with a terminal state constraint applied to the benchmark system of Section 5.3. This system consists of two carts connected by a spring as shown in Figure 2.4.2. As in Section 5.3, it is assumed that the masses are $m_1 = 1$ and $m_2 = 1$ and the spring constant k is treated as an uncertain parameter subject to the bound $0.5 \leq k \leq 2.0$. It is assumed that the variables x_1 and x_2 are available for measurement. From this, we obtain the following uncertain system of the form (6.4.4) which will be considered over the time interval $[0, 20]$.

$$\begin{aligned}
\dot{x} &= \begin{bmatrix} 0 & 0 & 1 & 0 \\ 0 & 0 & 0 & 1 \\ -1.25 & 1.25 & 0 & 0 \\ 1.25 & -1.25 & 0 & 0 \end{bmatrix} x + \begin{bmatrix} 0 \\ 0 \\ -0.75 \\ 0.75 \end{bmatrix} w + \begin{bmatrix} 0 \\ 0 \\ 0 \\ 1 \end{bmatrix} u, \\
y &= \begin{bmatrix} 1 & 0 & 0 & 0 \\ 0 & 1 & 0 & 0 \end{bmatrix} x + n(t), \\
w &= \Delta(t) \begin{bmatrix} 1 & -1 & 0 & 0 \end{bmatrix} x,
\end{aligned} \tag{6.4.17}$$

where

$$|\Delta(t)| \leq 1 \text{ and } \int_0^{20} 10^3 \|n(t)\|^2 dt \leq 10^{-4} x(0)^2.$$

We wish to construct a robust controller so that condition (6.4.5) is satisfied with $P_0 = 0.02 \times I$, $X_T = 0.2 \times I$ and $\gamma = 10^{-5}$. Choosing a value of $\tau = 18$

and solving Riccati equations (6.4.6) and (6.4.7), it is found that the conditions of Theorem 6.4.1 are satisfied. From these Riccati solutions, the controller (6.4.8) was constructed and applied to the system (6.4.17). Using the initial condition of $x(0) = [7\ 0\ 0\ 0]'$, the resulting closed loop system was simulated for different values of the uncertain parameter Δ. The results of these simulations are shown in Figure 6.4.1. For this initial condition, and the values of P_0, X_T and γ given above, condition (6.4.5) essentially amounts to the requirement $\|x(20)\| \leq 2.24$. From the simulations, it can be seen that although there is a large initial transient, this terminal state constraint condition is met for each value of the uncertain parameter.

For the sake of comparison, we now compare our results with those which could be obtained using existing methods. In particular, if the system were not uncertain and the full state was available for measurement, the problem could be solved as a finite-horizon linear quadratic regulator problem with cost functional

$$x(T)'X_T x(T) + \gamma \int_0^T \|u(t)\|^2 dt.$$

This problem is solved by using the state-feedback control law

$$u(t) = -\frac{1}{\gamma} B_1(t)'X(t)x(t)$$

where $X(\cdot)$ solves the Riccati equation

$$\begin{aligned} -\dot{X}(t) &= A(t)'X(t) + X(t)A(t) - X(t)B_1(t)B_1(t)'X(t)/\gamma; \\ X(T) &= X_T. \end{aligned}$$

If the full state is not available for measurement, then it can be estimated using a Kalman Filter based on the model

$$\begin{aligned} \dot{x}(t) &= A(t)x(t) + B_1(t)u(t); \\ y(t) &= C_2(t)x(t) + n(t) \end{aligned}$$

where the initial condition is regarded as a random variable and $n(t)$ is regarded as white noise. For the sake of this comparison, it assumed that the initial condition has a Gaussian probability distribution with zero mean and the following variance

$$\mathbf{E}[x(0)x(0)'] = P_0^{-1}.$$

Also, the noise $n(t)$ is assumed to have the covariance matrix as follows

$$\mathbf{E}[n(t)n(t)'] = N^{-1}.$$

The matrices P_0 and N are taken to be the same as above. The solution to this Kalman filtering problem obtained by solving the Riccati equation

$$\begin{aligned} \dot{Y}(t) &= A(t)Y(t) + Y(t)A(t)' - Y(t)C_2(t)'N(t)C_2(t)Y(t); \\ Y(0) &= P_0^{-1}. \end{aligned}$$

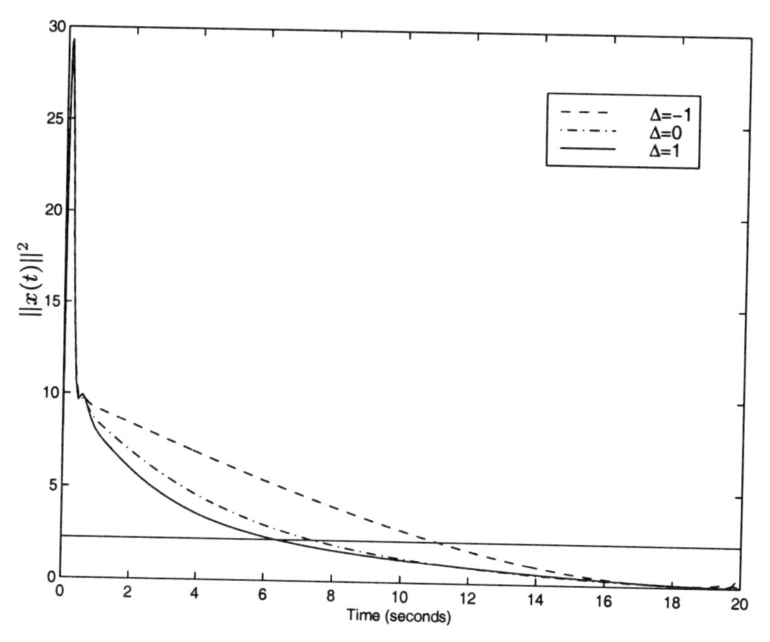

Figure 6.4.1. $||x||^2$ versus time for the robust controller (6.4.8).

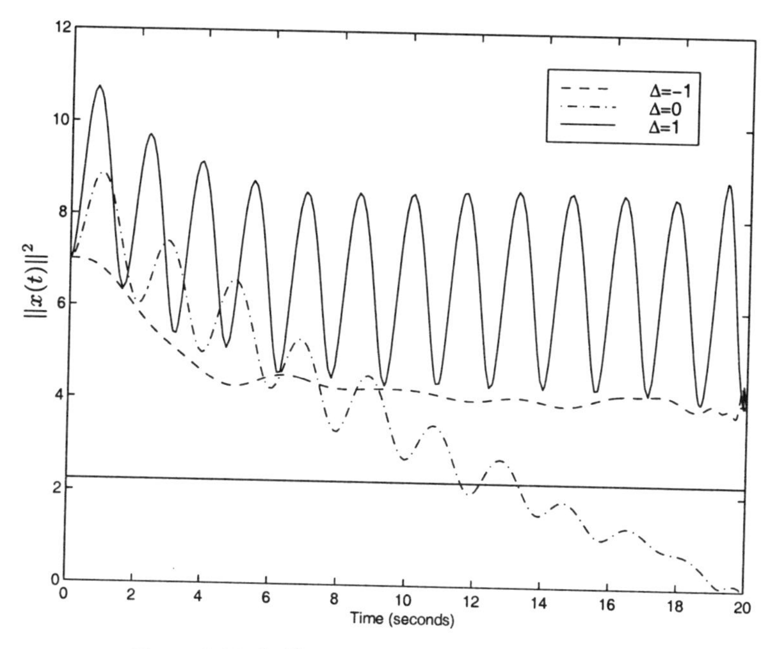

Figure 6.4.2. $||x||^2$ versus time for the LQG controller.

Combining this Kalman Filter with the above state-feedback controller leads to an LQG controller of the form (6.3.3) where

$$\begin{aligned} A_c(t) &= A(t) + B_1(t)C_c(t) - B_c(t)C_2(t); \\ B_c(t) &= Y(t)C_2(t)'N(t); \\ C_c(t) &= -B_1(t)'X(t)/\gamma; \\ D_c(t) &= 0. \end{aligned}$$

Such an LQG controller was constructed for the system (6.4.17) assuming $\Delta = 0$. Using an initial condition of $x(0) = [7\ 0\ 0\ 0]'$, the resulting closed loop system was simulated for different values of the uncertain parameter Δ. The results of these simulations are shown in Figure 6.4.2. From these simulations, it can be seen that although the LQG controller performs well for the nominal system ($\Delta = 0$), it does not satisfy the terminal state constraint for other values of the uncertain parameter. Thus we can see that our robust controller performs better than an LQG controller based purely on the nominal system.

6.5 Robust control with rejection of harmonic disturbances

6.5.1 *Problem statement*

Consider an uncertain system defined on the finite time interval $[0, T]$ as follows:

$$\begin{aligned} \dot{x}(t) &= A(t)x(t) + B_1(t)u(t) + B_2(t)w(t) + B_3(t)\nu(t); \\ z(t) &= C_1(t)x(t) + D_1(t)u(t); \\ y(t) &= C_2(t)x(t) + v(t) + D_2(t)\nu(t) \end{aligned} \tag{6.5.1}$$

where $x(t) \in \mathbf{R}^n$ is the *state*, $w(t) \in \mathbf{R}^p$ and $v(t) \in \mathbf{R}^l$ are the *uncertainty inputs*, $u(t) \in \mathbf{R}^m$ is the *control input*, $\nu(t) \in \mathbf{R}^N$ is the *harmonic disturbance input*, $z(t) \in \mathbf{R}^q$ is the *uncertainty output*, $y(t) \in \mathbf{R}^l$ is the *measured output*, $A(\cdot)$, $B_1(\cdot)$, $B_2(\cdot)$, $B_3(\cdot)$, $C_1(\cdot)$, $D_1(\cdot)$, $C_2(\cdot)$ and $D_2(\cdot)$ are bounded piecewise continuous matrix functions. As compared to the systems considered in the previous sections of this chapter, the system (6.5.1) allows for an additional disturbance input $\nu(\cdot)$.

Harmonic Disturbance

The harmonic disturbance in the above system is of the form

$$\nu(t) = \begin{bmatrix} a_1 \sin(\omega_1 t + \varphi_1) & \dots & a_N \sin(\omega_N t + \varphi_N) \end{bmatrix}' \tag{6.5.2}$$

where $\omega_1 > 0, \dots, \omega_N > 0$ are known constants. Also, $a_1, \dots, a_N, \varphi_1, \dots, \varphi_N$ are unknown constants. The set of all such inputs is denoted $\mathcal{N}$.

System Uncertainty

The uncertainty in the above system is represented by an equation of the form (6.4.2) where the following Integral Quadratic Constraint of the form (2.3.8) is satisfied:

Definition 6.5.1 *Let $d > 0$ be a given constant, $P_0 = P_0' > 0$ be a given positive-definite matrix, $Q(\cdot) = Q(\cdot)'$ and $N(\cdot) = N(\cdot)'$ be given bounded piecewise continuous weighting matrix functions associated with the system (6.5.1) and satisfying the condition: There exists a constant $\varepsilon > 0$ such that $Q(t) \geq \varepsilon I$ and $N(t) \geq \varepsilon I$ for all $t \in [0,T]$. We will consider uncertainty inputs of the form (6.4.2) and initial conditions $x(0)$ for the system (6.5.1) such that the following conditions hold: Given any control input $u(\cdot) \in \mathbf{L}_2[0,T]$ and any disturbance input $\nu(\cdot) \in \mathcal{N}$, then any corresponding solution to equations (6.5.1), (6.4.2) is defined on $[0,T]$ and*

$$\begin{aligned} x(0)'P_0x(0) + \int_0^T (w(t)'Q(t)w(t) + v(t)'N(t)v(t))dt \\ \leq d + \int_0^T \|z(t)\|^2 dt. \end{aligned} \tag{6.5.3}$$

The class of all such admissible uncertainty inputs and initial conditions is denoted Ξ.

The uncertainty description given above resembles that given in Definition 6.4.1. As in Section 6.4, this uncertainty description allows for $w(t)$ and $v(t)$ to depend dynamically on $x(t)$ and $u(t)$. Also, as in the previous section, the uncertain system (6.5.1), (6.5.3) allows for uncertainty satisfying a norm bound condition. However, in contrast to the problem considered in Section 6.4, in this section the initial condition is treated as an uncertain parameter. This is reflected by including the term $x(0)'P_0x(0)$ in the left-hand side of equation (6.5.3).

Let $X_T = X_T' \geq 0$ be a given matrix, $R(\cdot) = R(\cdot)'$ and $G(\cdot) = G(\cdot)'$ be given bounded piecewise continuous weighting matrix functions satisfying the condition: There exists a constant $\varepsilon > 0$ such that $R(t) \geq \varepsilon I$ and $G(t) \geq \varepsilon I$ for all $t \in [0,T]$. We will consider the following quadratic cost functional associated with the uncertain system (6.5.1), (6.5.2), (6.5.3)

$$\begin{aligned} J(x(\cdot), u(\cdot)) := x(T)'X_Tx(T) \\ + \int_0^T (x(t)'R(t)x(t) + u(t)'G(t)u(t))dt. \end{aligned} \tag{6.5.4}$$

We now consider a problem of robust control with harmonic disturbance rejection for the uncertain system (6.5.1), (6.5.2), (6.5.3). The class of controllers to be considered are finite dimensional time-varying linear output-feedback controllers of the form (6.3.3).

Definition 6.5.2 *Let $c_0 > 0, c_1 > 0, \ldots, c_N > 0$ be given positive constants associated with the system (6.5.1), (6.5.2), (6.5.3). The controller (6.3.3) is said to be a* robust controller with harmonic disturbance rejection , *if the following condition holds: the closed loop uncertain system (6.5.1), (6.5.2), (6.5.3), (6.3.3) satisfies the state constraint*

$$\sup_{\nu(\cdot)\in\mathcal{N},[x(0),w(\cdot),v(\cdot)]\in\Xi} \frac{J(x(\cdot),u(\cdot))}{c_0 + \sum_{i=1}^{N} c_i a_i^2} < 1. \tag{6.5.5}$$

An uncertain system (6.5.1), (6.5.2), (6.5.3) which admits a robust controller (6.3.3) with harmonic disturbance rejection is said to be robustly controllable *with harmonic disturbance rejection.*

6.5.2 *Design of a robust controller with harmonic disturbance rejection*

We now present the main result of this section. Also, we make a technical assumption which simplifies the solution to the robust control problem under consideration; e.g., see Section 3.2.2.

Assumption 6.5.1 *The matrix functions $C_1(\cdot)$ and $D_1(\cdot)$ satisfy the condition $C_1(\cdot)'D_1(\cdot) \equiv 0$.*

Let $\tau > 0$ be a given parameter. We will consider the following pair of Riccati differential equations

$$\begin{aligned} -\dot{X}(t) &= \hat{A}(t)'X(t) + X(t)\hat{A}(t) + \hat{C}_1(t)'\hat{C}_1(t) \\ &\quad + X(t)\left(\hat{B}_2(t)Q(t)^{-1}\hat{B}_2(t)' - \hat{B}_1(t)E(t)^{-1}\hat{B}_1(t)'\right)X(t); \end{aligned} \tag{6.5.6}$$

$$\begin{aligned} \dot{Y}(t) &= \hat{A}(t)Y(t) + Y(t)\hat{A}(t)' + \hat{B}_1(t)\dot{Q}(t)^{-1}\hat{B}_1(t)' \\ &\quad + Y(t)\left(\hat{C}_1(t)'\hat{C}_1(t) - \hat{C}_2(t)'N(t)\hat{C}_2(t)\right)Y(t) \end{aligned} \tag{6.5.7}$$

where

$$\hat{A}(t) = \begin{bmatrix} 0 & I & 0 \\ \Omega & 0 & 0 \\ B_3 & 0 & A \end{bmatrix}; \quad \hat{C}_1(t) = \begin{bmatrix} 0 & 0 & 0 \\ 0 & 0 & R^{\frac{1}{2}} \\ 0 & 0 & \sqrt{\tau}C_1 \end{bmatrix};$$

$$\hat{B}_1(t) = \begin{bmatrix} 0 \\ 0 \\ B_1 \end{bmatrix}; \quad \hat{B}_2(t) = \begin{bmatrix} 0 \\ 0 \\ \frac{1}{\sqrt{\tau}}B_2 \end{bmatrix};$$

$$\hat{D}_1(t) = \begin{bmatrix} G^{\frac{1}{2}} \\ 0 \\ \sqrt{\tau}D_1 \end{bmatrix}; \quad \hat{C}_2(t) = \begin{bmatrix} D_2 & 0 & C_2 \end{bmatrix};$$

$$\Omega = \mathrm{diag}\left[-\omega_1^2,\ -\omega_2^2,\ \ldots,\ -\omega_N^2\right];$$
$$E(\cdot) = \hat{D}_1(\cdot)'\hat{D}_1(\cdot). \tag{6.5.8}$$

Also, we introduce matrices $\hat{P}_0$ and $\hat{X}_T$ defined by

$$\hat{P}_0 = \begin{bmatrix} \Gamma_1 & 0 & 0 \\ 0 & \Gamma_2 & 0 \\ 0 & 0 & \tau P_0 \end{bmatrix}; \quad \hat{X}_T = \begin{bmatrix} 0 & 0 & 0 \\ 0 & 0 & 0 \\ 0 & 0 & X_T \end{bmatrix}$$

where $\Gamma_1 = \mathrm{diag}\left[c_1, \ldots, c_N\right]$ and $\Gamma_2 = \mathrm{diag}\left[\frac{c_1}{\omega_1^2}, \ldots, \frac{c_N}{\omega_N^2}\right]$.

Theorem 6.5.1 *Consider the uncertain system (6.5.1), (6.5.2), (6.5.3) and suppose that Assumption 6.5.1 is satisfied. Then the uncertain system (6.5.1), (6.5.2), (6.5.3) is robustly controllable with harmonic disturbance rejection if and only if there exists a constant $\tau \in (0, c_0/d]$ such that the following conditions hold:*

(i) There exists a solution $X(\cdot)$ to Riccati differential equation (6.5.6) defined on the interval $[0,T]$ with $X(T) = \hat{X}_T$. Furthermore, this solution satisfies $X(0) < \hat{P}_0$.

(ii) There exists a solution $Y(\cdot)$ to Riccati differential equation (6.5.7) defined on the interval $[0,T]$ with $Y(0) = \hat{P}_0^{-1}$. Furthermore, this solution satisfies $Y(t) > 0$ for all $t \in [0,T]$.

(iii) The spectral radius $\rho(Y(t)X(t))$ satisfies $\rho(Y(t)X(t)) < 1$ for all $t \in [0,T]$.

If conditions (i)-(iii) hold, then a controller of the form (6.3.3) with coefficient matrices

$$\begin{aligned} A_c(t) &= \hat{A}(t) + \hat{B}_2(t)C_c(t) - B_c(t)\hat{C}_2(t) \\ &\quad + \left(\hat{B}_2(t) - B_c(t)\right) Q(t)^{-1}\hat{B}_2(t)'X(t); \\ B_c(t) &= (I - Y(t)X(t))^{-1}\, Y(t)\hat{C}_2(t)'N(t); \\ C_c(t) &= -E(t)^{-1}\hat{B}_1(t)'X(t); \\ D_c(t) &= 0 \end{aligned} \tag{6.5.9}$$

is a robust controller with harmonic disturbance rejection.

Proof: *Necessity.* Suppose that there exists a controller of the form (6.3.3) such that condition (6.5.5) holds. Condition (6.5.5) implies that there exists a constant $\delta > 0$ such that

$$(1+\delta)J(x(\cdot), u(\cdot)) \le c_0 + \sum_{i=1}^{N} c_i a_i^2 \tag{6.5.10}$$

for all solutions to the system (6.5.1), (6.5.2), (6.5.3), (6.3.3). Now let $\mathcal{M}$ be the Hilbert space of all vector valued functions

$$\lambda(\cdot) := [x(\cdot), u(\cdot), w(\cdot), v(\cdot), \nu(\cdot)] \in \mathbf{L}_2[0, T]$$

connected by the equations (6.5.1), (6.3.3) and where $\nu(\cdot)$ is of the form (6.5.2). We also introduce continuous quadratic functionals $\mathcal{G}_0(\cdot)$ and $\mathcal{G}_1(\cdot)$ defined on $\mathcal{M}$ by the equations

$$\begin{aligned}
\mathcal{G}_0(\lambda(\cdot)) &:= (1+\delta)J(x(\cdot), u(\cdot)) - c_0 - \sum_{i=1}^{N} c_i a_i^2; \\
\mathcal{G}_1(\lambda(\cdot)) &:= d + \int_0^T (\|z(t)\|^2 - w(t)'Q(t)w(t) - v(t)'N(t)v(t))dt.
\end{aligned} \tag{6.5.11}$$

Since condition (6.5.10) holds for all solutions to the system (6.5.1), (6.3.3) with inputs $w(\cdot)$ and $v(\cdot)$ satisfying condition (6.5.3), then we have

$$\mathcal{G}_0(\lambda(\cdot)) \leq 0$$

for all $\lambda(\cdot) \in \mathcal{M}$ such that $\mathcal{G}_1(\lambda(\cdot)) \geq 0$. Since $d > 0$, if we let $w(t) \equiv 0$, $v(t) \equiv 0, \nu(t) \equiv 0$ and $x(0) = 0$, the corresponding vector function $\lambda(\cdot) \in \mathcal{M}$ is such that $\mathcal{G}_1(\lambda(\cdot)) > 0$. Hence, using the S-procedure Theorem 4.2.1, it follows that there exists a constant $\tau \geq 0$ such that the inequality

$$\mathcal{G}_0(\lambda(\cdot)) + \tau\mathcal{G}_1(\lambda(\cdot)) \leq 0 \tag{6.5.12}$$

is satisfied for all $\lambda(\cdot) \in \mathcal{M}$. That is

$$\begin{aligned}
&(1+\delta)x(T)'X_T x(T) - \sum_{i=1}^{N} c_i a_i^2 \\
&+ \int_0^T \begin{pmatrix} (1+\delta)[x(t)'R(t)x(t) + u(t)'G(t)u(t)] \\ +\tau[\|z(t)\|^2 - w(t)'Q(t)w(t) - v(t)'N(t)v(t)] \end{pmatrix} dt \\
&\leq c_0 - \tau d
\end{aligned} \tag{6.5.13}$$

Also, we can choose inputs $w(t) \equiv 0, v(t) \equiv 0, \nu(t) \equiv 0$ and initial condition $x(0) \neq 0$ such that $\mathcal{G}_0(\lambda(\cdot)) > 0$ for the corresponding solution to the system (6.5.1). Therefore, the corresponding vector function $\lambda(\cdot) \in \mathcal{M}$ is such that $\mathcal{G}_0(\lambda(\cdot)) > 0$. Referring to inequality (6.5.12), this implies that $\tau \neq 0$. However since $\tau \geq 0$, we can now conclude that $\tau > 0$.

Note that the disturbance signal $\nu(t)$ defined in (6.5.2) can be regarded as a solution to the equation

$$\ddot{\nu} = \Omega\nu.$$

Motivated by this fact, we introduce the augmented state vector $h = [\nu' \ \dot{\nu}' \ x']'$ and the augmented output vector $\hat{z} = [u'G^{\frac{1}{2}} \ x'R^{\frac{1}{2}} \ \sqrt{\tau}z']'$. Then consider the corresponding augmented system

$$\begin{aligned} \dot{h}(t) &= \hat{A}(t)h(t) + \tilde{B}_1(t)\hat{w}(t) + \hat{B}_2(t)u(t); \\ \hat{z}(t) &= \hat{C}_1(t)h(t) + \hat{D}_1(t)u(t); \\ y(t) &= \hat{C}_2(t)h(t) + \hat{D}_2(t)\hat{w}(t) \end{aligned} \tag{6.5.14}$$

where the coefficients are defined by (6.5.8) and

$$\begin{aligned} \hat{w}(t) &:= \sqrt{\tau}\begin{bmatrix} Q(t)^{\frac{1}{2}}w(t) \\ N(t)^{\frac{1}{2}}v(t) \end{bmatrix}; \ \tilde{B}_1(t) := \begin{bmatrix} \hat{B}_1(t)Q(t)^{-\frac{1}{2}} \ 0 \end{bmatrix}; \\ \hat{D}_2(t) &:= \begin{bmatrix} 0 \ N(t)^{-\frac{1}{2}} \end{bmatrix}. \end{aligned}$$

For this system, we let $\hat{w}(\cdot)$ be the disturbance input and let $\hat{z}(\cdot)$ be the controlled output. Using these definitions and the fact that $z(t)$ depends only on $x(t)$ and $u(t)$, it follows from inequality (6.5.13) that there exists a constant $\delta_1 \in (0, \delta)$ such that

$$\begin{aligned} &(1+\delta_1)h(T)'\hat{X}_T h(T) \\ &\quad + \int_0^T (1+\delta_1)[\|\hat{z}(t)\|^2 - \|\hat{w}(t)\|^2]dt - h(0)'\hat{P}_0 h(0) \\ &\leq c_0 - \tau d. \end{aligned} \tag{6.5.15}$$

If we consider this inequality with $\hat{z}(t) \equiv 0, \hat{w}(t) \equiv 0, h(t) \equiv 0$, it follows that $c_0 - \tau d \geq 0$. Hence, we have proved that $\tau \in (0, c_0/d]$. Furthermore, since the system (6.5.1), (6.3.3) is linear and the left side of the inequality (6.5.15) is quadratic, it follows from (6.5.15) that

$$\begin{aligned} &(1+\delta_1)h(T)'\hat{X}_T h(T) \\ &\quad + \int_0^T [(1+\delta_1)\|\hat{z}(t)\|^2 - \|\hat{w}(t)\|^2]dt - h(0)'\hat{P}_0 h(0) \leq 0. \end{aligned} \tag{6.5.16}$$

Indeed, if (6.5.16) does not hold for some $(\hat{w}(\cdot), h(0))$, then by scaling $(\hat{w}(\cdot), h(0))$, (6.5.15) would be violated since its right hand side is a constant. Inequality (6.5.16) then implies that the following condition holds:

$$J := \sup \left\{ \frac{h(T)'\hat{X}_T h(T) + \int_0^T \|\hat{z}(t)\|^2 dt}{h(0)'\hat{P}_0 h(0) + \int_0^T \|\hat{w}(t)\|^2 dt} \right\} < 1 \tag{6.5.17}$$

where the supremum is taken over all $h(0) \in \mathbf{R}^{n+2N}$ and $\hat{w}(\cdot) \in \mathbf{L}_2[0,\infty)$ such that

$$\|h(0)\|^2 + \int_0^T \|\hat{w}(t)\|^2 dt > 0.$$

That is, the controller (6.3.3) solves a H^∞ control problem defined by the system (6.5.14) and the H^∞ norm bound condition (6.5.17); see Section 3.2.2 and also [100]. Therefore, Theorem 3.2.6 implies that conditions (i)-(iii) are satisfied. This completes the proof of this part of the theorem.

Sufficiency. This proof follows along the same lines as the proof of the sufficiency part of Theorem 6.4.1.

■

Remark 6.5.1 It should be noted that the augmented system (6.5.14) is not stabilizable (even in the absence of uncertainty). This is due to the uncontrollable oscillatory modes corresponding to the disturbances signals (6.5.2). From this, it follows that the solutions to the Riccati equations (6.5.6), (6.5.7) will not converge to stabilizing steady state solutions in the limit as $T \to \infty$. Thus, the infinite-horizon case cannot be obtained directly from the results presented here. However, it may be possible to obtain a solution to the infinite-horizon case by partitioning the Riccati equation solutions according to the plant and disturbance dynamics in the augmented system (6.5.14).

6.6 Conclusions

This chapter has given solutions to finite-horizon state-feedback and output-feedback optimal guaranteed cost control problems for a class of time-varying uncertain linear systems with structured uncertainty. In Sections 6.2 and 6.3 the solution to these problems has been obtained by considering a new class of uncertain systems with structured uncertainty. This class of uncertain systems describes the structured uncertainties via averaged integral quadratic constraints. This uncertainty description is a variation of the previously considered integral quadratic constraint description of structured uncertainty. The principle difference between these two uncertainty descriptions relates to the characterization of the initial conditions on the uncertainty dynamics. The main advantage of the new uncertainty description is that it allows for a new S-procedure result to be used. This S-procedure result can be applied to a much wider class of problems than the S-procedure result of Yakubovich, Megretski and Treil.

Also in the robust control problem with a terminal constraint on the state, a corresponding integral quadratic constraint description of uncertainty has led to a controller exhibiting better robustness properties than the standard LQG controller.

The approach taken in Section 6.5 leads to a robust output-feedback controller guaranteeing an optimal level of robust performance in the presence of both noise input uncertainties and harmonic disturbances with known frequencies but uncertain magnitudes and phases.

7. Absolute stability, absolute stabilization and structured dissipativity

7.1 Introduction

In this chapter, we revisit the classical Lur'e-Postnikov approach to the stability of nonlinear control systems. The problems considered in this chapter include the problem of absolute stability and also its extensions such as the problem of absolute stabilization and the problem of structured dissipativity. As in previous chapters of this book, the approach taken here is to convert the problem under consideration into an equivalent H^∞ control problem. Such a conversion is accomplished by making use of the corresponding S-procedure theorem.

Popov's celebrated stability criterion concerning the absolute stability of systems containing an uncertain sector bounded nonlinearity is well known; e.g., see [161, 137]. An important feature of the Popov stability criterion is the fact that it is equivalent to the existence of a Lyapunov function of the Lur'e-Postnikov form; see [137]. This fact gives an indication as to the degree of conservatism involved in using the Popov criterion. In Section 7.2, we present a result which extends the Popov approach to the problem of robust stabilization. Towards this end, we define a notion of "absolute stabilizability with a Lyapunov function of the Lur'e-Postnikov form". This notion extends the absolute stability notion considered by Popov, to a corresponding stabilizability notion for output-feedback control. The main result of Section 7.2 is a necessary and sufficient condition for absolute stabilizability with a Lyapunov function of the Lur'e-Postnikov form. This condition is computationally tractable and involves the solutions of a pair of parameter dependent algebraic Riccati equations. These Riccati equations arise from a minimax problem with an uncertainty dependent cost functional. Furthermore, the controller synthesized on the basis of these Riccati equations leads to a closed loop system which is absolutely stable with an uncertainty dependent Lyapunov function of the Lur'e-Postnikov form. This result extends the quadratic stability and stabilizability approaches of Chapters 5 and 6 to allow for Lyapunov functions of the Lur'e-Postnikov form.

Note that in contrast to similar H^∞ control problems considered in previous chapters, the H^∞ control problem which arises in Section 7.2 is singular and is solved using perturbation techniques such as described in [102] and [172].

In Section 7.3, we consider connections between the problems of robust stability of nonlinear uncertain systems with structured uncertainty, the absolute stability of nonlinear uncertain systems, and the notion of dissipativity for nonlinear systems. In Chapter 1, we mentioned references concerned with the problem of stability for uncertain systems containing structured uncertainty. However, in these references, the underlying uncertain system is required to be linear. Section 7.3 extends these ideas into the realm of nonlinear systems. That is, we show using a corresponding *nonlinear* version of the S-procedure theorem, that the problem of robust stability for a nonlinear uncertain system with structured uncertainty is equivalent to the existence of a set of real scaling parameters such that a corresponding nonlinear uncertain system with unstructured uncertainty is robustly stable.

Also in Section 7.3, we introduce a notion of "structured dissipativity". This definition concerns a given nonlinear system and an associated collection of supply functions. One of the results presented in Section 7.3 shows that there is a direct connection between this notion of structured dissipativity and the absolute stability of an uncertain nonlinear system containing a number of uncertain nonlinearities. It is because of this connection that the term structured dissipativity is used.

The first main result presented in Section 7.3 shows that a nonlinear system has the structured dissipativity property if and only if it is dissipative with a single supply function which is a positive linear combination of the given supply functions. From this, the property of structured dissipativity is also shown to be equivalent to the satisfaction of a certain parameter dependent integral inequality. This connection is similar to the nonlinear version of the Kalman-Yakubovich-Popov Lemma given in [81].

The second main result presented in Section 7.3 concerns a given nonlinear uncertain system containing multiple structured uncertain nonlinearities each subject to an integral uncertainty constraint. These integral uncertainty constraints generalize the integral quadratic uncertainty constraints considered in previous chapters to the case of *nonlinear* uncertain systems. The result shows that a nonlinear uncertain system is absolutely stable if and only if the corresponding nonlinear system with associated supply functions (which defines the integral uncertainty constraints) has the structured dissipativity property. This result also leads to a sufficient condition for the stability of an interconnected nonlinear system.

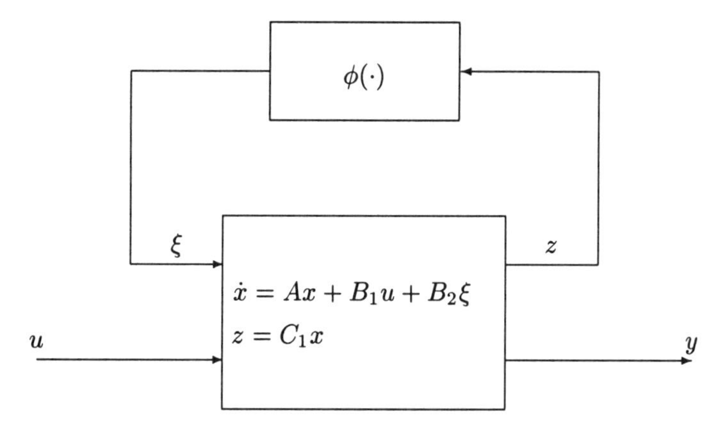

Figure 7.2.1. Uncertain system

7.2 Robust stabilization with a Lyapunov function of the Lur'e-Postnikov form

7.2.1 Problem statement

Consider the following time-invariant uncertain system of the form (2.3.3) with unstructured uncertainty:

$$\begin{aligned}\dot{x}(t) &= Ax(t) + B_1u(t) + B_2\xi(t);\\ \sigma(t) &= C_1x(t);\\ y(t) &= C_2x(t);\\ \xi(t) &= \phi(\sigma(t))\end{aligned} \tag{7.2.1}$$

where $x(t) \in \mathbf{R}^n$ is the *state*, $\sigma(t) \in \mathbf{R}$ is a *scalar uncertainty output*, $\xi(t) \in \mathbf{R}$ is a *scalar uncertainty input*, $u(t) \in \mathbf{R}^m$ is the *control input*, $y(t) \in \mathbf{R}^l$ is the *measured output* and $\phi(\cdot) : \mathbf{R} \to \mathbf{R}$ is an uncertain nonlinear mapping. This system is represented in the block diagram shown in Figure 7.2.1. We will suppose that the uncertain nonlinearity $\phi(\cdot)$ satisfies the following sector bound (e.g., see [137])

$$0 \leq \frac{\phi(\sigma)}{\sigma} \leq k \tag{7.2.2}$$

where $0 < k \leq \infty$ is a given constant associated with the system. We now consider a problem of absolute stabilization for the uncertain system (7.2.1), (7.2.2) via a causal linear output-feedback controller of the form

$$\begin{aligned}\dot{x}_c(t) &= A_cx_c(t) + B_cy(t);\\ u(t) &= C_cx_c(t) + D_cy(t).\end{aligned} \tag{7.2.3}$$

Note that the dimension of the controller state vector $x_c(t)$ may be arbitrary.

Let $h := \begin{bmatrix} x' & x_c' \end{bmatrix}'$ be the state vector of the closed loop system (7.2.1), (7.2.3). For the closed loop uncertain system (7.2.1), (7.2.2), (7.2.3), we will consider a Lyapunov function of the Lur'e-Postnikov form (e.g., see [120] and [137])

$$V(h) = h'Hh + \beta \int_0^\sigma \phi(\zeta)d\zeta \tag{7.2.4}$$

where $H > 0$ is a given matrix and $\beta > 0$ is a constant.

Definition 7.2.1 *The uncertain system (7.2.1), (7.2.2) is said to be* absolutely stabilizable with a Lyapunov function of the Lur'e-Postnikov form *if there exists a linear output-feedback controller of the form (7.2.3), a matrix $H > 0$ and constants $\beta > 0$ and $\epsilon_0 > 0$ such that the derivative $\dot{V}(h)$ of the function (7.2.4) along trajectories of the closed loop uncertain system (7.2.1), (7.2.2), (7.2.3) satisfies the following inequality:*

$$\dot{V}(h) \leq -\epsilon_0(\|h\|^2 + \phi^2(\sigma)). \tag{7.2.5}$$

Observation 7.2.1 If inequality (7.2.5) is satisfied, it follows immediately that the closed loop uncertain system (7.2.1), (7.2.2), (7.2.3) is absolutely stable; e.g., see [137]. This implies that there exist constants $c > 0$ and $\nu > 0$ such that

$$\|h(t)\| \leq c\|h(0)\|e^{-\nu t}$$

for all solutions to the closed loop system.

7.2.2 *Design of a robustly stabilizing controller*

We now present a result on robust stabilization via a Lyapunov function of the Lur'e-Postnikov form. This result is given in terms of the existence of solutions to a pair of parameter dependent algebraic Riccati equations. The Riccati equations under consideration are defined as follows: Let $\tau_0 \geq 0$ and $\delta > 0$ be given constants and introduce a constant α defined by the equation

$$\alpha := \frac{\tau_0}{k} - C_1 B_2. \tag{7.2.6}$$

Also, if $\alpha \neq 0$, we can introduce the matrices $\hat{A}, C_{\tau_0}, D_{\tau_0}, \hat{B}_1$ and E_1 defined by the equations

$$\begin{aligned}
C_{\tau_0} &= \frac{1}{2\alpha}(\tau_0 C_1 + C_1 A); \\
\hat{A} &= A + B_2 C_{\tau_0}; \\
D_{\tau_0} &= \frac{1}{2\alpha} C_1 B_1; \\
\hat{B}_1 &= B_1 + B_2 D_{\tau_0}; \\
E_1 &= D_{\tau_0}' D_{\tau_0}.
\end{aligned} \tag{7.2.7}$$

If $E_1 > 0$, we then consider the following Riccati equations

$$(\hat{A} - \hat{B}_1 E_1^{-1} D'_{\tau_0} C_{\tau_0})' X + X(\hat{A} - \hat{B}_1 E_1^{-1} D'_{\tau_0} C_{\tau_0}) + X(B_2 B_2' - \hat{B}_1 E_1^{-1} \hat{B}_1') X + C'_{\tau_0}(I - D_{\tau_0} E_1^{-1} D'_{\tau_0}) C_{\tau_0} = 0; \quad (7.2.8)$$

$$\hat{A} Y + Y \hat{A}' + Y(C'_{\tau_0} C_{\tau_0} - \frac{1}{\delta} C_2' C_2) Y + B_2 B_2' + \delta I = 0. \quad (7.2.9)$$

Theorem 7.2.1 *Consider the uncertain system (7.2.1), (7.2.2) and suppose that (A, B_2) is controllable, (A, C_1) is observable and $B_1' C_1' C_1 B_1 > 0$. Then the following statements are equivalent:*

(i) The nonlinear system (7.2.1), (7.2.2) is absolutely stabilizable with a Lyapunov function of the Lur'e-Postnikov form.

(ii) There exist constants $\tau_0 \geq 0$ and $\delta > 0$ such that if α and C_{τ_0} are defined by (7.2.6), (7.2.7), then $\alpha > 0$, the pair (A, C_{τ_0}) is observable and the Riccati equations (7.2.8), (7.2.9) have solutions $X > 0$ and $Y > 0$ such that the spectral radius of their product satisfies $\rho(XY) < 1$.

Furthermore, if condition (ii) is satisfied, then the system (7.2.1), (7.2.2) is absolutely stabilizable with a Lyapunov function of the Lur'e-Postnikov form via a linear controller of the form (7.2.3) with

$$\begin{aligned} A_c &= \hat{A} + \hat{B}_1 C_c - B_c C_2 + B_2 B_2' X + \delta X; \\ B_c &= \frac{1}{\delta}(I - YX)^{-1} Y C_2'; \\ C_c &= -E_1^{-1}(\hat{B}_1' X + D'_{\tau_0} C_{\tau_0}); \\ D_c &= 0. \end{aligned} \quad (7.2.10)$$

Proof: (i)⇒(ii). Suppose there exists a Lyapunov function of the form (7.2.4) with $H > 0, \beta > 0$ such that condition (7.2.5) holds. Let $V_0(h) := h'Hh$. Then the derivative of $V(h)$ along of trajectories of the system (7.2.1), (7.2.3) is given by

$$\dot{V}(h) = \dot{V}_0(h) + \beta \phi(\sigma) \dot{\sigma} = \dot{V}_0(h) + \beta \xi C_1 [Ax + B_1 u + B_2 \xi]. \quad (7.2.11)$$

Equations (7.2.1) and (7.2.3) imply that the control input u is a linear vector valued function of h and ξ. Moreover, the derivative $\dot{V}_0(h)$ of the quadratic form $V_0(h)$ along of trajectories of the system (7.2.1), (7.2.3) is a quadratic form in h and ξ. Hence, equation (7.2.11) implies that $\dot{V}(h)$ is a quadratic form in h and ξ. Therefore, quadratic functionals $\mathcal{G}_0(h, \xi)$ and $\mathcal{G}_1(h, \xi)$ can be defined on $\mathbf{R}^{n+1}$ as follows:

$$\begin{aligned} \mathcal{G}_0(h, \xi) &:= \frac{1}{\beta}[\dot{V}(h) + \epsilon_0(\|h\|^2 + \xi^2)]; \\ \mathcal{G}_1(h, \xi) &:= \sigma \xi - \frac{1}{k} \xi^2 \end{aligned}$$

where ϵ_0 is the constant from Definition 7.2.1 and $1/k := 0$ if $k = \infty$. Then condition (7.2.5) may be re-written as $\mathcal{G}_0(h,\xi) \leq 0$ if $\mathcal{G}_1(h,\xi) \geq 0$. Also, it is straightforward to verify that $\mathcal{G}_1(h,\xi) > 0$ for some $[h,\xi] \in \mathbf{R}^{n+1}$. Hence, Theorem 4.2.1 implies that there exists a constant $\tau \geq 0$ such that

$$\frac{1}{\beta}\dot{V}_0(h) + \tau(\sigma\xi - \frac{1}{k}\xi^2) + \xi C_1[Ax + B_1u + B_2\xi] \leq -\epsilon(\|h\|^2 + \xi^2) \quad (7.2.12)$$

with $\epsilon = \epsilon_0/\beta$. Now if $h(0) = 0$, then $V_0(h(0)) = 0$. Hence, it follows from (7.2.12) that

$$\begin{aligned}
&\frac{1}{\beta}[V_0(h(\infty)) - V_0(h(0))] \\
&\quad + \int_0^\infty \left[\tau\left(\sigma(t)\xi(t) - \frac{1}{k}\xi^2(t)\right) + \xi(t)C_1[Ax(t) + B_2\xi(t) + B_1u(t)]\right] dt \\
&\leq -\epsilon\int_0^\infty (\|h(t)\|^2 + \xi(t)^2)dt.
\end{aligned}$$

Furthermore since $H > 0$, this implies that

$$\begin{aligned}
&\int_0^\infty \left[\tau\left(\sigma(t)\xi(t) - \frac{1}{k}\xi^2(t)\right) + \xi(t)C_1[Ax(t) + B_1u(t) + B_2\xi(t)]\right] dt \\
&\leq -\epsilon\int_0^\infty (\|h(t)\|^2 + \xi(t)^2)dt \qquad (7.2.13)
\end{aligned}$$

for all $h(\cdot), u(\cdot), \sigma(\cdot)$ and $\xi(\cdot)$ in $\mathbf{L}_2[0,\infty)$ connected by (7.2.1), (7.2.3) and the initial condition $h(0) = 0$.

We now observe that equation (7.2.2) implies that there exists a constant $c_1 > 0$ such that

$$\sigma\xi - \frac{1}{k}\xi^2 \leq c_1(\|h\|^2 + \xi^2).$$

Also if we let $\varepsilon_1 := \varepsilon/(1 + c_1)$, it follows from (7.2.13) that given any $\tilde{\varepsilon} \in [0, \varepsilon_1]$, then

$$\begin{aligned}
&\int_0^\infty \begin{bmatrix} [\tau + \tilde{\varepsilon}]\left(\sigma(t)\xi(t) - \frac{1}{k}\xi^2(t)\right) \\ +\xi(t)C_1[Ax(t) + B_2\xi(t) + B_1u(t)] \end{bmatrix} dt \\
&\leq \int_0^\infty \left[\tilde{\varepsilon}\left(\sigma(t)\xi(t) - \frac{1}{k}\xi^2(t)\right) - \varepsilon(\|h(t)\|^2 + \xi(t)^2)\right] dt \\
&\leq -(\varepsilon - \tilde{\varepsilon}c_1)\int_0^\infty (\|h(t)\|^2 + \xi(t)^2)dt \\
&\leq -\varepsilon_1\int_0^\infty (\|h(t)\|^2 + \xi(t)^2)dt.
\end{aligned}$$

Hence,

$$\begin{aligned}
&\int_0^\infty \left[\tau_0\left(\sigma(t)\xi(t) - \frac{1}{k}\xi^2(t)\right) + \xi(t)C_1[Ax(t) + B_1u(t) + B_2\xi(t)]\right] dt \\
&\leq -\epsilon_1\int_0^\infty (\|h(t)\|^2 + \xi(t)^2)dt \qquad (7.2.14)
\end{aligned}$$

for all $\tau_0 \in [\tau, \tau + \epsilon_1]$.

We now observe that inequality (7.2.14) may be re-written as

$$\int_0^\infty (\sigma_0(t)\xi(t) - \alpha\xi^2(t))dt \leq -\epsilon_1 \int_0^\infty (\|h(t)\|^2 + \xi^2(t))dt \quad (7.2.15)$$

where α is defined by (7.2.6) and

$$\sigma_0(t) := \tau_0\sigma(t) + C_1[Ax(t) + B_1u(t)]. \quad (7.2.16)$$

From this, we will prove that $\alpha > 0$. Indeed, if $\alpha \leq 0$, then the inequality (7.2.15) implies

$$\int_0^\infty \sigma_0(t)\xi(t)dt \leq -\epsilon_1 \int_0^\infty \xi^2(t)dt. \quad (7.2.17)$$

Now let $\tilde{\xi}(s)$ denote the Laplace transform of the signal $\xi(t)$ and let $\tilde{\sigma}_0(s)$ denote the Laplace transform of the signal $\sigma_0(t)$. Also let $W(s)$ be the transfer function from $\xi(\cdot)$ to $\sigma_0(\cdot)$ in the system (7.2.1), (7.2.3), (7.2.16). Using Plancherel's Theorem, it follows from (7.2.17) that

$$\begin{aligned}
&\frac{1}{2}\int_{-\infty}^\infty \left(\tilde{\sigma}_0^*(j\omega)\tilde{\xi}(j\omega) + \tilde{\sigma}_0(j\omega)\tilde{\xi}^*(j\omega)\right) dt \\
&= \frac{1}{2}\int_{-\infty}^\infty (W^*(j\omega) + W(j\omega))\, \tilde{\xi}^*(j\omega)\tilde{\xi}(j\omega)d\omega \\
&\leq -\epsilon_1 \int_0^\infty \tilde{\xi}^*(j\omega)\tilde{\xi}(j\omega)d\omega.
\end{aligned}$$

However, this holds for all $\xi(\cdot) \in \mathbf{L}_2[0, \infty)$. Hence, we must have $\operatorname{Re} W(j\omega) \leq -\epsilon_1 < 0$ for all $\omega \in \mathbf{R}$. However, this is impossible, since $\lim_{\omega\to\infty} \operatorname{Re} W(j\omega) = 0$. Hence, we must have $\alpha > 0$.

Now consider the linear system

$$\begin{aligned}
\dot{x}(t) &= \hat{A}x(t) + \hat{B}_1u(t) + B_2w(t); \\
\sigma_1(t) &= C_{\tau_0}x(t) + D_{\tau_0}u(t); \\
y(t) &= C_2x(t)
\end{aligned} \quad (7.2.18)$$

where the matrices $\hat{A}$, $\hat{B}_1$, C_{τ_0} and D_{τ_0} are defined as in (7.2.7). With the substitution $w(t) = \xi(t) - \sigma_1(t)$, this system is equivalent to the system (7.2.1). Furthermore, by observing $\sigma_1 = \sigma_0/(2\alpha)$, it follows from (7.2.15) that

$$\begin{aligned}
\int_0^\infty (\sigma_1^2(t) - w^2(t))dt &= \int_0^\infty (2\sigma_1(t)\xi(t) - \xi^2(t))dt \\
&= \frac{1}{\alpha}\int_0^\infty (\sigma_0(t)\xi(t) - \alpha\xi^2(t))dt \\
&\leq -\frac{\epsilon_1}{\alpha}\int_0^\infty (\|h(t)\|^2 + \xi^2(t))dt. \quad (7.2.19)
\end{aligned}$$

We now observe that equations (7.2.1), (7.2.3) and (7.2.18) together imply that we can write

$$\begin{aligned}\sigma_1(t) &= C_{\tau_0}x(t) + D_{\tau_0}u(t)\\ &= C_{\tau_0}x(t) + D_{\tau_0}C_c x_c(t) + D_{\tau_0}D_c y(t)\\ &= C_{\tau_0}x(t) + D_{\tau_0}C_c x_c(t) + D_{\tau_0}D_c C_2 x(t).\end{aligned}$$

Hence, there exists a constant $c_2 > 0$ such that $\sigma_1^2 \le c_2(\|h\|^2 + \xi^2)$ for all h and ξ. From this inequality (7.2.19) implies

$$\int_0^\infty (\sigma_1^2(t) - w^2(t))dt \le -\varepsilon_2 \int_0^\infty \sigma_1^2(t)dt \tag{7.2.20}$$

where $\varepsilon_2 := \varepsilon_1/c_2 > 0$.

Inequality (7.2.20) implies that the controller (7.2.3) solves a standard H^∞ control problem. In this H^∞ control problem, the underlying linear system is described by the state equations (7.2.18) where $w(t) \in \mathbf{R}$ is the *disturbance input* and $\sigma_1(t) \in \mathbf{R}$ is the *controlled output*. The H^∞ norm bound requirement in this H^∞ control problem is

$$\sup_{w(\cdot)\in \mathbf{L}_2[0,\infty), x(0)=0} \frac{\int_0^\infty \sigma_1^2(t)dt}{\int_0^\infty w^2(t)dt} < 1. \tag{7.2.21}$$

Inequality (7.2.19) implies that condition (7.2.21) is satisfied. Furthermore, this condition holds for all $\tau_0 \in [\tau, \tau+\epsilon_1]$. Hence for all $\tau_0 \in [\tau, \tau+\epsilon_1]$, the controller (7.2.3) solves the H^∞ control problem defined by the system (7.2.18) and the norm bound condition (7.2.21). Using this fact, it follows from the first part of Theorem 3.2.4 (see also Theorem 3.3 of [147] and Theorem 2 of [172]) that there exist matrices $\tilde{X} > 0$ and F such that the following condition holds:

$$\begin{aligned}&(\hat{A} + \hat{B}_1F)'\tilde{X} + \tilde{X}(\hat{A} + \hat{B}_1F) + \tilde{X}B_2B_2'\tilde{X}\\ &+(C_{\tau_0} + D_{\tau_0}F)'(C_{\tau_0} + D_{\tau_0}F) < 0.\end{aligned} \tag{7.2.22}$$

Hence, $\hat{A} + \hat{B}_1F$ is stable and $\|C_{\tau_0} + D_{\tau_0}F)(sI - \hat{A} - \hat{B}_1F)^{-1}B_2\|_\infty < 1$. We now show that there exists a $\tau_0 \in [\tau, \tau + \epsilon_1]$ such that the pair $(\hat{A}, C_{\tau_0})$ is observable. We first establish that there exists a $\tau_0 \in [\tau, \tau + \epsilon_1]$ such that the pair (A, C_{τ_0}) is observable. Indeed, the observability matrix for the pair (A, C_{τ_0}) can be written as

$$N_{\tau_0} = \frac{\tau_0}{2\alpha}N + \frac{1}{2\alpha}NA \tag{7.2.23}$$

where

$$N := \begin{bmatrix} C_1 \\ C_1A \\ \vdots \\ C_1A^{n-1} \end{bmatrix}$$

is the observability matrix for the pair (A, C_1). However, we have assumed that the pair (A, C_1) is observable. Hence, the matrix N is nonsingular. From this, equation (7.2.23) implies that N_{τ_0} cannot be singular for all $\tau_0 \in [\tau, \tau + \epsilon_1]$. That is, there exists a $\tau_0 \in [\tau, \tau + \epsilon_1]$ such that the pair (A, C_{τ_0}) is observable. Furthermore since $\hat{A} = A + B_2 C_{\tau_0}$, the observability of the pair $(\hat{A}, C_{\tau_0})$ follows immediately from the observability of the pair (A, C_{τ_0}).

Using the observability of the pair $(\hat{A}, C_{\tau_0})$ and equation (7.2.22), it follows using a standard result in H^∞ control theory that the Riccati equation (7.2.8) has a solution $X > 0$ such that $X \leq \tilde{X}$; e.g., see Lemma 3.2.3 on page 70 or Theorem 3.4 of [147].

We now prove that the Riccati equation (7.2.9) has a solution $Y > 0$ such that $\rho(XY) < 1$. We again refer to the fact established above that for all $\tau_0 \in [\tau, \tau + \epsilon_1]$, the controller (7.2.3) solves the H^∞ control problem defined by the system (7.2.18) and the norm bound condition (7.2.21). This fact is now used in conjunction with the second part of Theorem 3.2.4. From this result, it follows that there exist matrices $\tilde{Y} > 0$ and L such that the following condition holds:

$$(\hat{A} + LC_2)\tilde{Y} + \tilde{Y}(\hat{A} + LC_2)' + \tilde{Y}C_{\tau_0}'C_{\tau_0}\tilde{Y} + B_2B_2' < 0. \qquad (7.2.24)$$

Hence, $\hat{A} + LC_2$ is stable and $\|C_{\tau_0}(sI - \hat{A} - LC_2)^{-1}B_2\|_\infty < 1$. Note that inequality (7.2.24) corresponds to the H^∞ control problem (7.2.18), (7.2.21) which is a singular H^∞ control problem. The approach taken to solve this singular H^∞ control problem is a perturbation approach such as contained in [141, 102, 172]. We first observe that equation (7.2.24) implies that

$$x'\left(\hat{A}\tilde{Y} + \tilde{Y}\hat{A}' + \tilde{Y}C_{\tau_0}'C_{\tau_0}\tilde{Y} + B_2B_2'\right)x < 0$$

for all non-zero $x \in \mathbf{R}^n$ such that $x'\tilde{Y}C_2'C_2\tilde{Y}x \leq 0$. Hence, Theorem 4.2.1 implies that there exists a constant $\mu \geq 0$ such that

$$\hat{A}\tilde{Y} + \tilde{Y}\hat{A}' - \mu\tilde{Y}C_2'C_2\tilde{Y} + \tilde{Y}C_{\tau_0}'C_{\tau_0}\tilde{Y} + B_2B_2' < 0.$$

Moreover, it is straightforward to verify that μ can be chosen so that $\mu > 0$. Also, we can re-write this inequality as

$$\hat{A}\tilde{Y} + \tilde{Y}\hat{A}' - \mu\tilde{Y}C_2'C_2\tilde{Y} + \tilde{Y}C_{\tau_0}'C_{\tau_0}\tilde{Y} + B_2B_2' + Q = 0.$$

where $Q > 0$. Now let the constant $\delta > 0$ be chosen sufficiently small so that $\delta I \leq Q$ and $1/\delta \geq \mu$. Then as in the proof of Theorem 2.1 of [141], this implies that the Riccati equation (7.2.9) has a solution $Y > 0$ such that $Y \leq \tilde{Y}$.

It remains to prove that the matrices X and Y satisfy the condition $\rho(XY) < 1$. Indeed, from Theorem 3.2.4, it follows that the matrices $\tilde{X}$ and $\tilde{Y}$ which satisfy equations (7.2.22) and (7.2.24) can also be chosen to satisfy the condition $\tilde{Y}^{-1} > \tilde{X}$. However, since we have $X \leq \tilde{X}$ and $Y \leq \tilde{Y}$, it follows that $\rho(XY) < 1$. This completes the proof of this part of the theorem.

(ii)⇒(i). Suppose there exist constants $\tau_0 \geq 0$ and $\delta > 0$ such that if α and C_{τ_0} are defined by (7.2.6) and (7.2.7), then $\alpha > 0$, the pair (A, C_{τ_0}) is observable and the Riccati equations (7.2.8) and (7.2.9) have solutions $X > 0$ and $Y > 0$ such that $\rho(XY) < 1$. Also, let the matrices $\hat{A}$, $\hat{B}_1$ and D_{τ_0} be defined as in (7.2.7).

We now consider a H^∞ control problem in which the underlying linear system is described by the state equations

$$\begin{aligned} \dot{x}(t) &= \hat{A}x(t) + \hat{B}_2\tilde{w}(t) + \hat{B}_1 u(t); \\ \sigma_1(t) &= C_{\tau_0}x(t) + D_{\tau_0}u(t); \\ y(t) &= C_2 x(t) + \hat{D}_{21}\tilde{w}(t) \end{aligned} \tag{7.2.25}$$

where $\hat{B}_2 := [B_2 \;\; 0 \;\; \sqrt{\delta}I]$ and $\hat{D}_{21} := [0 \;\; \sqrt{\delta}I \;\; 0]$. The H^∞ norm bound condition in this H^∞ control problem is as follows:

$$\sup_{\tilde{w}(\cdot)\in \mathbf{L}_2[0,\infty), x(0)=0} \frac{\int_0^\infty \sigma_1^2(t)dt}{\int_0^\infty \tilde{w}^2(t)dt} < 1. \tag{7.2.26}$$

Since $\hat{A} = A + B_2 C_{\tau_0}$, the observability of the pair (A, C_{τ_0}) implies the observability of the pair $(\hat{A}, C_{\tau_0})$. Also, the controllability of the pair (A, B_2) implies the controllability of the pair $(\hat{A}, B_2)$. Furthermore, this implies that the pair $(\hat{A}, \hat{B}_2)$ will be controllable. Hence, the H^∞ control problem defined by (7.2.25) and (7.2.26) satisfies all of the assumptions required for a nonsingular H^∞ control problem; e.g., see Section 3.2 and also [9, 147]. Therefore, using a standard result from H^∞ control theory, it follows that the controller (7.2.3), (7.2.10) solves this H^∞ control problem; e.g., see Theorem 3.2.3. From this, it follows immediately that the controller (7.2.3), (7.2.10) also solves the H^∞ control problem defined by the system (7.2.18) and H^∞ norm bound condition (7.2.21). Therefore, the closed loop system (7.2.18), (7.2.3), (7.2.10) is stable. Furthermore, there exists a constant $\epsilon_3 > 0$ such that

$$\int_0^\infty (\sigma_1^2(t) - w^2(t))dt \leq -\varepsilon_3 \int_0^\infty \|h(t)\|^2 dt \tag{7.2.27}$$

for all solutions $h(t)$ to the closed loop system (7.2.18), (7.2.3), (7.2.10) with initial condition $h(0) = 0$ and $w(\cdot) \in \mathbf{L}_2[0,\infty)$. However, inequality (7.2.2) implies that

$$0 \leq \xi^2 \leq k^2\sigma^2 = k^2|C_1 x|^2.$$

Hence, it follows from (7.2.27) that there exists a constant $\varepsilon_4 > 0$ such that

$$\int_0^\infty (\sigma_1^2(t) - w^2(t))dt \leq -\varepsilon_4 \int_0^\infty (\|h(t)\|^2 + \xi^2(t))dt. \tag{7.2.28}$$

Therefore, given any solution $h(t)$ to the closed loop system (7.2.1), (7.2.3), (7.2.10) with initial condition $h(0) = 0$ and $\xi(\cdot) \in \mathbf{L}_2[0,\infty)$, it follows from equations (7.2.6), (7.2.16), (7.2.18) and (7.2.28) that

$$\begin{aligned}\int_0^\infty &\left[\tau_0\left(\sigma(t)\xi(t) - \frac{1}{k}\xi^2(t)\right) + \xi(t)C_1[Ax(t) + B_2\xi(t) + B_1u(t)]\right] dt\\ &= \int_0^\infty (\sigma_0(t)\xi(t) - \alpha\xi^2(t))dt\\ &= \alpha\int_0^\infty (\sigma_1^2(t) - w^2(t))dt\\ &\leq -\varepsilon_4\alpha\int_0^\infty (\|h(t)\|^2 + \xi^2(t))dt.\end{aligned}$$

Hence, applying the Strict Bounded Real Lemma (Lemma 3.1.2) to the closed loop system (7.2.18), (7.2.3), (7.2.10), it follows that there exists a matrix $H > 0$ such that

$$\tilde{A}'H + H\tilde{A} + H\tilde{B}\tilde{B}'H + \tilde{C}'\tilde{C} < 0 \tag{7.2.29}$$

where

$$\begin{aligned}\tilde{A} &:= \begin{bmatrix} A + B_1D_cC_2 + B_2C_{\tau_0} + B_2D_{\tau_0}D_cC_2 & B_1C_c + B_2D_{\tau_0}C_c \\ B_cC_2 & A_c\end{bmatrix};\\ \tilde{B} &:= \begin{bmatrix} B_2 \\ 0 \end{bmatrix};\\ \tilde{C} &:= \begin{bmatrix} C_{\tau_0} + D_{\tau_0}D_cC_2 & D_{\tau_0}C_c\end{bmatrix}.\end{aligned}$$

Inequality (7.2.29) implies that there exists a constant $\delta > 0$ such that given any vector $h \in \mathbf{R}^{2n}$ and any constant $w \in \mathbf{R}$, then

$$\begin{aligned}-\delta\|h\|^2 &\geq 2h'H\tilde{A}h + h'H\tilde{B}\tilde{B}'Hh + h'\tilde{C}'\tilde{C}h\\ &= 2h'H(\tilde{A}h + \tilde{B}w) - 2h'H\tilde{B}w + h'H\tilde{B}\tilde{B}'Hh + \sigma_1^2\\ &= 2h'H(\tilde{A}h + \tilde{B}w) + (h'H\tilde{B} - w)^2 + \sigma_1^2 - w^2\\ &\geq 2h'H(\tilde{A}h + \tilde{B}w) + \sigma_1^2 - w^2\end{aligned} \tag{7.2.30}$$

where $\sigma_1 = \tilde{C}h$. Now, if we let $\xi = w + \sigma_1$ and

$$\bar{A} = \begin{bmatrix} A + B_1D_cC_2 & B_1C_c \\ B_cC_2 & A_c \end{bmatrix},$$

then it follows from (7.2.30) that

$$2h'H(\bar{A}h + \tilde{B}\xi) + \sigma_1^2 - w^2 < -\delta\|h\|^2. \tag{7.2.31}$$

We now write $h = [x' \;\; x_c']'$ and $u = D_cC_2x + C_cx_c$. Using this notation, it follows from equations (7.2.31), (7.2.6), (7.2.16) and (7.2.18) that

$$2\alpha h'H(\bar{A}h + \bar{B}\xi) + \tau_0(\sigma\xi - \frac{1}{k}\xi^2) + \xi C_1[Ax + B_1u + B_2\xi] \leq -\delta\alpha\|h\|^2. \tag{7.2.32}$$

We now consider the derivative $\dot{V}_0(h)$ of the quadratic form $V_0(h) = h'Hh$ along of trajectories of the closed loop system (7.2.1), (7.2.3), (7.2.10). Indeed, given any nonlinearity $\xi(t) = \phi(\sigma(t))$ satisfying the sector bound condition (7.2.2), then $\sigma\xi - \frac{1}{k}\xi^2 \geq 0$ and condition (7.2.32) implies that there exists a constant $\varepsilon_5 > 0$ such that

$$\frac{1}{\beta}\dot{V}_0(h) + \xi C_1[Ax + B_1 u + B_2\xi] \leq -\epsilon_5(\|h\|^2 + \xi^2) \tag{7.2.33}$$

where $\beta := 1/\alpha > 0$. Furthermore, since

$$\frac{d}{dt}\left[\int_0^\sigma \phi(\zeta)d\zeta\right] = \phi(\sigma)C_1[Ax + B_1 u + B_2\phi(\sigma)],$$

it follows from (7.2.33) that condition (7.2.5) must be satisfied. This completes the proof of the theorem.

■

Remark 7.2.1 It follows from the proof of the above theorem that if condition (ii) is satisfied for some $\delta^* > 0$, then it will also be satisfied for all $\delta \in (0, \delta^*]$. Thus, in order to test condition (ii), one would set $\delta > 0$ to a very small value and then perform a one parameter search over the parameter $\tau_0 \geq 0$. For each value of $\tau_0 \geq 0$, the Riccati equations (7.2.8), and (7.2.9) would be solved and the corresponding conditions on their solutions checked.

One of the key elements in the proof of the Theorem 7.2.1 was the use of a standard result on the output-feedback H^∞ control problem. It is straightforward to establish a state-feedback version of Theorem 7.2.1 using a corresponding result on state-feedback H^∞ control; see Theorem 3.2.2. Also, note that it should be possible to extend Theorem 7.2.1 so that some of the assumptions made on the uncertain system (7.2.1) can be removed. Indeed, we now prove a result concerning the case in which the assumption $B_1'C_1'C_1B_1 > 0$, does not hold. Instead we consider the alternative assumption that $C_1B_1 = 0$. Note that this assumption is satisfied in the example considered in Chapter 5.

We begin with a modification of Definition 7.2.1.

Definition 7.2.2 *Given the system (7.2.1) and a quadratic cost functional*

$$J(u(\cdot)) = \int_0^\infty (x'Rx + u'Gu)dt \tag{7.2.34}$$

where $R = R' > 0$ and $G = G' > 0$. A control law $u(t) = Kx(t)$ is said to define a guaranteed cost control with a Lyapunov function of the Lur'e-Postnikov form for the system (7.2.1), (7.2.2) and the cost functional (7.2.34) if there is a symmetric matrix $X > 0$ and a constant $\beta \geq 0$ such that

$$x'(R + K'GK)x + 2[(A + B_1K)x + B_2\xi]'[Xx + \beta C_1'\xi] < 0 \tag{7.2.35}$$

for all non-zero $x \in \mathbf{R}^n$ and all $\xi \in \mathbf{R}$ satisfying the sector condition

$$\frac{1}{k}\xi^2 - x'C_1'\xi \leq 0. \tag{7.2.36}$$

Remark 7.2.2 Note that the controller in this definition guarantees that

$$\sup_{\phi(\cdot)} J(Kx(\cdot)) \leq x'(0)(X + \beta k C_1'C_1)x(0),$$

where the supremum is taken over the set of all uncertainties satisfying the sector condition (7.2.2); e.g., see [26].

It is easy to establish that any guaranteed cost control for the system (7.2.1), (7.2.2) with a Lyapunov function of the Lur'e-Postnikov form will be an absolutely stabilizing control for this system with a Lyapunov function of the Lur'e-Postnikov form in the sense of Definition 7.2.1. Indeed, the state-feedback version of Definition 7.2.1 says that the uncertain system (7.2.1), (7.2.2) is absolutely stabilizable with a Lyapunov function of the Lur'e-Postnikov form if there exists a linear state feedback control law $u = Kx$, a matrix $X > 0$ and constants $\beta \geq 0$, $\epsilon_0 > 0$ such that

$$2[(A + B_1K)x + B_2\phi(Cx)]'[Xx + C_1'\beta\phi(C_1x)] < -\epsilon_0(\|x\|^2 + |\phi(C_1x)|^2) \tag{7.2.37}$$

for all $x \in \mathbf{R}^n$ and for any function $\phi(\cdot)$ satisfying the condition (7.2.2). Hence, if the control law $u = Kx$ is a guaranteed cost control with a Lyapunov function of the Lur'e-Postnikov form, then it follows from (7.2.35) that (7.2.37) holds for all uncertainty inputs satisfying the constraint (7.2.36). Thus, this control law is an absolutely stabilizing control with a Lyapunov function of the Lur'e-Postnikov form.

Conversely, if there exists a control law $u = Kx$ absolutely stabilizing the system (7.2.1) with a Lyapunov function of the Lur'e-Postnikov form, then for any positive-definite matrices R and G, there exists a constant $\delta > 0$ such that

$$\delta x'(R + K'GK)x + 2[(A + B_1K)x + B_2\xi]'(Xx + C_1'\beta\xi] < -\frac{\epsilon_0}{2}\|x\|^2 \tag{7.2.38}$$

for all $x \in \mathbf{R}^n$ and for all $\xi \in \mathbf{R}$ satisfying (7.2.36). Thus, given a cost functional of the form (7.2.34) corresponding to the chosen matrices R and G, there exists a constant $\delta > 0$ such that this control is a guaranteed cost control with a Lyapunov function of the Lur'e-Postnikov form corresponding to the matrix $\delta^{-1}X$ and the constant $\delta^{-1}\beta$. This observation establishes a connection between guaranteed cost control with a Lyapunov function of the Lur'e-Postnikov form, and absolutely stabilizing control with a Lyapunov function of the Lur'e-Postnikov form. This connection is analogous to the connection between quadratic guaranteed cost control and quadratically stabilizing control, which was established in Section 5.2.

The following theorem establishes a necessary and sufficient condition for a state-feedback controller to be a guaranteed cost controller with a Lyapunov function of the Lur'e-Postnikov form. In this case, the controller will also be an absolutely stabilizing controller with a Lyapunov function of the Lur'e-Postnikov form. The result is given under the technical assumption that $C_1 B_1 = 0$. This is in contrast to the technical assumption of Theorem 7.2.1 which requires that $B_1' C_1' C_1 B_1 > 0$.

Theorem 7.2.2 *Given a system (7.2.1) and a quadratic cost functional (7.2.34), suppose the pair* (A, B_1) *is stabilizable, the pairs* (A, C_1) *and* $(A, R^{1/2})$ *are detectable, and* $C_1 B_1 = 0$. *Then, the following statements are equivalent.*

(i) For some constants $\beta > 0$ *and* $\tau_0 \geq 0$, *there exists a positive-definite solution* X^+ *to the Riccati equation*

$$\begin{aligned}(A + &\frac{1}{2\alpha} B_2(\tau_0 C_1 + C_1 A) - B_1 G^{-1} B_1')' X \\ &+ X(A + \frac{1}{2\alpha} B_2(\tau_0 C_1 + C_1 A) - B_1 G^{-1} B_1') \\ &+ R + \frac{\beta}{2\alpha}(\tau_0 C_1 + C_1 A)'(\tau_0 C_1 + C_1 A) \\ &- X(B_1 G^{-1} B_1 - \frac{1}{2\alpha\beta} B_2 B_2') X = 0 \end{aligned} \quad (7.2.39)$$

such that the matrix

$$A + \frac{1}{2\alpha} B_2(\tau_0 C_1 + C_1 A) - B_1 G^{-1} B_1' X^+ + \frac{1}{2\alpha\beta} B_2 B_2' X^+ \quad (7.2.40)$$

is a stable matrix.

(ii) For the system (7.2.1), (7.2.2), there exists a state-feedback guaranteed cost controller with a Lyapunov function of the Lur'e-Postnikov form.

Furthermore, if condition (i) is satisfied and $X^+ > 0$ *is a solution to Riccati equation (7.2.39), then the state-feedback controller*

$$u(t) = Kx(t); \qquad K = -G^{-1} B_1' X^+ \quad (7.2.41)$$

is an absolutely stabilizing controller for the uncertain system (7.2.1), (7.2.2) with a Lyapunov function of the Lur'e-Postnikov form. Furthermore, this controller guarantees that

$$\sup J(Kx(\cdot)) \leq x'(0)(X^+ + \beta k C_1' C_1) x(0) \quad (7.2.42)$$

where the supremum is taken over the set of all uncertainties satisfying the sector condition (7.2.2).

Proof: (ii)⇒(i). Let $u = Kx$ be a state-feedback guaranteed cost controller with a Lyapunov function of the Lur'e-Postnikov form. Using Definition 7.2.2, inequality (7.2.35) implies that there exists a positive-definite matrix X and positive constants $\beta > 0$ and $\epsilon > 0$ such that

$$
\begin{aligned}
&x'(R + K'GK)x + 2[(A + B_1K)x + B_2\xi]'[Xx + \beta C_1\xi)] \\
&\leq -\epsilon(\|x\|^2 + \xi^2)
\end{aligned} \tag{7.2.43}
$$

for all $x \in \mathbf{R}^n$ and all $\xi \in \mathbf{R}$ satisfying condition (7.2.36). It is obvious that $\frac{1}{k}\xi^2 < \xi C_1 x$ for some $(x, \xi) \in \mathbf{R}^{n+1}$. Hence, the S-procedure Theorem 4.2.1, implies that there exists a constant $\tau \geq 0$ such that

$$
\begin{aligned}
&x'(R + K'GK)x + 2[(A + B_1K)x + B_2\xi]'[Xx + \beta C_1'\xi)] \\
&+2\beta\tau(x'C_1'\xi - \frac{1}{k}\xi^2) \leq -\epsilon(\|x\|^2 + \xi^2).
\end{aligned} \tag{7.2.44}
$$

for all nonzero $x \in \mathbf{R}^n$ and all $\xi \in \mathbf{R}$.

In a similar fashion to the proof of Theorem 7.2.1, we can replace the above inequality with the inequality

$$
\begin{aligned}
&x'(R + K'GK)x + 2[(A + B_1K)x + B_2\xi]'[Xx + \beta C_1'\xi)] \\
&+2\beta\tau_0(x'C_1'\xi - \frac{1}{k}\xi^2) \leq -\epsilon_1(\|x\|^2 + \xi^2)
\end{aligned} \tag{7.2.45}
$$

where $\epsilon_1 = \epsilon/(1 + c_1)$, and $\tau_0 \in [\tau, \tau + \epsilon_1]$. This inequality is a quadratic inequality with respect to ξ. The coefficient of the quadratic term in ξ on the left-hand side of (7.2.45) is equal to $-2\alpha\beta$, where the constant α is defined as in (7.2.6). By letting $x = 0$ in (7.2.45), we conclude that $\alpha > 0$. Hence, (7.2.45) and the assumption $C_1B_1 = 0$ imply that

$$
\begin{aligned}
&(A + B_1K)'X + X(A + B_1K) + R + K'GK \\
&+\frac{1}{2\alpha\beta}(\beta(A'C_1' + \tau_0C_1') + XB_2)(B_2'X + \beta(\tau_0C_1 + C_1A)) < 0.
\end{aligned}
$$

Now, using the notation

$$
\begin{aligned}
&\tilde{A} = A + \frac{1}{2\alpha}B_2(\tau_0C_1 + C_1A); \quad \tilde{B}_1 = B_1G^{-1/2}; \quad \tilde{B}_2 = (2\alpha\beta)^{-1/2}B_2; \\
&\tilde{C} = \begin{bmatrix} R^{1/2} \\ \left(\frac{\beta}{2\alpha}\right)^{1/2}(\tau_0C_1 + C_1A) \\ 0 \end{bmatrix}; \qquad \tilde{D} = \begin{bmatrix} 0 \\ 0 \\ I \end{bmatrix}; \\
&\tilde{K} = G^{1/2}K; \qquad \tilde{A}_K = \tilde{A} + \tilde{B}_1\tilde{K}
\end{aligned} \tag{7.2.46}
$$

this inequality becomes

$$
\tilde{A}_K'X + X\tilde{A}_K + \tilde{C}'\tilde{C} + \tilde{K}'\tilde{D}'\tilde{D}\tilde{K} + X\tilde{B}_2'\tilde{B}_2X < 0. \tag{7.2.47}
$$

Associated with this inequality is a state-feedback H^∞ control problem defined by the system

$$\dot{x} = \tilde{A}x + \tilde{B}_1 v + \tilde{B}_2 w;$$
$$\tilde{z} = \tilde{C}x + \tilde{D}v.$$

We verify now that this system satisfies the assumptions of Lemma 3.2.3 on page 70. Indeed, using the Strict Bounded Real Lemma (Lemma 3.1.2), it follows from (7.2.47) that the matrix $\tilde{A}_K$ is stable, and the state-feedback matrix $\tilde{K}$ leads to the satisfaction of the H^∞ norm bound condition

$$\sup_{w \in \mathbf{L}_2} \frac{\int_0^\infty \|\tilde{z}(t)\|^2 dt}{\int_0^\infty \|w(t)\|^2 dt} < 1.$$

This condition is equivalent to the corresponding H^∞ norm condition (3.2.26) required in Lemma 3.2.3. Also, since $R > 0$, it follows immediately that the pair $(\tilde{A}, \tilde{C})$ is detectable.

Now, Lemma 3.2.3 applied to (7.2.47), implies that the Riccati equation

$$\tilde{A}'\tilde{X} + \tilde{X}\tilde{A} - \tilde{X}(\tilde{B}_1\tilde{B}_1' - \tilde{B}_2\tilde{B}_2')\tilde{X} + \tilde{C}'\tilde{C} = 0$$

has a solution $X^+ \geq 0$ such that $X^+ < X$ and the matrix (7.2.40) is stable. Furthermore, since $R > 0$, then $\tilde{C}'\tilde{C} > 0$ and the above equation implies that $X^+ > 0$. Substituting (7.2.46) into this equation completes the proof.

(i)⇒(ii). This part of the theorem is similar to the proof of Theorem 5.5.1 on page 172. Also, the proof of the claim that the controller (7.2.41) is a guaranteed cost controller with a Lyapunov function of the Lur'e-Postnikov form, follows along the same lines as the proof of the corresponding claim in Theorem 5.5.1; see also [26].

■

7.3 Structured dissipativity and absolute stability for nonlinear uncertain systems

7.3.1 *Preliminary remarks*

In this section, we show that the use of integral uncertainty constraints allows us to extend the use of the S-procedure for structured uncertainties into the realm of nonlinear systems. Also, we introduce a new definition of absolute stability which extends the notion of absolute stability presented in Definition 3.5.2 to nonlinear uncertain systems; see also [256, 259, 177, 185]. A feature of this uncertainty model and corresponding definition of absolute stability is that it is closely related to the notion of dissipativity which arises in the modern theory of nonlinear systems; e.g., see [248, 79, 80, 81, 33]. Thus in this section, we establish a connection between the problems of robust stability for nonlinear uncertain systems with structured uncertainty, the absolute stability of nonlinear uncertain systems, and the notion of dissipativity for nonlinear systems.

The definition of dissipativity given in [248], concerns a given nonlinear system with an associated "supply function." This supply function represents the rate of "generalized energy" flow into the system. For a nonlinear system to be dissipative, it is required that there exists a suitable "storage function" for the system. This storage function must be found such that, over any given interval of time, the change in the storage function is less than the integral of the supply function. In this definition, the storage function acts as a measure of the generalized energy stored in the system. Thus, a dissipative system is one in which generalized energy is continually dissipated. The advantage of this general definition is that it includes such concepts as passivity and finite gain stability as special cases.

The study of dissipativity is particularly useful when investigating the stability of the feedback interconnection of two nonlinear systems; e.g., see [80]. Such a situation would arise when a given nonlinear system is subject to a single uncertain nonlinearity. In particular, for the case of a linear uncertain system containing a single uncertain nonlinearity, this leads to the standard small gain condition for robust stability.

A notion of *structured dissipativity* is presented in this section. This definition concerns a given nonlinear system and an associated *collection* of supply functions. In this case, we require the existence of a storage function such that over any interval of time, the system is dissipative with that storage function and at least one of the given supply functions. This notion is directly connected with the absolute stability of a nonlinear uncertain system containing a number of uncertain nonlinearities.

7.3.2 *Definitions*

Consider the following nonlinear system of the form (2.3.9):

$$\dot{x}(t) = g(x(t), \xi(t)) \tag{7.3.1}$$

where $x(t) \in \mathbf{R}^n$ is the *state* and $\xi(t) \in \mathbf{R}^m$ is the *input*. The set Ξ of all admissible inputs consists of all locally integrable vector functions from $\mathbf{R}$ to $\mathbf{R}^m$. Also, consider the associated set of supply rates introduced in Section 2.3.3:

$$w_1(x(t), \xi(t)), \;\; w_2(x(t), \xi(t)), \;\; \ldots, \;\; w_k(x(t), \xi(t)). \tag{7.3.2}$$

and corresponding integral functionals:

$$\begin{aligned}
\mathcal{G}_1\left(x(\cdot), \xi(\cdot)\right) &= \int_0^\infty w_1(x(t), \xi(t))dt; \\
\mathcal{G}_2\left(x(\cdot), \xi(\cdot)\right) &= \int_0^\infty w_2(x(t), \xi(t))dt; \\
&\vdots \\
\mathcal{G}_k\left(x(\cdot), \xi(\cdot)\right) &= \int_0^\infty w_k(x(t), \xi(t))dt.
\end{aligned} \tag{7.3.3}$$

We assume the system (7.3.1) and associated set of functions (7.3.3) satisfy Assumptions 4.4.1–4.4.3 of Section 4.4. That is, the following assumptions are satisfied.

Assumption 7.3.1 *The functions $g(\cdot,\cdot), w_1(\cdot,\cdot), \dots, w_k(\cdot,\cdot)$ are Lipschitz continuous.*

Assumption 7.3.2 *For all $\xi(\cdot) \in \mathbf{L}_2[0,\infty)$ and all initial conditions $x(0) \in \mathbf{R}^n$, the corresponding solution $x(\cdot)$ belongs to $\mathbf{L}_2[0,\infty)$ and the corresponding quantities $\mathcal{G}_1(x(\cdot),\xi(\cdot)), \dots, \mathcal{G}_k(x(\cdot),\xi(\cdot))$ are finite.*

Assumption 7.3.3 *Given any $\varepsilon > 0$, there exists a constant $\delta > 0$ such that the following condition holds: For any input function*

$$\xi_0(\cdot) \in \{\xi_0(\cdot) \in \mathbf{L}_2[0,\infty), \|\xi_0(\cdot)\|_2^2 \le \delta\}$$

and any $x_0 \in \{x_0 \in \mathbf{R}^n : \|x_0\| \le \delta\}$, let $x_0(t)$ denote the corresponding solution to (7.3.1) with initial condition $x_0(0) = x_0$. Then

$$|\mathcal{G}_j(x_0(\cdot),\xi_0(\cdot))| < \varepsilon$$

for $j = 0, 1, \dots, k$.

In addition, we assume the following.

Assumption 7.3.4 *The inequalities*

$$w_1(x,0) \le 0, \;\; w_2(x,0) \le 0, \;\; \dots, \;\; w_k(x,0) \le 0$$

are satisfied for all $x \in \mathbf{R}^n$.

Assumption 7.3.5 *For any $x_0 \in \mathbf{R}^n$, there exists a time $T > 0$ and an input $\xi_0(\cdot) \in \Xi$ such that for the solution $x(\cdot)$ to the system (7.3.1) with initial condition $x(0) = 0$ and the input $\xi_0(\cdot)$, we have $x(T) = x_0$.*

Note that Assumption 7.3.5 is a controllability type assumption on the system (7.3.1), and Assumption 7.3.3 is a stability type assumption.

Consider the system (7.3.1) with associated function $w(x(t),\xi(t))$. For this system, we have the following standard definition of dissipativity; e.g., see [248, 79, 80, 81, 33].

Definition 7.3.1 *A system (7.3.1) with supply rate $w(x(t),\xi(t))$ is said to be* dissipative *if there exists a nonnegative function $V(x_0) : \mathbf{R}^n \to \mathbf{R}$ called a* storage function, *such that $V(0) = 0$ and the following condition holds: Given any input $\xi(\cdot) \in \Xi$ and any corresponding solution to (7.3.1) with an interval of existence $[0, t_*)$ (that is, t_* is the upper time limit for which the solution exists), then*

$$V(x(t_0)) - V(x(0)) \le \int_0^{t_0} w(x(t),\xi(t))dt \qquad (7.3.4)$$

for all $t_0 \in [0, t_*)$.

We now introduce a notion of structured dissipativity. This definition extends the above definition to the case in which there are a number of supply rates associated with the system (7.3.1).

Definition 7.3.2 *A system (7.3.1) with supply rates (7.3.2) is said to have the* structured dissipativity property *if there exists a nonnegative storage function* $V(x_0) : \mathbf{R}^n \to \mathbf{R}$, *such that* $V(0) = 0$ *and the following condition holds: Given any input* $\xi(\cdot) \in \Xi$ *and any corresponding solution to (7.3.1) with an interval of existence* $[0, t_*)$, *then*

$$V(x(t_0)) - V(x(0)) \leq \max_{j=1,2,\ldots,k} \left\{ \int_0^{t_0} w_j(x(t), \xi(t)) dt \right\} \tag{7.3.5}$$

for all $t_0 \in [0, t_*)$.

Let $\delta > 0$ be a given constant. We introduce the following functions associated with the system (7.3.1)

$$\begin{aligned} w_1^\delta(x(t), \xi(t)) &:= w_1(x(t), \xi(t)) - \delta(\|x(t)\|^2 \\ &\qquad + \|g(x(t), \xi(t))\|^2 + \|\xi(t)\|^2); \\ w_2^\delta(x(t), \xi(t)) &:= w_2(x(t), \xi(t)) - \delta(\|x(t)\|^2 \\ &\qquad + \|g(x(t), \xi(t))\|^2 + \|\xi(t)\|^2); \\ &\vdots \\ w_k^\delta(x(t), \xi(t)) &:= w_k(x(t), \xi(t)) - \delta(\|x(t)\|^2 \\ &\qquad + \|g(x(t), \xi(t))\|^2 + \|\xi(t)\|^2). \end{aligned} \tag{7.3.6}$$

Definition 7.3.3 *The system (7.3.1) with supply rates (7.3.2) is said to have the* strict structured dissipativity property *if there exists a constant* $\delta > 0$ *such that the system (7.3.1) with supply rates (7.3.6) has the structured dissipativity property.*

7.3.3 A connection between dissipativity and structured dissipativity

The following theorem establishes a connection between dissipativity and structured dissipativity for nonlinear systems.

Theorem 7.3.1 *Consider the nonlinear system (7.3.1) with supply rates (7.3.2) and suppose that Assumptions 7.3.1–7.3.5 are satisfied. Then the following statements are equivalent:*

(i) The system (7.3.1) with supply rates (7.3.2) has the structured dissipativity property.

(ii) *There exist constants* $\tau_1 \geq 0, \tau_2 \geq 0, \ldots, \tau_k \geq 0$ *such that* $\sum_{j=1}^{k} \tau_j = 1$ *and the following condition holds: Given any input* $\xi(\cdot) \in \Xi$ *and the corresponding solution to the equation (7.3.1) with initial condition* $x(0) = 0$ *defined on an existence interval* $[0, t_*)$, *then*

$$\int_0^{t_0} w_\tau(x(t), \xi(t))dt \geq 0 \tag{7.3.7}$$

for all $t_0 \in [0, t_*)$ *where* $w_\tau(\cdot, \cdot)$ *is defined by*

$$w_\tau(x(t), \xi(t)) := \sum_{j=1}^{k} \tau_j w_j(x(t), \xi(t)). \tag{7.3.8}$$

(iii) *There exist constants* $\tau_1 \geq 0, \tau_2 \geq 0, \ldots, \tau_k \geq 0$ *such that* $\sum_{j=1}^{k} \tau_j = 1$ *and the system (7.3.1) is dissipative with supply rate (7.3.8).*

Proof: (i)⇒(ii). In order to prove this statement, we first establish the following proposition. For the system (7.3.1) satisfying Assumptions 7.3.1–7.3.3 , we define a set $\Omega \subset \mathbf{L}_2[0, \infty) \times \mathbf{L}_2[0, \infty)$ as follows: Ω is the set of pairs $\{x(\cdot), \xi(\cdot)\}$ such that $\xi(\cdot) \in \mathbf{L}_2[0, \infty)$ and $x(\cdot)$ is the corresponding solution to (7.3.1) with initial condition $x(0) = 0$; *c.f.* the set Ω defined in Theorem 4.4.1.

Proposition 1 *Consider the set* Ω *defined above. If the system (7.3.1) with supply rates (7.3.2) has the structured dissipativity property then there exist constants* $\tau_1 \geq 0, \ldots, \tau_k \geq 0$ *such that* $\sum_{j=1}^{k} \tau_j = 1$ *and condition*

$$\tau_1 \mathcal{G}_1(x(\cdot), \xi(\cdot)) + \tau_2 \mathcal{G}_2(x(\cdot), \xi(\cdot)) + \cdots + \tau_k \mathcal{G}_k(x(\cdot), \xi(\cdot)) \geq 0 \tag{7.3.9}$$

holds for all pairs $\{x(\cdot), \xi(\cdot))\} \in \Omega$.

In order to establish this proposition, we first prove that $\mathcal{G}_1(x(\cdot), \xi(\cdot)) \geq 0$ for all pairs $\{x(\cdot), \xi(\cdot)\} \in \Omega$ such that

$$\mathcal{G}_2(x(\cdot), \xi(\cdot)) < 0, \ldots, \mathcal{G}_k(x(\cdot), \xi(\cdot)) < 0. \tag{7.3.10}$$

Indeed, if condition (7.3.10) holds for a pair $\{x_0(\cdot), \xi_0(\cdot)\} \in \Omega$, then there exists a time $T \geq 0$ such that

$$\int_0^{t_0} w_2(x_0(t), \xi_0(t))dt < 0, \ldots, \int_0^{t_0} w_k(x_0(t), \xi_0(t))dt < 0 \tag{7.3.11}$$

for all $t_0 \geq T$. We now consider the corresponding storage function $V(\cdot)$. If $x_0(0) = 0$, then $V(x_0(0)) = 0$ and since $V(x_0(t_0)) \geq 0$, we have

$$V(x_0(t_0)) - V(x_0(0)) \geq 0.$$

Furthermore, this and condition (7.3.5) imply that

$$\max_{j=1,2,\ldots,k}\left\{\int_0^{t_0} w_j(x_0(t),\xi_0(t))dt\right\} \geq 0.$$

Therefore, if conditions (7.3.11) hold, then

$$\int_0^{t_0} w_1(x_0(t),\xi_0(t))dt \geq 0$$

for all $t_0 \geq T$. Hence, $\mathcal{G}_1(x_0(\cdot),\xi_0(\cdot)) \geq 0$. Now using the S-procedure Theorem 4.4.1, it follows that there exist constants $\tau_1 \geq 0, \tau_2 \geq 0, \ldots, \tau_k \geq 0$ such that $\sum_{j=1}^k \tau_j = 1$ and condition (7.3.9) is satisfied. This completes the proof of the proposition.

Proposition 2 *If there exist constants $\tau_1 \geq 0$, $\tau_2 \geq 0$, $\ldots$, $\tau_k \geq 0$ such that $\sum_{j=1}^k \tau_j = 1$ and condition (7.3.9) holds for all pairs $\{x(\cdot),\xi(\cdot)\} \in \Omega$, then condition (7.3.7), (7.3.8) holds for all $t_0 \geq 0$ and all pairs $\{x(\cdot),\xi(\cdot)\}$ such that $\xi(\cdot) \in \Xi$ and $x(\cdot)$ is the corresponding solution to (7.3.1) with initial condition $x(0) = 0$.*

Indeed, let $t_0 \geq 0$ be given. Also, consider an input $\xi_{t_0}(\cdot) \in \Xi$ such that $\xi_{t_0}(t) = 0$ for all $t \geq t_0$ and let $x_{t_0}(\cdot)$ be the corresponding solution to (7.3.1) with initial condition $x(0) = 0$. Since $\xi_{t_0}(\cdot) \in \mathbf{L}_2[0,\infty)$, Assumption 7.3.2 implies that $x_{t_0}(\cdot) \in \mathbf{L}_2[0,\infty)$. Also using Assumption 7.3.4, it follows that

$$\begin{aligned}\int_0^{t_0} w_\tau(x_{t_0}(t),\xi_{t_0}(t))dt &= \sum_{j=1}^k \tau_j \mathcal{G}_j(x_{t_0}(\cdot),\xi_{t_0}(\cdot)) \\ &\quad -\sum_{j=1}^k \tau_j \int_{t_0}^\infty w_j(x_{t_0}(t),0)dt \\ &\geq \sum_{j=1}^k \tau_j \mathcal{G}_j(x_{t_0}(\cdot),\xi_{t_0}(\cdot)).\end{aligned}$$

From this, condition (7.3.7) follows directly using condition (7.3.9). This completes the proof of the proposition and this part of the theorem.

(ii)⇒(iii). Suppose that condition (ii) holds and introduce the function $V(x_0)$: $\mathbf{R}^n \to \mathbf{R}$ defined by

$$V(x_0) := \sup_{x(0)=x_0,\xi(\cdot)\in\Xi,t_0\geq 0}\left\{-\int_0^{t_0} w_\tau(x(t),\xi(t))dt\right\} \tag{7.3.12}$$

where $w_\tau(\cdot,\cdot)$ is the function defined in (7.3.8). Using a standard approach from the theory of dissipative systems, we will prove that the system (7.3.1) is dissipative with supply rate $w_\tau(\cdot,\cdot)$ and storage function $V(\cdot)$; e.g., see [79, 80, 81].

Indeed, the condition $V(0) = 0$ follows from the definition of $V(\cdot)$ by using inequality (7.3.7). Furthermore, Assumption 7.3.4 implies that $V(x_0) \geq 0$ for all x_0. We now use condition (7.3.7) to prove that $V(x_0) < \infty$ for all x_0. Indeed, given any $x_0 \in \mathbf{R}^n$, it follows from Assumption 7.3.5 that there exists a time $T > 0$ and a function $\xi_0(\cdot) \in \Xi$ such that for the system (7.3.1) with initial condition $x(0) = 0$ and input $\xi_0(\cdot)$, we have $x(T) = x_0$. Also, let $\xi_1(\cdot) \in \Xi$ be given and define $\xi(\cdot) \in \Xi$ by

$$\xi(t) := \begin{cases} \xi_0(t) & \text{for } 0 \leq t < T; \\ \xi_1(t) & \text{for } T \leq t < \infty. \end{cases}$$

Now let $x(\cdot)$ be the solution to the system (7.3.1) with initial condition $x(0) = 0$ and input $\xi(\cdot)$. Then given any time $t_1 \geq T$, it follows from (7.3.7) that

$$\begin{aligned} 0 &\leq \int_0^{t_1} w_\tau(x(t), \xi(t))dt \\ &= \int_0^T w_\tau(x(t), \xi_0(t))dt + \int_T^{t_1} w_\tau(x(t), \xi_1(t))dt. \end{aligned}$$

Hence,

$$-\int_T^{t_1} w_\tau(x(t), \xi_1(t))dt \leq \int_0^T w_\tau(x(t), \xi_0(t))dt < \infty. \tag{7.3.13}$$

We now let $t_0 = t_1 - T \geq 0$, $\tilde{x}(t) = x(t+T)$, $\tilde{\xi}_1(t) = \xi_1(t+T)$, and use the time-invariance of the system (7.3.1) to conclude that equation (7.3.13) can be re-written as

$$-\int_0^{t_0} w_\tau(\tilde{x}(t), \tilde{\xi}_1(t))dt \leq \int_0^T w_\tau(x(t), \xi_0(t))dt < \infty. \tag{7.3.14}$$

Furthermore, we have $x(T) = x_0$ and hence $\tilde{x}(0) = x_0$. Thus, given any $t_0 \geq 0$ and any $\tilde{\xi}_1(\cdot) \in \Xi$, then $\tilde{x}(\cdot)$, the corresponding solution to the system (7.3.1) with initial condition $\tilde{x}(0) = x_0$ and input $\tilde{\xi}_1(\cdot)$ satisfies inequality (7.3.14). Hence, $V(x_0) < \infty$ for all x_0.

To complete the proof of this part of the theorem, it remains only to establish that the dissipativity condition (7.3.4) is satisfied. Indeed, given any $\xi(\cdot) \in \Xi$ and any $T \geq t_0$, then if $x(t)$ is the corresponding solution to (7.3.1) with initial condition $x(0)$, it follows from (7.3.12) that

$$\begin{aligned} V(x(0)) &\geq -\int_0^T w_\tau(x(t), \xi(t))dt \\ &= -\int_0^{t_0} w_\tau(x(t), \xi(t))dt - \int_{t_0}^T w_\tau(x(t), \xi(t))dt. \end{aligned}$$

Hence,

$$-\int_{t_0}^T w_\tau(x(t), \xi(t))dt - V(x(0)) \leq \int_0^{t_0} w_\tau(x(t), \xi(t))dt. \tag{7.3.15}$$

Now define $\xi_1(t) := \xi(t + t_0)$ for $t \in [0, T - t_0]$. Using the time invariance of the system (7.3.1), it follows that

$$-\int_{t_0}^{T} w_\tau(x(t), \xi(t))dt = -\int_{0}^{T-t_0} w_\tau(\tilde{x}(t), \xi_1(t))dt$$

where $\tilde{x}(\cdot)$ is the solution to (7.3.1) with initial condition $\tilde{x}(0) = x(t_0)$ and input $\xi_1(\cdot)$. Substituting this into (7.3.15), it follows that

$$-\int_{0}^{T-t_0} w_\tau(\tilde{x}(t), \xi_1(t))dt - V(x(0)) \leq \int_{0}^{t_0} w_\tau(x(t), \xi(t))dt$$

However, this holds for all $\xi_1(\cdot) \in \Xi$ and for all $T \geq t_0$. Thus using (7.3.12), we can conclude that the following dissipation inequality is satisfied:

$$V(x(t_0)) - V(x(0)) \leq \int_{0}^{t_0} w_\tau(x(t), \xi(t))dt.$$

(iii)⇒(i). If condition (iii) holds then there exists a function $V(x_0) \geq 0$ such that $V(0) = 0$ and

$$V(x(t_0)) - V(x(0)) \leq \sum_{j=1}^{k} \tau_j \int_{0}^{t_0} w_j(x(t), \xi(t))dt \tag{7.3.16}$$

for all solutions to the system (7.3.1) and all $t_0 \geq 0$. However, since $\tau_j \geq 0$ and $\sum_{j=1}^{k} \tau_j = 1$, inequality (7.3.16) implies

$$\begin{aligned} V(x(t_0)) - V(x(0)) &\leq \sum_{j=1}^{k} \tau_j \int_{0}^{t_0} w_j(x(t), \xi(t))dt \\ &\leq \max_{j=1,2,\ldots,k} \left\{ \int_{0}^{t_0} w_j(x(t), \xi(t))dt \right\}. \end{aligned}$$

That is, condition (7.3.5) is satisfied. This completes the proof of the theorem. ■

7.3.4 *Absolute stability for nonlinear uncertain systems*

In this section, we consider a nonlinear uncertain system defined by the state equations (7.3.1) and associated functions (7.3.2). In this case, $\xi(\cdot)$ is interpreted as a *structured uncertainty input*. The uncertainty in the system (7.3.1) is described by the following equation of the form (2.3.11):

$$\xi(t) = \phi(t, x(\cdot)) \tag{7.3.17}$$

It is assumed that this uncertainty satisfies the integral constraint defined in Definition 2.3.3 on page 26. That is, the admissible uncertainties are defined as in the following definition.

Definition 7.3.4 *An uncertainty of the form (7.3.17) is an admissible uncertainty for the system (7.3.1) if the following condition holds: Given any solution to equations (7.3.1), (7.3.17) with an interval of existence* $[0, t_*)$, *there exists a sequence* $\{t_i\}_{i=1}^{\infty}$ *and constants* $d_1 \geq 0, d_2 \geq 0, \ldots, d_k \geq 0$ *such that* $t_i \to t_*$, $t_i \geq 0$ *and*

$$\int_0^{t_i} w_j(x(t), \xi(t))dt \leq d_j \tag{7.3.18}$$

for all i *and for* $j = 1, 2, \ldots, k$. *Note that* t_* *and* t_i *may be equal to infinity.*

Further remarks concerning this class of uncertainties can be found in Section 2.3.3.

We now introduce a corresponding notion of absolute stability for the uncertain system (7.3.1), (7.3.17). This definition extends the standard definition of absolute stability to the case of nonlinear uncertain systems; e.g., see [256, 260, 259, 177, 185].

Definition 7.3.5 *Consider the uncertain system (7.3.1), (7.3.17). This uncertain system is said to be* absolutely stable *if there exists a nonnegative function* $W(\cdot) : \mathbf{R}^n \to \mathbf{R}$ *and a constant* $c > 0$ *such that the following conditions hold:*

(i) $W(0) = 0$.

(ii) For any initial condition $x(0) = x_0$ *and any uncertainty input* $\xi(\cdot) \in \mathbf{L}_2[0, \infty)$, *the system (7.3.1) has a unique solution which is defined on* $[0, \infty)$.

(iii) Given any admissible uncertainty for the uncertain system (7.3.1), (7.3.17), then any solution to equations (7.3.1), (7.3.17) satisfies $[x(\cdot), \dot{x}(\cdot), \xi(\cdot)] \in \mathbf{L}_2[0, \infty)$ *(hence,* $t_* = \infty$*) and*

$$\int_0^{\infty} (\|x(t)\|^2 + \|\dot{x}(t)\|^2 + \|\xi(t)\|^2)dt \leq W(x(0)) + c\sum_{j=1}^{k} d_j. \tag{7.3.19}$$

Remark 7.3.1 In the above definition, condition (7.3.19) requires that the $\mathbf{L}_2$ norm of any solution to (7.3.1) and (7.3.17) be bounded in terms of the norm of the initial condition and the 'measure of mismatch' d for this solution.

Observation 7.3.1 It follows from the above definition that for any admissible uncertainty, an absolutely stable uncertain system (7.3.1), (7.3.17) has the property that $x(t) \to 0$ as $t \to \infty$. Indeed, using the fact that $x(\cdot) \in \mathbf{L}_2[0, \infty)$ and $\dot{x}(\cdot) \in \mathbf{L}_2[0, \infty)$, it follows that $x(t) \to 0$ as $t \to \infty$.

Theorem 7.3.2 *Consider the nonlinear uncertain system (7.3.1), (7.3.17) with associated functions (7.3.2) and suppose that Assumptions 7.3.1–7.3.3, 7.3.4 and 7.3.5 are satisfied. Then the following statements are equivalent:*

(i) The uncertain system (7.3.1), (7.3.17) is absolutely stable.

(ii) There exist constants $\delta > 0, \tau_1 \geq 0, \tau_2 \geq 0, \ldots, \tau_k \geq 0$ such that $\sum_{j=1}^{k} \tau_j = 1$ and the following condition holds: Given any input $\xi(\cdot) \in \Xi$, let $x(\cdot)$ be the corresponding solution to equation (7.3.1) with initial condition $x(0) = 0$. Furthermore, suppose $x(\cdot)$ is defined on an existence interval $[0, t_*)$. Then

$$\int_0^{t_0} w_\tau(x(t), \xi(t))dt \geq \delta \int_0^{t_0} (\|x(t)\|^2 + \|\dot{x}(t)\|^2 + \|\xi(t)\|^2)dt \quad (7.3.20)$$

for all $t_0 \in [0, t_*)$ where $w_\tau(\cdot, \cdot)$ is defined by equation (7.3.8).

(iii) The system (7.3.1) with supply rate functions (7.3.2) has the strict structured dissipativity property.

(iv) There exist constants $\delta > 0, \tau_1 \geq 0, \tau_2 \geq 0, \ldots, \tau_k \geq 0$ such that $\sum_{j=1}^{k} \tau_j = 1$ and the system (7.3.1) is dissipative with supply rate

$$\begin{aligned} w_\tau^\delta(x(t), \xi(t)) := & \sum_{j=1}^{k} \tau_j w_j(x(t), \xi(t)) \\ & -\delta(\|x(t)\|^2 + \|g(x(t), \xi(t))\|^2 + \|\xi(t)\|^2). \end{aligned} \quad (7.3.21)$$

Proof: (i)⇒(ii). Let $\delta := 1/(kc)$ where c is the constant from Definition 7.3.5. Consider functionals defined by

$$\begin{aligned} \mathcal{G}_1^\delta(x(\cdot), \xi(\cdot)) &= \int_0^\infty w_1^\delta(x(t), \xi(t))dt; \\ \mathcal{G}_2^\delta(x(\cdot), \xi(\cdot)) &= \int_0^\infty w_2^\delta(x(t), \xi(t))dt; \\ &\vdots \\ \mathcal{G}_k^\delta(x(\cdot), \xi(\cdot)) &= \int_0^\infty w_k^\delta(x(t), \xi(t))dt \end{aligned} \quad (7.3.22)$$

where the functions $w_j^\delta(\cdot, \cdot)$ are as defined in (7.3.6). Also, let Ω be the set defined in Theorem 7.3.1. We will prove that if

$$\mathcal{G}_2^\delta(x(\cdot), \xi(\cdot)) < 0, \ \ldots, \ \mathcal{G}_k^\delta(x(\cdot), \xi(\cdot)) < 0$$

for some pair $\{x(\cdot), \xi(\cdot)\} \in \Omega$, then $\mathcal{G}_1^\delta(x(\cdot), \xi(\cdot)) \geq 0$. Indeed if $\mathcal{G}_1^\delta(x(\cdot), \xi(\cdot)) < 0$, we have

$$\int_0^\infty w_j(x(t), \xi(t))dt < \delta \int_0^\infty (\|x(t)\|^2 + \|g(x(t), \xi(t))\|^2 + \|\xi(t)\|^2)dt$$

for $s = 1, 2, \ldots, k$. Hence, there exist constants $d_1, d_2, \ldots, d_k$ such that

$$d_j < \delta \int_0^\infty (\|x(t)\|^2 + \|g(x(t), \xi(t))\|^2 + \|\xi(t)\|^2)dt \tag{7.3.23}$$

for $j = 1, 2, \ldots, k$ and the pair $\{x(\cdot), \xi(\cdot)\}$ satisfies the integral constraints (7.3.18) with $t_i = \infty$. However since $x(0) = 0$, Definition 7.3.5 implies that

$$\int_0^\infty (\|x(t)\|^2 + \|g(x(t), \xi(t))\|^2 + \|\xi(t)\|^2)dt \leq c \sum_{j=1}^k d_j. \tag{7.3.24}$$

Furthermore, since $\delta = 1/(kc)$, inequality (7.3.24) contradicts inequality (7.3.23). Hence, for any pair $\{x(\cdot), \xi(\cdot)\} \in \Omega$ such that $\mathcal{G}_2^\delta(x(\cdot), \xi(\cdot)) < 0, \ldots, \mathcal{G}_k^\delta(x(\cdot), \xi(\cdot)) < 0$, then $\mathcal{G}_1^\delta(x(\cdot), \xi(\cdot)) \geq 0$. Therefore, Theorem 4.4.1 implies that there exist constants $\tau_1 \geq 0, \tau_2 \geq 0, \ldots, \tau_k \geq 0$ such that $\sum_{j=1}^k \tau_j = 1$ and

$$\sum_{j=1}^k \tau_j \mathcal{G}_j^\delta(x(\cdot), \xi(\cdot)) \geq 0 \tag{7.3.25}$$

for all pairs $\{x(\cdot), \xi(\cdot)\} \in \Omega$. Furthermore, according to Proposition 2 in the proof of Theorem 7.3.1, condition (7.3.25) implies condition (7.3.20). This completes the proof of this part of the theorem.

(ii)⇒(iii), (iii)⇒(iv). These implications are immediate consequences of Theorem 7.3.1 and Definition 7.3.2.

(iv)⇒(i). Condition (iv) implies that there exist constants $\delta > 0, \tau_1 \geq 0, \ldots, \tau_k \geq 0$ and a nonnegative function $V(\cdot) : \mathbf{R}^n \to \mathbf{R}$ such that $V(0) = 0$ and

$$V(x(t_0)) - V(x(0)) \leq \int_0^{t_0} w_\tau^\delta(x(\cdot), \xi(\cdot))dt \tag{7.3.26}$$

for all solutions $x(\cdot)$ to the system (7.3.1), where the function $w_\tau^\delta(\cdot, \cdot)$ is defined as in equation (7.3.21). Therefore, if an input $\xi(\cdot)$ satisfies the integral constraint (7.3.18), it follows from (7.3.26) that

$$\int_0^{t_i} (\|x(t)\|^2 + \|\dot{x}(t)\|^2 + \|\xi(t)\|^2)dt \leq \frac{1}{\delta}(V(x(0)) + \sum_{j=1}^k \tau_j d_j) \tag{7.3.27}$$

for all i. However, it follows from Definition 7.3.4 that $t_i \to t_*$ where t_* is the upper limit of the existence interval of the solution $\{x(\cdot), \xi(\cdot)\}$. Hence, condition (7.3.27) implies that $t_* = \infty$ and the inequality (7.3.19) holds with $W(x_0) = \frac{1}{\delta} V(x_0)$ and with

$$c = \frac{1}{\delta} \max_{j=1,2,\ldots,k} \{\tau_j\}.$$

This completes the proof of the theorem.

■

Remark 7.3.2 For the case of linear uncertain systems with a single uncertainty and a quadratic storage function (Lyapunov function), the above theorem reduces to a well known result in the absolute stability literature; e.g., see [256, 259]. Furthermore in this case, condition (ii) is equivalent to the circle criterion. For the case of linear uncertain systems with structured uncertainty, the above theorem reduces to the result of [125].

Now consider the nonlinear interconnected system

$$
\begin{aligned}
\Sigma: \quad \dot{x}(t) &= g(x(t), \xi_1(t), \xi_2(t), \dots, \xi_k(t)); \\
y_1(t) &= z_1(x(t), \dot{x}(t)); \\
y_2(t) &= z_2(x(t), \dot{x}(t)); \\
&\vdots \\
y_k(t) &= z_k(x(t), \dot{x}(t)); \\
\\
\Sigma_1: \quad \dot{x}_1(t) &= g_1(x_1(t), y_1(t)); \\
\xi_1(t) &= h_1(x_1(t), \dot{x}_1(t)); \\
\\
\Sigma_2: \quad \dot{x}_2(t) &= g_2(x_2(t), y_2(t)); \\
\xi_2(t) &= h_2(x_2(t), \dot{x}_2(t)); \\
&\vdots \\
\Sigma_k: \quad \dot{x}_k(t) &= g_k(x_k(t), y_k(t)); \\
\xi_k(t) &= h_k(x_k(t), \dot{x}_k(t))
\end{aligned}
\tag{7.3.28}
$$

where $x(t) \in \mathbf{R}^n, x_1(t) \in \mathbf{R}^{n_1}, x_2(t) \in \mathbf{R}^{n_2}, \dots, x_k(t) \in \mathbf{R}^{n_k}$ are the state vectors; $\xi_1(t) \in \mathbf{R}^{m_1}, \xi_2(t) \in \mathbf{R}^{m_2}, \dots, \xi_k(t) \in \mathbf{R}^{m_k}, y_1(t) \in \mathbf{R}^{q_1}, y_2(t) \in \mathbf{R}^{q_2}, \dots, y_k(t) \in \mathbf{R}^{q_k}$ are the input-output vectors. A block diagram of this interconnected system is shown in Figure 7.3.1.

We assume the system (7.3.28) satisfies the following assumptions:

Assumption 7.3.6 *There exists a constant $c_0 > 0$ such that*

$$\|z_j(x(t), \dot{x}(t))\|^2 \leq c_0(\|x(t)\|^2 + \|\dot{x}(t)\|^2)$$

for $j = 1, 2, \dots, k$.

Assumption 7.3.7 *Suppose that $y_j(\cdot) \in \mathbf{L}_2[0, \infty)$ is a given input of the subsystem Σ_j and $\{x_j(\cdot), \xi_j(\cdot)\}$ is any corresponding solution. Then the condition $\xi_j(\cdot) \in \mathbf{L}_2[0, \infty)$ implies that $\{x(\cdot), \dot{x}(\cdot)\} \in \mathbf{L}_2[0, \infty)$.*

Note, Assumption 7.3.7 is an observability type assumption on each of the subsystems Σ_j.

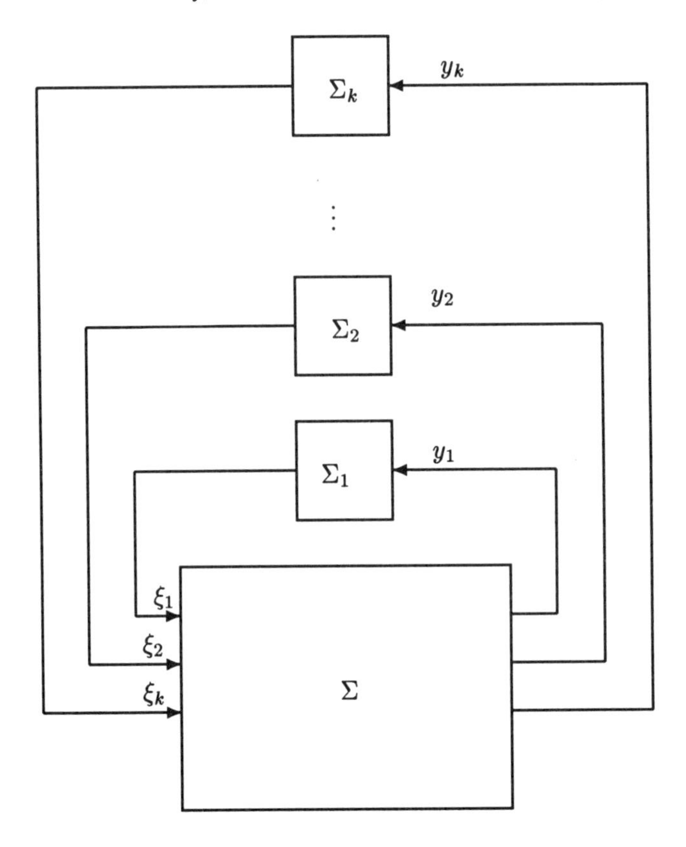

Figure 7.3.1. Interconnected nonlinear system

Definition 7.3.6 *The system (7.3.28) is said to be globally asymptotically stable if all of its solutions* $\{x(t), x_1(t), x_2(t), \ldots, x_k(t)\}$ *are defined on* $[0, \infty)$ *and tend to zero as* $t \to \infty$.

Associated with the system (7.3.28), we will consider the following functions

$$w_1(x_1(t), \xi_1(t)), \;\; w_2(x_2(t), \xi_2(t)), \;\; \ldots, \;\; w_k(x_k(t), \xi_k(t)). \tag{7.3.29}$$

Now we are in a position to present a stability criterion for the nonlinear interconnected system (7.3.28).

Theorem 7.3.3 *Consider the system (7.3.28) with associated functions (7.3.29) and suppose that the subsystem* Σ *satisfies Assumptions 7.3.1–7.3.5 and the subsystems* $\Sigma_1, \Sigma_2, \ldots, \Sigma_k$ *satisfy Assumptions 7.3.6 and 7.3.7. Also, suppose that the subsystem* Σ *with supply rates (7.3.29) has the strict structured dissipativity property and the subsystem* Σ_j *with supply rate* $-w_j(x_j(t), \xi_j(t))$ *is dissipative for* $j = 1, 2, \ldots, k$. *Then the system (7.3.28) is globally asymptotically stable.*

Proof: The dissipativity of the subsystem Σ_j implies that there exists a function $V_j(x_j) \geq 0$ such that

$$V_j(x_j(t_0)) - V_j(x_j(0)) \leq -\int_0^{t_0} w_j(x_j(t), \xi_j(t))dt.$$

Hence, any solution of the subsystem Σ_j satisfies the integral constraint

$$\int_0^{t_i} w_j(x_j(t), \xi_j(t))dt \leq d_j \tag{7.3.30}$$

with any $t_i \geq 0$ and with $d_j = V_j(x_j(0))$. Therefore, according to Theorem 7.3.2, the strict structured dissipativity of the system Σ implies that the system Σ with inputs $\xi_1(\cdot), \xi_2(\cdot), \ldots, \xi_k(\cdot)$ satisfying the integral constraints (7.3.30) is absolutely stable. Hence using Definition 7.3.6, it follows that $\{x(\cdot), \dot{x}(\cdot), \xi_1(\cdot), \xi_2(\cdot), \ldots, \xi_k(\cdot)\} \in \mathbf{L}_2[0, \infty)$. This and Assumption 7.3.6 imply that $\{y_1(\cdot), y_2(\cdot), \ldots, y_k(\cdot)\} \in \mathbf{L}_2[0, \infty)$. Furthermore, according to Assumption 7.3.7, if $\{y_j(\cdot), \xi_j(\cdot)\} \in \mathbf{L}_2[0, \infty)$ then $\{x_j(\cdot), \dot{x}_j(\cdot)\} \in \mathbf{L}_2[0, \infty)$. Using $\{x(\cdot), x_1(\cdot), x_2(\cdot), \ldots, x_k(\cdot)\} \in \mathbf{L}_2[0, \infty)$ and $\{\dot{x}(\cdot), \dot{x}_1(\cdot), \dot{x}_2(\cdot), \ldots, \dot{x}_k(\cdot)\} \in \mathbf{L}_2[0, \infty)$, it now follows that

$$\{x(t), x_1(t), x_2(t), \ldots, x_k(t)\} \to 0$$

as $t \to \infty$. This completes the proof of the theorem.

■

7.4 Conclusions

In this chapter, we have applied the idea of scaling to classical robust stability problems such as absolute stability and dissipativity for nonlinear systems. From the results presented, the reader may draw the following conclusions. Firstly, the H^∞ control based approach allows one to extend the methods of absolute stability to controller design problems. This leads to a corresponding notion of absolute stabilizability. Furthermore, the results presented lead to a possible approach to the problem of non-quadratic optimal guaranteed cost control.

Also, the results presented concerning the notion of structured dissipativity enable nonconservative dissipativity and absolute stability results to be obtained for complex interconnected systems. In particular, a small gain theorem type analysis becomes possible for systems incorporating more than two interconnected blocks.

8. Robust control of stochastic uncertain systems

8.1 Introduction

The results presented in this chapter extend the notions of minimax optimal control and absolute stabilization to the realm of stochastic uncertain systems. Some motivation for this extension was given in the Chapter 1 and in Section 2.4. In particular, in Subsections 2.4.1 and 2.4.2, some examples where given of uncertain systems which led naturally to descriptions in terms of stochastic processes. Also, Section 2.4 introduced definitions of stochastic uncertain system which provided a stochastic uncertain system framework for stochastic systems with multiplicative noise and additive noise. In this chapter, we address minimax optimal control problems for the stochastic uncertain systems introduced in Section 2.4.

The first problem to be considered in this chapter is the stochastic H^∞ control problem. The stochastic H^∞ control problem was originally solved in references [228, 227, 234], where a solution to the state-feedback stochastic H^∞ control problem was given in terms of certain modified Riccati equations, and in [82, 84] where LMI type results were obtained for the output-feedback case. In Section 8.2, we present the results of [228, 227, 234] on state-feedback stochastic H^∞ control. Also, Section 8.2 presents a stochastic version of Bounded Real Lemma.

The stochastic H^∞ control results presented in Section 8.2 play the same fundamental role in the solution to the corresponding state-feedback minimax optimal control problem given in Section 8.3 as deterministic H^∞ control theory plays in the derivation of the results of Chapter 5. In particular, the results presented in Section 8.2 enable us to obtain a tractable solution to a minimax guaranteed cost control problem for stochastic uncertain systems with multiplicative noise. This problem is addressed in Section 8.3. The stochastic uncertainty model considered in this section was defined in Section 2.4.1. Note that stochastic systems with multiplicative noise arise naturally in many control problems and are considered extensively in the literature; e.g., see [32, 56, 57, 83, 226, 264].

The results presented in Section 8.3 are closely related to those of Section 5.3. The main problem addressed in Section 8.3 is to find a static linear state-feedback controller yielding a given level of performance in the face of structured stochastic uncertainties. The class of controllers considered are static state-feedback controllers which absolutely stabilize the system in a stochastic sense. The construction of these controllers is based on the stabilizing solution to a *generalized* Riccati equation arising in the stochastic H^∞ control problem. The results presented in this section originally appeared in [228, 234].

One important feature which distinguishes the stochastic minimax control problem considered in Section 8.3 from the corresponding deterministic problem considered in Section 5.3 is as follows. The stochastic Ito equations, which give a mathematically rigorous description of the system dynamics, are non-autonomous equations. This is reflected by the fact that one cannot reduce a problem with initial condition $x(s) = h$, $s > 0$ to a problem with an initial condition imposed at time zero. This observation is important when the corresponding S-procedure result is considered.

Recall that it is possible to obtain some robust control results for systems with random parameters using the approach presented in Chapter 6. Indeed, the averaged uncertainty constraints of Chapter 6 were motivated by the fact that the model of a system and the system uncertainty might be obtained by taking a series of data measurements and using an averaging process. However, while the approach of Chapter 6 allows for a probabilistic characterization of the initial conditions on the uncertainty dynamics, it does not allow for stochastic noise processes acting on the system. This limitation in the problem formulation can be resolved by considering uncertain systems described by stochastic differential equations.

Our interest in stochastic versions of the results presented in Chapter 6 is motivated by a desire to develop a robust linear quadratic Gaussian (LQG) methodology based on stochastic modeling of the disturbance signals. In Section 8.4, we address the problem of finding an output-feedback minimax optimal controller for a stochastic uncertain system. This problem is considered over a finite time horizon. In fact, the main result of this section provides a complete solution to a problem of minimax LQG control on a finite time horizon. This result was originally presented in [230, 235] along with corresponding discrete-time results in [152, 153].

The approach used in Section 8.4 involves a version of the S-procedure theorem which is used to convert a constrained optimization problem into an unconstrained dynamic game type problem. Furthermore, the results presented in this section rely on the fact that this game-type problem also occurs in certain duality relations between risk-sensitive stochastic optimal control problems and stochastic dynamic games; e.g., see [39, 53, 170]. In Section 8.4, we use this duality together with the known solutions of corresponding risk-sensitive stochastic optimal control problems (e.g., see Section 3.3 and [12, 139]) to solve the minimax optimal control problem. In fact, the risk-sensitive optimal control problem that arises in Section 8.4 plays the same role as H^∞ control theory plays in the deterministic minimax optimal control problems of Sections 5.3, 6.2 and 6.3. However, this does not mean that stochastic minimax optimal control problems require different computational tools than deterministic minimax optimal control problems. Solutions to Linear Quadratic Exponentiated Gaussian risk-sensitive control problems exploit the same Riccati equations as those of deterministic H^∞ control theory. In particular, solutions to the stochastic minimax optimal control problems considered in this chapter are given in terms of appropriately scaled Riccati equations which are virtually the same as those arising in Sections 6.2 and 6.3.

The solution to an infinite-horizon version of the problem of Section 8.4 is presented in Section 8.5. The main result of Section 8.5 is a minimax LQG synthesis procedure based on a pair of algebraic Riccati equations arising in infinite-horizon risk-sensitive optimal control; see Section 3.3 and [139]. We show that solutions to a certain scaled infinite-horizon risk-sensitive control problem provide us with a controller which minimizes the worst case of a time-averaged cost functional. A version of this result was originally presented in [233].

Note, that the result of Section 8.5 improves considerably the result of reference [231]. In [231], a similar guaranteed cost control problem was considered for a class of uncertain systems in which the uncertainties were restricted to those that guaranteed Markov behaviour for the system. The requirement of Markov behaviour excluded from consideration many important classes of uncertainties such as dynamic uncertainties. Also, the model in [231] imposed a number of technical limitations and assumptions such as the existence of an invariant probability measure corresponding to the uncertain dynamics. These limitations are overcome in the results presented in Section 8.5.

8.2 H^∞ control of stochastic systems with multiplicative noise

In this section, we consider a stochastic version of the state-feedback H^∞ control problem associated with the system (2.4.1) of Section 2.4.1. We will use the notation defined in Section 2.4.1. Let $\{\Omega, \mathcal{F}, \mathcal{P}\}$ be a complete probability space and let $W_1(t)$ and $W_2(t)$ be two mutually independent Wiener processes in $\mathbf{R}^{q_1}$, $\mathbf{R}^{q_2}$ with covariance matrices Q_1 and Q_2, respectively. As in Section 2.4.1, let $\mathcal{F}_t$ denote the increasing sequence of Borel sub-σ-fields of $\mathcal{F}$, generated by $\{W_{1,2}(s), 0 \leq s < t\}$. Also, $\mathbf{E}$ and $\mathbf{E}\{\cdot|\mathcal{F}_t\}$ are the corresponding unconditional and conditional expectation operators respectively, where the latter is the expectation with respect to $\mathcal{F}_t$. Also, we shall consider the Hilbert spaces $\mathbf{L}_2(\Omega, \mathcal{F}_s, P)$, $\mathbf{L}_2(s, T; \mathbf{R}^n)$, $\mathbf{L}_2(s; \mathbf{R}^n)$ and $\mathbf{R}_Q^{n\times q}$ introduced in Section 2.4.1.

We consider a stochastic system driven by a control input $u(t) \in \mathbf{R}^{m_1}$ and an exogenous uncertainty input $w(t) \in \mathbf{R}^{m_2}$. The system is described by the stochastic differential Ito equation:

$$\begin{aligned} dx &= (Ax(t) + B_1u(t) + B_2w(t))dt + (Hx(t) + P_1u(t))dW_1(t) \\ &\quad + P_2w(t)dW_2(t); \\ z(t) &= Cx(t) + Du(t) \end{aligned} \tag{8.2.1}$$

where $x(t) \in \mathbf{R}^n$ is the state, and $z(t) \in \mathbf{R}^p$ is the uncertainty output. Here A, B_1, B_2, C, D are matrices of corresponding dimensions, and H, P_1, P_2 are linear bounded operators $\mathbf{R}^n \to \mathbf{R}^{n\times q_1}_{Q_1}, \mathbf{R}^{m_1} \to \mathbf{R}^{n\times q_1}_{Q_1}, \mathbf{R}^{m_2} \to \mathbf{R}^{n\times q_2}_{Q_2}$, respectively. In the sequel, we will use the adjoint operators H^*, P_1^*, P_2^* defined as follows:

$$\begin{aligned} &\langle H^*\Theta_1, x\rangle = \operatorname{tr}\Theta_1 Q_1 (Hx)', \quad \langle P_1^*\Theta_1, u\rangle = \operatorname{tr}\Theta_1 Q_1 (P_1 u)', \\ &\langle P_2^*\Theta_2, \xi\rangle = \operatorname{tr}\Theta_2 Q_2 (P_2\xi)' \\ &\forall x \in \mathbf{R}^n,\ u \in \mathbf{R}^{m_1},\ \xi \in \mathbf{R}^{m_2},\ \Theta_1 \in \mathbf{R}^{n\times q_1},\ \Theta_2 \in \mathbf{R}^{n\times q_2}. \end{aligned}$$

Definition 8.2.1 *Given the system (8.2.1), the associated stochastic H^∞ control problem is to find a matrix $K \in R^{m_1\times n}$ such that the state-feedback control law $u = Kx$ leads to the satisfaction of the following conditions:*

(i) The system

$$dx = (A + B_1K)xdt + (H + P_1K)xdW_1(t) \tag{8.2.2}$$

is exponentially stable in the mean square sense; i.e., *there exist constants $c > 0$, $\alpha > 0$ such that*

$$\mathbf{E}||x(t)||^2 \le ce^{-\alpha(t-s)}\mathbf{E}||h||^2.$$

As a consequence, the matrix $A + B_1K$ is stable;

(ii) The closed loop system corresponding to system (8.2.1) with feedback control law $u = Kx$,

$$\begin{aligned} dx &= [(A + B_1K)x(t) + B_2w(t)]dt \\ &\quad + (H + P_1K)x(t)dW_1(t) + P_2w(t)dW_2(t) \end{aligned} \tag{8.2.3}$$

satisfies the following stochastic H^∞-norm condition: There exists a constant $\varepsilon > 0$ such that

$$\begin{aligned} &\mathbf{E}\int_s^\infty \left(||(C + DK)x(t)||^2 - ||w(t)||^2\right) dt \\ &\le -\varepsilon\mathbf{E}\int_s^\infty ||w(t)||^2 dt \quad \text{for } x(s) = 0 \end{aligned}$$

for each $w \in \mathbf{L}_2(s; \mathbf{R}^{m_2})$.

Remark 8.2.1 The reader may observe that the system (8.2.1) is a time-varying system. Therefore, the term "H^∞ norm" does not apply to this system in a rigorous sense. However, we will see in the sequel that the class of stochastic systems with multiplicative noise under consideration manifest properties which are similar to those of deterministic systems in classical H^∞ control theory. In order to emphasize the parallels between the problem considered in this section and the problem of deterministic H^∞ control, we will refer to the problem considered in this section as a stochastic H^∞ control problem.

8.2.1 A stochastic differential game

In this subsection, we consider the linear quadratic stochastic differential game associated with the system (8.2.1). This game problem is similar to the one in the paper [227], where the special case of $P_1 = P_2 = 0$ was considered. The results presented below extend in a straightforward manner those of reference [227]. This extension, originally presented in [234], is of primary importance to the results of Section 8.3 and will be extensively used in the sequel.

As in [227, 234], the linear quadratic stochastic differential game considered in this subsection assumes that the underlying system has certain stability properties. In particular, we will require that the matrix A is stable. In the subsequent subsections, this assumption will be weakened; see Assumption 8.2.2 in the next subsection.

Assumption 8.2.1 *The linear system*

$$dx(t) = Ax(t)dt + HxdW_1(t), \qquad x(s) = h, \tag{8.2.4}$$

corresponding to (8.2.1) driven by the control input $u(\cdot) \equiv 0$ and uncertainty input $w(\cdot) \equiv 0$, is exponentially stable in the mean square sense; i.e., *there exist constants $c > 0$, $\alpha > 0$ such that*

$$\mathbf{E}\|x(t)\|^2 \leq ce^{-\alpha(t-s)}\mathbf{E}\|h\|^2. \tag{8.2.5}$$

As a consequence, the matrix A is stable.

Remark 8.2.2 In order to check Assumption 8.2.1, one can use Lyapunov arguments reducing the stability test to the problem of finding a feasible solution Y to the linear matrix inequality

$$A'Y + YA + H^*YH \leq -\bar{\epsilon}I, \qquad Y' = Y > 0,$$

for some $\bar{\epsilon} > 0$.

Consider the stochastic differential game defined by the system (8.2.1) and cost functional

$$\begin{aligned} \Im^{s,h}(u,w) &= \int_s^\infty \mathbf{E}\left\{F(x(t),u(t)) - \|w(t)\|^2\right\} dt; \qquad (8.2.6) \\ F(x,u) &:= \langle x, Rx + Qu\rangle + \langle u, Q'x + Gu\rangle \end{aligned}$$

where $R = R' \in R^{n\times n}$, $Q \in R^{n\times m_1}$, $G = G' \in R^{n_1\times m_1}$, $R \geq 0, G > 0$. In equation (8.2.6), $x(\cdot)$ denotes the solution to equation (8.2.1) with initial condition $x(s) = h$ and inputs $(u(\cdot), w(\cdot))$. In this game, $u(\cdot) \in \mathbf{L}_2(s, \mathbf{R}^{m_1})$ is the minimizing strategy, $w(\cdot) \in \mathbf{L}_2(s, \mathbf{R}^{m_2})$ is the maximizing strategy. Note, that Assumption 8.2.1 assures that any pair of inputs $(u(\cdot), w(\cdot)) \in \mathbf{L}_2(s, \mathbf{R}^{m_1}) \times \mathbf{L}_2(s, \mathbf{R}^{m_2})$ leads to a square integrable solution on $[s, \infty)$ for any initial condition; e.g., see Lemma B.2 in Appendix B and also [27]. This implies that the cost functional (8.2.6) is well defined on $\mathbf{L}_2(s, \mathbf{R}^{m_1}) \times \mathbf{L}_2(s, \mathbf{R}^{m_2})$. Also, in the sequel, we will consider the finite-horizon version of the cost functional (8.2.6) which is defined as follows:

$$\Im_T^{s,h}(u, w) = \int_s^T \mathbf{E}\left\{F(x(t), u(t)) - \|w(t)\|^2\right\} dt, \qquad (s \leq T < \infty). \tag{8.2.7}$$

Note that for $T = \infty$, $\Im_\infty^{s,h}(u, w) = \Im^{s,h}(u, w)$.

In the stochastic differential game considered in this subsection, we seek to find

$$V := \inf_{u(\cdot)\in\mathbf{L}_2(s,\mathbf{R}^{m_1})} \sup_{w\in\mathbf{L}_2(s,\mathbf{R}^{m_2})} \Im^{s,h}(u, w). \tag{8.2.8}$$

The derivation of a solution to the above differential game requires the following lemma. This result extends the Bounded Real Lemma to the case of stochastic systems with multiplicative noise. Note that a version of this stochastic Bounded Real Lemma was originally proved in [227] in the special case of $P_1 = 0, P_2 = 0$. The general result presented here was originally proved in [234]. Also, a similar result was obtained in [84].

Lemma 8.2.1 (Stochastic Bounded Real Lemma) *Suppose that Assumption 8.2.1 is satisfied. Also, suppose there exists a constant $\varepsilon_2 > 0$ such that*

$$\Im^{0,0}(0, w) \leq -\varepsilon_2 |||w|||^2 \qquad \forall w \in \mathbf{L}_2(0, \mathbf{R}^{m_2}). \tag{8.2.9}$$

Then the following conditions hold:

(a) For each $s \geq 0$ and $h \in \mathbf{L}_2(\Omega, \mathcal{F}_s, P)$, there exists a unique $w_2^s(\cdot) \in \mathbf{L}_2(s; \mathbf{R}^{m_2})$ such that

$$\Im^{s,h}(0, w_2^s(\cdot)) = \sup_{w(\cdot)\in\mathbf{L}_2(s,\mathbf{R}^{m_2})} \Im^{s,h}(0, w). \tag{8.2.10}$$

Also, there exists a unique optimal $w_{2T}^s(\cdot) \in \mathbf{L}_2(s, T; \mathbf{R}^{m_2})$ maximizing the cost function $\Im_T^{s,h}(0, w(\cdot))$. Moreover,

$$|||w_{2T}^s(\cdot)|||^2 \leq c\mathbf{E}\|h\|^2; \qquad \Im_T^{s,h}(0, w_{2T}^s(\cdot)) \leq c\mathbf{E}\|h\|^2 \tag{8.2.11}$$

for some $c > 0$ independent of $s \leq T \leq \infty$ and h.

(b) There exists a nonnegative-definite symmetric matrix X_2 such that

$$\sup_{w(\cdot)\in \mathbf{L}_2(s,\mathbf{R}^{m_2})} \Im^{s,h}(0,w) = \mathbf{E}\langle h, X_2 h\rangle. \tag{8.2.12}$$

(c) For all $T > 0$, there exists a unique symmetric nonnegative-definite solution $X_{2T}(\cdot)$ to the generalized Riccati equation

$$\begin{aligned} &\frac{dX_{2T}}{dt} + A'X_{2T} + X_{2T}A + H^*X_{2T}H + R \\ &\quad + X_{2T}B_2(I - P_2^*X_{2T}P_2)^{-1}B_2'X_{2T} = 0; \\ &X_{2T}(T) = 0. \end{aligned} \tag{8.2.13}$$

This solution satisfies the conditions

$$0 \le X_{2T}(s) \le X_2; \quad I - P_2^*X_{2T}(s)P_2 > 0.$$

The optimal input $w_{2T}^s(\cdot)$ maximizing the functional (8.2.7) can be expressed in the form of a feedback control law

$$w_{2T}^s(t) = (I - P_2^*X_{2T}(t)P_2)^{-1}B_2'X_{2T}x_{2T}^s(t), \tag{8.2.14}$$

where $x_{2T}^s(\cdot)$ is the optimal trajectory for the corresponding closed loop system

$$\begin{aligned} dx(t) &= (A + B_2(I - P_2^*X_{2T}(t)P_2)^{-1}B_2'X_{2T}(t))xdt + HxdW_1(t) \\ &\quad + P_2(I - P_2^*X_{2T}(t)P_2)^{-1}B_2'X_{2T}(t)xdW_2(t), \\ x(s) &= h. \end{aligned} \tag{8.2.15}$$

(d) The matrix X_2 satisfying (8.2.12) is also the minimal solution to the generalized Riccati equation

$$A'X_2 + X_2A + H^*X_2H + R + X_2B_2(I - P_2^*X_2P_2)^{-1}B_2'X_2 = 0, \tag{8.2.16}$$

such that

$$I - P_2^*X_2P_2 > 0. \tag{8.2.17}$$

Proof: The problem (8.2.10) is a stochastic control problem with a sign-indefinite integrand in the cost functional. Thus, it is natural to refer to the uncertainty input $w(\cdot)$ as a "control" input. A solution to this class of stochastic control problems has been given in [31]; see also [32], where the results of [31] were extended to the infinite-dimensional case. Assumption 8.2.1 and condition (8.2.9) are the conditions required by Theorem 1 of these references. Applying the result of [31, 32] to the control problem (8.2.10), it follows that the first part of claim

(a) of the lemma holds. The existence of a symmetric matrix X_2 satisfying condition (8.2.12) also follows from the above mentioned result of [31, 32]. From the inequality $\Im^{s,h}(0,0) \geq 0$, it follows that $X_2 \geq 0$. Hence, claim (b) of the lemma also holds.

The optimal control problem

$$\sup_{\zeta \in \mathbf{L}_2(s,T;\mathbf{R}^{m_2})} \Im_T^{s,h}(0,\zeta) \tag{8.2.18}$$

considered in the second part of claim (a) of the lemma can be solved by the same method as in [20, 90]. Reference [20] presents a solution to the standard finite-horizon stochastic optimal control problem. The optimal control problem (8.2.18) differs from the optimal control problem considered in reference [20] in that the integrand in (8.2.7) is sign-indefinite. Hence, one needs to extend the results of [20] to the case considered in this subsection. This extension can be carried out in a similar fashion to that used in the corresponding problem for deterministic time-varying systems [90].

We first note that the control problem (8.2.18) has a unique solution. Indeed, as in [20], it follows that the functional (8.2.7) is continuous. Also, as in [27, 31, 32], it follows using the Riesz Representation Theorem that

$$\Im_T^{s,h}(0,\zeta) = -\pi(\zeta,\zeta) + 2\Upsilon(h,\zeta) + \Im_T^{s,h}(0,0)$$

where $\Upsilon(\cdot,\cdot)$ and $\pi(\cdot,\cdot)$ are bilinear forms on the Hilbert product spaces $\mathbf{L}_2(\Omega,\mathcal{F}_s,P) \times \mathbf{L}_2(s,T;\mathbf{R}^{m_2})$ and $\mathbf{L}_2(s,T;\mathbf{R}^{m_2}) \times \mathbf{L}_2(s,T;\mathbf{R}^{m_2})$ respectively. We now show that the bilinear form $\pi(\cdot,\cdot)$ is coercive [116]. Let $\mathbf{\Phi}_{s,T}$ denote a linear bounded operator $\mathbf{L}_2(s,T;\mathbf{R}^{m_2}) \to \mathbf{L}_2(s,T;\mathbf{R}^n)$ mapping a given control input $\zeta(\cdot)$ into a corresponding solution to the system

$$dx = (Ax(t) + B_2\zeta(t))dt + Hx(t)dW_1(t) + P_2\zeta(t)dW_2(t); \quad x(s) = 0. \tag{8.2.19}$$

Then, for any $s_2 \leq s_1 \leq T_1 \leq T_2$ and $\zeta(\cdot) \in \mathbf{L}_2(s_1,T_1,\mathbf{R}^{m_2})$,

$$|||R^{1/2}\mathbf{\Phi}_{s_1,T_1}\zeta|||^2 = |||R^{1/2}\mathbf{\Phi}_{s_2,T_1}\tilde{\zeta}|||^2 \leq |||R^{1/2}\mathbf{\Phi}_{s_2,T_2}\tilde{\zeta}|||^2$$

where $\tilde{\zeta}(\cdot) \in \mathbf{L}_2(s_2,T_2;\mathbf{R}^{m_2})$ is the extension of $\zeta(\cdot)$ to $[s_2;T_2]$ obtained by defining $\tilde{\zeta}(\cdot)$ to be zero on $[s_2,s_1]$ and $[T_1,T_2]$. From this inequality and from (8.2.9), it follows that

$$-\pi(\zeta,\zeta) = |||R^{1/2}\mathbf{\Phi}_{s,T}\zeta|||^2 - |||\zeta|||^2 \leq -\varepsilon_2|||\zeta|||^2 \tag{8.2.20}$$

for all $0 \leq s < T \leq \infty$. That is, the bilinear form $\pi(\cdot,\cdot)$ is coercive. Then, Theorem 1.1 of [116] implies that the control problem (8.2.18) has a unique solution which is characterized by the equation

$$(\mathbf{\Phi}_{s,T}^* R\mathbf{\Phi}_{s,T} - \mathcal{I})\zeta_T^s = -\mathbf{\Phi}_{s,T}^* Rx_{s,h},$$

where $x_{s,h}(\cdot)$ is the solution to (8.2.4) with the initial condition $x(s) = h$, and $\mathcal{I}$ denotes the identity operator. From (8.2.20), the operators $R^{1/2}\Phi_{s,T}$ and $(\mathcal{I} - \Phi_{s,T}^* R \Phi_{s,T})^{-1}$ are uniformly bounded. Hence, the conditions (8.2.11) are satisfied. This establishes the second part of claim (a) of the lemma.

As in Theorem 6.1 of [20], one can prove that the above facts imply the existence of a unique nonnegative-definite solution to Riccati equation (8.2.13). This proof requires the following proposition.

Proposition 1 *Let $\hat{X}$ be a given nonnegative-definite symmetric matrix such that*

$$I - P_2^* \hat{X} P_2 \geq \lambda I; \ \ \|(I - P_2^* \hat{X} P_2)^{-1}\| \leq 1/\lambda.$$

For any $\tilde{X}$ such that

$$\|\tilde{X} - \hat{X}\| \leq \lambda/(2\|P_2\|_Q^2),$$

the matrix $I - P_2^* \tilde{X} P_2$ is positive-definite and hence invertible. Also,

$$\|(I - P_2^* \tilde{X} P_2)^{-1}\| \leq 2/\lambda.$$

The proof of this proposition follows using the same arguments as those used in proving the corresponding fact in [20]. From the condition of the proposition, one has

$$\|P_2^*(\tilde{X} - \hat{X})P_2)\| \leq \lambda/2. \tag{8.2.21}$$

Thus,

$$\begin{aligned} w'(I - P_2^* \tilde{X} P_2)w &= w'(I - P_2^* \hat{X} P_2)w - w' P_2^* (\tilde{X} - \hat{X}) P_2 w \\ &\geq \lambda \|w\|^2 - \frac{\lambda}{2}\|w\|^2; \end{aligned}$$

i.e., $I - P_2^* \tilde{X} P_2 \geq \frac{\lambda}{2} I$. Thus, the proposition follows.

Let $\alpha > 0$ be a given constant. Consider a constant $\lambda > 0$ and a matrix valued function $\mathcal{P}(t)$, $t \in [T - \alpha, T]$, such that

$$0 \leq \mathcal{P}(t) = \mathcal{P}'(t); \ \ I - P_2^* \mathcal{P}(t) P_2 \geq \lambda I$$

and hence

$$\sup_t \|(I - P_2^* \mathcal{P}(t) P_2)^{-1}\| \leq 1/\lambda.$$

As in [20], we also consider the set of matrix valued functions

$$\kappa_{\mathcal{P}}^{\alpha} := \left\{ \tilde{X}(t), t \in [T - \alpha, T] : \|\tilde{X} - \mathcal{P}\| \leq \frac{\lambda}{2\|P_2\|_Q^2} \right\}.$$

We define a distance measure on the set $\kappa_{\mathcal{P}}^{\alpha}$ by

$$\rho(\hat{X}, \tilde{X}) := \sup_{t \in [T-\alpha, T]} \|\tilde{X}(t) - \hat{X}(t)\|.$$

With this distance measure, the set $\kappa_{\mathcal{P}}^{\alpha}$ becomes a closed subset of the Banach space of bounded continuous matrix valued functions. Let the mapping $\varphi(\cdot)$ be defined as follows:

$$\varphi(Z) := A'Z + ZA + R + H^*ZH + ZB_2(I - P_2^*ZP_2)^{-1}B_2'Z.$$

It follows from Proposition 1 that this mapping is well defined on $\kappa_{\mathcal{P}}^{\alpha}$ and that $\|\varphi(\tilde{X}(t))\| \leq c_{\mathcal{P}}$. Moreover, since $\mathcal{P}$ has been chosen as described above, then

$$\sup_{\mathcal{P}} c_{\mathcal{P}} \leq c_{\lambda}.$$

Also, as in Proposition 1, it can be readily established by calculating the derivative of $\varphi(Z)$ with respect to Z, there exists a constant $c_k > 0$ such that if $\hat{X}(\cdot)$ and $\tilde{X}(\cdot)$ are in $\kappa_{\mathcal{P}}^{\alpha}$ with $\mathcal{P}$ self-adjoint, nonnegative-definite and such that $I - P_2^*\mathcal{P}(t)P_2 \geq \lambda I$, then

$$\|\varphi(\tilde{X}(t)) - \varphi(\hat{X}(t))\| \leq c_k\|\tilde{X}(t) - \hat{X}(t)\|.$$

This observation allows us to establish that the mapping

$$\mathcal{Q}(\tilde{X}(\cdot))(t) := \mathcal{P} + \int_t^T \varphi(\tilde{X}(\theta))d\theta, \qquad \forall \tilde{X} \in \kappa_{\mathcal{P}}^{\alpha},$$

is a contraction on $\kappa_{\mathcal{P}}^{\alpha}$, provided α is chosen to be sufficiently small. The reader is referred to [20, Theorem 6.1] for details.

Let $\mathcal{P}(\cdot) = 0$. This choice agrees with the above conditions on the function $\mathcal{P}(\cdot)$. From the Contraction Mapping Theorem, the contraction $\mathcal{Q}$ has a unique fixed point in $\kappa_{\mathcal{P}\equiv 0}^{\alpha}$. Denoting this point $X_{2T}(\cdot)$, we observe that $X_{2T}(\cdot)$ satisfies equation (8.2.13) on $[T-\alpha, T]$. Also from Proposition 1, for a certain $\lambda_1 > 0$, $I - P_2^*X_{2T}P_2 \geq \lambda_1 I$ and hence $I - P_2^*X_{2T}P_2$ is boundedly invertible. As in Lemma 2.2 of [90], this fact implies that

$$\sup_{\zeta \in \mathbf{L}_2(s,T,\mathbf{R}^{m_2})} \Im_T^{s,h}(0,\zeta) = \mathbf{E}\langle h, X_{2T}(s)h\rangle.$$

for any $s \in [T-\alpha, T]$. Also, we see from the above equation that

$$0 \leq X_{2T}(s) \leq X_2.$$

Consequently, the operator X_{2T} is uniformly bounded on $[T-\alpha, T]$. Furthermore, it is straightforward to verify using equation (8.2.13) that the unique optimal control can be expressed in the feedback form

$$w_{2T}^s(t) = (I - P_2^*X_{2T}(t)P_2)^{-1}B_2'X_{2T}x_{2T}^s(t).$$

Claim (c) of the lemma now follows by iterating the procedure described above a finite number of steps with $\mathcal{P}(t) \equiv X_{2T}(T - (i-1)\alpha)$ at the i-th step.

To prove claim (d), first note that it is straightforward to show $X_{2T}(s)$ is monotonically increasing in T. Hence, for all $s \geq 0$, there exists a matrix

$$\bar{X}_2(s) = \lim_{T\to\infty} X_{2T}(s).$$

Hence $0 \leq \bar{X}_2(s) \leq X_2$ for all $s \geq 0$. Consequently, $I - P_2^* \bar{X}_2(s) P_2 \geq 0$. Also, letting T approach infinity in equation (8.2.13), it follows that $\bar{X}_2(s)$ satisfies this equation for all $s \in [0, \infty)$ such that $I - P_2^* \bar{X}_2(s) P_2$ is nonsingular.

For all $s \in [0, \infty)$ such that $I - P_2^* \bar{X}_2(s) P_2$ is singular, we must have that either $\frac{d}{ds}\bar{X}_2(s) = \infty$ or $\bar{X}_2(s)$ is not differentiable. Moreover, it follows from the second part of claim (a) that

$$\Im_T^{s,h}(0, w_2^s(\cdot)) \leq \mathbf{E}\langle h, X_{2T}(s)h\rangle.$$

Letting T approach infinity, it follows from this inequality and (8.2.10), (8.2.12) that

$$\Im^{s,h}(0, w_2^s(\cdot)) \leq \mathbf{E}\langle h, \bar{X}_2(s)h\rangle \leq \mathbf{E}\langle h, X_2 h\rangle = \Im^{s,h}(0, w_2^s(\cdot)).$$

Hence, $\bar{X}_2(s) = X_2$, and $\frac{d}{ds}\bar{X}_2(s) = 0$ for all $s \in [0, \infty)$. This implies, that the matrix $I - P_2^* X_2 P_2$ is nonsingular. That is, condition (8.2.17) is satisfied.

■

Remark 8.2.3 As in [90], it can be shown that the matrix X_2 satisfying conditions (8.2.12), (8.2.16) and (8.2.17) is such that the system

$$\begin{aligned} dx(t) &= (A + B_2(I - P_2^* X_2 P_2)^{-1} B_2' X_2) x dt + H x dW_1(t) \\ &\quad + P_2 (I - P_2^* X_2 P_2)^{-1} B_2' X_2 x dW_2(t), \quad x(s) = h \end{aligned} \tag{8.2.22}$$

is exponentially mean-square stable. Note that in the particular case where $H = 0$, $P_2 = 0$, equation (8.2.16) becomes a Riccati equation with a stable A matrix. Hence in this case, this equation has a nonnegative-definite stabilizing solution satisfying condition (8.2.17).

The problem of solving a generalized ARE is known to be a challenging problem; e.g., see reference [166] and references therein. Reference [166] presents a numerical algorithm based on homotopy methods which solves a general class of perturbed Riccati equations. The generalized algebraic Riccati equation (8.2.16) is virtually the same as those in [166]. Hence, a useful approach in solving (8.2.16) would be to apply the method of [166]. Also, it is worth noting that the solution X_2 to equation (8.2.16) and inequality (8.2.17) necessarily satisfies the linear matrix inequalities

$$\begin{bmatrix} -A'X_2 - X_2 A - R - H^* X_2 H & X_2 B_2 \\ B_2' X_2 & I - P_2^* X_2 P_2 \end{bmatrix} \geq 0, \qquad X_2 \geq 0.$$

Hence, the desired matrix X_2 exists only if the above LMIs are feasible. Numerical aspects of solving generalized algebraic Riccati equations will be discussed in Section 8.2.3. In particular, in Section 8.2.3, we give a useful empirical algorithm for solving the above generalized Riccati equations.

Theorem 8.2.1 *Suppose that Assumption 8.2.1 is satisfied and there exists a constant $\varepsilon_2 > 0$ such that condition (8.2.9) holds. Also, assume that there exists a constant $\varepsilon_1 > 0$ such that*

$$F(x,u) > \varepsilon_1 \|u\|^2 \qquad \forall x \in R^n, u \in R^{m_1}. \tag{8.2.23}$$

Then the following conditions hold:

(a) For each $s \geq 0$ and $h \in \mathbf{L}_2(\Omega, \mathcal{F}_s, P)$, there exists a unique saddle point (minimax pair) for the cost functional (8.2.6) in $\mathbf{L}_2(s, R^{m_1}) \times \mathbf{L}_2(s, R^{m_2})$.

(b) There exists a unique nonnegative-definite symmetric solution $X \in \mathbf{R}^{n\times n}$ to the generalized game type algebraic Riccati equation

$$\begin{aligned} &A'X + XA + H^*XH + R + XB_2(I - P_2^*XP_2)^{-1}B_2'X \\ &\quad -(XB_1 + H^*XP_1 + Q)(G + P_1^*XP_1)^{-1}(XB_1 + H^*XP_1 + Q)' \\ &= 0. \end{aligned} \tag{8.2.24}$$

such that $I - P_2^*XP_2 > 0$ and the upper value of the differential game (8.2.8) is given by

$$V = \mathbf{E}\langle h, Xh\rangle. \tag{8.2.25}$$

(c) The minimax pair can be expressed in the feedback form

$$u = F_1 x; \qquad w = F_2 x$$

where

$$\begin{aligned} F_1 &= -(G + P_1^*XP_1)^{-1}(XB_1 + H^*XP_1 + Q)'; \\ F_2 &= (I - P_2^*XP_2)^{-1}B_2'X. \end{aligned}$$

(d) The stochastic closed loop system

$$\begin{aligned} dx &= (A + B_1F_1 + B_2F_2)x + (H + P_1F_1)x dW_1(t) \\ &\quad + P_2F_2 x dW_2(t); \\ x(s) &= h, \end{aligned} \tag{8.2.26}$$

is exponentially mean square stable. As a consequence, the matrix $A + B_1F_1 + B_2F_2$ is stable.

Proof: Note that condition (8.2.23) of the theorem implies that

$$\Im^{0,0}(u(\cdot), 0) \geq \varepsilon_1 \|u(t)\|_2^2 \qquad \forall u(\cdot) \in \mathbf{L}_2([0,\infty) \times \Omega; R^{n_1}). \tag{8.2.27}$$

Conditions (8.2.27) and (8.2.9) are convexity-concavity conditions which guarantee the existence of a unique minimax pair for the cost functional (8.2.6); e.g., see

[10]. Indeed, as in [31, 32], using the Riesz Representation Theorem, we obtain the following representation for the cost functional (8.2.6):

$$\Im^{s,h}(u,w) = \left\langle\!\!\left\langle \begin{pmatrix} u \\ w \end{pmatrix}, \mathcal{G}\begin{pmatrix} u \\ w \end{pmatrix} \right\rangle\!\!\right\rangle + 2\left\langle\!\!\left\langle \begin{pmatrix} u \\ w \end{pmatrix}, \begin{pmatrix} g_{1s}h \\ -g_{2s}h \end{pmatrix} \right\rangle\!\!\right\rangle + \Im^{s,h}(0,0).$$

Here the symbol $\langle\!\langle \cdot,\cdot \rangle\!\rangle$ denotes the inner product in $\mathbf{L}_2(s,\mathbf{R}^{m_1}) \times \mathbf{L}_2(s,\mathbf{R}^{m_2})$. Furthermore, it follows from the Riesz Representation Theorem that $\mathcal{G}$ is a linear bounded self-adjoint operator

$$\mathcal{G} : \mathbf{L}_2(s,\mathbf{R}^{m_1}) \times \mathbf{L}_2(s,\mathbf{R}^{m_2}) \to \mathbf{L}_2(s,\mathbf{R}^{m_1}) \times \mathbf{L}_2(s,\mathbf{R}^{m_2})$$

which has the following structure

$$\mathcal{G} = \begin{pmatrix} \tilde{G} & \tilde{\Psi} \\ \tilde{\Psi}^* & -\tilde{\Upsilon} \end{pmatrix}.$$

The block components $\tilde{G}$, $\tilde{\Upsilon}$, $\tilde{\Psi}$ in this operator are linear bounded operators mapping $\mathbf{L}_2(s,\mathbf{R}^{m_1}) \to \mathbf{L}_2(s,\mathbf{R}^{m_1})$, $\mathbf{L}_2(s,\mathbf{R}^{m_2}) \to \mathbf{L}_2(s,\mathbf{R}^{m_2})$ and $\mathbf{L}_2(s,\mathbf{R}^{m_2}) \to \mathbf{L}_2(s,\mathbf{R}^{m_1})$ respectively. Also, the operators $\tilde{G}$, $\tilde{\Upsilon}$ are self-adjoint operators, and the operators g_{1s}, g_{2s} are linear bounded operators $\mathbf{L}_2(\Omega,\mathcal{F}_s,P) \to \mathbf{L}_2(s,\mathbf{R}^{m_1})$, $\mathbf{L}_2(\Omega,\mathcal{F}_s,P) \to \mathbf{L}_2(s,\mathbf{R}^{m_2})$ respectively. Furthermore, conditions (8.2.27), (8.2.9) imply that the operators $\tilde{G}$, $\tilde{\Upsilon}$ are strictly positive-definite. Then, it follows from the results of [10], that there exists a unique saddle point of the functional (8.2.6) on $\mathbf{L}_2(s,\mathbf{R}^{m_1}) \times \mathbf{L}_2(s,\mathbf{R}^{m_2})$.

Proposition 2 *There exists a symmetric matrix X satisfying equation (8.2.25).*

To prove this proposition, we first note that the conditions $\tilde{G} > 0$, $\tilde{\Upsilon} > 0$ imply that the operators $\tilde{G} + \tilde{\Psi}\tilde{\Upsilon}^{-1}\tilde{\Psi}^*$, $\tilde{\Upsilon} + \tilde{\Psi}^*\tilde{G}^{-1}\tilde{\Psi}$ are boundedly invertible. Now, letting

$$\begin{aligned} \hat{u}^s &= (\tilde{G} + \tilde{\Psi}\tilde{\Upsilon}^{-1}\tilde{\Psi}^*)^{-1}(\tilde{\Psi}\tilde{\Upsilon}^{-1}g_{2s} - g_{1s})h, \\ \hat{w}^s &= -(\tilde{\Upsilon} + \tilde{\Psi}^*\tilde{G}^{-1}\tilde{\Psi})^{-1}(\tilde{\Psi}^*\tilde{G}^{-1}g_{1s} + g_{2s})h. \end{aligned} \tag{8.2.28}$$

and observing that this definition implies $(\hat{u}^s, \hat{w}^s) \in \mathbf{L}_2(s,\mathbf{R}^{m_1}) \times \mathbf{L}_2(s,\mathbf{R}^{m_2})$, it is straightforward to verify by completing the squares, that

$$\begin{aligned} \Im^{s,h}(u(\cdot),w(\cdot)) &= \left\langle\!\!\left\langle \begin{pmatrix} u - \hat{u}^s \\ w - \hat{w}^s \end{pmatrix}, \mathcal{G}\begin{pmatrix} u - \hat{u}^s \\ w - \hat{w}^s \end{pmatrix} \right\rangle\!\!\right\rangle \\ &\quad - \left\langle\!\!\left\langle \begin{pmatrix} \hat{u}^s \\ \hat{w}^s \end{pmatrix}, \mathcal{G}\begin{pmatrix} \hat{u}^s \\ \hat{w}^s \end{pmatrix} \right\rangle\!\!\right\rangle + \Im^{s,h}(0,0). \end{aligned}$$

To prove the above result, the following operator identities are used

$$\begin{aligned} &\tilde{G}^{-1}\tilde{\Psi}(\tilde{\Upsilon} + \tilde{\Psi}^*\tilde{G}^{-1}\tilde{\Psi})^{-1} = (\tilde{G} + \tilde{\Psi}\tilde{\Upsilon}^{-1}\tilde{\Psi}^*)^{-1}\tilde{\Psi}\tilde{\Upsilon}^{-1}; \\ &\mathcal{I} - (\tilde{\Upsilon} + \tilde{\Psi}^*\tilde{G}^{-1}\tilde{\Psi})^{-1}\tilde{\Psi}^*\tilde{G}^{-1}\tilde{\Psi} = (\tilde{\Upsilon} + \tilde{\Psi}^*\tilde{G}^{-1}\tilde{\Psi})^{-1}\tilde{G}. \end{aligned}$$

Then, the saddle point is given by the following equations

$$u^s = u^s(h) = \hat{u}^s; \qquad w^s = w^s(h) = \hat{w}^s. \tag{8.2.29}$$

The subsequent proof of equation (8.2.25) follows along the lines of the proof of the corresponding result in linear quadratic stochastic optimal control [31, 32]. Let $\mathfrak{U}^s$ and $\mathfrak{W}^s$ be the operators defined by equation (8.2.28) and mapping the space $\mathbf{L}_2(\Omega, \mathcal{F}_s, P)$ into the spaces $\mathbf{L}_2(s, \mathbf{R}^{m_1})$ and $\mathbf{L}_2(s, \mathbf{R}^{m_2})$, respectively. From this definition, it follows that $\mathfrak{U}^s$ and $\mathfrak{W}^s$ are linear bounded operators. Also, let $\hat{x}^s(\cdot)$ be a solution to equation (8.2.1) corresponding to the minimax pair $(\hat{u}^s(\cdot), \hat{w}^s(\cdot))$ and satisfying the initial condition $x(s) = h$. Assumption 8.2.1 and Lemma B.2 of Appendix B imply that the operator $\mathfrak{X}^s$ defined by the equation $\hat{x}^s(\cdot) = \mathfrak{X}^s$ is also a linear bounded operator $\mathbf{L}_2(\Omega, \mathcal{F}_s, P)$ to $\mathbf{L}_2(s, \mathbf{R}^n)$. We now define the operator

$$\mathcal{X}_s := (\mathfrak{X}^s)^* R \mathfrak{X}^s + (\mathfrak{X}^s)^* Q \mathfrak{U}^s + (\mathfrak{U}^s)^* Q \mathfrak{X}^s + (\mathfrak{U}^s)^* G \mathfrak{U}^s - (\mathfrak{W}^s)^* G \mathfrak{W}^s. \tag{8.2.30}$$

From this definition, the operator $\mathcal{X}_s$ is a linear bounded self-adjoint operator $\mathbf{L}_2(\Omega, \mathcal{F}_s, P) \to \mathbf{L}_2(\Omega, \mathcal{F}_s, P)$. Also,

$$\inf_{u(\cdot) \in \mathbf{L}_2(s, \mathbf{R}^{m_1})} \sup_{w \in \mathbf{L}_2(s, \mathbf{R}^{m_2})} \mathfrak{I}^{s,h}(u, w) = \langle\!\langle \mathcal{X}_s h, h \rangle\!\rangle. \tag{8.2.31}$$

For a given set $\Lambda \in \mathcal{F}_s$, we now define subspaces $\mathbf{L}^\Lambda$ and $\mathbf{L}_{2s}^{\Lambda,q}$ of the Hilbert spaces $\mathbf{L}_2(\Omega, \mathcal{F}_s, P)$ and $\mathbf{L}_2(s, \mathbf{R}^q)$ respectively as follows:

$$\begin{aligned} \mathbf{L}^\Lambda &:= \{h \in \mathbf{L}_2(\Omega, \mathcal{F}_s, P) : h(\omega) = 0 \text{ if } \omega \notin \Lambda \text{ a.s.}\}; \\ \mathbf{L}_{2s}^{\Lambda,q} &:= \{\lambda(\cdot) \in \mathbf{L}_2(s, \mathbf{R}^q) : p(\cdot, \omega) \equiv 0 \text{ if } \omega \notin \Lambda \text{ a.s.}\} \end{aligned}$$

where q is one of the numbers n, m_1 or m_2 and $\lambda(\cdot)$ is one of the processes $x(\cdot)$, $u(\cdot)$ or $w(\cdot)$, respectively. The orthogonal projection mapping onto these subspaces is denoted π_Λ. Obviously,

$$[\pi_\Lambda h](\omega) = \chi_\Lambda(\omega) h(\omega); \quad [\pi_\Lambda \lambda](\cdot, \omega) = \chi_\Lambda(\omega) \lambda(\cdot, \omega) \quad \text{a.s.}$$

where $\chi_\Lambda(\omega)$ denotes the indicator function of the set Λ. We now show that the projection π_Λ commutes with the operator $\mathcal{X}_s$.

Let $h \in \mathbf{L}^\Lambda$, $u(\cdot) \in \mathbf{L}_{2s}^{\Lambda,m_1}$ and $w(\cdot) \in \mathbf{L}_{2s}^{\Lambda,m_2}$. Then, it follows from the definitions of the Bochner integral [265] and the Ito integral [117] that $x(\cdot)$, the corresponding solution to equation (8.2.1), belongs to $\mathbf{L}_{2s}^{\Lambda,n}$. This observation implies that the projection mapping π_Λ commutes with the operators $\tilde{G}$, $\tilde{\Upsilon}$, $\tilde{\Psi}$, $\mathcal{G}$, g_{1s} and g_{2s}. Furthermore, by pre and postmultiplying the equations

$$\pi_\Lambda \tilde{G} = \tilde{G} \pi_\Lambda; \quad \pi_\Lambda \tilde{\Upsilon} = \tilde{\Upsilon} \pi_\Lambda$$

by $\tilde{G}^{-1}$ and $\tilde{\Upsilon}^{-1}$ respectively, we show that the operators $\tilde{G}^{-1}$ and $\tilde{\Upsilon}^{-1}$ also commute with π_Λ. Thus, it follows from equations (8.2.28) and (8.2.30) that

$$\begin{aligned} \pi_\Lambda \mathfrak{U}^s &= \mathfrak{U}^s \pi_\Lambda; \qquad \pi_\Lambda \mathfrak{W}^s = \mathfrak{W}^s \pi_\Lambda; \\ \pi_\Lambda \mathfrak{X}^s &= \mathfrak{X}^s \pi_\Lambda; \qquad \pi_\Lambda \mathcal{X}^s = \mathcal{X}^s \pi_\Lambda. \end{aligned} \tag{8.2.32}$$

That is, the projection π_Λ commutes with the operator $\mathcal{X}_s$ defined by equation (8.2.30).

The above properties of the projection π_Λ allow us to show that there exists an $\mathcal{F}_s$-measurable mapping $X_s(\omega) : \Omega \to \mathbf{R}^{n\times n}$ such that

$$\begin{aligned}&\langle g(\omega), X_s(\omega)h(\omega)\rangle \\ &= \langle g(\omega), (\mathcal{X}_s h)(\omega)\rangle \\ &= \mathbf{E}\left\{\int_s^\infty \left(\begin{array}{l}\langle x^s(g), Rx^s(h) + Qu^s(h)\rangle \\ +\langle u^s(g), Q'x^s(h) + Gu^s(h)\rangle - \langle w^s(g), w^s(h)\rangle\end{array}\right) dt \Big| \mathcal{F}_s\right\}\end{aligned} \tag{8.2.33}$$

where $x^s(h)$, $x^s(g)$ are solutions to equation (8.2.1) corresponding to the minimax pairs $(u^s(h), w^s(h))$ and $(u^s(g), w^s(g))$, satisfying the initial conditions $x(s) = h$, $x(s) = g$, respectively. This claim follows from the observation that equations (8.2.30) and (8.2.32) imply the following equation

$$\begin{aligned}&\int_\Lambda \langle g(\omega), (\mathcal{X}_s h)(\omega)\rangle P(d\omega) \\ &= \int_\Lambda \left\{\int_s^\infty \left(\begin{array}{l}\langle \mathfrak{X}^s g, (R\mathfrak{X}^s + Q\mathfrak{U}^s)h\rangle + \langle \mathfrak{U}^s g, (Q'\mathfrak{X}^s + G\mathfrak{X}^s)h\rangle \\ -\langle \mathfrak{W}^s g, \mathfrak{W}^s h\rangle\end{array}\right) dt\right\} P(d\omega).\end{aligned} \tag{8.2.34}$$

Indeed, choose h and g to be constant with probability one. For such initial conditions, the right hand side of equation (8.2.34) defines an absolutely continuous function $\mathcal{F}_s \to \mathbf{R}$. Using the Radon-Nikodym Theorem [265], it follows that there exists an $\mathcal{F}_s$-measurable Radon-Nikodym derivative $\mathfrak{x}_s(\omega, h, g)$ of this function with respect to the probability measure P. Furthermore, from the right hand side of equation (8.2.34), the function $\mathfrak{x}_s(\omega, h, g)$ is a bilinear function of h and g and hence, there exists an $\mathcal{F}_s$-measurable matrix $X_s(\omega)$ such that

$$\langle X_s(\omega)h, g\rangle := \mathfrak{x}_s(\omega, h, g).$$

Also, since the $\mathcal{F}_s$-measurable set Λ in equation (8.2.34) was chosen arbitrarily, this leads to the following equation

$$\begin{aligned}&\langle g, X_s(\omega)h\rangle \\ &= \mathbf{E}\left\{\int_s^\infty \left(\begin{array}{l}\langle \mathfrak{X}^s g, (R\mathfrak{X}^s + Q\mathfrak{U}^s)h\rangle \\ +\langle \mathfrak{U}^s g, (Q'\mathfrak{X}^s + G\mathfrak{X}^s)h\rangle - \langle \mathfrak{W}^s g, \mathfrak{W}^s h\rangle\end{array}\right) dt \Big| \mathcal{F}_s\right\}\end{aligned} \tag{8.2.35}$$

which holds for any a.s. constant h and g.

Let Λ_1 and Λ_2 be two $\mathcal{F}_s$-measurable sets. By multiplying the above identity by $\chi_{\Lambda_1\cap\Lambda_2}$ and taking into account equation (8.2.32) on the right hand side, it follows that equation (8.2.35) remains valid for initial conditions of the form $h = h_0\chi_{\Lambda_1}$, $g = g_0\chi_{\Lambda_2}$ where h_0, g_0 are two constant vectors. From this, equation (8.2.35) is readily extended to hold for any simple functions h, g. Also, since the set of simple functions is dense in $\mathbf{L}_2(\Omega, \mathcal{F}_s, P)$, we conclude that equation (8.2.35) holds for all $h, g \in \mathbf{L}_2(\Omega, \mathcal{F}_s, P)$. Thus, equation (8.2.33) follows.

To complete the proof, we establish that $X_s(\omega) = X_0(\Gamma_s\omega)$ a.s. where $\{\Gamma_s, s \geq 0\}$ is the translation semigroup generated by the Wiener process $(W_1(t), W_2(t))$; see [45] and Appendix B. Indeed, let $h \in \mathbf{L}_2(\Omega, \mathcal{F}_0, P)$ be given. Consider the minimax pairs for the differential games associated with the functionals $\Im^{0,h}$ and $\Im^{s,h_s}$ where $h_s(\omega) = h(\Gamma_s\omega)$. Using Lemma B.1, it follows that the minimax pairs for these two differential games are connected as follows:

$$\begin{aligned} u^s(h_s)(t+s,\omega) &= u^0(h)(t,\Gamma_s\omega); \\ w^s(h_s)(t+s,\omega) &= w^0(h)(t,\Gamma_s\omega); \\ x^s(h_s)(t+s,\omega) &= x^0(h)(t,\Gamma_s\omega). \end{aligned}$$

Using this property in equation (8.2.33) for $h_s = h(\Gamma_s\omega)$ and $g_s = g(\Gamma_s\omega)$ leads to the following equation

$$\begin{aligned} &\langle g_s(\omega), X_s(\omega)h_s(\omega)\rangle \\ &= \mathbf{E}\left\{\int_0^\infty \left(\begin{array}{l} \langle x^s(g_s)(t+s,\omega), Rx^s(h_s)(t+s,\omega)\rangle \\ +\langle x^s(g_s)(t+s,\omega), Qu^s(h_s)(t+s,\omega)\rangle \\ +\langle u^s(g_s)(t+s,\omega), Q'x^s(h_s)(t+s,\omega) \\ +\langle u^s(g_s)(t+s,\omega), Gu^s(h_s)(t+s,\omega)\rangle \\ -\langle w^s(g_s)(t+s,\omega), w^s(h_s)(t+s,\omega)\rangle \end{array}\right) dt \Big| \mathcal{F}_s \right\} \\ &= \mathbf{E}\left\{\int_0^\infty \left(\begin{array}{l} \langle x^0(g)(t,\Gamma_s\omega), Rx^0(h)(t,\Gamma_s\omega) \\ +\langle x^0(g)(t,\Gamma_s\omega), Qu^0(h)(t,\Gamma_s\omega)\rangle \\ +\langle u^0(g)(t,\Gamma_s\omega), Q'x^0(h)(t,\Gamma_s\omega) \\ +\langle u^0(g)(t,\Gamma_s\omega), Gu^0(h)(t,\Gamma_s\omega)\rangle \\ -\langle w^0(g)(t,\Gamma_s\omega), w^0(h)(t,\Gamma_s\omega)\rangle \end{array}\right) dt \Big| \mathcal{F}_s \right\}. \end{aligned}$$

We now use Proposition 1 of Appendix B. Using equation (B.3) along with the above equations, this leads to the conclusion that

$$\begin{aligned} &\langle g_s(\omega), X_s(\omega)h_s(\omega)\rangle \\ &= \mathbf{E}\left\{\int_0^\infty \left(\begin{array}{l} \langle x^0(g)(t), Rx^0(h)(t) + Qu^0(h)(t)\rangle \\ +\langle u^0(g)(t), Q'x^0(h)(t) + Gu^0(h)(t)\rangle \\ -\langle w^0(g)(t), w^0(h)(t)\rangle \end{array}\right) dt \Big| \mathcal{F}_0 \right\} (\Gamma_s\omega) \\ &= \langle g(\Gamma_s\omega), X_0(\Gamma_s\omega)h(\Gamma_s\omega)\rangle. \end{aligned}$$

That is,

$$X_s(\omega) = X_0(\Gamma_s\omega) \quad \text{a.s.}$$

for all $s \geq 0$. We now recall that $\mathcal{F}_0$ is the minimum σ-algebra with respect to which the sets $\{W_1(0) \in \Lambda_1, W_2(0) \in \Lambda_2\}$ are measurable. This fact implies that $\mathcal{F}_0$-measurable random variables are constant almost everywhere on Ω. Hence,

$\mathcal{F}_0$-measurable random variables are invariant with respect to the metrically transitive transformation Γ_s; see Appendix B and also [45]. Thus, we arrive at the conclusion that

$$X_s(\omega) = X_0(\omega) := X \quad \text{a.s.}$$

for all $s \geq 0$. Together with equation (8.2.31), this leads to the existence of a matrix X satisfying (8.2.25). Obviously, the matrix X is symmetric, since the operator $\mathcal{X}_s$ is self-adjoint and the matrix $X_s(\omega)$ is symmetric. Thus, the proposition follows.

The proof of the claim that the operator X satisfying (8.2.25) also satisfies equation (8.2.24) follows using the same arguments as those used in proving Theorem 3.4 of [90]. This proof is based on some well known facts in linear quadratic stochastic optimal control theory concerning the existence of nonnegative-definite solutions to the generalized Riccati equation

$$\begin{aligned} \frac{dX_{1T}}{dt} &+ A'X_{1T} + X_{1T}A + H^*X_{1T}H + R \\ &-(X_{1T}B_1 + H^*X_{1T}P_1 + Q)(G + P_1^*X_{1T}P_1)^{-1} \\ &\times(X_{1T}B_1 + H^*X_{1T}P_1 + Q)' = 0; \\ X_{1T}(T) &= 0 \end{aligned} \tag{8.2.36}$$

and the Riccati equation (8.2.13). Note that under the conditions of the theorem, the existence of a nonnegative-definite solution to equation (8.2.13) is guaranteed by Lemma 8.2.1.

Equation (8.2.36) is a Riccati differential equation corresponding to the following standard stochastic optimal control problem

$$\inf_{u \in \mathbf{L}_2([s,T]\times\Omega; R^{m_1})} \Im_T^{s,h}(u, 0). \tag{8.2.37}$$

For the particular case $Q = 0$, the solution to the optimal control problem (8.2.37) can be found for example, in references [20, 249]. Note that under condition (8.2.23) of the theorem, the results of [20] are readily extended to the case $Q \neq 0$. The extension of Theorem 6.1 of reference [20] to the case $Q \neq 0$ implies that there exists a unique nonnegative-definite symmetric bounded solution $X_{1T}(t)$ to equation (8.2.36) and that the feedback control law

$$u_{1T} = -(G + P_1^*X_{1T}P_1)^{-1}(B_1'X_{1T} + P_1^*X_{1T}H + Q')x$$

solves the optimal control problem (8.2.37).

Now consider the stochastic differential game associated with (8.2.1) and the cost functional (8.2.7):

$$\inf_{u(\cdot)\in \mathbf{L}_2(s,T,\mathbf{R}^{m_1})} \sup_{w(\cdot)\in \mathbf{L}_2(s,T,\mathbf{R}^{m_2})} \Im_T^{s,h}(u, w). \tag{8.2.38}$$

Under conditions (8.2.23) and (8.2.9), stochastic counterparts to Theorems 3.1 and 3.2 of [90] can be established.

Proposition 3 *If conditions (8.2.23), (8.2.9) are satisfied, then the game problem (8.2.38) has a unique saddle point $(u_T(\cdot), w_T(\cdot))$. Furthermore, there exists a unique nonnegative-definite solution to the Riccati equation*

$$\begin{aligned}\frac{dX_T(s)}{ds} &+ A'X_T + X_TA + H^*X_TH + R - (X_TB_1 + H^*X_TP_1 + Q) \\ &\times (G + P_1^*X_TP_1)^{-1}(X_TB_1 + H^*X_TP_1 + Q)' \\ &+ X_TB_2(I - P_2^*X_TP_2)^{-1}B_2'X_T = 0; \qquad (8.2.39) \\ X_T(T) &= 0\end{aligned}$$

and the saddle point of the game (8.2.38) is characterized by the feedback control laws

$$\begin{aligned}&u_T = F_{1T}x; \quad w_T = F_{2T}x; \\ &F_{1T} := -(G + P_1^*X_TP_1)^{-1}(B_1'X_T + P_1^*X_TH + Q'); \qquad (8.2.40) \\ &F_{2T} := (I - P_2^*X_TP_2)^{-1}B_2'X_T. \qquad (8.2.41)\end{aligned}$$

The existence of a unique saddle point of the differential game (8.2.38) follows from the same concavity convexity arguments as those used above and can be proved using the main result of [10].

Suppose that a solution X_T to equation (8.2.39) exists on the interval $[T-\alpha, T]$. Let $s \in [T-\alpha, T]$ be given. Then, applying Ito's formula to the quadratic form $x(t)'X_T(t)x(t)$, where $x(t)$ satisfies equation (8.2.1), we obtain

$$\begin{aligned}\Im_T^{s,h}(u,w) = &\ \mathbf{E}h'X_T(s)h \\ &-\mathbf{E}\int_s^T \|(I - P_2^*X_TP_2)^{1/2}(w(t) - F_{2T}x(t))\|^2 dt \\ &+\mathbf{E}\int_s^T \|(G + P_1^*X_TP_1)^{1/2}(u(t) - F_{1T}x(t))\|^2 dt\end{aligned}$$

where the matrices F_{1T} and F_{2T} are defined by equations (8.2.40) and (8.2.41). Hence, we conclude that the pair (8.2.40), (8.2.41) satisfies the saddle point condition:

$$\Im_T^{s,h}(u_T, w) \le \Im_T^{s,h}(u_T, w_T) = \mathbf{E}h'X_T(s)h \le \Im_T^{s,h}(u, w_T). \qquad (8.2.42)$$

It remains to prove that equation (8.2.39) has a solution as required. We can prove this claim in the same fashion as in the proof of Lemma 8.2.1 by choosing α sufficiently small. Indeed as in the proof Lemma 8.2.1, one can consider the mapping

$$\begin{aligned}\tilde{\varphi}(\tilde{X}(t)) := &\ A'\tilde{X}(t) + \tilde{X}(t)A + R + H^*\tilde{X}(t)H \\ &-(\tilde{X}(t)B_1 + H^*\tilde{X}(t)P_1 + Q)(G + P_1^*\tilde{X}(t)P_1)^{-1} \\ &\times(B_1'\tilde{X}(t) + P_1^*\tilde{X}(t)H + Q') \\ &+\tilde{X}(t)B_2(I - P_2^*\tilde{X}(t)P_2)^{-1}B_2'\tilde{X}(t)\end{aligned}$$

defined on the closed set $\kappa_{\mathcal{P}}^{\alpha}$. For any $\mathcal{P}$ chosen as in the proof of Lemma 8.2.1, we note that since $G + P_1^* \mathcal{P} P_1 \geq G \geq \bar{\lambda} I$, then $\|(G + P_1^* \mathcal{P} P_1)^{-1}\| \leq 1/\bar{\lambda}$. As in Proposition 1 (see also [20, Theorem 6.1]), we have that $\|(G + P_1^* \tilde{X} P_1)^{-1}\| \leq 2/\bar{\lambda}$ for any $\tilde{X} \in \kappa_{\mathcal{P}}^{\alpha}$. This observation allows us to conclude that the mapping $\tilde{\varphi}$ is well-defined on $\kappa_{\mathcal{P}}^{\alpha}$ and that the mapping

$$\tilde{\mathcal{Q}}(\tilde{X}(\cdot))(t) := \mathcal{P} + \int_t^T \tilde{\varphi}(\tilde{X}(\theta))d\theta$$

is a contraction on $\kappa_{\mathcal{P}}^{\alpha}$ provided α is chosen sufficiently small. Then, fixed point arguments lead us to the conclusion that there exists a bounded solution $X_T(t)$ to equation (8.2.39) on $[T-\alpha, T]$, and also $I - P_2^* X_T(t) P_2 > 0$. Using (8.2.42) with this solution, we have that

$$\mathbf{E}h' X_{1T} h = \Im_T^{s,h}(u_{1T}, 0) \leq \Im_T^{s,h}(u_T, 0) \leq \Im_T^{s,h}(u_T, w_T) = \mathbf{E}h' X_T h.$$

Also,

$$\mathbf{E}h' X_T h \leq \Im_T^{s,h}(0, w_T) \leq \Im_T^{s,h}(0, w_{2T}) = \mathbf{E}h' X_{2T} h.$$

Hence $0 \leq X_{1T}(s) \leq X_T(s) \leq X_{2T}(s) \leq X_2$ for all $s \in [T-\alpha, T]$. Thus, by partitioning the interval $[0, T]$ into subintervals not longer than α and iterating the above fixed point procedure, we can establish the existence of a unique global solution to Riccati equation (8.2.39). Since the game under consideration (8.2.38) has a unique saddle point, it follows from (8.2.42) that this saddle point is given by equations (8.2.40) and (8.2.41). This completes the proof of Proposition 3.

The remainder of the proof of Theorem 8.2.1 follows using arguments similar to those used in proving Theorem 3.4 in [90]. Consider a pair of inputs (8.2.40) and (8.2.41) extended to the interval $[T, \infty)$ by defining $(u_T(\cdot), w_T(\cdot))$ to be zero on this interval. This pair is again denoted by $(u_T(\cdot), w_T(\cdot))$. By optimality we have

$$\Im_T^{s,h}(u_T, w^s) \leq \Im_T^{s,h}(u_T, w_T) \leq \Im_T^{s,h}(u^s, w_T) \leq \Im^{s,h}(u^s, w^s). \quad (8.2.43)$$

Also, we have already shown that $X_T(s) \leq X_2$. This bound on $X_T(s)$ holds for any $s < T$.

On the other hand, for any pair of times $T_1 < T_2$, let $\hat{w}(\cdot)$ denote the input $w_{T_1}(\cdot)$ extended to the interval $[T_1, T_2]$ by defining $\hat{w}(t)$ to be zero on this interval. Also, let $\hat{x}(\cdot)$ denote the corresponding solution to (8.2.1) driven by the control input $u_{T_2}(\cdot)$ and uncertainty input $\hat{w}(\cdot)$. Then we have

$$\begin{aligned}
\mathbf{E}\langle h, X_{T_2}(s)h\rangle &= \Im_{T_2}^{s,h}(u_{T_2}, w_{T_2}) \\
&\geq \Im_{T_2}^{s,h}(u_{T_2}, \hat{w}) \\
&= \Im_{T_1}^{s,h}(u_{T_2}, w_{T_1}) + \mathbf{E}\int_{T_1}^{T_2} F(\hat{x}(t), u_{T_2}(t))dt \\
&> \Im_{T_1}^{s,h}(u_{T_2}, w_{T_1}) \geq \Im_{T_1}^{s,h}(u_{T_1}, w_{T_1}) = \mathbf{E}\langle h, X_{T_1}(s)h\rangle.
\end{aligned}$$

This shows that $X_T(s)$ is monotonically increasing in T. Hence, for all $s \geq 0$, the limit $\bar{X}(s) = \lim_{T\to\infty} X_T(s)$ exists. Consequently, $\bar{X}(s) \leq X_2$ for all $s \geq 0$. This inequality and inequality (8.2.16) from Lemma 8.2.1 together imply that $I - P_2^* \bar{X}(s) P_2 > 0$. This in turn implies that there exist bounded limits for $F_{1T}(s)$, $F_{2T}(s)$ as $T \to \infty$. These limits are denoted $\bar{F}_1(s)$ and $\bar{F}_2(s)$ respectively. If we pass to the limit as $T \to \infty$ in equation (8.2.39), it then follows that $\frac{dX_T(s)}{ds} \to \frac{d\bar{X}(s)}{ds}$. This implies that $X_T(s) \to \bar{X}(s)$, $F_{1T}(s) \to \bar{F}_1(s)$, $F_{2T}(s) \to \bar{F}_2(s)$ as $T \to \infty$, uniformly on each interval $[0, T_0]$. From this fact, it follows that $x_T(\cdot) \to \bar{x}(\cdot)$ strongly in $C([s, T_0], \mathbf{L}_2(\Omega, \mathcal{F}, P, R^n))$. Here $\bar{x}(\cdot)$ is the solution to the system

$$\begin{aligned} dx &= (A + B_1\bar{F}_1(t) + B_2\bar{F}_2(t))xdt + (H + P_1\bar{F}_1(t))xdW_1(t) \\ &\quad + P_2\bar{F}_1(t)xdW_2(t); \quad x(s) = h. \end{aligned}$$

The above convergence result follows from Theorem 5.2 on page 118 of [62].

The strong convergence of $x_T(\cdot)$ to $\bar{x}(\cdot)$ implies that $u_T(\cdot) \to \bar{u}(\cdot) = \bar{F}_1(\cdot)\bar{x}(\cdot)$ strongly in $C([s, T_0], \mathbf{L}_2(\Omega, \mathcal{F}, P, R^{m_1}))$ as $T \to \infty$. Also, $w_T(\cdot) \to \bar{w}(\cdot) = \bar{F}_2(\cdot)\bar{x}(\cdot)$ strongly in $C([s, T_0], \mathbf{L}_2(\Omega, \mathcal{F}, P, R^{m_2}))$ as $T \to \infty$, for any $T_0 > s$.

Also, we have

$$\varepsilon_1 |||u_T|||^2 \leq \Im_T^{s,h}(u_T, 0) \leq \Im_T^{s,h}(u_T, w_T) \leq \Im^{s,h}(u^s, w^s) = \mathbf{E}\langle h, Xh \rangle.$$

This implies that the sequence $\{u_T(\cdot)\}$ is bounded in $\mathbf{L}_2(s, \mathbf{R}^{m_1})$. Furthermore, the inequality

$$0 \leq \Im_T^{s,h}(0, w_T) \leq \Im^{s,h}(0, w_T)$$

implies that for any constant $\nu > 0$,

$$|||w_T|||^2 \leq (1 + \nu)||R^{1/2}\Phi_{s,\infty}||^2 |||w_T|||^2 + \left(1 + \frac{1}{\nu}\right) |||R^{1/2}x_{s,h}|||^2.$$

From (8.2.20), there exists a constant $\nu > 0$ such that

$$1 - (1 + \nu)||R^{1/2}\Phi_{s,\infty}||^2 > 0.$$

With this choice of ν, we obtain

$$|||w_T|||^2 \leq c_3 |||R^{1/2}x_{s,h}|||^2$$

where $c_3 > 0$ is a constant independent of T. Hence, the sequence $\{w_T(\cdot)\}$ is bounded in $\mathbf{L}_2(s, \mathbf{R}^{m_2})$. Thus, one can extract subsequences, again denoted by $\{u_T(\cdot)\}$, $\{w_T(\cdot)\}$, such that $(u_T(\cdot), w_T(\cdot)) \to (\tilde{u}(\cdot), \tilde{w}(\cdot)) \in \mathbf{L}_2(s, \mathbf{R}^{m_1}) \times \mathbf{L}_2(s, \mathbf{R}^{m_2})$ weakly as $T \to \infty$. This leads us to the conclusion that $\bar{u}(\cdot) \equiv \tilde{u}(\cdot)$ and $\bar{w}(\cdot) \equiv \tilde{w}(\cdot)$, which implies that $(\bar{u}(\cdot), \bar{w}(\cdot)) \in \mathbf{L}_2(s, \mathbf{R}^{m_1}) \times \mathbf{L}_2(s, \mathbf{R}^{m_2})$.

Letting $T \to \infty$ in (8.2.43), we obtain

$$\Im^{s,h}(\bar{u}, w^s) \leq \mathbf{E}\langle h, \bar{X}(s)h \rangle \leq \Im^{s,h}(u^s, \bar{w}) \leq \Im^{s,h}(u^s, w^s) \leq \Im^{s,h}(\bar{u}, w^s).$$

Thus, we have $\mathbf{E}\langle h, \bar{X}(s)h\rangle = \Im^{s,h}(u^s, w^s) = \mathbf{E}\langle h, Xh\rangle$ and therefore

$$(u^s(\cdot), w^s(\cdot)) \equiv (\bar{u}(\cdot), \bar{w}(\cdot)).$$

That is, the required feedback representation of the saddle point is satisfied. To see that the matrix X satisfies equation (8.2.24), one needs to pass to the limit as $T \to \infty$ in equation (8.2.39). Also, we see that $X = \lim_{T\to\infty} X_T$ is a minimal solution to equation (8.2.24) and this solution satisfies condition (8.2.25). It is clear that the solution satisfying these conditions is unique.

The claim that the system (8.2.26) is exponentially mean-square stable follows from the fact that for linear stochastic systems, stochastic $\mathbf{L}_2$-stability is equivalent to stochastic exponential mean-square stability; e.g., see [89].

■

Remark 8.2.4 In the particular case of $H = P_1 = P_2 = 0$, equation (8.2.24) reduces to the Riccati equation arising in deterministic state-feedback H^∞ control. In the particular case in which $B_2 = 0$, $P_2 = 0$, this equation reduces to a generalized Riccati equation arising in linear-quadratic stochastic optimal control; see [20, 249].

8.2.2 *Stochastic H^∞ control with complete state measurements*

In this subsection, we consider a stochastic H^∞ control problem related to the system (8.2.1). The results presented in this subsection extend corresponding results of [227] to our more general setup. From now on, we no longer assume that the system (8.2.4) is exponentially stable. That is, we no longer assume that Assumption 8.2.1 is satisfied, and hence we do not require the *à priori* stability of the matrix A in equation (8.2.1). Instead, we will use a property which is the stochastic counterpart of detectability.

Assumption 8.2.2 *Assume $D'D > 0$ and let $\tilde{R} := C'(I - D(D'D)^{-1}D')C \geq 0$. There exists a matrix $N \in R^{n\times m}$ such that the matrix $A - B_1(D'D)^{-1}D'C - N\tilde{R}^{1/2}$ is stable,* i.e., $\|\exp((A - B_1(D'D)^{-1}D'C - N\tilde{R}^{1/2})t)\| \leq ae^{-\alpha t}$, *and*

$$\frac{a^2}{\alpha}\|H - P_1(D'D)^{-1}D'C\|^2_{Q_1} < 1. \tag{8.2.44}$$

Note that the exponential stability of the nominal closed loop system (8.2.2) is a sufficient condition for the system (8.2.3) to have solutions lying in $\mathbf{L}_2(s; \mathbf{R}^n)$ for any $w(\cdot) \in \mathbf{L}_2(s; \mathbf{R}^{m_2})$.

Theorem 8.2.2 *Suppose Assumption 8.2.2 is satisfied. Then the stochastic H^∞ control problem in Definition 8.2.1 has a solution if and only if there exists a minimal nonnegative-definite symmetric solution X to the generalized algebraic*

Riccati equation

$$\begin{aligned} &A'X + XA + H^*XH + C'C + XB_2(I - P_2^*XP_2)^{-1}B_2'X \\ &\quad -(XB_1 + H^*XP_1 + C'D)(D'D + P_1^*XP_1)^{-1}(B_1'X + P_1^*XH + D'C) \\ &= 0, \end{aligned} \tag{8.2.45}$$

such that $I - P_2^*XP_2 > 0$ and the stochastic system

$$\begin{aligned} dx = &(A - B_1(D'D + P_1^*XP_1)^{-1}(B_1'X + P_1^*XH + D'C) \\ &+B_2(I - P_2^*XP_2)^{-1}B_2'X)xdt \\ &+(H + P_1(D'D + P_1^*XP_1)^{-1}(B_1'X + P_1^*XH + D'C))xdW_1 \\ &+P_2(I - P_2^*XP_2)^{-1}B_2'XxdW_2(t) \end{aligned} \tag{8.2.46}$$

is exponentially stable in the mean square sense. If this condition is satisfied, then a corresponding stabilizing controller which solves the stochastic H^∞ problem is given by

$$K = -(D'D + P_1^*XP_1)^{-1}(B_1'X + P_1^*XH + D'C). \tag{8.2.47}$$

Conversely, if the stochastic H^∞ control problem in Definition 8.2.1 has a solution, then the solution to the Riccati equation (8.2.45) satisfies the condition:

$$\mathbf{E}\langle h, Xh\rangle \leq \sup_{w \in \mathbf{L}_2(s;\mathbf{R}^{m_2})} \mathbf{E}\int_s^{+\infty} \left(\|Cx(t) + Du(t)\|^2 - \|w(t)\|^2\right) dt \tag{8.2.48}$$

for any state-feedback controller $u(\cdot)$ such that the closed loop system corresponding to this controller has $\mathbf{L}_2$-summable solutions for all $w \in \mathbf{L}_2(s;\mathbf{R}^{m_2})$, and the supremum on the right-hand side of (8.2.48) is finite.

Proof: The following proof extends arguments used in the proof of the corresponding result in [227]. Indeed, by substituting

$$v = u + (D'D)^{-1}D'Cx \tag{8.2.49}$$

into equation (8.2.1), the stochastic system under consideration becomes

$$\begin{aligned} dx &= (\tilde{A}x(t) + B_1v(t) + B_2w(t))dt + (\tilde{H}x(t) + P_1v(t))dW_1(t) \\ &\qquad + P_2w(t)dW_2(t); \\ \tilde{z} &= \tilde{C}x + \tilde{D}v. \end{aligned} \tag{8.2.50}$$

Here the following notation is used

$$\begin{aligned} &\tilde{A} = A - B_1(D'D)^{-1}D'C; \\ &\tilde{H} = H - P_1(D'D)^{-1}D'C; \\ &\tilde{C} = \begin{bmatrix} \tilde{R}^{1/2} \\ 0 \end{bmatrix}; \quad \tilde{D} = \begin{bmatrix} 0 \\ D \end{bmatrix} \end{aligned} \tag{8.2.51}$$

and the matrix $\tilde{R}$ is defined as in Assumption 8.2.2. Note that a matrix K solves the stochastic H^∞ problem defined in Section 8.2.2 if and only if the matrix $\tilde{K} = K + (D'D)^{-1}D'C$ solves a corresponding stochastic H^∞ problem associated with the system (8.2.50). The stochastic H^∞ control problem associated with the system (8.2.50) is defined in the same fashion as the original H^∞ control problem; i.e, given the system (8.2.50), find a matrix $\tilde{K} \in R^{m_1 \times n}$ such that the state feedback control law $v = \tilde{K}x$ satisfies the following conditions:

(i') The stochastic system

$$dx = (\tilde{A} + B_1\tilde{K})xdt + (\tilde{H} + P_1\tilde{K})xdW_1(t) \tag{8.2.52}$$

is exponentially stable in the mean square sense and the matrix $\tilde{A} + B_1\tilde{K}$ is stable;

(ii') The closed loop system corresponding to the system (8.2.50) with state-feedback control law $v = \tilde{K}x$:

$$\begin{aligned} dx &= [(\tilde{A} + B_1\tilde{K})x(t) + B_2w(t)]dt + (\tilde{H} + P_1\tilde{K})x(t)dW_1(t) \\ &\quad + P_2w(t)dW_2(t) \end{aligned} \tag{8.2.53}$$

satisfies the following stochastic H^∞-norm condition: There exists a constant $\tilde{\varepsilon} > 0$ such that

$$\mathbf{E}\int_s^{+\infty} \left(\|(\tilde{C} + \tilde{D}\tilde{K})x(t)\|^2 - \|w(t)\|^2\right) dt \le -\tilde{\varepsilon}\mathbf{E}\int_s^{+\infty} \|w(t)\|^2 dt \tag{8.2.54}$$

for all $w(\cdot) \in \mathbf{L}_2(s; \mathbf{R}^{m_2})$ where $x(s) = 0$.

Note that the problem (i'), (ii') is simpler than the original problem because we have $\tilde{C}'\tilde{D} = 0$ and $\tilde{D}'\tilde{D} = D'D > 0$ in this case. Also, by Assumption 8.2.2, there exists a matrix N such that the matrix $\tilde{A} - N\tilde{C}$ is stable and

$$\|\exp((\tilde{A} - N\tilde{C})t)\| \le ae^{-\alpha t}; \qquad \frac{a^2}{\alpha}\|\tilde{H}\|_{Q_1}^2 < 1. \tag{8.2.55}$$

This observation implies that the stochastic system

$$dy = (\tilde{A} - N\tilde{C})ydt + \tilde{H}ydW_1(t); \qquad y(s) = y_0 \in \mathbf{L}_2(\Omega, \mathcal{F}_s, P), \tag{8.2.56}$$

is exponentially mean square stable. In the particular case when $\tilde{H} = 0$, this is equivalent to the detectability of the pair $(\tilde{C}, \tilde{A})$.

We first solve the problem (i'), (ii').

Lemma 8.2.2

(a) *If the state-feedback stochastic H^∞ control problem (i'), (ii') has a solution, then there exists a minimal nonnegative-definite symmetric solution X to the generalized algebraic Riccati equation*

$$\begin{aligned} &\tilde{A}'X + X\tilde{A} + \tilde{H}^*X\tilde{H} + \tilde{C}'\tilde{C} + XB_2(I - P_2^*XP_2)^{-1}B_2'X \\ &\quad -(XB_1 + \tilde{H}^*XP_1)(\tilde{D}'\tilde{D} + P_1^*XP_1)^{-1}(B_1'X + P_1^*X\tilde{H}) \\ &= 0 \end{aligned} \tag{8.2.57}$$

such that $I - P_2^*XP_2 > 0$ and the stochastic system

$$dx = (\tilde{A} + B_1\tilde{F}_1 + B_2\tilde{F}_2)xdt + (H + P_1\tilde{F}_1)xdW_1(t) + P_2\tilde{F}_2xdW_2(t), \tag{8.2.58}$$

is exponentially stable in the mean-square sense where

$$\begin{aligned} \tilde{F}_1 &= -(\tilde{D}'\tilde{D} + P_1^*XP_1)^{-1}(B_1'X + P_1^*X\tilde{H}); \\ \tilde{F}_2 &= (I - P_2^*XP_2)^{-1}B_2'Xx. \end{aligned} \tag{8.2.59}$$

Also,

$$\mathbf{E}\langle h, Xh\rangle \leq \sup_{w \in \mathbf{L}_2(s;\mathbf{R}^{m_2})} \mathbf{E}\int_s^{+\infty} \left(\|\tilde{C}x(t) + \tilde{D}\hat{v}(t)\|^2 - \|w(t)\|^2\right) dt \tag{8.2.60}$$

for any state-feedback controller $\hat{v}(\cdot)$ such that the closed loop system corresponding to equation (8.2.50) and this controller has $\mathbf{L}_2$-summable solutions for all $w(\cdot) \in \mathbf{L}_2(s;\mathbf{R}^{m_2})$ and the supremum on the left-hand side of (8.2.60) is finite.

(b) *Conversely, if there exists a nonnegative-definite symmetric solution X to equation (8.2.57) such that $I - P_2^*XP_2 > 0$ and the stochastic system (8.2.58) is exponentially stable, then the state-feedback stochastic H^∞ control problem (i'), (ii') has a solution. The corresponding stabilizing feedback control law which solves this stochastic H^∞ problem is given by*

$$\tilde{K} = -(\tilde{D}'\tilde{D} + P_1^*XP_1)^{-1}(B_1'X + P_1^*X\tilde{H}). \tag{8.2.61}$$

Proof: We will use the following notation:

$$\tilde{A}_K = \tilde{A} + B_1\tilde{K}; \ \tilde{C}_K = \tilde{C} + \tilde{D}\tilde{K}; \ \tilde{H}_K = \tilde{H} + P_1\tilde{K}.$$

(b)⇒(a). Let X be a nonnegative-definite symmetric solution to equation (8.2.57) such that the stochastic system (8.2.58) is exponentially mean square stable. Let $\tilde{K}$ be given by (8.2.61). We wish to establish that X and $\tilde{K}$ satisfy conditions (i'), (ii').

We first prove the stability of the system (8.2.52). Note that using (8.2.61), equation (8.2.57) is equivalent to the equation

$$\tilde{A}_K'X + X\tilde{A}_K + \tilde{H}_K^* X \tilde{H}_K + \tilde{C}'\tilde{C} + \tilde{K}'(\tilde{D}'\tilde{D})\tilde{K} \\ +XB_2(I - P_2^* X P_2)^{-1}B_2'X = 0. \tag{8.2.62}$$

This implies that

$$\tilde{A}_K'X + X\tilde{A}_K + \tilde{H}_K^* X \tilde{H}_K + \tilde{C}'\tilde{C} + \tilde{K}'(\tilde{D}'\tilde{D})\tilde{K} \leq 0. \tag{8.2.63}$$

We now proceed as in [89, Lemma 4.6]. Letting $x(t)$ be the solution to equation (8.2.52), we first note that from (8.2.63)

$$\mathbf{E}\int_0^\infty \left(\|\tilde{C}x(t)\|^2 + \|\tilde{D}\tilde{K}x(t)\|^2\right) dt \leq \mathbf{E}h'Xh < \infty,$$

and hence $\tilde{C}x(\cdot)$ and $\tilde{K}x(\cdot)$ are square integrable (the latter holds since $\tilde{D}'\tilde{D} > 0$). Also, note that

$$\|\tilde{H} + P_1\tilde{K}\|_{Q_1}^2 \leq (1+\nu)\|\tilde{H}\|_{Q_1}^2 + (1 + \frac{1}{\nu})\|P_1\tilde{K}\|_{Q_1}^2$$

for all $\nu > 0$. From (8.2.55), it follows that ν can be chosen to be sufficiently small in order to guarantee

$$\frac{(1+\nu)a^2}{\alpha}\|\tilde{H}\|_{Q_1}^2 < 1. \tag{8.2.64}$$

This observation allows us to conclude that $\mathbf{E}\|x(t)\|^2 \in \mathbf{L}_1[0,\infty)$. The proof of this fact follows along the same lines as the proof of Lemma 4.6 in [89]. This implies that (i') holds.

We will now prove that condition (ii') is also satisfied. From (8.2.62), we have

$$\mathbf{E}\langle x(t), Xx(t)\rangle + \mathbf{E}\int_0^t \{\|\tilde{C}_K x(t)\|^2 - \|w\|^2\}dt \\ = -\mathbf{E}\int_0^t \|(I - P_2^* X P_2)^{1/2}(w - (I - P_2^* X P_2)^{-1}B_2'Xx)\|^2 dt \tag{8.2.65}$$

where $x(\cdot)$ is the solution to (8.2.53) corresponding to the initial condition $x(0) = 0$.

Note that the substitution $w = \zeta + (I - P_2^* X P_2)^{-1}B_2'Xx$ into equation (8.2.53) leads to the following stochastic system

$$dx = ((\tilde{A}_K + B_2(I - P_2^* X P_2)^{-1}B_2'X)x + B_2\zeta(t))dt + \tilde{H}_K x dW_1(t) \\ +P_2(\zeta + (I - P_2^* X P_2)^{-1}B_2'Xx)dW_2; \quad x(0) = 0. \tag{8.2.66}$$

In particular, the input $\zeta(t) = 0$ corresponds to the stable system (8.2.58). This implies (e.g., see Lemma B.2 in Appendix B and also [27]) that the solutions of equation (8.2.66) satisfy the condition $|||x||| \leq c_0|||\zeta|||$, where $c_0 > 0$ is a constant independent of $\zeta(\cdot)$. That is, the mapping $\zeta(\cdot) \to x(\cdot)$ generated by equation (8.2.66) is a bounded mapping $\mathbf{L}_2(0;\mathbf{R}^{m_2}) \to \mathbf{L}_2(0;\mathbf{R}^n)$. Therefore, the mapping $\zeta(\cdot) \to w(\cdot) = \zeta(\cdot) + (I - P_2XP_2)^{-1}B_2'Xx(\cdot)$ is also a bounded mapping $\mathbf{L}_2(0;\mathbf{R}^{m_2}) \to \mathbf{L}_2(0;\mathbf{R}^{m_2})$. Thus, there exists a constant $c > 0$ such that

$$|||w(\cdot)||| \leq c|||\zeta(\cdot)||| \quad \forall \zeta \in \mathbf{L}_2(0;\mathbf{R}^{m_2}).$$

Note, the restriction of any solution to the stochastic system (8.2.66) to an interval $[0,t]$ is equal to the restriction to $[0,t]$ of the corresponding solution to the stochastic system (8.2.53). Since, $X \geq 0$ and $I - P_2^*XP_2 > 0$, then (8.2.65) implies

$$\begin{aligned}\mathbf{E}\int_0^\infty \{\|(\tilde{C}+\tilde{D}\tilde{K})x(t)\|^2 - \|w\|^2\}dt &\leq -\epsilon\mathbf{E}\int_0^\infty \|\zeta\|^2 dt \\ &\leq -\frac{\epsilon}{c^2}\mathbf{E}\int_0^\infty \|w\|^2 dt.\end{aligned}$$

(a)⇒(b). Given a matrix $\tilde{K}$ satisfying conditions (i') and (ii'), we wish to prove that there exists a solution to equation (8.2.57) stabilizing the system (8.2.58).

Let $\delta \in (0,\bar{\delta}]$ be a given constant and consider the stochastic differential game of the form (8.2.8) associated with the system

$$\begin{aligned} dx &= (\tilde{A}_K x + B_1 v + B_2 w)dt + (\tilde{H}_k x + P_1 v)dW_1(t) + P_2 w dW_2(t); \\ \tilde{z} &= \tilde{C}_K x + \tilde{D}v \end{aligned} \tag{8.2.67}$$

and the cost functional

$$\Im_\delta^{s,h}(v,w) = \int_s^{+\infty} \mathbf{E}\left\{\|\tilde{C}_K x(t) + \tilde{D}v(t)\|^2 + \delta\|v(t)\|^2 - \|w(t)\|^2\right\} dt. \tag{8.2.68}$$

It follows from condition (i') that the system (8.2.52) is stable and hence the system (8.2.67) satisfies Assumption 8.2.1. Also, it follows from (ii') that the system (8.2.53) satisfies condition (8.2.9) of Lemma 8.2.1 and Theorem 8.2.1 with $\varepsilon_2 = \tilde{\varepsilon}$. In particular, Lemma 8.2.1 defines a matrix X_2 such that $I - P_2^*X_2P_2 > 0$. Note that this matrix is independent of δ.

Condition (8.2.23) of Theorem 8.2.1 is also satisfied with $\varepsilon_1 = \delta$. It follows from this theorem that for each $\delta \in (0,\bar{\delta}]$, the equation

$$\begin{aligned} &\tilde{A}_K'X + X\tilde{A}_K + \tilde{H}_K^*X\tilde{H}_K + \tilde{C}_K\tilde{C}_K + XB_2(I - P_2^*XP_2)^{-1}B_2'X \\ &\quad -(XB_1 + \tilde{H}_K^*XP_1 + \tilde{C}_K'\tilde{D})(\tilde{D}'\tilde{D} + \delta I + P_1^*XP_1)^{-1} \\ &\quad \times(B_1'X + P_1^*X\tilde{H}_K + \tilde{D}'\tilde{C}_K) \\ &= 0 \end{aligned} \tag{8.2.69}$$

has a minimal nonnegative-definite symmetric solution X^δ such that the system

$$\begin{aligned} dx = {} & (\tilde{A}_K + B_1F_{1,\delta} + B_2F_{2,\delta})xdt + (\tilde{H}_K + P_1F_{1,\delta})xdW_1(t) \\ & + P_2F_{2,\delta}xdW_2(t) \end{aligned} \tag{8.2.70}$$

is exponentially mean square stable. In equation (8.2.70),

$$F_{1,\delta} = -(\tilde{D}'\tilde{D} + \delta I + P_1^* X^\delta P_1)^{-1}(B_1' X^\delta + P_1^* X^\delta \tilde{H}_K + \tilde{C}_K'\tilde{D}),$$
$$F_{2,\delta} = (I - P_2^* X^\delta P_2)^{-1} B_2' X^\delta.$$

Also, it follows from the proof of Theorem 8.2.1 that $X^\delta \leq X_2$ and hence $I - P_2^* X^\delta P_2 > 0$.

As in Theorem 4.2 of reference [90], it follows that there exists a matrix

$$X := \lim_{\delta \downarrow 0} X^\delta, \ X = X' \geq 0$$

which satisfies equation (8.2.57). To verify this, we use the fact that $\tilde{C}'\tilde{D} = 0$. Indeed, note that since X_2 is independent of δ, then $X \leq X_2$ and hence $I - P_2^* X P_2 > 0$. Now, letting $\delta \downarrow 0$ in (8.2.69), we obtain:

$$\begin{aligned}
0 = \; & \tilde{A}_K' X + X\tilde{A}_K + \tilde{H}_K^* X \tilde{H}_K + \tilde{C}_K \tilde{C}_K \\
& -(XB_1 + \tilde{H}_K^* X P_1 + \tilde{C}_K'\tilde{D})(\tilde{D}'\tilde{D} + P_1^* X P_1)^{-1} \\
& \times(B_1' X + P_1^* X \tilde{H}_K + \tilde{D}'\tilde{C}_K) + XB_2(I - P_2^* X P_2)^{-1} B_2' X \\
= \; & \tilde{A}' X + X\tilde{A} + \tilde{H}^* X \tilde{H} + \tilde{C}'\tilde{C} \\
& -(XB_1 + \tilde{H}^* X P_1)(\tilde{D}'\tilde{D} + P_1^* X P_1)^{-1}(B_1' X + P_1^* X \tilde{H}) \\
& +XB_2(I - P_2^* X P_2)^{-1} B_2' X \\
& +(XB_1 + \tilde{H}^* X P_1)\tilde{K} + \tilde{K}'(XB_1 + \tilde{H}^* X P_1)' \\
& +\tilde{K}'(\tilde{D}'\tilde{D} + P_1^* X P_1)\tilde{K} \\
& -(XB_1 + \tilde{H}^* X P_1)\tilde{K} - \tilde{K}'(XB_1 + \tilde{H}^* X P_1)' \\
& -\tilde{K}'(\tilde{D}'\tilde{D} + P_1^* X P_1)\tilde{K} \\
= \; & \tilde{A}' X + X\tilde{A} + \tilde{H}^* X \tilde{H} + \tilde{C}'\tilde{C} \\
& -(XB_1 + \tilde{H}^* X P_1)(\tilde{D}'\tilde{D} + P_1^* X P_1)^{-1}(B_1' X + P_1^* X \tilde{H}) \\
& +XB_2(I - P_2^* X P_2)^{-1} B_2' X.
\end{aligned}$$

That is, equation (8.2.57) is satisfied. Also, the following limits exist:

$$\begin{aligned}
F_{1,K} &= \lim_{\delta \downarrow 0} F_{1,\delta} \\
&= -\tilde{K} - (\tilde{D}'\tilde{D} + P_1^* X P_1)^{-1}(B_1' X + P_1' X \tilde{H}) = \tilde{F}_1 - \tilde{K}; \\
F_{2,K} &= \lim_{\delta \downarrow 0} F_{2,\delta} = (I - P_2^* X P_2)^{-1} B_2' X = \tilde{F}_2.
\end{aligned}$$

Note that X is defined as the limit of a sequence of minimal solutions and therefore, it represents the minimal solution to (8.2.57).

We now show that the system (8.2.58) is exponentially mean-square stable. Consider solutions $x_\delta(\cdot)$ and $x(\cdot)$ to equations (8.2.70) and (8.2.58), respectively and corresponding to the initial conditions $x_\delta(s) = x(s) = h$. Note that

$$\begin{aligned}
0 \leq \Im_\delta^{s,h}(F_{1,\delta} x_\delta, 0) \leq \mathbf{E} h' X^\delta h &= \Im_\delta^{s,h}(F_{1,\delta} x_\delta, F_{2,\delta} x_\delta) \\
&\leq \Im^{s,h}(0, F_{2,\delta} x_\delta) \\
&\leq \mathbf{E} h' X_2 h \qquad (8.2.71)
\end{aligned}$$

where the matrix X_2 is as defined above. Hence, as in the proof of Theorem 8.2.1, $F_{2,\delta}x_\delta(\cdot)$ is bounded. Furthermore, it follows from (8.2.71) that $\Im_\delta^{s,h}(F_{1,\delta}x_\delta, F_{2,\delta}x_\delta)$ is bounded. Since $F_{2,\delta}x_\delta(\cdot)$ is bounded, this together with $\tilde{C}'\tilde{D} = 0$ and $\tilde{D}'\tilde{D} > 0$, implies that $\tilde{C}x_\delta(\cdot)$ and $(\tilde{K} + F_{1,\delta})x_\delta(\cdot)$ are bounded on the corresponding spaces $\mathbf{L}_2[s,\infty)$. Hence, one can extract subsequences which have weak limit points in the corresponding $\mathbf{L}_2$-spaces. These limits are $\mathbf{L}_2$-summable functions on $[s,\infty) \times \Omega$. On the other hand, as in the proof of Theorem 8.2.1, for any $T_0 > s$, $x_\delta(\cdot) \to x(\cdot)$ in $C([s,T_0]; \mathbf{L}_2(\Omega, \mathcal{F}, P, R^n))$. Therefore, for any $T_0 > s$, the restrictions of functions $\tilde{C}x(\cdot)$, $(\tilde{K}+F_{1,K})x(\cdot) = \tilde{F}_1 x(\cdot)$ and $\tilde{F}_2 x(\cdot)$ to the interval $[s,T_0]$ are equal to the restrictions to $[s,T_0]$ of the corresponding weak limit points. Hence $\tilde{C}x(\cdot)$, $\tilde{F}_1 x(\cdot)$, $\tilde{F}_2 x(\cdot)$ are square integrable functions on $[s,\infty) \times \Omega$.

We now re-write the system (8.2.58) in the following form

$$\begin{aligned} dx &= ((\tilde{A} - N\tilde{C})x + B_1\tilde{F}_1 x(t) + B_2\tilde{F}_2 x(t) + N\tilde{C}x(t))dt \\ &\quad +(\tilde{H} + P_1\tilde{F}_1)x dW_1(t) + P_2\tilde{F}_2 x dW_2(t) \end{aligned}$$

where N is the matrix defined in Assumption 8.2.2. Since the system (8.2.56) is stable and the inputs $\tilde{F}_1 x(\cdot)$, $\tilde{F}_2 x(\cdot)$, and $\tilde{C}x(\cdot)$ are square integrable on $[s,\infty) \times \Omega$, then $x(\cdot) \in \mathbf{L}_2(s; \mathbf{R}^n)$. Hence, the stochastic system (8.2.58) is exponentially mean square stable.

To show that (8.2.60) holds, we consider a finite-horizon version of the functional (8.2.68):

$$\Im_{\delta,T}^{s,h}(v,w) = \int_s^T \mathbf{E}\left\{\|\tilde{C}_K x(t) + \tilde{D}v(t)\|^2 + \delta\|v(t)\|^2 - \|w(t)\|^2\right\} dt$$

with $v(\cdot) \in \mathbf{L}_2(s,T; \mathbf{R}^{m_1})$ and $w(\cdot) \in \mathbf{L}_2(s,T; \mathbf{R}^{m_2})$. As in the infinite-horizon case, it follows from Proposition 3 (see also [90, Theorem 4.1]), that there exists a matrix $X_T = \lim_{\delta \downarrow 0} X_T^\delta$, satisfying the Riccati differential equation

$$\begin{aligned} &\frac{dX_T(s)}{ds} + \tilde{A}_K' X_T + X_T\tilde{A}_K + \tilde{H}_K^* X_T \tilde{H}_K + \tilde{C}_K'\tilde{C}_K \\ &\quad - (X_T B_1 + \tilde{H}_K^* X_T P_1 + \tilde{C}_K'\tilde{D})(\tilde{D}'\tilde{D} + P_1^* X_T P_1)^{-1} \\ &\quad \times (X_T B_1 + \tilde{H}_K^* X_T P_1 + \tilde{C}_K'\tilde{D})' \\ &\quad + X_T B_2 (I - P_2^* X_T P_2)^{-1} B_2' X_T = 0; \quad X_T(T) = 0. \end{aligned} \tag{8.2.72}$$

Let $\hat{v}(\cdot)$ be any state-feedback controller such that the closed loop system obtained when this controller is applied to (8.2.50) has $\mathbf{L}_2$-summable trajectories for all $\mathbf{L}_2$-summable uncertainty inputs. Then, the corresponding closed loop system obtained when the state-feedback controller $v(\cdot) = \hat{v}(\cdot) - \tilde{K}x$ is applied to

(8.2.67) also has $\mathbf{L}_2$-summable trajectories for all $\mathbf{L}_2$-summable uncertainty inputs. Hence, if $w(\cdot)$ is an $\mathbf{L}_2$-summable uncertainty input, $w(\cdot) \in \mathbf{L}_2(s,T;\mathbf{R}^{m_2})$, and $\tilde{w}(\cdot)$ denotes its extension to the interval $[T,\infty)$ obtained by setting $\tilde{w}(\cdot)$ to zero on $[T,\infty)$, then (8.2.72) implies

$$\begin{aligned}
&\mathbf{E}\langle h, X_T(s)h\rangle \\
&\leq \sup_{w\in\mathbf{L}_2(s,T;\mathbf{R}^{m_2})} \int_s^T \mathbf{E}\left\{\|\tilde{C}_K x(t) + \tilde{D}v(t)\|^2 - \|w(t)\|^2\right\} dt \\
&\leq \sup_{w\in\mathbf{L}_2(s,T;\mathbf{R}^{m_2})} \int_s^\infty \mathbf{E}\left\{\|\tilde{C}_K x(t) + \tilde{D}v(t)\|^2 - \|\tilde{w}(t)\|^2\right\} dt \\
&\leq \sup_{w\in\mathbf{L}_2(s;\mathbf{R}^{m_2})} \int_s^\infty \mathbf{E}\left\{\|\tilde{C}_K x(t) + \tilde{D}v(t)\|^2 - \|w(t)\|^2\right\} dt \\
&= \sup_{w\in\mathbf{L}_2(s;\mathbf{R}^{m_2})} \int_s^\infty \mathbf{E}\left\{\|\tilde{C} x(t) + \tilde{D}\hat{v}(t)\|^2 - \|w(t)\|^2\right\} dt.
\end{aligned}$$

Also, note that $X_T \to X$ as $T \to \infty$ since X is a minimal solution. Thus, (8.2.60) holds.

■

We now are in position to complete the proof of Theorem 8.2.2. Indeed, by letting

$$K = \tilde{K} - (D'D)^{-1}D'C$$

and using the notation (8.2.51), equations (8.2.57), (8.2.58) are transformed into equations (8.2.45), (8.2.46) respectively. Also, the feedback matrix (8.2.61) is transformed into the matrix (8.2.47). Hence the theorem follows from Lemma 8.2.2.

■

8.2.3 *Illustrative example*

In this section, we present an example illustrating Theorem 8.2.2. The system to be controlled consists of two carts as shown in Figure 2.4.2. We will regard the disturbance $w(\cdot)$ as a time-varying external disturbance which acts on the right cart. A control force $u(\cdot)$ acts on the left cart. The spring constant k has a specified nominal value of $k_0 = 1.25$. However, this value can vary and is considered uncertain, $k \in [0.5, 2]$. We assume that the masses of carts are $m_1 = m_2 = 1$. Then, the system is described by the state equations

$$\begin{aligned}
\dot{x} &= (A + F(k(t) - k_0)S)x + B_1 u + B_2 w(t); \qquad (8.2.73)\\
z &= Cx + Du
\end{aligned}$$

where $x = [x_1 \; x_2 \; \dot{x}_1 \; \dot{x}_2]' \in \mathbf{R}^4$, and

$$A = \begin{bmatrix} 0 & 0 & 1 & 0 \\ 0 & 0 & 0 & 1 \\ -1.25 & 1.25 & 0 & 0 \\ 1.25 & -1.25 & 0 & 0 \end{bmatrix}; \quad B_1 = \begin{bmatrix} 0 \\ 0 \\ 0 \\ 1 \end{bmatrix}; \quad B_2 = \begin{bmatrix} 0 \\ 0 \\ 1 \\ 0 \end{bmatrix};$$

$$C = \begin{bmatrix} 1 & 0 & 0 & 0 \\ 0 & 1 & 0 & 0 \\ 0 & 0 & 0 & 0 \end{bmatrix}; \quad D = \begin{bmatrix} 0 \\ 0 \\ 1 \end{bmatrix};$$

$$F = \begin{bmatrix} 0 \\ 0 \\ -1 \\ 1 \end{bmatrix}; \quad S = \begin{bmatrix} 1 & -1 & 0 & 0 \end{bmatrix}. \tag{8.2.74}$$

A state-feedback controller corresponding to the nominal value of k can be found that guarantees $\|\mathfrak{W}_{wz}(s)\|_\infty \leq 2.5$, where $\mathfrak{W}_{wz}(s)$ is the closed loop nominal transfer function

$$\mathfrak{W}_{wz}(s) = (C + DK)(sI - (A + B_1 K))^{-1} B_2.$$

However, this H^∞ norm bound holds only for the nominal value of k. Now, suppose it is required to construct a controller that guarantees the disturbance attenuation bound $\gamma^2 = 4$ in the face of variations in the value of k.

Let $\Delta(t) = k(t) - k_0$. In order to apply results of this section, we assume that for all $t \geq 0$, $\Delta(t)$ is the Gaussian white noise process with zero mean and $\mathbf{E}\Delta^2(t) = \sigma^2$. We can then choose the value of the parameter σ such that k obeys the bound $0.5 \leq k \leq 2$ with a sufficiently high probability. For example, for $\sigma = \bar{\sigma} = 0.25$, we have $P(|k(t) - k_0| \leq 0.75) \geq 0.997$. This probability is increasing as $\sigma^2 \downarrow 0$. This model of the spring constant variation leads to the following stochastic system model of the form (8.2.1):

$$\begin{aligned} dx &= (Ax + B_1 u + B_2 w)dt + HxdW_1(t); \\ z &= Cx + Du. \end{aligned} \tag{8.2.75}$$

In equation (8.2.75), A, B_1, B_2, C and D are given by (8.2.74), W_1 is a scalar Wiener process with identity covariance and $H = \bar{\sigma} FS$. Note that $C'D = 0$ in this particular example. We now apply Theorem 8.2.2 to this system.

Note that equation (8.2.75) satisfies Assumption 8.2.2 with

$$N = \begin{bmatrix} 317.2223 & 0.0039 & 0 \\ 0.0039 & 317.2223 & 0 \\ 314.9827 & 1.2451 & 0 \\ 1.2451 & 314.9827 & 0 \end{bmatrix}.$$

For this value of N, we have $a = e^{0.4}$, $\alpha = 1$, and $(a^2/\alpha)\|H\|^2 = 0.5564 < 1$.

The process of designing a controller based on Theorem 8.2.2 involves solving equation (8.2.45). As mentioned in Remark 8.2.3 on page 255, Riccati equations of the form (8.2.45) can be solved using homotopy methods; e.g., see [166]. For this particular example, we found the following procedure to work well. Consider the auxiliary equation

$$A'X + XA + \mu H'XH + C'C - X\left(B_1(D'D)^{-1}B_1 - \frac{1}{\gamma^2}B_2B_2'\right)X = 0. \tag{8.2.76}$$

In the particular case when $\mu = 1$, this equation corresponds to equation (8.2.45). Also, consider a series of the form

$$X(\mu) = \sum_{j=0}^{\infty} \mu^j X_j. \tag{8.2.77}$$

The formal substitution of this series into (8.2.76) leads to a standard game type algebraic Riccati equation for X_0:

$$A'X_0 + X_0A + C'C - X_0\left(B_1(D'D)^{-1}B_1' - \frac{1}{\gamma^2}B_2B_2'\right)X_0 = 0 \tag{8.2.78}$$

and an infinite sequence of Lyapunov equations:

$$\begin{aligned}
&\tilde{A}'X_1 + X_1\tilde{A} + H'X_0H = 0; \\
&\tilde{A}'X_j + X_j\tilde{A} + H'X_{j-1}H \\
&\quad - \sum_{\nu=1}^{j-1} X_\nu\left(B_1(D'D)^{-1}B_1' - \frac{1}{\gamma^2}B_2B_2'\right)X_{j-\nu} = 0; \quad (8.2.79) \\
&\qquad j = 2, 3, \ldots
\end{aligned}$$

Here

$$\tilde{A} = A - B_1(D'D)^{-1}B_1'X_0 - \frac{1}{\gamma^2}B_2B_2'X_0.$$

Note that in this example, the pair (A, B_1) is controllable and the pair (C, A) is observable. This implies that there is at most one positive-definite solution to (8.2.78) such that the matrix $\tilde{A}$ is stable. If there exists such a stabilizing solution to (8.2.78), then the Lyapunov equations (8.2.79) have unique solutions for $j = 2, 3 \ldots$. If the sequence of truncations $\sum_{j=0}^{m} \mu^j X_j$ has a nonnegative-definite limit as $m \to \infty$, then this limit must satisfy equation (8.2.76). Although there are no results available which guarantee the success of the above algorithm, it has been found to work well in practice.

Using the above algorithm, the following matrix is obtained by summing the first 20 terms of the series (8.2.77)

$$\hat{X} = \begin{bmatrix} 2.8456 & -2.0816 & -0.6195 & -0.8547 \\ -2.0816 & 7.1554 & 5.6625 & 3.7458 \\ -0.6195 & 5.6625 & 6.9848 & 3.3103 \\ -0.8547 & 3.7458 & 3.3103 & 3.1986 \end{bmatrix}.$$

This matrix is found to satisfy equation (8.2.45) with an absolute accuracy of $\sim 10^{-7}$. Moreover, it can readily be verified that $\hat{A}'M + M\hat{A} + H'MH < 0$, where M is the positive definite symmetric solution to the Lyapunov equation

$$\hat{A}'M + M\hat{A} = -I; \qquad \hat{A} := A - B_1(D'D)^{-1}B_1'\hat{X} + \frac{1}{\gamma^2}B_2B_2'\hat{X}.$$

Hence, the Lyapunov function $V(x) = x'Mx$ can be used to establish the exponential stability of the system (8.2.46) with $X = \hat{X}$. The corresponding controller is

$$u = Kx; \quad K = \begin{bmatrix} 0.8547 & -3.7458 & -3.3103 & -3.1986 \end{bmatrix}. \tag{8.2.80}$$

To verify the properties of this controller, the closed loop system (8.2.73), (8.2.80) was simulated with the input

$$w(t) = \begin{cases} 1, & t \leq 2, \\ 0, & t \geq 2. \end{cases} \tag{8.2.81}$$

Plots of the system outputs corresponding to three chosen realizations of $\Delta(t)$ are shown in Figure 8.2.1. It is of interest to note that we obtain

$$\frac{\int_0^\infty \|z(t)\|^2 dt}{\int_0^\infty \|w(t)\|^2 dt} \approx 8.77 > \gamma^2$$

in the case $\Delta(t) \equiv -0.75$. This in no way contradicts Theorem 8.2.2, since this theorem guarantees the bound of $\gamma^2 = 4$ on the ratio of expected values of the above $\mathbf{L}_2$-norms.

8.3 Absolute stabilization and minimax optimal control of stochastic uncertain systems with multiplicative noise

8.3.1 The stochastic guaranteed cost control problem

Consider an uncertain stochastic system of the form (2.4.1) described by the stochastic differential Ito equation:

$$\begin{aligned} dx &= (Ax(t) + B_1u(t) + B_2\xi(t))dt + (Hx(t) + P_1u(t))dW_1(t) \\ &\quad + P_2\xi(t)dW_2(t); \\ z(t) &= Cx(t) + Du(t) \end{aligned} \tag{8.3.1}$$

where $x(t) \in \mathbf{R}^n$ is the state, $u(t) \in \mathbf{R}^{m_1}$ is the control input, $z(t) \in \mathbf{R}^p$ is a vector assembling all uncertainty outputs and $\xi(t) \in \mathbf{R}^{m_2}$ is a vector assembling all uncertainty inputs.

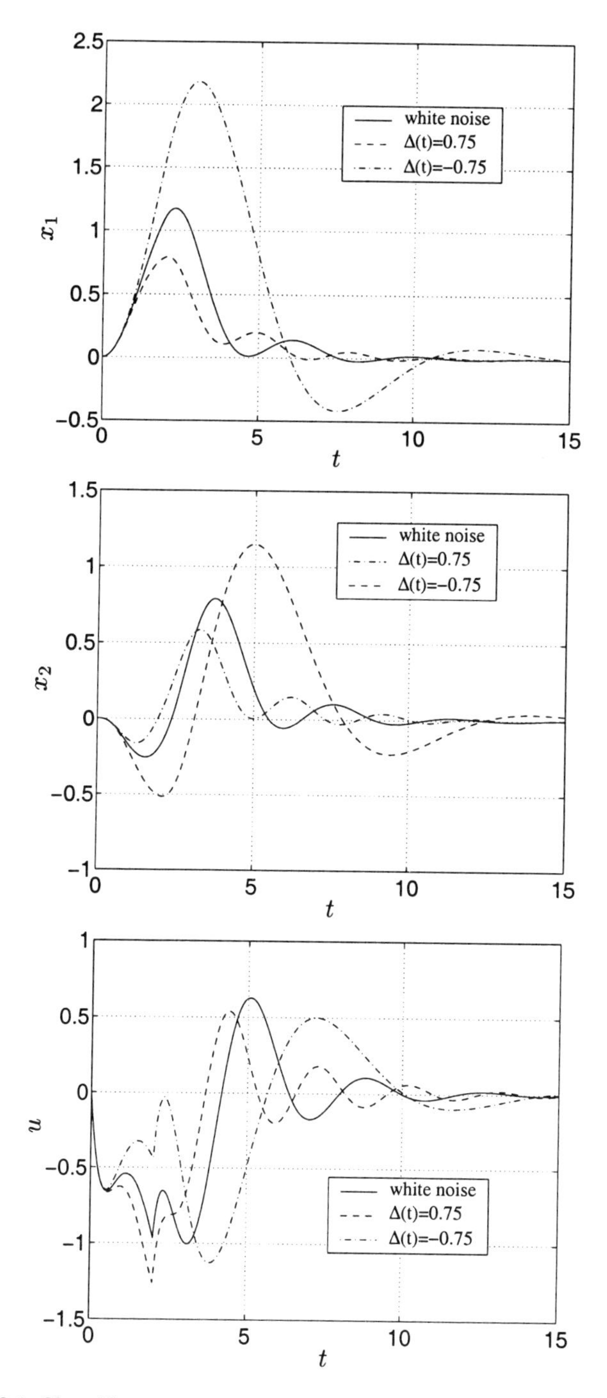

Figure 8.2.1. Closed loop system outputs versus time for three realizations of $\Delta(t)$.

The uncertainty in the above system (8.3.1) is described by an equation of the form

$$\xi(t) := \phi(t, x(\cdot), u(\cdot)). \tag{8.3.2}$$

This uncertainty is required to satisfy a stochastic integral quadratic constraint similar to that in Definition 2.4.1.

Definition 8.3.1 *Let $\bar{R} \geq 0$, $\bar{G} > 0$, $S > 0$ be given matrices. Then an uncertainty of the form (8.3.2) is said to be admissible if the following conditions are satisfied.*

(i) For any $s \geq 0$, if $u(\cdot) \in \mathbf{L}_2(s, T; R^{m_1})$ then there exists a unique solution to (8.3.1), (8.3.2) in $\mathbf{L}_2(s, T; R^n)$;

(ii) There exists a sequence $\{t_i\}_{i=1}^{\infty}$ such that $t_i > s$, $t_i \to \infty$ as $j \to \infty$ and the following condition holds. If $u(\cdot) \in \mathbf{L}_2([s, t_i]; R^{m_1})$ and $x(\cdot) \in \mathbf{L}_2([s, t_i]; R^n)$, then $\xi(\cdot) \in \mathbf{L}_2([s, t_i]; R^{m_2})$, and

$$\int_s^{t_i} \mathbf{E}\langle \xi(t), \bar{G}\xi(t)\rangle dt \leq \mathbf{E}\langle x(s), Sx(s)\rangle + \int_s^{t_i} \mathbf{E}\langle z(t), \bar{R}z(t)\rangle dt. \tag{8.3.3}$$

We use the notation $\Xi(\bar{R}, \bar{G}, S)$ to denote the set of admissible uncertainties. However, we will write Ξ wherever it produces no confusion.

Observation 8.3.1 Observe that the trivial uncertainty $\xi(\cdot) \equiv 0$ satisfies the above constraint. We refer to the system corresponding to this uncertainty as the *nominal* system.

An uncertain system involving structured uncertainty arises when the plant contains several uncertain feedback loops. This situation can be described by the decomposition of the uncertainty output vector z and the uncertainty input vector ξ into a number of subvectors as follows:

$$z = [z_1', \ldots, z_k']'; \quad \xi = [\xi_1', \ldots, \xi_k']'.$$

This in turn induces a corresponding block decomposition of the matrices C, D, B_2, and P_2 in equation (8.3.1); see equation (2.4.1) in Section 2.4.1. In this case, Definition 8.3.1 should then be modified in order to account for this uncertainty structure. Indeed, in this case, the corresponding uncertainty description requires that each uncertainty loop satisfies its own stochastic integral quadratic constraint of the form (8.3.3)

$$\int_s^{t_i} \mathbf{E}\langle \xi_j(t), \bar{G}_j\xi_j(t)\rangle dt \leq \mathbf{E}\langle x(s), S_j x(s)\rangle + \int_s^{t_i} \mathbf{E}\langle z_j(t), \bar{R}_j z_j(t)\rangle dt; \quad j = 1, \ldots, k \tag{8.3.4}$$

where $\bar{R}_j \geq 0$, $\bar{G}_j > 0$ and $S_j > 0$; e.g., see Definition 2.3.2. Then, for any constants $\tau_1 > 0, \ldots, \tau_k > 0$, we replace all of the above constraints by a single stochastic integral quadratic constraint of the form (8.3.3):

$$\int_s^{t_i} \mathbf{E}\langle \xi(t), \bar{G}_\tau \xi(t) \rangle dt \leq \mathbf{E}\langle x(s), S_\tau x(s) \rangle + \int_s^{t_i} \mathbf{E}\langle z(t), \bar{R}_\tau z(t) \rangle \quad (8.3.5)$$

where $\bar{R}_\tau$ and $\bar{G}_\tau$ are the block diagonal matrices

$$\bar{R}_\tau = \operatorname{diag}[\tau_1 \bar{R}_1, \ldots, \tau_k \bar{R}_k], \quad \bar{G}_\tau = \operatorname{diag}[\tau_1 \bar{G}_1, \ldots, \tau_k \bar{G}_k]$$

and $S_\tau = \sum_{j=1}^k \tau_j S_j$.

The main problem addressed in this subsection is to find a minimax optimal static linear state-feedback controller for the system (8.3.1). We now set up this problem.

Let $R \in \mathbf{R}^{n \times n}$, $G \in \mathbf{R}^{m_1 \times m_1}$, $R' = R > 0$, $G' = G > 0$, be given matrices. Associated with the uncertain system (8.3.1), (8.3.3), consider the cost functional

$$J^{s,h}(u(\cdot)) = \int_s^\infty \mathbf{E}\left(\langle x(t), Rx(t) \rangle + \langle u(t), Gu(t) \rangle\right) dt \quad (8.3.6)$$

where $x(t)$ is the solution to (8.3.1), (8.3.2) satisfying the initial condition $x(s) = h$.

Definition 8.3.2 *Given a constant $\gamma > 0$ and cost functional (8.3.6), the state feedback controller*

$$u^0 = Kx \quad (8.3.7)$$

is said to be a guaranteed cost controller *for the uncertain system (8.3.1), (8.3.3) with cost functional (8.3.6) and initial condition h, if it satisfies the following conditions:*

(i) This controller stabilizes the nominal system; i.e., *the resulting closed loop nominal system*

$$dx(t) = [A + B_1 K]x(t)dt + [H + P_1 K]x(t)dW_1(t), \quad x(s) = h, \quad (8.3.8)$$

is exponentially stable in mean square sense and hence there exist constants $c > 0$, $\alpha > 0$ such that

$$\mathbf{E}\|x(t)\|^2 \leq c\mathbf{E}\|h\|^2 e^{-\alpha(t-s)}.$$

(ii) For all $s > 0$, $h \in \mathbf{L}_2(\Omega, \mathcal{F}_s, P)$ and admissible uncertainty input (8.3.2), the corresponding solution to the closed loop uncertain system (8.3.1),

(8.3.3), (8.3.7)

$$\begin{aligned} dx(t) &= ([A + B_1K]x(t) + B_2\xi(t))\,dt + [H + P_1K]x(t)dW_1(t) \\ &\quad + P_2\xi(t)dW_2(t); \\ \begin{pmatrix} z \\ u \end{pmatrix} &= \begin{bmatrix} C + DK \\ K \end{bmatrix} x; \qquad x(s) = h \end{aligned} \tag{8.3.9}$$

is in $\mathbf{L}_2(s, \mathbf{R}^n)$. *Furthermore, this implies the corresponding control input and uncertainty input satisfy* $u(\cdot) \in \mathbf{L}_2(s, \mathbf{R}^{m_1})$ *and* $\xi(\cdot) \in \mathbf{L}_2(s, \mathbf{R}^{m_2})$.

(iii) *The corresponding value of the cost functional (8.3.6) is bounded by the constant* γ *for all admissible uncertainties:*

$$\sup_{\xi(\cdot)\in\Xi} J^{s,h}(u(\cdot)) \leq \gamma. \tag{8.3.10}$$

Note, that in the case of the structured uncertainty, the above definition remains essentially the same but with the replacement of the constraint (8.3.3) by the constraints (8.3.4).

Let $\mathcal{K}$ denote the set of guaranteed cost controllers of the form (8.3.7). Note that the constant γ in Definition 8.3.2 defines the required level of robust performance for the closed loop system. In the (non-optimal) guaranteed cost control problem, it is sufficient to construct any controller satisfying condition (8.3.10). In this section, we address an optimal guaranteed cost control problem in which we seek to find a controller guaranteeing the minimum upper bound on the worst-case performance of the closed loop uncertain system:

$$\inf_{u(\cdot)\in\mathcal{K}} \sup_{\xi\in\Xi} J^{s,h}(u(\cdot)) \tag{8.3.11}$$

Note, that in this minimax optimization problem, the admissible uncertainties $\xi(\cdot)$ are those which satisfy the constraint (8.3.3) (or the constraints (8.3.4)). Thus, the optimization problem (8.3.11) is a constrained minimax optimization problem. A controller solving the constrained minimax optimization problem (8.3.11) will be referred to as a minimax optimal guaranteed cost controller. The derivation of a solution to the problem (8.3.11) relies on absolute stabilization results presented in the next subsection.

8.3.2 *Stochastic absolute stabilization*

In this subsection, we address to the problem of absolute stabilization via a static state-feedback controller. We construct a stabilizing controller of the form (8.3.7) which leads to a closed loop uncertain system (8.3.1), (8.3.3), (8.3.7) which is absolutely stable in the following sense.

Definition 8.3.3 *A controller of the form (8.3.7) is said to absolutely stabilize the stochastic uncertain system (8.3.1), (8.3.3) if the following conditions hold:*

(i) The nominal closed loop system is stable. That is, for any initial condition random variable $x(s) = h \in \mathbf{L}_2(\Omega, \mathcal{F}_s, P)$, the system (8.2.2) is exponentially mean square stable.

(ii) There exists a constant $c > 0$, independent of the initial condition, such that for any admissible uncertainty $\xi(\cdot)$, the corresponding solution to the closed loop system (8.3.1), (8.3.2), (8.3.7) belongs to $\mathbf{L}_2(s, \mathbf{R}^n)$. Furthermore, the corresponding uncertainty input $\xi(\cdot)$ belongs to $\mathbf{L}_2(s, \mathbf{R}^{m_2})$ and

$$|||x(\cdot)|||^2 + |||\xi(\cdot)|||^2 \leq c\mathbf{E}\|h\|^2. \tag{8.3.12}$$

Given matrices $R \in \mathbf{R}^{n \times n}$, $G \in \mathbf{R}^{m_1 \times m_1}$, $R' = R > 0$ and $G' = G > 0$, consider the following generalized algebraic Riccati equation of the form (8.2.45):

$$\begin{aligned} &A'X + XA + H^*XH + \bar{C}'\bar{C} + X\bar{B}_2(I - \bar{P}_2^*X\bar{P}_2)^{-1}\bar{B}_2'X \\ &\quad -(XB_1 + H^*XP_1 + \bar{C}'\bar{D})(\bar{D}'\bar{D} + P_1^*XP_1)^{-1}(B_1'X + P_1^*XH + \bar{D}'\bar{C}) \\ &= 0 \end{aligned} \tag{8.3.13}$$

where

$$\begin{aligned} &\bar{C} = \begin{bmatrix} R^{1/2} \\ 0 \\ \bar{R}^{1/2}C \end{bmatrix}; \quad \bar{D} = \begin{bmatrix} 0 \\ G^{1/2} \\ \bar{R}^{1/2}D \end{bmatrix}; \\ &\bar{B}_2 = B_2\bar{G}^{-1/2}; \quad \bar{P}_2 = P_2\bar{G}^{-1/2}. \end{aligned} \tag{8.3.14}$$

Also, associated with the stochastic uncertain system (8.3.1), (8.3.3), is a cost functional of the form (8.3.6).

Theorem 8.3.1 *Suppose Assumption 8.2.2 is satisfied and there exists a minimal nonnegative-definite stabilizing solution X to the generalized Riccati equation (8.3.13);* i.e., *X is such that $I - \bar{P}_2^*X\bar{P}_2 > 0$ and the system*

$$\begin{aligned} dx &= \begin{bmatrix} A - B_1(\bar{D}'\bar{D} + P_1^*XP_1)^{-1}(B_1'X + P_1^*XH + \bar{D}'\bar{C}) \\ \bar{B}_2(I - \bar{P}_2^*X\bar{P}_2)^{-1}\bar{B}_2'X \end{bmatrix} xdt \\ &\quad + \left[H - P_1(\bar{D}'\bar{D} + P_1^*XP_1)^{-1}(B_1'X + P_1^*XH + \bar{D}'\bar{C})\right] xdW_1(t) \\ &\quad + \bar{B}_2(I - \bar{P}_2^*X\bar{P}_2)^{-1}\bar{B}_2'XxdW_2(t) \end{aligned} \tag{8.3.15}$$

is exponentially stable in mean square sense. In particular, the matrix

$$A - B_1(\bar{D}'\bar{D} + P_1^*XP_1)^{-1}(B_1'X + P_1^*XH + \bar{D}'\bar{C}) + \bar{B}_2(I - \bar{P}_2^*X\bar{P}_2)^{-1}\bar{B}_2'X$$

is stable. Then the controller

$$\begin{aligned} u^0(t) &= Kx(t); \\ K &= -(\bar{D}'\bar{D} + P_1^*XP_1)^{-1}(B_1'X + P_1^*XH + \bar{D}'\bar{C}) \end{aligned} \tag{8.3.16}$$

is an absolutely stabilizing controller for uncertain system (8.3.1), (8.3.3). Furthermore, the corresponding value of the cost functional (8.3.6) satisfies the bound

$$\sup_{\xi\in\Xi} J^{s,h}(u^0(\cdot)) \leq \mathbf{E}\langle h, (X+S)h\rangle. \tag{8.3.17}$$

Proof: Consider a stochastic differential game defined by the system

$$\begin{aligned} dx(t) &= [Ax(t) + B_1u(t) + \bar{B}_2w(t)]dt + (Hx(t) + P_1u(t))dW_1(t) \\ &\quad + \bar{P}_2w(t)dW_2(t) \end{aligned} \tag{8.3.18}$$

and the cost functional

$$\begin{aligned} \bar{\Im}^{s,h}(u(\cdot), w(\cdot)) &= J^{s,h}(u(\cdot)) + \mathbf{E}\int_s^\infty \left(\langle z(t), \bar{R}z(t)\rangle - \|w\|^2\right) dt \\ &= \mathbf{E}\int_s^\infty (\|\bar{C}x + \bar{D}u\|^2 - \|w\|^2)dt \end{aligned} \tag{8.3.19}$$

where the matrices $\bar{B}_2$, $\bar{P}_2$, $\bar{C}$ and $\bar{D}$ are defined as in (8.3.14). Under Assumption 8.2.2, the conditions of the sufficiency part of Theorem 8.2.2 are satisfied for the system (8.3.18) and cost functional (8.3.19). This leads to the conclusion that the controller (8.3.16) solves the stochastic H^∞ control problem associated with (8.3.18), (8.3.19). That is, the system (8.2.2) with K is defined by equation (8.3.16), is exponentially stable in the mean square sense. As a consequence, the matrix $A + B_1K$ is stable. Furthermore, the controller (8.3.16) satisfies the following stochastic H^∞-norm bound condition: There exists a constant $\varepsilon > 0$ such that

$$\mathbf{E}\int_0^\infty \left(\|(\bar{C} + \bar{D}K)x(t)\|^2 - \|w(t)\|^2\right) dt \leq -\varepsilon\mathbf{E}\int_0^\infty \|w(t)\|^2 dt \tag{8.3.20}$$

for $x(0) = 0$ and for all $w(\cdot) \in \mathbf{L}_2(0, \mathbf{R}^{m_2})$, where $x(\cdot)$ is the solution to the stochastic system

$$\begin{aligned} dx(t) &= ((A + B_1K)x(t) + \bar{B}_2w(t))dt + (H + P_1K)x(t)dW_1(t) \\ &\quad + \bar{P}_2w(t)dW_2(t), \end{aligned} \tag{8.3.21}$$

obtained by the substitution of (8.3.16) into equation (8.3.18). This conclusion shows that the system (8.3.21) and cost functional (8.3.19) satisfy the conditions of Lemma 8.2.1. Using this lemma, we obtain

$$\bar{\Im}^{s,h}(u^0(\cdot), w(\cdot)) \leq \mathbf{E}\langle h, X_Kh\rangle \quad \forall w(\cdot) \in \mathbf{L}_2(0, \mathbf{R}^{m_2}) \tag{8.3.22}$$

where the matrix X_K is the minimal solution to the following Riccati equation of the form (8.2.16):

$$\begin{aligned} &(A + B_1K)'X_K + X_K(A + B_1K) + (\bar{C} + \bar{D}K)'(\bar{C} + \bar{D}K) \\ &\quad + X_K\bar{B}_2(I - \bar{P}_2^*X_K\bar{P}_2)^{-1}\bar{B}_2'X_K + (H + P_1K)^*X_K(H + P_1K) = 0; \\ &I - \bar{P}_2^*X_K\bar{P}_2 > 0. \end{aligned} \tag{8.3.23}$$

Moreover, it is straightforward to verify that equation (8.3.13) can be transformed into the following equation

$$\begin{aligned}&(A+B_1K)'X+X(A+B_1K)+(H+P_1K)^*X(H+P_1K)\\&\quad+(\bar{C}+\bar{D}K)'(\bar{C}+\bar{D}K)+X\bar{B}_2(I-\bar{P}_2^*X\bar{P}_2)^{-1}\bar{B}_2'X=0;\\&I-\bar{P}_2^*X\bar{P}_2>0\end{aligned} \tag{8.3.24}$$

which is the same as equation (8.3.23). Thus, the minimal solutions to these equations are equal: $X = X_K$. Consequently, one can replace X_K in (8.3.22) by the solution to the algebraic Riccati equation (8.3.13) to obtain:

$$\Im^{s,h}(u^0(\cdot),w(\cdot)) \le \mathbf{E}\langle h, Xh\rangle. \tag{8.3.25}$$

The subsequent proof requires the use of Theorem 3 of reference [32]. This result, referred to as the stochastic counterpart to the Kalman-Yakubovich lemma, establishes the existence of a solution to a certain linear matrix inequality related to a sign-indefinite linear quadratic stochastic optimal control problem. We wish to apply this result to the following "optimal control problem"

$$\sup_{w(\cdot)} \Im^{s,h}(u^0, w(\cdot))$$

with underlying system (8.3.21) where the matrix K is given by (8.3.16). The conditions under which one can apply Theorem 3 in [32] are virtually the same as the conditions of Lemma 8.2.1.

Note that Theorem 3 in [32] requires the cost functional satisfy a certain coercivity condition; e.g., see[32, 116]. In our case, the fact that this condition is satisfied follows in a straightforward way from the stochastic H^∞ norm bound condition (8.3.20). Also note that Assumption 3 of reference [32] holds in our case, since we deal with finite dimensional equations and operators. Since we have verified the conditions of Lemma 8.2.1 for the system (8.3.21) and functional (8.3.19), then the application of Theorem 3 in [32] implies the existence of a symmetric matrix M and positive constant ϵ satisfying the inequality

$$\begin{aligned}&-2x'M[(A+B_1K)x+\bar{B}_2w]-x'(H+P_1K)^*M(H+P_1K)x\\&\quad-w'\bar{P}_2^*M\bar{P}_2w-\|(\bar{C}+\bar{D}K)x\|^2+\|w\|^2\\&\ge \epsilon(\|x\|^2+\|w\|^2)\end{aligned} \tag{8.3.26}$$

for all $x \in \mathbf{R}^n$ and $w \in \mathbf{R}^{m_2}$. This inequality is equivalent to a linear matrix inequality which can be re-written in the following standard form:

$$\left[\begin{array}{c|c} -\left(\begin{array}{l}(A+B_1K)'M+M(A+B_1K)\\+(H+P_1K)^*M(H+P_1K)\\+(\bar{C}+\bar{D}K)'(\bar{C}+\bar{D}K)\end{array}\right) & -M\bar{B}_2\\ \hline -\bar{B}_2'M & I-\bar{P}_2M\bar{P}_2\end{array}\right]>0.$$

Recall that the nominal closed loop system (8.2.2) corresponding to the system (8.3.21) driven by the uncertainty input $\xi(t) \equiv 0$ is exponentially stable in the mean square sense. Hence, $\mathbf{E}\|x(t)\|^2 \to 0$ as $t \to \infty$ in this case. Therefore, it follows from (8.3.26) that

$$\mathbf{E}\langle h, Mh\rangle \geq \epsilon_1 \mathbf{E}\int_s^\infty \|x(t)\|^2 dt$$

and hence $M > 0$. To establish this fact, we have used Ito's formula along the solution to the system (8.2.2) on the interval $[s, t]$, with initial condition $x(s) = h$, and then taken the limit $t \to, \infty$.

Now let $\xi(\cdot)$ be an uncertainty input, $\xi(\cdot) \in \Xi$, and define

$$w(t) = \begin{cases} \bar{G}^{1/2}\xi(t) = \bar{G}^{1/2}\phi(t, x(\cdot)|_s^t, Kx(\cdot)|_s^t) & \text{if } t \in [s, t_i]; \\ 0 & \text{if } t > t_i. \end{cases}$$

where t_i is defined as in Definition 8.3.1. Also, let $x(t)$ be the corresponding solution to (8.3.21) satisfying the initial condition $x(s) = h$. Then it follows from inequality (8.3.26) that

$$\begin{aligned} & -\mathbf{E}\langle x(t_i), Mx(t_i)\rangle + \mathbf{E}\langle x(s), Mx(s)\rangle \\ & + \mathbf{E}\int_s^{t_i} (-\|(\bar{C} + \bar{D}K)x(t)\|^2 + \|w(t)\|^2)dt \\ & \geq \epsilon \mathbf{E}\int_s^{t_i} (\|x(t)\|^2 + \langle \xi(t), \bar{G}\xi(t)\rangle)dt. \end{aligned} \tag{8.3.27}$$

Ito's formula is used to derive (8.3.27) from (8.3.26). Hence, using (8.3.3) and the nonnegativeness of R and G, we obtain

$$\begin{aligned} & -\mathbf{E}\langle x(t_i), Mx(t_i)\rangle + \mathbf{E}\langle x(s), Mx(s)\rangle + \mathbf{E}\langle x(s), Sx(s)\rangle \\ & \geq -\mathbf{E}\langle x(t_i), Mx(t_i)\rangle + \mathbf{E}\langle x(s), Mx(s)\rangle \\ & \quad -\mathbf{E}\int_s^{t_i} (\langle x(t), Rx(t)\rangle + \langle Kx(t), GKx(t)\rangle)dt \\ & \quad -\mathbf{E}\int_s^{t_i} \left(\langle z(t), \bar{R}z(t)\rangle + \langle \xi(t), \bar{G}\xi(t)\rangle\right) dt \\ & \geq \epsilon_1 \mathbf{E}\int_s^{t_i} (\|x(t)\|^2 + \langle \xi(t), \bar{G}\xi(t)\rangle)dt. \end{aligned} \tag{8.3.28}$$

Since we have established that $M > 0$ and $x(s) = h$, then (8.3.28) implies

$$\epsilon_1 \mathbf{E}\int_s^{t_i} (\|x(t)\|^2 + \langle \xi(t), \bar{G}\xi(t)\rangle)dt \leq \mathbf{E}\langle h, (M + S)h\rangle. \tag{8.3.29}$$

Thus, it follows that the expression on the left hand side of (8.3.29) is uniformly bounded with respect to t_i. Therefore, (8.3.29) implies $x(\cdot) \in \mathbf{L}_2(s, \mathbf{R}^n)$ for any admissible $\xi(\cdot)$. Also, using the fact that $\bar{G} > 0$, it follows that $\xi(\cdot) \in \mathbf{L}_2(s, \mathbf{R}^{m_2})$ for all admissible uncertainties $\xi(\cdot)$. Therefore, for any admissible uncertainty $\xi(\cdot)$, the input $w(\cdot) = \bar{G}^{1/2}\xi(\cdot)$ is an admissible disturbance input in the stochastic H^∞ control problem defined by the system (8.3.18) and cost functional (8.3.19). For this uncertainty input, it follows from (8.3.19) and (8.3.25) that

$$\begin{aligned} J^{s,h}(u^0) &\leq -\int_0^\infty \mathbf{E}\left(\langle z(t), \bar{R}z(t)\rangle - \langle \xi(t), \bar{G}\xi(t)\rangle\right) dt + \mathbf{E}\langle h, Xh\rangle \\ &\leq \mathbf{E}\langle h, (X+S)h\rangle. \end{aligned}$$

■

Remark 8.3.1 We have shown that the generalized algebraic Riccati equation (8.3.13) can be transformed into an algebraic Riccati equation of the form (8.2.16). Hence, as in the case of equation (8.2.16), a possible approach to solving the algebraic Riccati equation (8.3.13) is to apply homotopy methods; e.g., see Remark 8.2.3, Section 8.2.3 and also [166].

8.3.3 State-feedback minimax optimal control

In this subsection, we assume that the uncertainty in the system is structured and each uncertainty satisfies its own stochastic integral quadratic constraint of the form (8.3.4). Given constants $\tau_1 > 0, \ldots, \tau_k > 0$, we replace all of these constraints by a single scaled stochastic integral quadratic constraint (8.3.5). For the stochastic uncertain system (8.3.1) where the uncertainty (8.3.2) satisfies this structured uncertainty constraint, we solve the corresponding minimax optimal guaranteed cost control problem. The main result of this subsection, Theorem 8.3.2, shows that the problem of finding the minimax optimal guaranteed cost controller can be reduced to a finite dimensional optimization problem. In this subsection, we assume that the system (8.3.1) satisfies Assumption 8.2.2.

Let $\mathcal{T}$ denote the set of vectors $\tau \in \mathbf{R}^k_+$, such that the corresponding Riccati equation (8.3.13) has a nonnegative-definite stabilizing solution X_τ. That is

$$\mathcal{T} := \left\{ \tau \in \mathbf{R}^k \;:\; \begin{array}{l} \tau_1 > 0, \ldots, \tau_k > 0, \text{ and Riccati equation (8.3.13)} \\ \text{with } \bar{R} = \bar{R}_\tau, \bar{G} = \bar{G}_\tau, \text{ has a stabilizing solution} \\ X_\tau \geq 0, \text{ such that } \bar{G}_\tau - P_2^* X_\tau P_2 > 0. \end{array} \right\} \tag{8.3.30}$$

Theorem 8.3.2

(i) Given $\gamma > 0$ and initial condition h, there exists a guaranteed cost controller for the uncertain system (8.3.1), (8.3.3) if and only if the set $\mathcal{T}$ is non-empty.

(ii) Suppose the set $\mathcal{T}$ is non-empty. Then

$$\inf_{u(\cdot)\in\mathcal{K}} \sup_{\xi\in\Xi} J^{s,h}(u(\cdot)) = \inf_{\tau\in\mathcal{T}} \mathbf{E}\left[\langle h, (X_\tau + S_\tau)h\rangle\right]. \tag{8.3.31}$$

Furthermore, suppose $\tau^* \in \mathcal{T}$ attains the infimum on the right-hand side of equation (8.3.31). Then the corresponding controller $u^0_{\tau^*}(\cdot)$ defined by equation (8.3.16) with $X = X_{\tau^*}$, is the state-feedback minimax optimal guaranteed cost controller which minimizes the worst case of the cost functional (8.3.19) in the constrained stochastic optimization problem (8.3.11) subject to the stochastic integral quadratic constraint (8.3.4). This controller also absolutely stabilizes the stochastic uncertain system (8.3.1).

The proof of this theorem is similar to the proof of the main result of Section 5.3. As in Section 5.3, we use the S-procedure to reduce our constrained optimization problem to a problem without constraints. However, in contrast to the results presented in Section 5.3, the system (8.3.1) is non-autonomous due to the stochastic perturbations. This leads us to consider a collection of specially constructed shift operators in order to satisfy conditions of the S-procedure. This construction involves metrically transitive transformations of stochastic processes which were used in Section 8.2. The basic properties of these transformations are given in Appendix B.

As in Section 5.3, we begin with a result which shows that a stabilizing guaranteed cost controller for the uncertain system (8.3.1), (8.3.2) exists if and only if the set $\mathcal{T}$ defined in (8.3.30) is non-empty.

Lemma 8.3.1 *Given a positive constant γ and an initial condition $x(0) = h$, the following statements are equivalent:*

(i) There exists a guaranteed cost controller of the form (8.3.7) such that the closed loop uncertain system (8.3.1), (8.3.3), (8.3.7) with the cost functional (8.3.6) satisfies the guaranteed cost bound

$$\sup_{\xi \in \Xi} J^{s,h}(u(\cdot)) < \gamma. \tag{8.3.32}$$

(ii) There exists a $\tau \in \mathcal{T}$ such that

$$\mathbf{E}\langle h, (X_\tau + S_\tau)h\rangle < \gamma. \tag{8.3.33}$$

Proof: (i)$\Rightarrow$(ii). First observe that since (8.3.7) corresponds to a guaranteed cost controller, the nominal closed loop system (8.3.8) is exponentially stable. This implies that for any $\xi(\cdot) \in \mathbf{L}_2(s; \mathbf{R}^{m_2})$, $x(\cdot)$ the corresponding solution to (8.3.9) belongs to $\mathbf{L}_2(s; \mathbf{R}^n)$. As a consequence, the corresponding uncertainty output $z(\cdot)$ is in $\mathbf{L}_2(s; \mathbf{R}^p)$ and the corresponding control input $\hat{u}(\cdot)$ is in $\mathbf{L}_2(s; \mathbf{R}^{m_1})$. Here, $\hat{u}(\cdot)$ denotes the control input corresponding to the guaranteed cost controller (8.3.7).

Note, that (8.3.32) implies the existence of a constant $\varepsilon > 0$ such that

$$(1+\varepsilon)J^{s,h}(u_\phi(\cdot)) \le \gamma - \varepsilon \quad \forall \xi \in \Xi(\bar{R}_1, \bar{G}_1, S_1, \ldots, \bar{R}_k, \bar{G}_k, S_k), \tag{8.3.34}$$

where $u_\phi(\cdot)$ is the control input generated by the closed loop uncertain system (8.3.9), (8.3.2).

It follows from the above observation that the following quadratic functionals are well defined on $\mathbf{L}_2(s;\mathbf{R}^{m_2})$:

$$\begin{aligned} \mathcal{G}_0(\xi(\cdot)) &:= -(1+\varepsilon)J^{s,h}(\hat{u}(\cdot))+\gamma-\varepsilon; \\ \mathcal{G}_j(\xi(\cdot)) &:= \mathbf{E}\int_0^\infty \left(\langle z_j(t),\bar{R}_j z_j(t)\rangle - \langle \xi_j(t),\bar{G}_j\xi_j(t)\rangle\right) dt + \mathbf{E}\langle h, S_j h\rangle; \\ & \quad j=1,2,\ldots,k. \end{aligned} \tag{8.3.35}$$

Here, the functions $z_j(\cdot)$ correspond to the closed loop system (8.3.9) with uncertainty input $\xi(\cdot)\in\mathbf{L}_2(s;\mathbf{R}^{m_2})$. Furthermore, since (8.3.7) corresponds to a guaranteed cost controller, then for any admissible uncertainty input $\xi(\cdot)$, we have $\xi(\cdot)\in\mathbf{L}_2(s;\mathbf{R}^{m_2})$ and $z(\cdot)\in\mathbf{L}_2(s;\mathbf{R}^{p})$. Therefore, since $t_i\to\infty$ as $j\to\infty$, it follows from (8.3.4) that

$$\mathcal{G}_j(\xi(\cdot))\geq 0 \quad \forall\xi\in\Xi(\bar{R}_1,\bar{G}_1,S_1,\ldots,\bar{R}_k,\bar{G}_k,S_k) \qquad (j=1,\ldots,k). \tag{8.3.36}$$

We now verify that the quadratic functionals $\mathcal{G}_0,\mathcal{G}_1,\ldots,\mathcal{G}_k$ form an S-system in the terminology of Definition 4.3.1; *i.e.,* we show that there exists a sequence of bounded linear operators $\mathbf{T}_i:\mathbf{L}_2(s;\mathbf{R}^{m_2})\to\mathbf{L}_2(s;\mathbf{R}^{m_2})$, $i=1,2,\ldots$, such that this sequence converges weakly to zero in $\mathcal{L}(\mathbf{L}_2(s;\mathbf{R}^{m_2}))$ [1] and

$$J^{s,0}(u_i(\cdot))\to J^{s,0}(u(\cdot)); \quad \mathcal{G}_j(\xi_i(\cdot))\to\mathcal{G}_j(\xi(\cdot)) \quad \text{as } j\to\infty \tag{8.3.37}$$

for all $\xi(\cdot)\in\mathbf{L}_2(s;\mathbf{R}^{m_2})$ and $j=1,2,\ldots,k$. Here $u_i(\cdot)$ corresponds to the closed loop system (8.3.9) with uncertainty input $\mathbf{T}_i\xi(\cdot)$ and zero initial condition. We choose the operators $\mathbf{T}_i$ as follows: For each $\xi(\cdot)\in\mathbf{L}_2(s;\mathbf{R}^{m_2})$,

$$\xi_i(t,\omega)=(\mathbf{T}_i\xi)(t,\omega):=\begin{cases}0 & \text{if } s\leq t<t_i;\\ \xi(t-t_i,\Gamma_{t_i}\omega), & \text{if } t\geq t_i\end{cases} \tag{8.3.38}$$

where t_i is the sequence defined as in Definition 8.3.1, $t_i\to\infty$ as $j\to\infty$. Also, Γ_{t_i}, $i=1,2,\ldots$, are metrically transitive transformations from the translation semigroup generated by the Wiener process (W_1,W_2); see Appendix B for details. Given any two functions $\xi_1(\cdot),\xi_2(\cdot)\in\mathbf{L}_2(s;\mathbf{R}^{m_2})$, the Cauchy-Schwartz inequality gives

$$\begin{aligned} \left|\mathbf{E}\int_s^\infty \langle(\mathbf{T}_i\xi_1)(t),\xi_2(t)\rangle dt\right|^2 &= \left|\mathbf{E}\int_{t_i}^\infty \langle\xi_1(t-t_i,\Gamma_{t_i}\omega),\xi_2(t)\rangle dt\right|^2 \\ &\leq \mathbf{E}\int_s^\infty \|\xi_1(t)\|^2 dt\,\mathbf{E}\int_{t_i}^\infty \|\xi_2\|^2 dt. \end{aligned}$$

Here, we have used the fact that the transformations Γ_{t_i} preserve probability measures, and $\mathbf{E}f(x(\Gamma_{t_i}\omega))=\mathbf{E}f(x)$ for any Borel-measurable vector function f.

[1] $\mathcal{L}(Z)$ where Z is a Banach space, denotes the space of linear bounded operators $Z\to Z$.

However, the term $\mathbf{E}\int_{t_i}^{\infty}\|\xi_2\|^2 dt$ in the above inequality tends to 0 as $j \to \infty$. This implies $\mathbf{T}_i \to 0$ weakly.

Next, observe that Lemma B.1 in Appendix B (see also Lemma 2 of [27]) states that the solution to the closed loop system (8.3.9) with zero initial condition and uncertainty input $\mathbf{T}_i\xi$ is such that

$$
\begin{aligned}
x_i(t,\omega) &= \begin{cases} 0 & \text{if } s \le t < t_i; \\ x(t-t_i, \Gamma_{t_i}\omega) & \text{if } t \ge t_i; \end{cases} \\
u_i(t,\omega) &= \begin{cases} 0 & \text{if } s \le t < t_i; \\ u(t-t_i, \Gamma_{t_i}\omega) & \text{if } t \ge t_i; \end{cases} \\
z_i(t,\omega) &= \begin{cases} 0 & \text{if } s \le t < t_i; \\ z(t-t_i, \Gamma_{t_i}\omega) & \text{if } t \ge t_i; \end{cases}
\end{aligned}
\tag{8.3.39}
$$

Then, the substitution of (8.3.38) and (8.3.39) into (8.3.35) gives (8.3.37). Thus, we conclude that the above family of functionals forms an S-system. Furthermore, since $S_j > 0$, then for given non-zero initial condition $x(0) = h$, a zero uncertainty input $\xi(\cdot) \equiv 0$ gives $\mathcal{G}_j(0) > 0$ for all $j = 1, 2, \ldots, k$. Also, it follows from (8.3.34) and (8.3.36) that the condition $\mathcal{G}_1(\xi(\cdot)) \ge 0, \ldots, \mathcal{G}_k(\xi(\cdot)) \ge 0$ implies that $\mathcal{G}_0(\xi(\cdot)) \ge 0$.

We have now satisfied all of the conditions of the S-procedure Theorem 4.3.1 on page 112. Therefore, for a given non-zero initial condition $x(0) = h$, there exist constants $\tau_1 \ge 0, \ldots, \tau_k \ge 0$ such that

$$
\mathcal{G}_0(\xi) \ge \sum \tau_j \mathcal{G}_j(\xi) \quad \forall \xi(\cdot) \in \mathbf{L}_2(s; \mathbf{R}^{m_2}). \tag{8.3.40}
$$

We now use the constants τ_j to define a functional

$$
J_{\varepsilon,\tau}^{s,h}(\xi(\cdot)) := \varepsilon J^{s,h}(\hat{u}(\cdot)) + \bar{\Im}^{s,h}(\hat{u}(\cdot), \bar{G}_\tau^{1/2}\xi(\cdot)) \tag{8.3.41}
$$

where the coefficients of $\bar{\Im}^{s,h}$ in (8.3.19) are defined as in (8.3.14):

$$
\begin{aligned}
\bar{\Im}^{s,h}(u(\cdot), \bar{G}_\tau^{1/2}\xi(\cdot)) &= J^{s,h}(u(\cdot)) \\
&\quad + \mathbf{E}\int_s^{\infty} \left(\langle z(t), \bar{R}_\tau z(t) \rangle - \langle \xi(t), \bar{G}_\tau \xi(t) \rangle \right) dt.
\end{aligned}
$$

It follows from (8.3.40) that

$$
J_{\varepsilon,\tau}^{s,h}(\xi(\cdot)) \le \gamma - \varepsilon - \mathbf{E}\langle h, S_\tau, h \rangle \quad \forall \xi \in \mathbf{L}_2(s; \mathbf{R}^{m_2}). \tag{8.3.42}
$$

Using the same arguments as in Section 5.3 (see Propositions 2 and 3 in the proof of Lemma 5.3.1) it then follows that

$$
J_{\varepsilon,\tau}^{s,0}(\xi(\cdot)) \le 0 \quad \forall \xi \in \mathbf{L}_2(s; \mathbf{R}^{m_2}). \tag{8.3.43}
$$

Also $\tau_j > 0$ for all $j = 1, \ldots, k$.

We now show that if the constants $\tau_1, \ldots, \tau_k$ are chosen as in (8.3.40), then the generalized Riccati equation (8.3.13) with $\bar{R} = \bar{R}_\tau$ and $\bar{G} = \bar{G}_\tau$, has a nonnegative-definite stabilizing solution X_τ such that $\bar{G}_\tau - P_2^* X_\tau P_2 > 0$. In order to establish this fact, first observe that (8.3.43) implies

$$\bar{\Im}^{s,0}(\hat{u}, \bar{G}^{1/2}\xi(\cdot)) \leq -\varepsilon J^{s,0}(\hat{u}(\cdot)) \leq 0 \qquad \forall \xi \in \mathbf{L}_2(s; \mathbf{R}^{m_2}). \tag{8.3.44}$$

Therefore, since $R > 0$, $G > 0$ and $\bar{G}_\tau > 0$, it follows that there exists a constant $\alpha > 0$ such that

$$\begin{aligned}\mathbf{E}\int_s^\infty & \left(\|\bar{C}x(t) + \bar{D}\hat{u}(t)\|^2 - \|w(t)\|^2\right) dt \\ & < -\alpha \mathbf{E}\int_s^\infty \|w(t)\|^2 dt \quad \forall w(\cdot) \in \mathbf{L}_2(s; \mathbf{R}^{m_2})\end{aligned} \tag{8.3.45}$$

Here, $x(\cdot)$ and $\hat{u}(\cdot)$ are the solution and the control input defined by equation (8.3.9) in which $\xi(\cdot) = \bar{G}^{-1/2}w(\cdot)$, with initial condition $x(0) = h = 0$. Indeed, suppose (8.3.45) does not hold; *i.e.*, there exists a sequence $\{w^l(\cdot)\}_{l=1}^\infty \subset \mathbf{L}_2(s; \mathbf{R}^{m_2})$ such that

$$\lim_{l\to\infty} \frac{\mathbf{E}\int_s^\infty \|\bar{C}x^l(t) + \bar{D}\hat{u}^l(t)\|^2 dt}{|||w^l(t)|||^2} = 1. \tag{8.3.46}$$

Since (8.3.9) is linear with respect to $\xi(\cdot)$ and (8.3.41) is a quadratic functional with respect to $\xi(\cdot)$, it is clear that we can choose this sequence such that $|||w^l(\cdot)||| = 1$. Then, (8.3.44) and (8.3.46) together imply $x^l(\cdot) \to 0$ strongly since $R > 0$. Indeed, from (8.3.44) we have

$$\bar{\Im}^{s,0}(\hat{u}^l, w^l(\cdot)) \leq -\varepsilon_1 \mathbf{E}\int_0^\infty \|x^l(t)\|^2 dt \leq 0. \tag{8.3.47}$$

Also, from (8.3.46), the expression on the left hand side of (8.3.47) tends to zero as $l \to \infty$. Hence $x^l(\cdot) \to 0$ strongly. Therefore, we have $|||\bar{C}x^l(\cdot) + \bar{D}\hat{u}^l(\cdot)||| \to 0$ which contradicts (8.3.46) since $|||w^l(\cdot)||| = 1$. Thus (8.3.45) must hold.

Inequality (8.3.45) implies that the given guaranteed cost controller solves the stochastic H^∞ control problem corresponding to the system (8.3.1) and cost functional (8.3.19). Hence using Theorem 8.2.2, we conclude that there exists a nonnegative-definite symmetric stabilizing solution X_τ to Riccati equation (8.3.13). This solution is the minimal nonnegative-definite solution to (8.3.13) and

$$\bar{G}_\tau - P_2^* X_\tau P_2 = \bar{G}_\tau^{1/2}(I - \bar{P}_2^* X_\tau \bar{P}_2)\bar{G}_\tau^{1/2} > 0.$$

Thus, $\tau \in \mathcal{T}$. Moreover, inequality (8.2.48) established in Theorem 8.2.2 and condition (8.3.42) together imply

$$\begin{aligned}\mathbf{E}\langle h, X_\tau h\rangle &\leq \sup_{w(\cdot)} \bar{\Im}^{s,h}(\hat{u}(\cdot), w(\cdot)) \\ &\leq \sup_{\xi(\cdot)} J_{\varepsilon,\tau}^{s,h}(\xi(\cdot)) \\ &\leq \gamma - \varepsilon - \mathbf{E}\langle h, S_\tau h\rangle.\end{aligned} \tag{8.3.48}$$

Hence, condition (8.3.33) holds.

(ii)⇒(i). This part of proof follows immediately from Theorem 8.3.1.

■

Proof of Theorem 8.3.2: The theorem follows immediately from Lemma 8.3.1. In particular, note that the absolute stabilizing properties of the minimax optimal guaranteed cost controller were established in the second part of the proof of Lemma 8.3.1 where we referred to Theorem 8.3.1.

■

8.4 Output-feedback finite-horizon minimax optimal control of stochastic uncertain systems with additive noise

In this section, we consider stochastic uncertain systems with additive noise. As mentioned previously, our aim in this section and the subsequent section is to present a robust version of LQG control. To this end, we adopt the stochastic uncertain system framework introduced in Section 2.4.2.

Let $T > 0$ be a constant which will denote the finite time horizon. Also, let $(\Omega, \mathcal{F}, P)$ be a complete probability space defined as in Section 2.4.2. On this probability space, a p-dimensional standard Wiener process $W(\cdot)$ and a Gaussian random variable $x_0 \colon \Omega \to \mathbf{R}^n$ with mean $\check{x}_0$ and variance $Y_0 > 0$ are defined as described in Section 2.4.2. We suppose $p = r + l$ and recall that the first r entries of the vector process $W(\cdot)$ correspond to the system noise and the last l entries correspond to the measurement noise. As described in Section 2.4.2, the space Ω is endowed with a filtration $\{\mathcal{F}_t, t \geq 0\}$. This filtration has been completed by including all sets of P-probability zero.

8.4.1 *Definitions*

Stochastic nominal system

On the probability space defined in Section 2.4.2, we consider the system and measurement dynamics driven by a noise input $W(\cdot)$ and a control input $u(\cdot)$. These dynamics are described by the following stochastic differential equation of the form (2.4.6):

$$\begin{aligned} dx(t) &= (A(t)x(t) + B_1(t)u(t))dt + B_2(t)dW(t); \quad x(0) = x_0; \\ z(t) &= C_1(t)x(t) + D_1(t)u(t); \\ dy(t) &= C_2(t)x(t)dt + D_2(t)dW(t); \quad y(0) = y_0. \end{aligned} \tag{8.4.1}$$

In the above system, $x(t) \in \mathbf{R}^n$ is the state, $u(t) \in \mathbf{R}^m$ is the control input, $z(t) \in \mathbf{R}^q$ is the uncertainty output and $y(t) \in \mathbf{R}^l$ is the measured output. All coefficients in equations (8.4.1) are assumed to be sufficiently smooth matrix valued functions defined on $[0, T]$. Also, we assume that there exists a constant $d_0 > 0$ such that

$$\Gamma := D_2(t)D_2'(t) \geq d_0 I \quad \text{for all } t \geq 0.$$

Let $\{\mathcal{Y}_t, t \geq 0\}$ be the filtration generated by the observation process $y(\cdot)$:

$$\mathcal{Y}_t = \sigma\{y(s), s \leq t\}.$$

In the sequel, we will consider control inputs $u(\cdot)$ adapted to the filtration $\{\mathcal{Y}_t, t \geq 0\}$. That is, attention will be restricted to output-feedback controllers of the form (2.4.7):

$$u(t) = \mathcal{K}(t, y(\cdot)|_0^t). \tag{8.4.2}$$

Here $\mathcal{K}(\cdot)$ is a non-anticipative function such that for any $t \geq 0$, the function $\mathcal{K}(t, \cdot)\colon C([0,T], \mathbf{R}^l) \to \mathbf{R}^m$ is $\mathcal{Y}_t$-measurable. Also, we assume that the function $\mathcal{K}(t, y)$ is piecewise continuous in t and Lipschitz continuous in y.

It is assumed that the closed loop system (8.4.1), (8.4.2) has a unique strong solution. We will see that the controller solving the robust control problem considered in this section is a linear output-feedback controller. For this class of controllers, the above assumption will be automatically satisfied.

Stochastic uncertain systems

In this section, we adopt the uncertainty description given in Definition 2.4.2. This definition defines the stochastic uncertainty using a set $\mathcal{P}$ of probability measures consisting of all the probability measures Q such that $Q(\Lambda) = P(\Lambda)$ for $\Lambda \in \mathcal{F}_0$ and having finite relative entropy between the probability measure Q and the reference probability measure P. Given any two probability measures Q and P, the relative entropy $h(Q\|P)$ is defined by equation (A.1) on page 419. From the properties of relative entropy, it was observed in Section 2.4.2 that the set $\mathcal{P}$ is convex.

We now define the relative entropy constraint uncertainty description. This definition is a specialization of Definition 2.4.2 of Section 2.4.2. Let M be a given nonnegative-definite symmetric weighting matrix. Also, let $S \in \mathbf{R}^{n \times n}$ be a given positive-definite symmetric matrix.

Definition 8.4.1 *A probability measure $Q \in \mathcal{P}$ is said to define an admissible uncertainty if the following* stochastic uncertainty constraint *is satisfied:*

$$\frac{1}{2}\mathbf{E}^Q\left[x'(T)Mx(T) + \int_0^T \|z(t)\|^2 dt + x'(0)Sx(0)\right] - h(Q\|P) \geq 0. \tag{8.4.3}$$

In (8.4.3), $x(\cdot)$, $z(\cdot)$ are defined by equations (8.4.1).

We let Ξ denote the set of probability measures defining admissible uncertainties. Elements of Ξ are also called admissible probability measures.

Note that the set Ξ is non-empty. Indeed, $P \in \Xi$ since the conditions $M \geq 0$, $S > 0$ and the fact that $h(P\|P) = 0$ imply the constraint (8.4.3) is strictly satisfied in this case; e.g., see [53].

The minimax optimal control problem

Associated with the system (8.4.1), consider a cost functional J^Q of the form

$$J^Q(u(\cdot)) = \mathbf{E}^Q\left[\frac{1}{2}x'(T)M_T x(T) + \frac{1}{2}\int_0^T (x'(t)R(t)x(t) + u'(t)G(t)u(t))\,dt\right]. \tag{8.4.4}$$

In (8.4.4), M_T is a nonnegative-definite symmetric matrix of dimension $n \times n$. Also, $R(\cdot)$ and $G(\cdot)$ are positive-definite symmetric matrix valued functions such that $R(t) \in \mathbf{R}^{n\times n}$ and $G(t) \in \mathbf{R}^{m\times m}$.

We are concerned with an optimal control problem associated with the system (8.4.1), the cost functional (8.4.4) and the constraints (8.4.3). In this optimal control problem, we seek to find a control law $u(\cdot)$ of the form (8.4.2) minimizing the worst case value of the cost functional J^Q:

$$\inf_{u(\cdot)} \sup_{Q\in\Xi} J^Q(u(\cdot)). \tag{8.4.5}$$

Here the maximizing player input is an admissible probability measure $Q \in \Xi$ satisfying the constraint (8.4.3).

In Section 2.4.2, we observed that relative entropy uncertainty constraints such as the constraint (8.4.3) describe the admissible uncertainty inputs as those uncertainty inputs which are in a suitable relation to the corresponding system state and applied control input. The resulting set Ξ of admissible probability measures therefore depends on the controller $u(\cdot)$. This fact leads to the observation that the minimizing player imposes restrictions on the choice of strategy available to the maximizing player in the minimax optimal control problem (8.4.5). This observation reveals a significant difference between standard game type minimax optimal control problems and the minimax optimal control problem (8.4.5) related to worst case LQG controller design. To emphasize this difference, we refer to the minimax optimal control problem (8.4.5) as a *constrained* minimax optimal control problem. Further discussion of such constrained minimax optimal control problems is given in Section 2.4.2.

The derivation of a solution to the constrained minimax optimal control problem (8.4.5) proceeds in two steps. In Section 8.4.2, a connection between the constrained minimax optimal control problem (8.4.5) and a similar minimax optimal control problem without constraints will be established. This unconstrained minimax optimal control problem is defined in terms of the system (8.4.1) and the following augmented cost functional

$$J_\tau^Q(u(\cdot)) :=$$
$$J^Q(u(\cdot)) + \tau \mathbf{E}^Q\left[\frac{1}{2}x'(T)Mx(T) + \frac{1}{2}\int_0^T \|z(t)\|^2 dt + \frac{1}{2}x'(0)Sx(0)\right]$$
$$-\tau h(Q\|P) \tag{8.4.6}$$

where $\tau > 0$ is a given positive constant and $Q \in \mathcal{P}$. In this unconstrained minimax optimal control problem, we seek to minimize the quantity $\sup_{Q\in\mathcal{P}} J_\tau^Q(u(\cdot))$.

The second theoretical result which will be exploited in this section is the duality relationship between free energy and relative entropy established in [39]; see also Appendix A and [53, 152]. Associated with the system (8.4.1), we consider the following risk-sensitive cost functional of the form (3.3.10):

$$\Im_{\tau,T}(u(\cdot))$$
$$= 2\tau \log \mathbf{E}\left\{ \exp\left(\frac{1}{2\tau}\left[x'(T)M_\tau x(T) + \int_0^T F_\tau(x(t), u(t))dt\right]\right)\right\} \tag{8.4.7}$$

where

$$\begin{aligned} F_\tau(x,u) &:= x'R_\tau(t)x + 2x'\Upsilon_\tau(t)u + u'G_\tau(t)u; \quad (8.4.8)\\ R_\tau(t) &:= R(t) + \tau C_1'(t)C_1(t);\\ G_\tau(t) &:= G(t) + \tau D_1'(t)D_1(t);\\ \Upsilon_\tau(t) &:= \tau C_1'(t)D_1(t);\\ M_\tau &:= M_T + \tau M. \end{aligned} \tag{8.4.9}$$

By applying the duality result of Lemma A.1 to the system (8.4.1) and the cost functional (8.4.7), it follows that for each admissible control $u(\cdot)$,

$$\sup_{Q\in\mathcal{P}} J_\tau^Q(u(\cdot)) = \frac{1}{2}\left[\Im_{\tau,T}(u(\cdot)) + \tau \mathbf{E}x'(0)Sx(0)\right]; \tag{8.4.10}$$

see Corollary 3.1 and Remark 3.2 of [39]. Note that in deriving (8.4.10), we have used the fact that the probability measures Q and P coincide on $\mathcal{F}_0$-measurable sets, and hence $\mathbf{E}^Q x'(0)Sx(0) = \mathbf{E}x'(0)Sx(0)$. From the duality relationship (8.4.10), one can also conclude that the minimizing controller $u(\cdot)$ solving the above unconstrained stochastic game is equal to the controller solving the following risk-sensitive optimal control problem

$$\inf_{u(\cdot)} \Im_{\tau,T}(u(\cdot)). \tag{8.4.11}$$

In this risk-sensitive optimal control problem, an output-feedback controller of the form (8.4.2) is sought to minimize the cost functional (8.4.7). A mathematically rigorous definition of the class of admissible controllers in the optimal control problem (8.4.11) was given in Section 3.3.2. In Section 8.4.3, we will present a version of this definition which is adapted to the problem (8.4.11). Also in Section 3.3.2, a solution to a risk-sensitive control problem of the form (8.4.11) was presented; see Theorem 3.3.1. It is important to note that the optimal LEQG controller derived in Theorem 3.3.1 is expressed in the form of a linear state-feedback control law combined with a state estimator; see equations (3.3.19), (3.3.20). Hence, linking the result of Theorem 3.3.1 with those on the connection between a constrained minimax optimal control problem and an unconstrained stochastic game, and using the duality relation (8.4.10), we arrive at a solution to the original output-feedback minimax optimal control problem (8.4.5).

An equivalent form for the constrained minimax optimal control problem

To obtain an equivalent form for the constrained minimax optimal control problem (8.4.5), recall that in Section 2.4.2, we characterize each probability measure $Q \in \mathcal{P}$ in terms of a stochastic process $(\xi(t), \mathcal{F}_t)$, $t \in [0,T]$. This leads us to derive an equivalent representation for the stochastic uncertain system (8.4.1). Also, it should be noted that the corresponding equivalent form of the relative entropy constraint (8.4.3) is

$$\mathbf{E}^Q \int_0^T \|\xi(t)\|^2 dt \leq \mathbf{E}^Q \left[x'(0)Sx(0) + x'(T)Mx(T) + \int_0^T \|z(t)\|^2 dt \right]; \tag{8.4.12}$$

see also (2.4.20).

Note that the constrained minimax optimal control problem (8.4.5) can be rewritten with $\xi(\cdot)$ as the maximizer:

$$\inf_{u(\cdot)} \sup_{\xi(\cdot)} \mathbf{E}^Q \left[\frac{1}{2} x'(T)M_T x(T) + \frac{1}{2} \int_0^T (x'(t)R(t)x(t) + u'(t)G(t)u(t))\, dt \right]. \tag{8.4.13}$$

Here $x(\cdot)$ is the weak solution to equation (2.4.19) driven by the uncertainty input $\xi(\cdot)$ as described in Section 2.4.2.

8.4.2 *Finite-horizon minimax optimal control with stochastic uncertainty constraints*

In the previous subsection, we set up the constrained minimax optimal control problem (8.4.5). As in references [151, 153, 229, 235], we will show that this constrained optimal control problem can be converted into an unconstrained optimal control problem. This unconstrained optimal control problem is defined in terms of the system (8.4.1) and cost functional (8.4.6). In this unconstrained optimal control problem, we seek to minimize the quantity $\sup_{Q \in \mathcal{P}} J_T^Q(u(\cdot))$. Let V_τ

denote the upper value of the stochastic game:

$$V_\tau = \inf_{u(\cdot)} \sup_{Q\in\mathcal{P}} J_\tau^Q(u(\cdot)) \tag{8.4.14}$$

where the infimum is with respect to controllers $u(\cdot)$ of the form (8.4.2). Also, we define a set $\mathcal{T} \subset \mathbf{R}$ as

$$\mathcal{T} := \{\tau > 0 : V_\tau < \infty\}. \tag{8.4.15}$$

The following result establishes a relationship between the constrained and unconstrained optimal control problems defined above. This result requires that the system (8.4.1) satisfies the following assumption.

Assumption 8.4.1 *For any controller $u(\cdot)$ of the form (8.4.2),*

$$\sup_{Q\in\mathcal{P}} J^Q(u(\cdot)) = \infty. \tag{8.4.16}$$

In the sequel, we will show that this assumption amounts to a standard controllability assumption with respect to the input $W(\cdot)$ and an observability assumption with respect to the cost functional (8.4.4). As in [153], we need this assumption to rule out the case where $\tau = 0$.

Theorem 8.4.1 *If Assumption 8.4.1 is satisfied, then the following conclusions can be made.*

(i) The value (8.4.5) is finite if and only if the set $\mathcal{T}$ is non-empty.

(ii) If the set $\mathcal{T}$ is non-empty, then

$$\inf_{u(\cdot)} \sup_{Q\in\Xi} J^Q(u(\cdot)) = \inf_{\tau\in\mathcal{T}} V_\tau. \tag{8.4.17}$$

Furthermore, suppose $\tau_* \in \mathcal{T}$ attains the infimum on the right hand side of equation (8.4.17). Also, let $u^*(\cdot)$ be an optimal control input for the game (8.4.14) corresponding to $\tau = \tau_*$. That is, $u^*(\cdot)$ is a control input satisfying the equation

$$\sup_{Q\in\mathcal{P}} J_{\tau_*}^Q(u^*) = V_{\tau_*} = \inf_{\tau\in\mathcal{T}} V_\tau \tag{8.4.18}$$

Then, $u^*(\cdot)$ is a minimax optimal control solving the constrained stochastic optimal control problem (8.4.5).

To prove the above theorem, we need a result concerning the following *worst case performance problem*. Consider the following stochastic system on the probability space $(\Omega, \mathcal{F}, P)$, which is obtained by substituting the control input defined by equation (8.4.2), into equation (8.4.1):

$$\begin{aligned} dx(t) &= \mathfrak{A}(t, x(\cdot)|_0^t)dt + B_2 dW(t); \quad x(0) = x_0 \\ z &= \mathfrak{C}(t, x(\cdot)|_0^t). \end{aligned} \tag{8.4.19}$$

As in the case of the system (8.4.1), an equivalent representation of this system on a probability space $(\Omega, \mathcal{F}, Q)$, $Q \in \mathcal{P}$ can be obtained using the associated uncertainty input $\xi(\cdot)$. For the system (8.4.19) and the cost functional (8.4.4), consider the optimization problem

$$\sup_{Q \in \Xi} J^Q(\mathcal{K}(\cdot))$$

where $\mathcal{K}(\cdot)$ is the given controller of the form (8.4.2). In the above optimization problem, one seeks to evaluate the worst case performance for the given controller $\mathcal{K}$.

Let $\tilde{V}_\tau$ denote the optimum value in the corresponding unconstrained optimization problem

$$\tilde{V}_\tau := \sup_{Q \in \mathcal{P}} J_\tau^Q(\mathcal{K}(\cdot))$$

and define the set $\tilde{\mathcal{T}}$ as follows:

$$\tilde{\mathcal{T}} := \{\tau > 0 : \tilde{V}_\tau < \infty\}. \tag{8.4.20}$$

Lemma 8.4.1 *Suppose Assumption 8.4.1 is satisfied. Then the following conclusions hold.*

(i) The value $\sup_{Q \in \Xi} J^Q(\mathcal{K}(\cdot))$ *is finite if and only if the set* $\tilde{\mathcal{T}}$ *is non-empty.*

(ii) If $\tilde{\mathcal{T}} \neq \varnothing$, *then*

$$\sup_{Q \in \Xi} J^Q(\mathcal{K}(\cdot)) = \min_{\tau \in \tilde{\mathcal{T}}} \tilde{V}_\tau. \tag{8.4.21}$$

Proof: The proof of this lemma is similar to the proof of the corresponding result in [151, 229]. We first prove part (i). That is, we prove that the condition $\tilde{\mathcal{T}} \neq \varnothing$ is a necessary and sufficient condition for $\sup_{Q \in \Xi} J^Q(\mathcal{K}(\cdot)) < \infty$.

Sufficiency. Suppose the set $\tilde{\mathcal{T}}$ is non-empty and let $\tau \in \tilde{\mathcal{T}}$ be given. Then, the corresponding optimum value of the functional $J_\tau^Q(\mathcal{K}(\cdot))$ is finite; *i.e.*, $\tilde{V}_\tau < \infty$. Also, it follows from the definition of $\tilde{V}_\tau$, that

$$\begin{aligned}
&J^Q(\mathcal{K}(\cdot)) \\
&\quad + \tau \left\{ \mathbf{E}^Q \left[\frac{1}{2} x'(T) M x(T) + \frac{1}{2} \int_0^T \|z(t)\|^2 dt + \frac{1}{2} x'(0) S x(0) \right] - h(Q\|P) \right\} \\
&\leq \tilde{V}_\tau
\end{aligned}$$

for any $Q \in \Xi \subseteq \mathcal{P}$. From this and (8.4.3), it follows that

$$J^Q(\mathcal{K}(\cdot)) \leq \tilde{V}_\tau < \infty \quad \forall Q \in \Xi. \tag{8.4.22}$$

Necessity. To prove this part of the lemma, we use the S-procedure. Suppose that $\sup_{Q\in\Xi} J^Q(\mathcal{K}(\cdot))$ is finite. In order to establish that the set $\tilde{\mathcal{T}}$ is non-empty, we introduce the notation

$$c := \sup_{Q\in\Xi} J^Q(\mathcal{K}(\cdot)). \tag{8.4.23}$$

Also, a functional of the form (4.6.1) is defined as follows:

$$g_1(x(\cdot)|_0^T, x_0, \omega) := \tfrac{1}{2}x'(T)Mx(T) + \tfrac{1}{2}\int_0^T \|z(t)\|^2 dt + \tfrac{1}{2}x'(0)Sx(0)$$

where $x(\cdot)$ is the solution to (8.4.19) corresponding to the given initial condition $x(0) = x_0$ and $z(\cdot)$ is the corresponding uncertainty output.

We now define the corresponding functional $\mathcal{G}_1$ as in (4.6.2) and define the functional $\mathcal{G}_0$ as

$$\mathcal{G}_0(Q) := -J^Q(\mathcal{K}(\cdot)) + c. \tag{8.4.24}$$

Note that given any $Q \in \mathcal{P}$ such that $\mathcal{G}_1(Q) \geq 0$, it follows from (8.4.3) that $Q \in \Xi$. Hence, it follows from (8.4.24) and (8.4.23) that $\mathcal{G}_0(Q) \geq 0$. Thus, the condition of Theorem 4.6.1 is satisfied. Furthermore since $S > 0$, it follows that if we choose $Q = P$, then $\mathcal{G}_1(Q) > 0$. Hence, the second part of Theorem 4.6.1 implies that there exist a constant $\tau \geq 0$ such that for any $Q \in \mathcal{P}$

$$\begin{aligned}&-J^Q(\mathcal{K}(\cdot)) + c \\ &\geq \tau\mathbf{E}^Q\left[\frac{1}{2}x'(T)Mx(T) + \frac{1}{2}\int_0^T \|z(t)\|^2 dt + \frac{1}{2}x'(0)Sx(0)\right] - \tau h(Q\|P).\end{aligned}$$

Therefore,

$$\begin{aligned}&J^Q(\mathcal{K}(\cdot)) + \tau\left\{\frac{1}{2}\mathbf{E}^Q\left[x'(T)Mx(T) + \int_0^T \|z(t)\|^2 dt\right] - h(Q\|P)\right\} \\ &\leq c - \frac{1}{2}\tau\mathbf{E}^Q x'(0)Sx(0).\end{aligned} \tag{8.4.25}$$

Also, condition (8.4.25) implies

$$\sup_{Q\in\mathcal{P}} J_\tau^Q(\mathcal{K}(\cdot)) \leq c.$$

Thus,

$$\tilde{V}_\tau \leq c = \sup_{Q\in\Xi} J^Q(\mathcal{K}(\cdot)). \tag{8.4.26}$$

Moreover, it follows from Assumption 8.4.1 that

$$\sup_{Q\in\mathcal{P}} J^Q(\mathcal{K}(\cdot)) = \infty.$$

This fact ensures $\tau \neq 0$. Hence $\tau > 0$ and therefore the set $\tilde{\mathcal{T}}$ is non-empty. This completes the proof of condition (i).

To establish condition (ii) in the statement of the lemma, we simply note that if the set $\tilde{\mathcal{T}}$ is non-empty, then (8.4.22) and (8.4.26) imply (8.4.21).

■

Proof of Theorem 8.4.1: We first prove part (i) of the theorem.

Sufficiency. Suppose $\mathcal{T} \neq \varnothing$ and let $\tau \in \mathcal{T}$ be given. It follows from the definition of the set $\mathcal{T}$ that given any $\varepsilon > 0$, there exists a controller $\mathcal{K}_\varepsilon(\cdot)$ such that

$$\sup_{Q \in \mathcal{P}} J_\tau^Q(\mathcal{K}_\varepsilon(\cdot)) \leq V_\tau + \varepsilon.$$

Now observe that the closed loop system defined by the controller $\mathcal{K}_\varepsilon(\cdot)$ and the system (8.4.1) is a system of the form of (8.4.19). Hence, it follows from (8.4.20) that $\tau \in \tilde{\mathcal{T}}_{\mathcal{K}_\varepsilon}$, where $\tilde{\mathcal{T}}_{\mathcal{K}_\varepsilon}$ denotes the set $\tilde{\mathcal{T}}$ defined in Lemma 8.4.1 corresponding to this closed loop system. Moreover, Lemma 8.4.1 implies that

$$\sup_{Q \in \Xi_{\mathcal{K}_\varepsilon}} J^Q(\mathcal{K}_\varepsilon(\cdot)) = \min_{\tau \in \tilde{\mathcal{T}}_{\mathcal{K}_\varepsilon}} \left[\sup_{Q \in \mathcal{P}} J_\tau^Q(\mathcal{K}_\varepsilon(\cdot)) \right] \leq V_\tau + \varepsilon. \quad (8.4.27)$$

Hence, taking the infimum with respect to admissible controllers $\mathcal{K}(\cdot)$, it follows from (8.4.27) that

$$\inf_{\mathcal{K}} \sup_{Q \in \Xi} J^Q(\mathcal{K}(\cdot)) \leq V_\tau. \quad (8.4.28)$$

Also, it follows from the definition of the functional $J^Q(u)$, that

$$\inf_{\mathcal{K}} \sup_{Q \in \Xi} J^\xi(\mathcal{K}(\cdot)) \geq 0.$$

Thus, this value is finite.

Necessity. To prove necessity, we suppose that

$$\tilde{c} := \inf_{u(\cdot)} \sup_{Q \in \Xi} J^Q(u(\cdot)) < \infty.$$

Then, there exists a controller $\hat{u}(\cdot)$ of the form (8.4.2) and a positive constant $\tilde{\varepsilon}$ such that the corresponding closed loop system satisfies

$$\sup_{Q \in \Xi} J^Q(\hat{u}(\cdot)) < \tilde{c} + \tilde{\varepsilon}. \quad (8.4.29)$$

Now, as in the proof of Lemma 8.4.1, we apply the S-procedure Theorem 4.6.1. Indeed, Theorem 4.6.1 implies the existence of a constant $\tau \geq 0$ such that for any $Q \in \mathcal{P}$

$$\sup_{Q \in \mathcal{P}} J_\tau^Q(\hat{u}(\cdot)) \leq \tilde{c} + \tilde{\varepsilon}.$$

Hence,

$$V_\tau = \inf_{u(\cdot)} \sup_{Q \in \mathcal{P}} J_\tau^Q(u(\cdot)) \leq \tilde{c} + \tilde{\varepsilon} < \infty. \tag{8.4.30}$$

Therefore, $V_\tau < \infty$. Also, using Assumption 8.4.1 and a proof similar to that of Lemma 8.4.1, it follows that $\tau > 0$. Thus, $\tau \in \mathcal{T}$. This completes the proof of the first part of the theorem.

We now prove part (ii) of the theorem. Equation (8.4.17) will be established as in the proof of the corresponding result in [151, 229]. In order to achieve this, we first note that for a given controller $\tilde{u}(\cdot)$ of the form (8.4.2), Lemma 8.4.1 implies

$$\sup_{Q \in \Xi} J^Q(\tilde{u}(\cdot)) = \inf_{\tau \in \tilde{\mathcal{T}}_{\tilde{u}}} \left(\sup_{Q \in \mathcal{P}} J_\tau^Q(\tilde{u}(\cdot)) \right) \tag{8.4.31}$$

for the closed loop system (8.4.19). Note that on the left hand side of this equation, the infimum with respect to $Q \in \Xi$ is taken over the set Ξ corresponding to the controller $\tilde{u}(\cdot)$; see Remark 2.4.4 on page 41.

We now choose $\tilde{u}(\cdot) = \hat{u}(\cdot)$ where $\hat{u}(\cdot)$ is the controller defined by condition (8.4.29). Substituting (8.4.31), with $\tilde{u}(\cdot) = \hat{u}(\cdot)$, into inequality (8.4.29) and using the definition of the constant $\tilde{c}$, it follows that

$$\inf_{u(\cdot)} \sup_{Q \in \Xi} J^Q(u(\cdot)) + \tilde{\varepsilon} \geq \inf_{\tau \in \tilde{\mathcal{T}}_{\hat{u}}} \left(\sup_{Q \in \mathcal{P}} J_\tau^Q(\hat{u}(\cdot)) \right).$$

The infimum with respect to $u(\cdot)$ on the left hand side of this inequality is over the set of admissible controllers. Hence, taking the infimum with respect to $u(\cdot)$, it follows that

$$\inf_{u(\cdot)} \sup_{Q \in \Xi} J^Q(u(\cdot)) \geq \inf_{u(\cdot)} \inf_{\tau \in \tilde{\mathcal{T}}_u} \left(\sup_{Q \in \mathcal{P}} J_\tau^Q(u(\cdot)) \right).$$

Furthermore, using the definitions of $\tilde{\mathcal{T}}_u$ and $\mathcal{T}$, it follows from the sufficiency part of (i) that $\tilde{\mathcal{T}}_u \subset \mathcal{T}$. Therefore,

$$\begin{aligned} \inf_{u(\cdot)} \sup_{Q \in \Xi} J^Q(u(\cdot)) &\geq \inf_{u(\cdot)} \inf_{\tau \in \mathcal{T}} \left(\sup_{Q \in \mathcal{P}} J_\tau^Q(u(\cdot)) \right) \\ &= \inf_{\tau \in \mathcal{T}} \inf_{u(\cdot)} \left(\sup_{Q \in \mathcal{P}} J_\tau^Q(u(\cdot)) \right) \\ &= \inf_{\tau \in \mathcal{T}} V_\tau. \end{aligned}$$

Also, it follows from (8.4.28) that

$$\inf_{u(\cdot)} \sup_{Q \in \Xi} J^Q(u(\cdot)) \leq \inf_{\tau \in \mathcal{T}} V_\tau.$$

From this, equation (8.4.17) follows.

Now let $\tau_* \in \mathcal{T}$ be such that condition (8.4.18) is satisfied. Also, let Ξ_* denote the class of admissible probability measures satisfying the uncertainty constraint (8.4.3) corresponding to the control input $u^*(\cdot)$ defined by equation (8.4.18). Hence,

$$\sup_{Q \in \Xi_*} J^Q(u^*) \geq \inf_{u(\cdot)} \sup_{Q \in \Xi} J^Q(u).$$

Moreover, the substitution of the control input $u^*(\cdot)$ into the functional (8.4.6) implies

$$J^Q(u^*) \leq J^Q_{\tau_*}(u^*) = \inf_{\tau \in \mathcal{T}} V_\tau = \inf_{u(\cdot)} \sup_{Q \in \Xi} J^Q(u)$$

for all $Q \in \Xi_*$. Hence, $\sup_{Q \in \Xi_*} J^Q(u^*) \leq \inf_{u(\cdot)} \sup_{Q \in \Xi} J^Q(u)$. This completes the proof of the theorem.

■

Remark 8.4.1 Note that in general, the optimization problem which occurs on the right hand side of (8.4.17) may be non-convex. However, since this optimization problem involves only a single parameter, it can be solved using standard line search methods.

8.4.3 Design of a finite-horizon minimax optimal controller

Using Theorem 8.4.1, we will now convert the original constrained minimax optimal control problem into a risk-sensitive optimal control problem. This will be achieved by using the duality relationship between free energy and relative entropy established in [39]; see Lemma A.1 and also [53, 152, 153].

Associated with the system (8.4.1), consider a cost functional (8.4.7). As mentioned above, the result of reference [39] applied to the system (8.4.1) and cost functional (8.4.7) (see Corollary 3.1 and Remark 3.2 of [39]), implies that

$$V_\tau = \frac{1}{2}\left[\inf_{u(\cdot)} \Im_{\tau,T}(u(\cdot)) + \tau \mathbf{E} x'(0) S x(0)\right] \tag{8.4.32}$$

Furthermore, the minimizing controller $u(\cdot)$ for the stochastic game (8.4.14) is the same as the output-feedback controller solving the stochastic risk-sensitive optimal control problem (8.4.11). Hence, linking the results of Section 3.3.2 with those of Theorem 8.4.1 and the duality relation (8.4.32), we arrive at a solution to the original minimax optimal control problem (8.4.5). This solution is given in terms of a pair of the Riccati differential equations. In order to apply the results of Section 3.3.2, we require the following assumption to be satisfied. This assumption adapts Assumption 3.3.1 to the case being considered in this section. We first note that one condition of Assumption 3.3.1 is automatically satisfied in the case considered in this section. Indeed, using the notation

$$\tilde{C}_1 := \begin{bmatrix} \frac{1}{\sqrt{\tau}}R^{1/2} \\ 0 \\ C_1 \end{bmatrix}; \quad \tilde{D}_1 := \begin{bmatrix} 0 \\ \frac{1}{\sqrt{\tau}}G^{1/2} \\ D_1 \end{bmatrix}; \tag{8.4.33}$$

we obtain

$$R_\tau - \Upsilon_\tau G_\tau^{-1}\Upsilon_\tau' = \tau\tilde{C}_1'(I - \tilde{D}_1(\tilde{D}_1'\tilde{D}_1)^{-1}\tilde{D}_1')\tilde{C}_1.$$

This implies that $R_\tau - \Upsilon_\tau G_\tau^{-1}\Upsilon_\tau' \geq 0$ for all $t \in [0,T]$.

Assumption 8.4.2 *There exists a constant $\tau > 0$ such that the following conditions hold:*

(i) The Riccati differential equation

$$\begin{aligned} \dot{Y} &= (A - B_2 D_2'\Gamma^{-1}C_2)Y + Y(A - B_2 D_2'\Gamma^{-1}C_2)' \\ &\quad -Y(C_2'\Gamma^{-1}C_2 - \frac{1}{\tau}R_\tau)Y + B_2(I - D_2'\Gamma^{-1}D_2)B_2'; \\ Y(0) &= Y_0 \end{aligned} \tag{8.4.34}$$

has a symmetric solution $Y : [0,T] \to \mathbf{R}^{n\times n}$ such that $Y(t) \geq c_0 I$ for some $c_0 > 0$ and for all $t \in [0,T]$.

(ii) The Riccati differential equation

$$\begin{aligned} \dot{X} &+ X(A - B_1 G_\tau^{-1}\Upsilon_\tau') + (A - B_1 G_\tau^{-1}\Upsilon_\tau')'X \\ &+(R_\tau - \Upsilon_\tau G_\tau^{-1}\Upsilon_\tau') - X(B_1 G_\tau^{-1}B_1' - \frac{1}{\tau}B_2 B_2')X = 0; \\ X(T) &= M_\tau \end{aligned} \tag{8.4.35}$$

has a symmetric nonnegative definite solution $X : [0,T] \to \mathbf{R}^{n\times n}$.

(iii) For each $t \in [0,T]$,

$$\rho(Y(t)X(t)) < \tau \qquad \forall t \in [0,T], \tag{8.4.36}$$

where $\rho(\cdot)$ denotes the spectral radius of a matrix.

The class of admissible controllers now can be defined as in Section 3.3.2. For any constant τ satisfying Assumption 8.4.2, consider estimator state equations of the form (3.3.16) and (3.3.17) where the matrices R_τ, Υ_τ, G_τ and M_τ replace the matrices R, Υ, G and M, respectively. The class of admissible controllers considered in Section 3.3.2 are controllers of the form (8.4.2) with a function $\mathcal{K}(t,y)$ which is piecewise continuous in t and Lipschitz continuous in y. Also, these controllers are required to satisfy the causality condition defined in Section 3.3.2.

As observed in [139], the class of admissible controllers includes the class of linear controllers. Thus, if the controller which attains the infimum of the functional (8.4.7) in the risk-sensitive optimal control problem (8.4.11) is linear, then this controller will necessarily be admissible.

As in Section 3.3.2, the solutions to Riccati differential equations (8.4.34), (8.4.35) will define the optimal controller in the risk-sensitive optimal control problem. Indeed, for any constant τ satisfying Assumption 8.4.2, consider the controller state equations

$$\begin{aligned} d\check{x}(t) = & \left[A + \frac{1}{\tau} Y R_\tau - (Y C_2' + B_2 D_2')\Gamma^{-1} C_2\right] \check{x} dt \\ & -(B_1 + \frac{1}{\tau} Y \Upsilon_\tau) G_\tau^{-1} (B_1' X + \Upsilon_\tau')(I - \frac{1}{\tau} Y X)^{-1} \check{x} dt \\ & +(Y C_2' + B_2 D_2')\Gamma^{-1} dy(t); \quad \check{x}(0) = \check{x}_0 \end{aligned} \tag{8.4.37}$$

obtained from (3.3.16) with the substitution of the feedback control law

$$u^*(t) = -G_\tau^{-1}(t)(B_1'(t)X(t) + \Upsilon_\tau'(t))[I - \frac{1}{\tau} Y(t)X(t)]^{-1} \check{x}(t). \tag{8.4.38}$$

The following lemma is a straightforward consequence of the output-feedback risk-sensitive optimal control result of Theorem 3.3.1 and the duality result of Lemma A.1.

Lemma 8.4.2 *Suppose Assumption 8.4.2 is satisfied. Then the set $\mathcal{T}$ defined in (8.4.15) is non-empty. Furthermore, for any τ satisfying Assumption 8.4.2,*

$$\begin{aligned} V_\tau = & \\ & \frac{1}{2}\left[\check{x}_0' X(0)(I - \frac{1}{\tau} Y_0 X(0))^{-1} \check{x}_0 + \tau \mathbf{E}\{x'(0) S x(0)\}\right] \\ & -\frac{1}{2}\tau \log |I - \frac{1}{\tau} M_\tau Y(T)| \\ & +\frac{1}{2}\int_0^T \mathrm{tr} \begin{bmatrix} Y(t) R_\tau + (Y(t) C_2'(t) + B_2(t) D_2'(t))\Gamma^{-1}(t) \\ \times (C_2(t) Y(t) + D_2(t) B_2'(t)) X(t)(I - \frac{1}{\tau} Y(t) X(t))^{-1} \end{bmatrix} dt. \end{aligned} \tag{8.4.39}$$

Proof: Theorem 3.3.1 (see also Theorem 2 of reference [139]) implies that the controller (8.4.38) is an optimal controller for the risk-sensitive optimal control problem (8.4.11) associated with the system (8.4.1) and the cost functional (8.4.7). Also,

$$\begin{aligned} & \check{x}_0' X(0)(I - \frac{1}{\tau} Y_0 X(0))^{-1} \check{x}_0 - \tau \log |I - \frac{1}{\tau} M_\tau Y(T)| \\ & + \int_0^T \mathrm{tr} \begin{bmatrix} Y(t) R_\tau + (Y(t) C_2'(t) + B_2(t) D_2'(t))\Gamma^{-1}(t) \\ \times (C_2(t) Y(t) + D_2(t) B_2'(t)) X(t)(I - \frac{1}{\tau} Y(t) X(t))^{-1} \end{bmatrix} dt \end{aligned} \tag{8.4.40}$$

is the optimal value of the cost functional (8.4.7). It follows from the duality formula (8.4.10) that the controller (8.4.38) is the optimal minimizing strategy in the stochastic game (8.4.14) associated with the cost functional (8.4.6) and the system (8.4.1). Hence, if a constant τ satisfies Assumption 8.4.2, then it also belongs to $\mathcal{T}$. Equation (8.4.39) follows from (8.4.32) and (8.4.40).

■

Using Lemma 8.4.2, the main result of this section now follows.

Theorem 8.4.2

(i) Suppose there exists a constant $\tau > 0$ such that Assumption 8.4.2 is satisfied. Then the set $\mathcal{T}$ is non-empty. Furthermore, suppose $\tau_* \in \mathcal{T}$ attains the infimum on the right-hand side of equation (8.4.39). Then, the corresponding control input $u^*(\cdot)$ defined by (8.4.38) is the output-feedback controller minimizing the worst case of the cost functional (8.4.4) in the constrained stochastic minimax optimization problem (8.4.5) subject to the stochastic integral quadratic constraint (8.4.3).

(ii) Conversely, suppose that Assumption 8.4.1 is satisfied. Also suppose that for all $\tau > 0$, either equation (8.4.34) or equation (8.4.35) has a conjugate point[2] in $[0, T]$, or condition (8.4.36) fails to hold. Then

$$\sup_{Q\in\Xi} J(u(\cdot)) = \infty \tag{8.4.41}$$

for any controller $u(\cdot)$ defined by state equations of the form

$$\begin{aligned} d\check{x}(t) &= A_c(t)\check{x}(t)dt + B_c(t)dy(t), \qquad \check{x}(0) = \check{x}_0 \\ u(t) &= K(t)\check{x}(t), \end{aligned} \tag{8.4.42}$$

where $A_c(\cdot)$, $B_c(\cdot)$ and $K(\cdot)$ are matrix valued functions of appropriate dimension.

Proof: Part (i) of the theorem is a straightforward consequence of Lemma 8.4.2 and Theorem 8.4.1.

To prove part (ii) of the theorem, a corresponding result from finite-horizon deterministic H^∞ control theory is used. Indeed, the assumption of this part of the theorem is the same as the condition required in the third part of Theorem 5.6 of [9]. It follows from this result that for any controller of the form $u = \mathcal{K}(t, y(\cdot)|_0^t)$ generated by the state equations (8.4.42), then

$$\sup_{\bar{\xi}(\cdot)\in\mathbf{L}_2[0,T]} \frac{1}{2}\left[\begin{array}{l} \bar{x}'(T,\omega)M_\tau\bar{x}(T,\omega) \\ +\int_0^T \left(F_\tau(\bar{x}(t,\omega),\bar{u}(t,\omega)) - \tau\|\bar{\xi}(t)\|^2\right)dt \end{array}\right] = \infty$$

[2]That is, one of these Riccati equations has a finite escape time; e.g., see [9].

for all $\omega \in \Omega$, where $\bar{x}(\cdot,\omega)$ is the solution to the state equations

$$\begin{aligned}
&\dot{\bar{x}}(t,\omega) = A(t)\bar{x}(t,\omega) + B_1(t)\bar{u}(t,\omega) + B_2\bar{\xi}(t);\\
&\bar{x}(0,\omega) = x_0(\omega);\\
&\dot{\bar{y}}(t,\omega) = C_2(t)\bar{x}(t,\omega) + D_2\bar{\xi}(t);\\
&\bar{y} = 0
\end{aligned} \tag{8.4.43}$$

and

$$\bar{u}(t,\omega) = \mathcal{K}(t, \bar{y}(\cdot,\omega)|_0^t). \tag{8.4.44}$$

Hence,

$$\sup_{\bar{\xi}(\cdot)\in\mathbf{L}_2[0,T]} \frac{1}{2}\mathbf{E}\left[\bar{x}'(T)M_\tau\bar{x}(T) + \int_0^T F_\tau(\bar{x}(\cdot),\bar{u}(\cdot)) - \tau\|\bar{\xi}(t)\|^2 dt\right] = \infty.$$

That is, for any $\epsilon > 0$, there exists a $\bar{\xi}_\epsilon(\cdot)$ such that $\bar{\xi}_\epsilon(\cdot) \in \mathbf{L}_2[0,T]$ a.s. and

$$\begin{aligned}
&\frac{1}{2}\mathbf{E}\left[\bar{x}_\epsilon'(T)M_\tau\bar{x}_\epsilon(T) + \int_0^T \left(F_\tau(\bar{x}_\epsilon(\cdot),\bar{u}_\epsilon(\cdot)) - \tau\|\bar{\xi}_\epsilon(t)\|^2\right) dt + \tau x_0' S x_0\right]\\
&> \frac{1}{\epsilon}
\end{aligned} \tag{8.4.45}$$

where $\bar{x}_\epsilon(\cdot,\omega)$ is the solution to the system (8.4.43) corresponding to uncertainty inputs $\bar{\xi}_\epsilon(\cdot)$ and $\bar{\delta}_\epsilon(\cdot)$. Also, $\bar{u}_\epsilon(\cdot,\omega)$ is the control input corresponding to the output-feedback controller (8.4.44).

Now consider a probability measure $Q_\epsilon \in \mathcal{P}$ corresponding to the *deterministic* uncertainty input $\bar{\xi}_\epsilon(\cdot)$. The construction of such a probability measure is as discussed in Section 2.4.2. Also, recall that in the probability space $(\Omega, \mathcal{F}, Q_\epsilon)$, the dynamics of the original system (8.4.1) are described by the following stochastic differential equations of the form (2.4.19):

$$\begin{aligned}
dx(t) &= (A(t)x(t) + B_1(t)\mathcal{K}(t, y(\cdot)|_0^t) + B_2(t)\bar{\xi}_\epsilon(t))dt\\
&\qquad + B_2(t)d\tilde{W}(t); \quad x(0) = x_0;\\
z(t) &= C_1(t)x(t) + D_1(t)\mathcal{K}(t, y(\cdot)|_0^t);\\
dy(t) &= (C_2(t)x(t) + D_2(t)\bar{\xi}_\epsilon(t))dt + D_2(t)d\tilde{W}(t); \quad y(0) = 0;
\end{aligned} \tag{8.4.46}$$

where $\tilde{W}(\cdot)$ is a Wiener process. Hence, using the fact that the operator generated by equation (8.4.42) is a linear operator, for any $t > 0$ we have

$$\begin{aligned}
x(t) &= \bar{x}_\epsilon(t) + \tilde{x}(t);\\
y(t) &= \bar{y}_\epsilon(t) + \tilde{y}(t);\\
u(t) &= \bar{u}_\epsilon(t) + \tilde{u}(t)
\end{aligned}$$

where

$$
\begin{aligned}
d\tilde{x}(t) &= (A(t)\tilde{x}(t) + B_1(t)\mathcal{K}(t,\tilde{y}(\cdot)|_0^t)dt \\
&\qquad\qquad + B_2(t)d\tilde{W}(t); \quad \tilde{x}(0) = 0; \\
\tilde{z}(t) &= C_1(t)\tilde{x}(t) + D_1\mathcal{K}(t,\tilde{y}(\cdot)|_0^t); \\
d\tilde{y}(t) &= C_2(t)\tilde{x}(t)dt + D_2(t)d\tilde{W}(t); \quad \tilde{y}(0) = 0
\end{aligned} \tag{8.4.47}
$$

and $\tilde{u}(t) = \mathcal{K}(t,\tilde{y}(\cdot)|_0^t)$. Note that since the initial condition in equation (8.4.47) is $\tilde{x}(0) = 0$, then $\check{x}_0 = \mathbf{E}\tilde{x}(0) = 0$ in the state estimator (8.4.42). This state estimator determines the control input $\tilde{u}(\cdot)$.

Also, note that x_0 and $\tilde{W}(\cdot)$ are independent. Hence, equations (8.4.43) and (8.4.47) imply that $\bar{x}_\epsilon(\cdot)$ and $\tilde{x}(\cdot)$ are independent. Consequently $\bar{y}_\epsilon(\cdot)$ and $\tilde{y}(\cdot)$ are also independent. Furthermore, it follows from the above observation concerning the choice of the initial condition in the definition of $\tilde{u}(t)$ that the processes $\bar{u}_\epsilon(\cdot)$ and $\tilde{u}(\cdot)$ are independent. This implies the following identities:

$$
\begin{aligned}
\mathbf{E}^{Q_\epsilon}x'(T)M_\tau x(T) &= \mathbf{E}^{Q_\epsilon}\bar{x}_\epsilon'(T)M_\tau\bar{x}_\epsilon(T) + \mathbf{E}^{Q_\epsilon}\tilde{x}'(T)M_\tau\tilde{x}(T); \\
\mathbf{E}^{Q_\epsilon}F_\tau(x(t),u(t)) &= \mathbf{E}^{Q_\epsilon}F_\tau(\bar{x}_\epsilon(t),\bar{u}_\epsilon(t)) + \mathbf{E}^{Q_\epsilon}F_\tau(\tilde{x}(t),\tilde{u}(t)).
\end{aligned}
$$

Now observe that since the random variable on the left-hand side of equation (8.4.45) is $\mathcal{F}_0$-measurable, then the expectation $\mathbf{E}$ in (8.4.45) can be replaced with the expectation $\mathbf{E}^{Q_\epsilon}$. Therefore, it follows from the above identities and from (8.4.45) that for any $\epsilon > 0$, there exists a probability measure $Q \in \mathcal{P}$ such that

$$
J_\tau^{Q_\epsilon}(u(\cdot)) > \frac{1}{\epsilon}.
$$

Hence, we conclude that given a control $u(\cdot)$ of the form (8.4.42), then

$$
\sup_{Q\in\mathcal{P}} J_\tau^Q(u(\cdot)) = \infty \tag{8.4.48}
$$

for any $\tau > 0$.

Now suppose that part (ii) of the theorem does not hold. That is, we suppose that there exists a linear controller $u(\cdot)$ of the form (8.4.42) such that

$$
\sup_{Q\in\Xi} J^Q(u(\cdot)) < \infty.
$$

Using this condition and Assumption 8.4.1, it follows from Lemma 8.4.1 that there exists a constant $\tau > 0$ such that

$$
\sup_{Q\in\mathcal{P}} J_\tau^Q(u(\cdot)) < \infty.
$$

However, this contradicts (8.4.48). Thus the second part of the theorem follows.

■

The first part of Theorem 8.4.2 leads to a practical method for solving minimax optimal LQG design problems. Furthermore, note that the second part of the theorem is also important. Indeed, if one fails to find a $\tau > 0$ satisfying Assumption 8.4.2, then for the uncertain system under consideration, the optimal worst case LQG design problem does not admit a solution in terms of a linear output-feedback controller of the form (8.4.42).

Recall that Assumption 8.4.1 is a controllability type assumption. The following lemma illustrates a connection between Assumption 8.4.1 and the controllability property of the system (8.4.1).

Consider a linear controller $u(\cdot)$ of the form (8.4.42) and introduce the notation:

$$\bar{A} = \begin{bmatrix} A & B_1 K \\ B_c C_2 & A_c \end{bmatrix}; \quad \bar{B} = \begin{bmatrix} B_2 \\ B_c D_2 \end{bmatrix}; \quad \bar{R} = \begin{bmatrix} R & 0 \\ 0 & K'GK \end{bmatrix}. \tag{8.4.49}$$

Lemma 8.4.3 *Suppose a linear controller $u(\cdot)$ of the form (8.4.42) is such that the pair $(\bar{A}, \bar{B})$ is controllable. Then for this controller, condition (8.4.16) of Assumption 8.4.1 is satisfied.*

Proof: Under the condition of the lemma, for any matrix Φ of suitable dimension, the Lyapunov differential equation

$$\dot{\Pi} + (\bar{A} + \bar{B}\Phi)'\Pi + \Pi(\bar{A} + \bar{B}\Phi) + \bar{R} = 0; \quad \Pi(T) = 0 \tag{8.4.50}$$

has a nonnegative-definite solution on the interval $[0, T]$.

Let $\bar{\mathcal{P}}$ be the subset of $\mathcal{P}$ which includes the probability measures Q satisfying the following additional condition. On the probability space $(\Omega, \mathcal{F}, Q)$, the closed loop system corresponding to the uncertain system (8.4.1) and the given controller (8.4.42), is a system of the form

$$\bar{x} = (\bar{A} + \bar{B}\Phi)\bar{x}dt + \bar{B}d\tilde{W}; \quad \bar{x} := [x(\cdot)' \, \breve{x}(\cdot)']'.$$

Then, using Itô's formula and equation (8.4.50), one obtains

$$\begin{aligned} J^Q(u(\cdot)) &= \mathbf{E}^Q \int_0^T \bar{x}'(t)\bar{R}\bar{x}(t)dt \\ &= \mathbf{E}^Q \bar{x}'(0)\Pi\bar{x}(0) + \int_0^T \operatorname{tr} \bar{B}(t)\bar{B}'(t)\Pi(t)dt. \end{aligned}$$

Thus,

$$\sup_{Q \in \mathcal{P}} J^Q(u(\cdot)) \geq \sup_{Q \in \bar{\mathcal{P}}} J^Q(u(\cdot)) = \infty$$

and hence, the lemma follows. ■

8.5 Output-feedback infinite-horizon minimax optimal control of stochastic uncertain systems with additive noise

8.5.1 Definitions

In this section, we address an infinite-horizon minimax optimal control problem. This problem may be regarded as the extension to the infinite-horizon case, of the problem considered in Section 8.4. An important feature of the minimax optimal control problem to be considered, is that it involves a time averaged cost functional. Furthermore, we will introduce a definition of absolute stabilizability which accounts for the fact that the underlying system is subject to additive noise disturbances and its solutions do not necessarily belong to $\mathbf{L}_2[0,\infty)$.

The problem formulation will be as described in Sections 2.4.3 and 8.4. Let $(\Omega, \mathcal{F}, P)$ be the complete probability space constructed as in Sections 2.4.3 and 8.4. That is, $\Omega = \mathbf{R}^n \times \mathbf{R}^l \times C([0,\infty), \mathbf{R}^p)$, the probability measure P is defined as the product of a given probability measure on $\mathbf{R}^n \times \mathbf{R}^l$ and the standard Wiener measure on $C([0,\infty), \mathbf{R}^p)$. Also, let $W(\cdot)$ be a p-dimensional standard Wiener process, $x_0 : \Omega \to \mathbf{R}^n$ be a Gaussian random variable with mean $\check{x}_0$ and non-singular covariance matrix Y_0, and $\{\mathcal{F}_t, t \geq 0\}$ be a complete filtration as defined in Section 2.4.3. Also, the random variable x_0 and the Wiener process $W(\cdot)$ are assumed to be stochastically independent on $(\Omega, \mathcal{F}, P)$.

On the probability space $(\Omega, \mathcal{F}, P)$, we consider the system and measurement dynamics driven by the noise input $W(\cdot)$ and a control input $u(\cdot)$. These dynamics are described by the following stochastic differential equations of the form (2.4.25):

$$\begin{aligned} dx(t) &= (Ax(t) + B_1 u(t))dt + B_2 dW(t); \quad x(0) = x_0; \\ z(t) &= C_1 x(t) + D_1 u(t); \\ dy(t) &= C_2 x(t)dt + D_2 dW(t); \quad y(0) = 0. \end{aligned} \tag{8.5.1}$$

In these equations, all of the coefficients are assumed to be constant matrices of appropriate dimensions. Thus, the system (8.5.1) can be viewed as a time-invariant version of the nominal system (8.4.1). In particular, the variables x, u, z, and y have the same meanings as in Section 8.4. Also, as in Section 8.4, we assume that

$$\Gamma := D_2 D_2' > 0.$$

In the minimax optimal control problem to be considered, attention will be restricted to the following linear output-feedback controllers of the form (2.4.26):

$$\begin{aligned} d\hat{x} &= A_c \hat{x} + B_c dy(t); \\ u &= K\hat{x} \end{aligned} \tag{8.5.2}$$

where $\hat{x} \in \mathbf{R}^{\hat{n}}$ is the state of the controller and $A_c \in \mathbf{R}^{\hat{n}\times\hat{n}}$, $K \in \mathbf{R}^{m\times\hat{n}}$, and $B_c \in \mathbf{R}^{\hat{n}\times q}$. Let $\mathcal{U}$ denote this class of linear controllers. Note that the controller

(8.5.2) is adapted to the filtration $\{\mathcal{Y}_t, t \geq 0\}$ generated by the observation process y. The closed loop nominal system corresponding to controller (8.5.2) is described by a linear Ito differential equation of the form

$$\begin{aligned} d\bar{x} &= \bar{A}\bar{x}dt + \bar{B}dW(t); \\ z &= \bar{C}\bar{x}; \\ u &= \begin{bmatrix} 0 & K \end{bmatrix} \bar{x} \end{aligned} \tag{8.5.3}$$

and is considered on the probability space $(\Omega, \mathcal{F}, P)$. Note that the controller (8.5.2) and the closed loop system (8.5.3) represent a time-invariant version of the controller (8.4.42) and the corresponding closed loop system considered in Section 8.4. In equation (8.5.3), we therefore use the notation (8.4.49) assuming that the corresponding matrices are time-invariant. Also, we use the notation

$$\bar{C} = \begin{bmatrix} C_1 & D_1 K \end{bmatrix}.$$

Stochastic uncertain system

As in Section 8.4, we refer to the system (8.5.1) as the nominal system. Also, we will consider perturbations of this stochastic nominal system in terms of uncertain martingales as discussed in Section 2.4.3. The class of all such perturbation martingales is denoted $\mathcal{M}$; see Section 2.4.3. Associated with any martingale $\zeta(\cdot) \in \mathcal{M}$ and time $T > 0$ is a corresponding probability measure Q^T defined by

$$Q^T(d\omega) = \zeta(T)P^T(d\omega); \tag{8.5.4}$$

see (2.4.27). It was observed in Section 2.4.3 that the set $\mathcal{M}$ is convex. Also, an alternative representation for the perturbation probability measures in terms of an associated stochastic process $\xi(\cdot)$ and a Wiener process $\tilde{W}(\cdot)$ was discussed in Section 2.4.3. This led to an equivalent representation of the stochastic uncertain system on the probability space $(\Omega, \mathcal{F}_T, Q^T)$; see equation (2.4.29).

For the class of stochastic uncertain systems under consideration, the class of admissible uncertain probability measures was defined in Section 2.4.3; see Definition 2.4.3 on page 47. The class of admissible uncertainties is described in Section 2.4.3 in terms of the relative entropy functional $h(\cdot\|P^T)$, $T > 0$. Here, P^T denotes the restriction of the reference probability measure P to $(\Omega, \mathcal{F}_T)$.

Also, in Section 2.4.3, the above uncertainty class was shown to include a class of linear time-invariant uncertainties defined in terms of a standard H^∞ norm bound. Note that Definition 2.4.3 in Section 2.4.3 exploits a sequence of continuous positive martingales $\zeta_i(\cdot)$ in order to define a bound on the uncertainty. Recall that for any $T > 0$, the random variables $\zeta_i(T)$ were required to converge weakly to $\zeta(T)$ in $\mathbf{L}_1(\Omega, \mathcal{F}_T, P^T)$. This fact implies that for each $T > 0$, the sequence of probability measures $\{Q_i^T\}_{i=1}^\infty$ corresponding to the martingales $\zeta_i(\cdot)$ converges to the probability measure Q^T corresponding to the given martingale $\zeta(\cdot)$ in the following sense: For any bounded $\mathcal{F}_T$-measurable random variable η,

$$\lim_{i\to\infty} \int_\Omega \eta Q_i^T(d\omega) = \int_\Omega \eta Q^T(d\omega); \tag{8.5.5}$$

see equation (2.4.31). The above convergence of probability measures was denoted $Q_i^T \Rightarrow Q^T$, as $i \to \infty$. It was previously noted that the property $Q_i^T \Rightarrow Q^T$ implies that the sequence of probability measures Q_i^T converges weakly to the probability measure Q^T; see Remark 2.4.5.

We now repeat Definition 2.4.3 for ease of reference.

Definition 8.5.1 *Let d be a given positive constant. A martingale $\zeta(\cdot) \in \mathcal{M}$ is said to define an admissible uncertainty if there exists a sequence of continuous positive martingales $\{\zeta_i(t), \mathcal{F}_t, t \geq 0\}_{i=1}^{\infty} \subset \mathcal{M}$ which satisfies the following conditions:*

(i) For each i, $h(Q_i^T \| P^T) < \infty$ for all $T > 0$;

(ii) For all $T > 0$, $Q_i^T \Rightarrow Q^T$ as $i \to \infty$;

(iii) The following stochastic uncertainty constraint *is satisfied: For any sufficiently large $T > 0$ there exists a constant $\delta(T)$ such that* $\lim_{T\to\infty} \delta(T) = 0$ *and* [3]

$$\inf_{T'>T} \frac{1}{T'} \left[\frac{1}{2} \mathbf{E}^{Q_i^{T'}} \int_0^{T'} \|z(t)\|^2 dt - h(Q_i^{T'} \| P^{T'}) \right] \geq -d + \delta(T) \quad (8.5.6)$$

for all $i = 1, 2, \ldots$. In (8.5.6), the uncertainty output $z(\cdot)$ is defined by equation (8.5.1) considered on the probability space $(\Omega, \mathcal{F}_T, Q_i^T)$.

We let Ξ denote the set of martingales $\zeta(\cdot) \in \mathcal{M}$ corresponding to admissible uncertainties. Elements of Ξ are also called admissible martingales.

Further discussion concerning this definition can be found in Section 2.4.3. In particular, Section 2.4.3 gives some examples of uncertainties which lead to the satisfaction of condition (8.5.6). Also, note that the set Ξ is non-empty. Indeed, the martingale $\zeta(t) \equiv 1$ belongs to Ξ, since the constraint (8.5.6) is strictly satisfied in this case. This follows from the condition $d > 0$ and the fact that $h(P^T \| P^T) = 0$; see Section 2.4.3 for details.

In the sequel, we will use the following notation. Let $\mathcal{P}_T$ be the set of probability measures Q^T on $(\Omega, \mathcal{F}_T)$ such that $h(Q^T \| P^T) < \infty$. Also, the notation $\mathcal{M}_\infty$ will denote the set of martingales $\zeta(\cdot) \in \mathcal{M}$ such that $h(Q^T \| P^T) < \infty$ for all $T > 0$. Such a set $\mathcal{M}_\infty$ is convex. Indeed, for any two martingales $\zeta_1(\cdot), \zeta_2(\cdot) \in \mathcal{M}_\infty$ and a constant $0 < a < 1$, the measure Q^T corresponding to the martingale $a\zeta_1(\cdot) + (1-a)\zeta_2(\cdot)$ is as follows:

$$Q^T(d\omega) = (a\zeta_1(T) + (1-a)\zeta_2(T))P^T(d\omega) = aQ_1^T(d\omega) + (1-a)Q_2^T(d\omega)$$

[3] In this section, it will be convenient for us to use a constant d in the constraint (8.5.6) rather than the constant $d/2$ which was used in Definition 2.4.3.

for any $T > 0$, where Q_1^T, Q_2^T are probability measures corresponding to the martingales $\zeta_1(\cdot)$, $\zeta_2(\cdot)$. Obviously, Q^T is a probability measure. Also,

$$h(Q^T \| P^T) \leq a h(Q_1^T \| P^T) + (1-a) h(Q_2^T \| P^T) < \infty.$$

The above inequality follows from the fact that the functional $h(\cdot \| P^T)$ is convex. Hence, $Q^T \in \mathcal{P}_T$ for all $T > 0$ and $a\zeta_1(\cdot) + (1-a)\zeta_2(\cdot) \in \mathcal{M}_\infty$.

Note that the martingales $\zeta_i(\cdot)$ from Definition 8.5.1 belong to $\mathcal{M}_\infty$.

8.5.2 *Absolute stability and absolute stabilizability*

An important issue in any optimal control problem on an infinite time interval concerns the stabilizing properties of the optimal controller. For example, a critical issue addressed in Sections 5.3 and 8.3 was to prove the absolutely stabilizing property of the optimal control schemes presented in those sections. In this section, the systems under consideration are subject to additive noise. The solutions of such systems do not necessarily belong to $\mathbf{L}_2[0, \infty)$. Hence, we need a definition of absolute stabilizability which properly accounts for this feature of the systems under consideration.

Definition 8.5.2 *A controller of the form (8.4.2) is said to be an* absolutely stabilizing output-feedback controller *for the stochastic uncertain system (8.5.1), (8.5.6), if the process $x(\cdot)$ defined by the closed loop system corresponding to this controller satisfies the following condition. There exist constants $c_1 > 0$, $c_2 > 0$ such that for any admissible martingale $\zeta(\cdot) \in \Xi$,*

$$\limsup_{T\to\infty} \frac{1}{T}\left[\mathbf{E}^{Q^T} \int_0^T \left(\|x(t)\|^2 + \|u(t)\|^2\right) dt + h(Q^T \| P^T)\right] \leq c_1 + c_2 d. \quad (8.5.7)$$

The property of absolute stability is defined as a special case of Definition 8.5.2 corresponding to $u(\cdot) \equiv 0$. In this case, the system (8.5.1) becomes a system of the form

$$\begin{aligned} dx(t) &= Ax(t)dt + B_2 dW(t); \\ z(t) &= C_1 x(t). \end{aligned} \quad (8.5.8)$$

Definition 8.5.3 *The stochastic uncertain system corresponding to the state equations (8.5.8) with uncertainty satisfying the relative entropy constraint (8.5.6) is said to be* absolutely stable *if there exist constants $c_1 > 0$, $c_2 > 0$ such that for any admissible uncertainty $\zeta(\cdot) \in \Xi$,*

$$\limsup_{T\to\infty} \frac{1}{T}\left[\mathbf{E}^{Q^T} \int_0^T \|x(t)\|^2 dt + h(Q^T \| P^T)\right] \leq c_1 + c_2 d. \quad (8.5.9)$$

In the sequel, the following properties of absolutely stable systems will be used.

Lemma 8.5.1 *Suppose the stochastic nominal system (8.5.8) is mean square stable;* i.e.,

$$\limsup_{T\to\infty} \frac{1}{T}\mathbf{E}^{Q^T} \int_0^T \|x(t)\|^2 dt < \infty. \tag{8.5.10}$$

Also, suppose the pair (A, B_2) is stabilizable. Then, the matrix A must be stable.

Proof: Since the stochastic nominal system (8.5.8) satisfies condition (8.5.10, then for any vector y,

$$\limsup_{T\to\infty} \frac{1}{T} \int_0^T y'\mathbf{E}\left[x(t)x'(t)\right] y dt < \infty. \tag{8.5.11}$$

We will prove that the stability of the matrix A follows from condition (8.5.11). This proof is by contradiction.

Suppose that the matrix A is not stable and therefore, it has a left eigenvalue λ such that $\mathrm{Re}\lambda \geq 0$. Consider the left eigenvector y of the matrix A corresponding to the eigenvalue λ. Hence, $y'A = y'\lambda$. Since the pair (A, B_2) is stabilizable, it follows that $y'B_2 \neq 0$ and consequently,

$$y'B_2B_2'y > 0. \tag{8.5.12}$$

We now consider two cases:

Case 1. $\mathrm{Re}\lambda > 0$. In this case, we obtain the following bound on the quantity $y'\mathbf{E}\left[x(t)x'(t)\right] y$:

$$\begin{aligned} y'\mathbf{E}\left[x(t)x'(t)\right] y &= y'e^{At}\mathbf{E}\left[x(0)x'(0)\right] e^{A't}y + \int_0^t y'e^{A(t-s)}B_2B_2'e^{A'(t-s)}yds \\ &= e^{2\mathrm{Re}\lambda t}y'\mathbf{E}\left[x(0)x'(0)\right] y + \int_0^t e^{2\mathrm{Re}\lambda(t-s)}y'B_2B_2'yds \\ &\geq \frac{e^{2\mathrm{Re}\lambda t}-1}{2\mathrm{Re}\lambda}y'B_2B_2'y. \end{aligned}$$

Thus for any $T > 0$,

$$\frac{1}{T}\int_0^T y'\mathbf{E}\left[x(t)x'(t)\right] y dt \geq y'B_2B_2'y \cdot \frac{1}{T}\int_0^T \frac{e^{2\mathrm{Re}\lambda t}-1}{2\mathrm{Re}\lambda}dt. \tag{8.5.13}$$

Case 2. $\mathrm{Re}\lambda = 0$. In this case, we obtain the following bound on the quantity $y'\mathbf{E}\left[x(t)x'(t)\right] y$:

$$\begin{aligned} y'\mathbf{E}\left[x(t)x'(t)\right] y &= y'e^{At}\mathbf{E}\left[x(0)x'(0)\right] e^{A't}y + \int_0^t y'e^{A(t-s)}B_2B_2'e^{A'(t-s)}yds \\ &= y'\mathbf{E}\left[x(0)x'(0)\right] y + \int_0^t y'B_2B_2'yds \\ &\geq t \cdot y'B_2B_2'y. \end{aligned}$$

Thus for any $T > 0$,

$$\frac{1}{T}\int_0^T y'\mathbf{E}\left[x(t)x'(t)\right]y\,dt \geq y'B_2B_2'y \cdot \frac{1}{T}\frac{T^2}{2} = y'B_2B_2'y\frac{T}{2}. \tag{8.5.14}$$

Since $y'B_2B_2'y > 0$, the expressions on the right-hand side of inequalities (8.5.13) and (8.5.14) both approach infinity as $T \to \infty$. That is in both cases,

$$\limsup_{T\to\infty} \frac{1}{T}\int_0^T y'\mathbf{E}\left[x(t)x'(t)\right]y\,dt = \infty.$$

This yields the desired contradiction with (8.5.11).

■

Lemma 8.5.2 *Suppose that the stochastic uncertain system (8.5.8), (8.5.6) is absolutely stable and the pair (A, B_2) is stabilizable. Then for any nonnegative-definite matrix R, there exists a positive constant $\tau > 0$ such that the Riccati equation*

$$A'\Pi + \Pi A + R + \tau C_1'C_1 + \frac{1}{\tau}\Pi B_2B_2'\Pi = 0 \tag{8.5.15}$$

has a nonnegative-definite stabilizing solution.

Proof: Since the stochastic uncertain system is absolutely stable, it follows that the stochastic nominal system must be mean square stable (the martingale $\zeta(t) \equiv 1$ is an admissible uncertainty). Hence, it follows from Lemma 8.5.1 that the matrix A is stable. Now consider the absolutely stable uncertain system (8.5.8), (8.5.6). We consider a real rational uncertainty transfer function $\Delta(s) \in H^\infty$ satisfying condition (2.4.36) and having a state-space realization of the form

$$\begin{aligned}\dot{\eta} &= \Sigma\eta + \Theta z; \\ \xi &= \Psi\eta + \Phi z\end{aligned} \tag{8.5.16}$$

with a stable matrix Σ. It was shown in Lemma 2.4.2 that such H^∞ norm-bounded uncertainty gives rise to a martingale $\zeta(t)$ which satisfies the conditions of Definition 8.5.1. The assumptions of Lemma 2.4.2 are satisfied in the case considered in this lemma since the augmented system consisting of the system (8.5.8) and the system (8.5.16) is linear; see also Remark 2.4.7. Thus, the martingale $\zeta(t)$ corresponding to the chosen H^∞ norm-bounded uncertainty belongs to the set Ξ and hence, the stability condition (8.5.9) holds for this particular uncertainty. From this fact, it follows that since the system (8.5.16) is stable, then condition (8.5.9) implies that the augmented stochastic system

$$d\begin{bmatrix} x \\ \eta \end{bmatrix} = \begin{bmatrix} A + B_2\Phi C_1 & B_2\Psi \\ \Theta C_1 & \Sigma \end{bmatrix}\begin{bmatrix} x \\ \eta \end{bmatrix}dt + \begin{bmatrix} B_2 \\ 0 \end{bmatrix}d\tilde{W}(t) \tag{8.5.17}$$

is mean square stable. We now prove that the pair

$$\left(\begin{bmatrix} A + B_2\Phi C_1 & B_2\Psi \\ \Theta C_1 & \Sigma \end{bmatrix}, \begin{bmatrix} B_2 \\ 0 \end{bmatrix} \right) \tag{8.5.18}$$

is stabilizable. Indeed, let the complex number λ be such that $\mathrm{Re}\lambda \geq 0$ and there exists a vector $[y', \theta']$ such that the following equations hold:

$$y'(A + B_2\Phi C_1 - \lambda I) + \theta'\Theta C_1 = 0; \tag{8.5.19a}$$

$$y'B_2\Psi + \theta'(\Sigma - \lambda I) = 0; \tag{8.5.19b}$$

$$y'B_2 = 0. \tag{8.5.19c}$$

It is straightforward to verify that conditions (8.5.19b) and (8.5.19c) along with the stability of the matrix Σ imply that $\theta = 0$. Hence, it follows from (8.5.19a) and (8.5.19c) that

$$y'(A - \lambda I) = 0.$$

Since the pair (A, B_2) is assumed to be stabilizable, then the above equation along with equation (8.5.19c) implies that $y = 0$. Thus for any λ such that $\mathrm{Re}\lambda \geq 0$, equations (8.5.19) will be satisfied only if the vector $[y', \theta']'$ is zero. This implies that the pair (8.5.18) is stabilizable.

We have shown that the system (8.5.17) satisfies the conditions of Lemma 8.5.1. From Lemma 8.5.1, it follows that the matrix

$$\begin{bmatrix} A + B_2\Phi C_1 & B_2\Psi \\ \Theta C_1 & \Sigma \end{bmatrix}$$

is stable. This fact implies that the closed loop system corresponding to the system

$$\begin{aligned} \frac{d\tilde{x}}{dt} &= A\tilde{x} + B_2\tilde{\xi}; \\ \tilde{z} &= C_1\tilde{x}; \\ \tilde{\xi} &= \Delta(s)\tilde{z} \end{aligned} \tag{8.5.20}$$

and the given uncertainty $\Delta(s)$ with the state-space representation (8.5.16), is internally stable. Recall that $\Delta(s)$ can be chosen to be any stable transfer function satisfying the condition $\|\Delta(s)\|_\infty \leq 1$. Hence, we can apply the standard small gain theorem to the uncertain system (8.5.20), (8.5.16); e.g., see [271]. From the small gain theorem, it follows that the system (8.5.20) satisfies the following H^∞ norm bound:

$$\|C_1(sI - A)^{-1}B_2\|_\infty < 1.$$

That is, there exists a positive constant β such that

$$\|C_1(sI - A)^{-1}B_2\|_\infty \leq 1 - \beta. \tag{8.5.21}$$

Furthermore, it was shown that the matrix A is stable. Hence, there exists a positive constant α such that

$$\sup_{\omega \in \mathbf{R}} \|(j\omega I - A)^{-1} B_2\| = \|(sI - A)^{-1} B_2\|_\infty < \alpha. \tag{8.5.22}$$

We now choose the constant τ such that

$$\tau > \left[\frac{\alpha \|R^{1/2}\|}{\beta} \right]^2. \tag{8.5.23}$$

With this choice of the constant τ, we obtain:

$$\begin{aligned} &\left\| \begin{bmatrix} \frac{1}{\sqrt{\tau}} R^{1/2} \\ C_1 \end{bmatrix} (sI - A)^{-1} B_2 \right\|_\infty \\ &\leq \frac{1}{\sqrt{\tau}} \|R^{1/2}\| \cdot \|(sI - A)^{-1} B_2\|_\infty + \|C_1 (sI - A)^{-1} B_2\|_\infty \\ &\leq \frac{\alpha}{\sqrt{\tau}} \|R^{1/2}\| + 1 - \beta \\ &< 1. \end{aligned}$$

The stability of the matrix A and the above H^∞ norm condition are the conditions of the Strict Bounded Real Lemma; see Lemma 3.1.2 and also [147]. From the Strict Bounded Real Lemma, it follows that the Riccati equation (8.5.15) has a nonnegative-definite stabilizing solution.

■

Infinite-horizon minimax optimal control problem

Associated with the stochastic uncertain system (8.5.1), (8.5.6), consider a cost functional $J(\cdot)$ of the form

$$J(u(\cdot), \zeta(\cdot)) = \limsup_{T \to \infty} \frac{1}{2T} \mathbf{E}^{Q^T} \int_0^T F(x(t), u(t)) dt, \tag{8.5.24}$$

defined on solutions $x(\cdot)$ to (8.5.1). In (8.5.24),

$$F(x, u) := x' R x + u' G u,$$

and R and G are positive-definite symmetric matrices, $R \in \mathbf{R}^{n \times n}$, $G \in \mathbf{R}^{m \times m}$. Also, in equation (8.5.24), Q^T is the probability measure corresponding to the martingale $\zeta(\cdot)$; see equation (8.5.4).

In this section, we are concerned with a minimax optimal control problem associated with the system (8.5.1), cost functional (8.5.24) and the uncertainty set Ξ. In this problem, we seek to find a controller $u^*(\cdot)$ of the form (8.5.2) which minimizes the worst case value of the cost functional J in the face of uncertainty

$\zeta(\cdot) \in \Xi$ satisfying the constraint (8.5.6). That is, we consider the minimax optimal control problem:

$$\sup_{\zeta(\cdot)\in\Xi} J(u^*(\cdot),\zeta(\cdot)) = \inf_{u(\cdot)\in\mathcal{U}} \sup_{\zeta(\cdot)\in\Xi} J(u(\cdot),\zeta(\cdot)). \tag{8.5.25}$$

As in Section 8.4, the derivation of a solution to the above minimax optimal control problem relies on a duality relationship between free energy and relative entropy given in Lemma A.1; see also [39, 53, 153]. Associated with the system (8.5.1), consider the parameter dependent risk-sensitive cost functional (*c.f.* (8.4.7))

$$\Im_{\tau,T}(u(\cdot)) := \frac{2\tau}{T} \log \mathbf{E}\left\{\exp\left(\frac{1}{2\tau}\int_0^T F_\tau(x(t),u(t))dt\right)\right\} \tag{8.5.26}$$

where $\tau > 0$ is a given constant and

$$\begin{aligned} F_\tau(x,u) &:= x'R_\tau x + 2x'\Upsilon_\tau u + u'G_\tau u; &(8.5.27)\\ R_\tau &:= R + \tau C_1'C_1;\\ G_\tau &:= G + \tau D_1'D_1;\\ \Upsilon_\tau &:= \tau C_1'D_1. &(8.5.28)\end{aligned}$$

Note that these matrices are time-invariant versions of the corresponding matrices defined in Section 8.4; see equations (8.4.8) and (8.4.9). We will apply the duality result of Lemma A.1 of Appendix A; also, see [39]. When applied to the system (8.5.1) and the risk-sensitive cost functional (8.5.26) (see Corollary 3.1 and Remark 3.2 of [39]), this result states that for each admissible controller $u(\cdot)$,

$$\sup_{Q^T\in\mathcal{P}_T} J_{\tau,T}(u(\cdot),Q^T) = \frac{1}{2}\Im_{\tau,T}(u(\cdot)) \tag{8.5.29}$$

where

$$J_{\tau,T}(u(\cdot),Q^T) := \frac{1}{T}\left[\frac{1}{2}\mathbf{E}^{Q^T}\int_0^T F_\tau(x(t),u(t))dt - \tau h(Q^T\|P^T)\right] \tag{8.5.30}$$

is a parameter dependent augmented cost functional formed from the cost functional (8.5.24) and the relative entropy constraint (8.5.6); *c.f.* equation (8.4.6).

The use of the above duality result is a key step that enables us to replace the minimax optimal control problem by a risk-sensitive optimal control problem. Hence, we will be interested in constructing an output-feedback controller of the form (8.5.2) solving the following stochastic risk-sensitive optimal control problem

$$\inf_{u(\cdot)\in\mathcal{U}} \lim_{T\to\infty} \Im_{\tau,T}(u(\cdot)). \tag{8.5.31}$$

Recall that a risk-sensitive control problem of this form was considered in Section 3.3.3. There, a rigorous definition of the class of admissible controllers in the control problem (8.5.31) was given. This class of controllers, originally introduced in reference [139], includes linear output-feedback controllers. Furthermore, the risk-sensitive optimal controller derived in [139] was a linear output-feedback controller of the form (8.5.2). This fact leads us to restrict the attention to the class of controllers $\mathcal{U}$.

Also, recall that in the finite-horizon case, the risk-sensitive optimal controller is expressed as the combination of a linear state-feedback controller and a state estimator. In Section 8.4, the duality result (8.4.10) allowed us to prove that this controller is the solution to the original minimax optimal control problem (8.4.5). Since the functional (8.5.31) is a limit of the corresponding cost functional (8.5.26) defined on the finite time interval $[0, T]$, one might expect that the duality relation (8.5.29) would lead to a corresponding output-feedback controller solving the minimax optimal control problem (8.5.25). This idea is the basis of our approach to be presented in the next subsection.

8.5.3 A connection between risk-sensitive optimal control and minimax optimal control

In this subsection, we present results establishing a connection between the risk-sensitive optimal control problem (8.5.31) and the minimax optimal control problem (8.5.25).

For a given constant $\tau > 0$, let V_τ denote an optimal value of the risk-sensitive control problem (8.5.31); *i.e.*,

$$\begin{aligned} V_\tau &:= \inf_{u(\cdot)\in\mathcal{U}} \lim_{T\to\infty} \frac{1}{2}\Im_{\tau,T}(u(\cdot)) \\ &= \inf_{u(\cdot)} \lim_{T\to\infty} \frac{\tau}{T} \log \mathbf{E}\left\{\exp\left[\frac{1}{2\tau}\int_0^T F_\tau(x(s),u(s))ds\right]\right\}. \end{aligned}$$

Theorem 8.5.1 *Suppose that for a given $\tau > 0$, the risk-sensitive control problem (8.5.31) admits an optimal controller $u_\tau(\cdot) \in \mathcal{U}$ of the form (8.5.2) which guarantees a finite optimal value: $V_\tau < \infty$. Then, this controller is an absolutely stabilizing controller for the stochastic uncertain system (8.5.1) satisfying the relative entropy constraint (8.5.6). Furthermore,*

$$\sup_{\zeta(\cdot)\in\Xi} J(u_\tau(\cdot),\zeta(\cdot)) \le V_\tau + \tau d. \tag{8.5.32}$$

Proof: It follows from the condition of the theorem that

$$V_\tau = \lim_{T\to\infty} \frac{\tau}{T} \log \mathbf{E}\left\{\exp\left[\frac{1}{2\tau}\int_0^T F_\tau(x(s),u_\tau(s))ds\right]\right\} < \infty$$

where $u_\tau(\cdot)$ is the risk-sensitive optimal controller of the form (8.5.2) corresponding to the given τ. We wish to prove that this risk-sensitive optimal controller satisfies condition (8.5.7) of Definition 8.5.2.

Using the duality result (8.5.29), we obtain

$$\lim_{T\to\infty} \sup_{Q^T\in\mathcal{P}_T} \frac{1}{T}\left[\frac{1}{2}\mathbf{E}^{Q^T}\int_0^T F_\tau(x(s),u_\tau(s))ds - \tau h(Q^T\|P^T)\right] = V_\tau. \tag{8.5.33}$$

Equation (8.5.33) implies that for any sufficiently large $T > 0$, one can choose a sufficiently small positive constant $\hat{\delta} = \hat{\delta}(T) > 0$, such that $\lim_{T\to\infty}\hat{\delta}(T) = 0$ and

$$\sup_{Q^{T'}\in\mathcal{P}_{T'}} \frac{1}{T'}\left[\frac{1}{2}\mathbf{E}^{Q^{T'}}\int_0^{T'} F_\tau(x(s),u_\tau(s))ds - \tau h(Q^{T'}\|P^{T'})\right] \le V_\tau + \hat{\delta}(T) \tag{8.5.34}$$

for all $T' > T$. Thus, for the chosen constants $T > 0$ and $\hat{\delta}(T) > 0$ and for all $T' > T$,

$$\frac{1}{T'}\left[\frac{1}{2}\mathbf{E}^{Q^{T'}}\int_0^{T'} F_\tau(x(s),u_\tau(s))ds - \tau h(Q^{T'}\|P^{T'})\right] \le V_\tau + \hat{\delta}(T)$$

for any $Q^{T'} \in \mathcal{P}_{T'}$. Furthermore if $Q^{T'} \in \mathcal{P}_{T'}$ for all $T' > T$, then

$$\sup_{T'>T} \frac{1}{T'}\left[\frac{1}{2}\mathbf{E}^{Q^{T'}}\int_0^{T'} F_\tau(x(s),u_\tau(s))ds - \tau h(Q^{T'}\|P^{T'})\right] \le V_\tau + \hat{\delta}(T). \tag{8.5.35}$$

Let $\zeta(\cdot) \in \Xi$ be a given admissible uncertainty martingale and let $\zeta_i(\cdot)$ be a corresponding sequence of martingales as in Definition 8.5.1. Recall, that the corresponding probability measures Q_i^T belong to the set $\mathcal{P}_T$. Hence each probability measure Q_i^T satisfies condition (8.5.35); *i.e.*,

$$\sup_{T'>T} \frac{1}{T'}\left[\frac{1}{2}\mathbf{E}^{Q_i^{T'}}\int_0^{T'} F_\tau(x(s),u_\tau(s))ds - \tau h(Q_i^{T'}\|P^{T'})\right] \le V_\tau + \hat{\delta}(T). \tag{8.5.36}$$

Note that in condition (8.5.36), $\hat{\delta}(T)$ and T are the constants which are independent of i.

Since $F(x,u) \ge 0$ and $\tau > 0$, then condition (8.5.36) implies

$$\sup_{T'>T} \frac{1}{T'}\left[\frac{1}{2}\mathbf{E}^{Q_i^{T'}}\int_0^{T'} \|z(s)\|^2 ds - h(Q_i^{T'}\|P^{T'})\right] < \infty.$$

From this, it follows from (8.5.36) that for each integer i,

$$\begin{aligned}
&\sup_{T'>T} \frac{1}{2T'} \mathbf{E}^{Q_i^{T'}} \int_0^{T'} F(x(s), u_\tau(s))ds \\
&\quad + \tau \inf_{T'>T} \frac{1}{T'} \left[\frac{1}{2} \mathbf{E}^{Q_i^{T'}} \int_0^{T'} \|z(s)\|^2 ds - h(Q_i^{T'} \| P^{T'}) \right] \\
&\leq \sup_{T'>T} \frac{1}{T'} \left[\frac{1}{2} \mathbf{E}^{Q_i^{T'}} \int_0^{T'} \left(F(x(s), u_\tau(s)) + \tau \|z(s)\|^2 \right) ds - \tau h(Q_i^{T'} \| P^{T'}) \right] \\
&\leq V_\tau + \hat{\delta}(T).
\end{aligned}$$

This implies

$$\begin{aligned}
&\sup_{T'>T} \frac{1}{2T'} \mathbf{E}^{Q_i^{T'}} \int_0^{T'} F(x(s), u_\tau(s))ds \\
&\quad \leq V_\tau + \hat{\delta}(T) - \tau \inf_{T'>T} \frac{1}{T'} \left[\frac{1}{2} \mathbf{E}^{Q_i^{T'}} \int_0^{T'} \|z(s)\|^2 ds - h(Q_i^{T'} \| P^{T'}) \right] \\
&\quad \leq V_\tau + \tau d + \hat{\delta}(T) - \tau \delta(T).
\end{aligned} \tag{8.5.37}$$

The derivation of the last line of inequality (8.5.37) uses the fact that the probability measure Q_i^T satisfies condition (8.5.6). Also, note that in condition (8.5.37), the constants $\hat{\delta}(T)$, $\delta(T)$ and T are independent of i and $T' > T$.

We now let $i \to \infty$ in inequality (8.5.37). This leads to the following proposition.

Proposition 1 *For any admissible uncertainty* $\zeta(\cdot) \in \Xi$,

$$\sup_{T'>T} \frac{1}{2T'} \mathbf{E}^{Q^{T'}} \int_0^{T'} F(x(s), u_\tau(s))ds \leq V_\tau + \tau d + \hat{\delta}(T) - \tau \delta(T). \tag{8.5.38}$$

To establish this proposition, consider the space $\mathbf{L}_1(\Omega, \mathcal{F}_{T'}, P^{T'})$ endowed with the topology of weak convergence of random variables, where $T' > T$. We denote this space $\mathbf{L}_1^w$. Define the functional

$$\phi(\nu) := \frac{1}{2T'} \mathbf{E} \left[\nu \int_0^{T'} F(x(s), u_\tau(s))ds \right] \tag{8.5.39}$$

mapping $\mathbf{L}_1^w$ into the space of extended reals $\mathfrak{R} = \mathbf{R} \cup \{-\infty, \infty\}$. Also, consider a sequence of functionals mapping $\mathbf{L}_1^w \to \mathfrak{R}$ defined by

$$\phi_N(\nu) := \frac{1}{2T'} \mathbf{E} \left[\nu \int_0^{T'} F_N(x(s), u_\tau(s))ds \right]; \quad N = 1, 2, \ldots \tag{8.5.40}$$

where each function $F_N(\cdot)$ is defined as follows:

$$F_N(x,u) := \begin{cases} F(x,u) & \text{if } F(x,u) \leq N; \\ N & \text{if } F(x,u) > N. \end{cases}$$

Note that from the above definition, the sequence $\phi_N(\nu)$ is monotonically increasing in N for each ν. Also, we note that for any $N > 0$,

$$P\left(\frac{1}{2T'}\int_0^{T'} F_N(x(s), u_\tau(s))ds \leq N\right) = 1.$$

Hence, if $\nu_i \to \nu$ weakly, then $\phi_N(\nu_i) \to \phi_N(\nu)$. That is, each functional $\phi_N(\cdot)$ is continuous on the space $\mathbf{L}_1^w$. Therefore, the functional

$$\phi(\nu) = \lim_{N\to\infty} \phi_N(\nu)$$

is lower semi-continuous; e.g., see Theorem 10 on page 330 of [164]. Now, let $\nu = \zeta(T')$ be the Radon-Nikodym derivative of the probability measure $Q^{T'}$ and let $\nu_i = \zeta_i(T')$ be the Radon-Nikodym derivative of the probability measure $Q_i^{T'}$. Then, the fact that $\zeta_i(T') \to \zeta(T')$ weakly implies

$$\begin{aligned} \frac{1}{2T'}\mathbf{E}^{Q^{T'}}\int_0^{T'} F(x(s), u_\tau(s))ds &\leq \liminf_{i\to\infty} \frac{1}{2T'}\mathbf{E}^{Q_i^{T'}}\int_0^{T'} F(x(s), u_\tau(s))ds \\ &\leq V_\tau + \tau d + \hat{\delta}(T) - \tau\delta(T). \end{aligned} \tag{8.5.41}$$

Since the constants on the right hand side of (8.5.41) are independent of $T' > T$, condition (8.5.38) of the proposition now follows. This completes the proof of the proposition.

Note that from the above proposition, equation (8.5.32) of the theorem follows. Indeed, for any $\zeta(\cdot) \in \Xi$, Proposition 1 and the fact that $\hat{\delta}(T), \delta(T) \to 0$ as $T \to \infty$ together imply

$$\begin{aligned} J(u_\tau(\cdot), \zeta(\cdot)) &= \limsup_{T\to\infty} \frac{1}{2T}\mathbf{E}^{Q^T}\int_0^T F(x(s), u_\tau(s))ds \\ &= \lim_{T\to\infty} \sup_{T'>T} \frac{1}{2T'}\mathbf{E}^{Q^{T'}}\int_0^{T'} F(x(s), u_\tau(s))ds \\ &\leq V_\tau + \tau d. \end{aligned} \tag{8.5.42}$$

From condition (8.5.42), equation (8.5.32) of the theorem follows.

We now establish the absolute stabilizing property of the risk-sensitive optimal controller $u_\tau(\cdot)$. Indeed, since the matrices R and G are positive-definite, then inequality (8.5.38) implies

$$\limsup_{T\to\infty} \frac{1}{T}\mathbf{E}^{Q^T}\int_0^T \left(\|x(s)\|^2 + \|u_\tau(s)\|^2\right) ds \leq \alpha\left(V_\tau + \tau d\right) \tag{8.5.43}$$

where α is a positive constant which depends only on R and G.

To complete the proof, it remains to prove that there exist constants $c_1, c_2 > 0$, such that

$$\limsup_{T\to\infty} \frac{1}{T} h(Q^T \| P^T) < c_1 + c_2 d. \tag{8.5.44}$$

To this end, we note that for any sufficiently large T and for all $T' > T$, the constraint (8.5.6) implies

$$\frac{1}{T'} h(Q_i^{T'} \| P^{T'}) \leq \frac{1}{2T'} \mathbf{E}^{Q_i^{T'}} \int_0^{T'} \|z(s)\|^2 ds + d - \delta(T) \tag{8.5.45}$$

for all $i = 1, 2, \ldots$. We now observe that condition (8.5.37) implies that for all $T' > T$

$$\frac{1}{2T'} \mathbf{E}^{Q_i^{T'}} \int_0^{T'} \|z(s)\|^2 ds \leq \bar{c}(V_\tau + \tau d + \hat{\delta}(T) - \tau\delta(T)), \tag{8.5.46}$$

where $\bar{c}$ is a positive constant determined only by the matrices R, G, C_1 and D_1. From conditions (8.5.45), (8.5.46), Remark 2.4.5 and the fact that the relative entropy functional is lower semi-continuous, it follows that

$$\begin{aligned}
\frac{1}{T'} h(Q^{T'} \| P^{T'}) &\leq \liminf_{i\to\infty} \frac{1}{T'} h(Q_i^{T'} \| P^{T'}) \\
&\leq \bar{c}(V_\tau + \hat{\delta}(T)) + (\bar{c}\tau + 1)d - (\bar{c}\tau + 1)\delta(T)
\end{aligned}$$

for any $T' > T$. This inequality and the fact that $\delta(T) \to 0$ and $\hat{\delta}(T) \to 0$ as $T \to \infty$, together imply

$$\limsup_{T\to\infty} \frac{1}{T} h(Q^T \| P^T) \leq \bar{c}V_\tau + (\bar{c}\tau + 1)d.$$

Combining this condition and inequality (8.5.43), we obtain condition (8.5.7), where the constants c_1, c_2 are defined by V_τ, τ, α, $\bar{c}$, and hence independent of $\zeta(\cdot) \in \Xi$.

■

To formulate conditions under which a converse to Theorem 8.5.1 holds, we consider the closed loop system corresponding to the system (8.5.1) and a linear time-invariant output-feedback controller of the form (8.5.2). Recall that the closed loop nominal system corresponding to controller (8.5.2) is described by the linear Ito differential equation (8.5.3) on the probability space $(\Omega, \mathcal{F}, P)$. In the sequel, we will consider the class of linear controllers of the form (8.5.2) satisfying the following assumptions: the matrix $\bar{A}$ is stable, the pair $(\bar{A}, \bar{B})$ is controllable and the pair $(\bar{A}, \bar{R})$ is observable. Here, we use the notation (8.4.49) assuming that the corresponding matrices are time-invariant. Also, let $\mathfrak{D}_0$ be the set of all linear functions $\phi(\bar{x}) = \Phi\bar{x}$ such that the matrix $\bar{A} + \bar{B}\Phi$ is stable. Note that the pair $(\bar{A} + \bar{B}\Phi, \bar{B})$ is controllable since the pair $(\bar{A}, \bar{B})$ is controllable. Under these assumptions, the Markov process generated by the linear system

$$d\bar{x}_\phi(t) = (\bar{A} + \bar{B}\Phi)\bar{x}_\phi(t)dt + \bar{B}dW(t) \tag{8.5.47}$$

has a unique invariant probability measure ν^ϕ on $\mathbf{R}^{n+\hat{n}}$; e.g., see [267]. It is shown in [267] that the probability measure ν^ϕ is a Gaussian probability measure.

Lemma 8.5.3 *For every function $\phi(\cdot) \in \mathfrak{D}_0$, there exists a martingale $\zeta(\cdot) \in \mathcal{M}_\infty$ such that for any $T > 0$, the process*

$$\tilde{W}(t) = W(t) - \int_0^t \Phi\bar{x}(s)ds \tag{8.5.48}$$

is a Wiener process with respect to $\{\mathcal{F}_t, t \in [0,T]\}$ and the probability measure Q^T corresponding to the martingale $\zeta(\cdot)$. In equation (8.5.48), $\bar{x}(\cdot)$ is the solution to the nominal closed loop system (8.5.3) with initial probability distribution ν^ϕ.

Furthermore, the equation

$$d\bar{x} = (\bar{A} + \bar{B}\Phi)\bar{x}dt + \bar{B}d\tilde{W}(t) \tag{8.5.49}$$

considered on the probability space $(\Omega, \mathcal{F}_T, Q^T)$, admits a stationary solution $\bar{x}_\zeta(\cdot)$ such that ν^ϕ is an invariant probability measure for this solution:

$$Q^T(\bar{x}_\zeta(t) \in d\bar{x}) = \nu^\phi(d\bar{x}). \tag{8.5.50}$$

Proof: Let ν^ϕ be the Gaussian invariant probability measure corresponding to a given $\phi(\cdot) \in \mathfrak{D}_0$. Consider a stochastic process $\bar{x}(t)$ defined by equation (8.5.3) and having initial probability distribution ν^ϕ; *i.e.*, $P(x(0) \in d\bar{x}) = \nu^\phi(d\bar{x})$. Since the probability measure ν^ϕ is Gaussian, there exists a constant $\delta_0 > 0$ such that

$$\mathbf{E}\exp(\delta_0\|\bar{x}(0)\|^2) = \int \exp(\delta_0\|\bar{x}\|^2)\nu^\phi(d\bar{x}) < \infty.$$

Hence, it follows from the multidimensional version of Theorem 4.7 on page 137 of [117] that for any $T > 0$,

$$\sup_{t \le T} \mathbf{E}\exp(\delta_0\|\bar{x}(t)\|^2) < \infty.$$

This condition implies that the constant $\delta_1 = \delta_0/\|\Phi\|^2$ is such that

$$\begin{aligned}
&\sup_{t \le T} \mathbf{E}\exp(\delta_1\|\Phi\bar{x}(t)\|^2) \\
&\quad < \sup_{t \le T} \mathbf{E}\exp(\delta_1\|\Phi\|^2\|\bar{x}(t)\|^2) \\
&\quad < \sup_{t \le T} \mathbf{E}\exp(\delta_0\|\bar{x}(t)\|^2) \\
&\quad < \infty.
\end{aligned}$$

Furthermore,

$$\mathbf{E}\exp\left(\frac{1}{2}\int_0^T \|\Phi\bar{x}(t)\|^2 dt\right) < \infty; \tag{8.5.51}$$

see Example 3 on page 220 of [117]. Also, since the process $\bar{x}(t)$ is defined by a linear system with constant coefficients, then

$$P\left(\int_0^T \|\Phi\bar{x}(t)\|^2 dt < \infty\right) = 1. \tag{8.5.52}$$

Conditions (8.5.51) and (8.5.52) are the conditions of Lemma 2.4.1 which shows that the random process $\tilde{W}(\cdot)$ defined by equation (8.5.48) is a Wiener process with respect to $\{\mathcal{F}_t, t \in [0,T]\}$ and the probability measure Q^T defined in Lemma 2.4.1. Also, as in [53, 39],

$$h(Q^T\|P^T) = \frac{1}{2}\mathbf{E}^{Q^T}\int_0^T \|\Phi\bar{x}(t)\|^2 dt < \infty$$

for each $T < \infty$ since $\bar{x}(t)$ is a solution to the linear system (8.5.3). Thus, $Q^T \in \mathcal{P}_T$ for all $T > 0$. Hence, $\zeta(\cdot) \in \mathcal{M}_\infty$.

We now consider the system (8.5.49) on the probability space $(\Omega, \mathcal{F}_T, Q^T)$ with initial distribution ν^ϕ. Also, consider the system (8.5.47) on the probability space $(\Omega, \mathcal{F}_T, P^T)$ with initial distribution ν^ϕ. Since the probability distributions of the Wiener process $\tilde{W}$ considered on the uncertainty probability space $(\Omega, \mathcal{F}_T, Q^T)$ are the same as those of the Wiener process W considered on the probability space $(\Omega, \mathcal{F}_T, P^T)$, it then follows from Proposition 3.10 on page 304 of [96] that the stochastic process $\bar{x}_\zeta(\cdot)$ defined by equation (8.5.49) and the corresponding stochastic process $x_\phi(\cdot)$ defined by equation (8.5.47) have the same probability distribution under their respective probability measures. In particular, from the fact that ν^ϕ is the invariant probability measure of the Markov process $x_\phi(\cdot)$ it follows that

$$Q^T(\bar{x}_\zeta(t) \in d\bar{x}) = P^T(\bar{x}_\phi(t) \in d\bar{x}) = \nu^\phi(d\bar{x}). \tag{8.5.53}$$

The lemma now follows from equation (8.5.53).

■

We now present a converse to Theorem 8.5.1.

Theorem 8.5.2 *Suppose that there exists a controller $u^*(\cdot) \in \mathcal{U}$ such that the following conditions are satisfied:*

(i) $\sup_{\zeta \in \Xi} J(u^*(\cdot), \zeta(\cdot)) < c < \infty.$

(ii) The controller $u^*(\cdot)$ is an absolutely stabilizing controller such that the corresponding closed loop matrix $\bar{A}$ is stable, the pair $(\bar{A}, \bar{B})$ is controllable and the pair $(\bar{A}, \bar{R})$ is observable.

Then, there exists a constant $\tau > 0$ such that the corresponding risk-sensitive optimal control problem (8.5.31) has a solution which guarantees a finite optimal value. Furthermore,

$$V_\tau + \tau d < c. \tag{8.5.54}$$

Proof: Given a controller $u^*(\cdot) \in \mathcal{U}$ satisfying the conditions of the theorem, this controller also satisfies the condition

$$\sup_{\zeta \in \Xi} \liminf_{T\to\infty} \frac{1}{2T} \mathbf{E}^{Q^T} \int_0^T F(x(s), u^*(s)) ds < c. \tag{8.5.55}$$

Also, since the closed loop system is absolutely stable, then there exists a sufficiently small $\bar{\varepsilon} > 0$ such that

$$\begin{aligned} &\liminf_{T\to\infty} \frac{1}{2T} \mathbf{E}^{Q^T} \int_0^T F(x(s), u^*(s)) ds \\ &\quad + \bar{\varepsilon} \liminf_{T\to\infty} \frac{1}{2T} \mathbf{E}^{Q^T} \int_0^T \|x(t)\|^2 dt \le c - \bar{\varepsilon} \end{aligned} \tag{8.5.56}$$

for all $\zeta(\cdot) \in \Xi$.

Consider the functionals

$$\begin{aligned} \mathcal{G}_0(\zeta(\cdot)) &:= c - \bar{\varepsilon} - \liminf_{T\to\infty} \frac{1}{2T} \mathbf{E}^{Q^T} \int_0^T F(x(s), u^*(s)) ds \\ &\quad - \bar{\varepsilon} \liminf_{T\to\infty} \frac{1}{2T} \mathbf{E}^{Q^T} \int_0^T \|x(t)\|^2 dt; \\ \mathcal{G}_1(\zeta(\cdot)) &:= -d - \liminf_{T\to\infty} \frac{1}{T} \left[\frac{1}{2} \mathbf{E}^{Q^T} \int_0^T \|z(s)\|^2 ds - h(Q^T \| P^T) \right]. \end{aligned} \tag{8.5.57}$$

Note that since the controller $u^*(\cdot)$ is an absolutely stabilizing controller, then both of these functionals is well defined on the set Ξ.

Now consider a martingale $\zeta(\cdot) \in \mathcal{M}_\infty$ such that

$$\mathcal{G}_1(\zeta(\cdot)) \le 0. \tag{8.5.58}$$

This condition implies that the martingale $\zeta(\cdot)$ satisfies the conditions of Definition 8.5.1 with $\zeta_i(\cdot) = \zeta(\cdot)$. Indeed, condition (i) of Definition 8.5.1 is satisfied since $\zeta(\cdot) \in \mathcal{M}_\infty$. Condition (ii) is trivial in this case. Also, let $\delta(T)$ be any function chosen to satisfy the conditions $\lim_{T\to\infty} \delta(T) = 0$ and

$$\inf_{T'>T} \frac{1}{T'} \left[\frac{1}{2} \mathbf{E}^{Q^T} \int_0^{T'} \|z(s)\|^2 ds - h(Q^{T'} \| P^{T'}) \right] \ge -d + \delta(T)$$

for all sufficiently large $T > 0$. The existence of such a function $\delta(T)$ follows from condition (8.5.58). Then condition (8.5.6) of Definition 8.5.1 is also satisfied. Thus, condition (8.5.58) implies that each martingale $\zeta(\cdot) \in \mathcal{M}_\infty$ satisfying this condition is an admissible uncertainty martingale. That is $\zeta(\cdot) \in \Xi$. From condition (8.5.56), it follows that $\mathcal{G}_0(\zeta(\cdot)) \geq 0$.

We have now shown that the satisfaction of condition (8.5.56) implies that the following condition is satisfied:

$$\text{If } \mathcal{G}_1(\zeta(\cdot)) \leq 0, \text{ then } \mathcal{G}_0(\zeta(\cdot)) \geq 0. \tag{8.5.59}$$

Furthermore, the set of martingales satisfying condition (8.5.58) has an interior point $\zeta(t) \equiv 1$; see the remark following Definition 8.5.1. Also, it follows from the properties of the relative entropy functional, that the functionals $\mathcal{G}_0(\cdot)$ and $\mathcal{G}_1(\cdot)$ are convex. We have now verified all of the conditions needed to apply Lemma 4.6.1. Indeed, the second part of Lemma 4.6.1 implies that there exists a constant $\tau_0 \geq 0$ such that

$$\mathcal{G}_0(\zeta(\cdot)) + \tau_0 \mathcal{G}_1(\zeta(\cdot)) \geq 0 \tag{8.5.60}$$

for all $\zeta(\cdot) \in \mathcal{M}_\infty$. We now show that the conditions of the theorem guarantee that $\tau_0 > 0$.

Proposition 2 *In inequality (8.5.60), $\tau_0 > 0$.*

Consider the system (8.5.49) where $\phi(\bar{x}) := \Phi\bar{x}$ belongs to $\mathfrak{D}_0$. From Lemma 8.5.3, the corresponding martingale $\zeta_\Phi(\cdot)$ belongs to the set $\mathcal{M}_\infty$.

Now consider the quantity

$$\liminf_{T\to\infty} \frac{1}{2T} \mathbf{E}^{Q_\Phi^T} \int_0^T F(x(t), u^*(t))dt = \liminf_{T\to\infty} \frac{1}{2T} \mathbf{E}^{Q_\Phi^T} \int_0^T \bar{x}(t)'\bar{R}\bar{x}(t)dt.$$

Here, Q_Φ^T is the probability measure corresponding to the martingale $\zeta_\Phi(\cdot)$, $\bar{x}(\cdot)$ is the solution to the corresponding system (8.5.1) considered on the probability space $(\Omega, \mathcal{F}_T, Q_\Phi^T)$. Also, consider the Lyapunov equation

$$(\bar{A} + \bar{B}\Phi)'\Pi + \Pi(\bar{A} + \bar{B}\Phi) + \bar{R} = 0. \tag{8.5.61}$$

Since the matrix $\bar{A}+\bar{B}\Phi$ is stable, then this matrix equation admits a nonnegative-definite solution Π. Using Ito's formula, it is straightforward to show that equation (8.5.61) leads to the inequality

$$\begin{aligned}
&\frac{1}{T}\mathbf{E}^{Q_\Phi^T} \int_0^T \bar{x}(t)'\bar{R}\bar{x}(t)dt \\
&= -\frac{1}{T}\mathbf{E}^{Q_\Phi^T}\bar{x}(T)'\Pi\bar{x}(T) + \frac{1}{T}\mathbf{E}^{Q_\Phi^T}\bar{x}(0)'\Pi\bar{x}(0) + \operatorname{tr}\bar{B}\bar{B}'\Pi \\
&\geq -\frac{1}{T}\mathbf{E}^{Q_\Phi^T}\bar{x}(T)'\Pi\bar{x}(T) + \operatorname{tr}\bar{B}\bar{B}'\Pi.
\end{aligned} \tag{8.5.62}$$

Let α be a positive constant such that all of the eigenvalues of the matrix $\bar{A}+\bar{B}\Phi$ are located in the half-plane Re$s \leq -\alpha$. Then, we obtain

$$
\begin{aligned}
&\frac{1}{T}\mathbf{E}^{Q_\Phi^T}\bar{x}(T)'\Pi\bar{x}(T) \\
&\leq \frac{1}{T}\mathbf{E}^{Q_\Phi^T}\bar{x}'(0)e^{T(\bar{A}+\bar{B}\Phi)'}\Pi e^{T(\bar{A}+\bar{B}\Phi)}\bar{x}(0) \\
&\quad +\frac{1}{T}\int_0^T e^{(T-s)(\bar{A}+\bar{B}\Phi)'}\bar{B}'\Pi\bar{B}e^{(T-s)(\bar{A}+\bar{B}\Phi)}ds \\
&\leq \frac{\|\Pi\|}{T}\mathbf{E}^{Q_\Phi}\|\bar{x}(0)\|^2 e^{-2\alpha T} + \frac{\|\Pi\|\|\bar{B}\|^2}{T}\frac{1-e^{-2\alpha T}}{2\alpha}.
\end{aligned}
$$

Hence,

$$
\lim_{T\to\infty}\frac{1}{T}\mathbf{E}^{Q_\Phi^T}\bar{x}(T)'\Pi\bar{x}(T) = 0.
$$

Using this result, we let T approach infinity in (8.5.62). This leads to the following condition:

$$
\liminf_{T\to\infty}\frac{1}{T}\mathbf{E}^{Q_\Phi^T}\int_0^T \bar{x}(t)'\bar{R}\bar{x}(t)dt \geq \operatorname{tr}\bar{B}\bar{B}'\Pi.
$$

This condition implies that

$$
\begin{aligned}
&\sup_{\varsigma(\cdot)\in\mathcal{M}_\infty}\liminf_{T\to\infty}\frac{1}{2T}\mathbf{E}^{Q^T}\int_0^T F(x(t),u^*(t))dt \\
&\geq \sup_{\Phi:\bar{A}+\bar{B}\Phi \text{ is stable}}\frac{1}{2}\operatorname{tr}\bar{B}\bar{B}'\Pi = \infty.
\end{aligned} \tag{8.5.63}
$$

Using (8.5.63), the proposition follows. Indeed, suppose that $\tau_0 = 0$. Then condition (8.5.60) implies that

$$
\begin{aligned}
&\sup_{\varsigma(\cdot)\in\mathcal{M}_\infty}\left[\begin{array}{l}\liminf_{T\to\infty}\frac{1}{2T}\mathbf{E}^{Q^T}\int_0^T F(x(t),u^*(t))dt \\ +\bar{\varepsilon}\liminf_{T\to\infty}\frac{1}{2T}\mathbf{E}^{Q^T}\int_0^T \|x(t)\|^2 dt\end{array}\right] \\
&\leq c-\bar{\varepsilon}<\infty.
\end{aligned} \tag{8.5.64}
$$

Inequality (8.5.64) leads to a contradiction with condition (8.5.63). From this, it follows that $\tau_0 > 0$.

Proposition 3 *The Riccati equation*

$$
\bar{A}'\Pi + \Pi\bar{A} + \bar{R} + \tau_0\bar{C}'\bar{C} + \frac{1}{\tau_0}\Pi\bar{B}\bar{B}'\Pi = 0 \tag{8.5.65}
$$

admits a positive-definite stabilizing solution.

We first note that the pair $(\bar{A}, \bar{R} + \tau_0 \bar{C}'\bar{C})$ is observable, since the pair $(\bar{A}, \bar{R})$ is observable. Hence, if $\Pi \geq 0$ satisfies equation (8.5.65), then $\Pi > 0$. Thus, it is sufficient to prove that equation (8.5.65) admits a nonnegative-definite stabilizing solution. This is true if and only if the following bound on the H^∞ norm of the corresponding transfer function is satisfied:

$$\|\mathcal{H}_{\tau_0}(s)\|_\infty \leq 1, \tag{8.5.66}$$

where

$$\mathcal{H}_{\tau_0}(s) := \begin{bmatrix} \frac{1}{\sqrt{\tau_0}} \bar{R}^{1/2} \\ \bar{C} \\ \frac{\sqrt{\bar{\varepsilon}}}{\sqrt{\tau_0}} I \end{bmatrix} (sI - \bar{A})^{-1} \bar{B};$$

see Lemma 5 and Theorem 5 of [247].

In order to prove the above claim, we note that condition (8.5.60) implies that for any martingale $\zeta \in \mathcal{M}_\infty$,

$$\begin{aligned} &\liminf_{T\to\infty} \frac{1}{T} \mathbf{E}^{Q^T} \int_0^T \frac{1}{2\tau_0} \bar{x}(t)' \bar{R} \bar{x}(t) dt \\ &+ \frac{\bar{\varepsilon}}{\tau_0} \liminf_{T\to\infty} \frac{1}{2T} \mathbf{E}^{Q^T} \int_0^T \|x(t)\|^2 dt \\ &+ \liminf_{T\to\infty} \frac{1}{T} \left[\frac{1}{2} \mathbf{E}^{Q^T} \int_0^T \|z(s)\|^2 ds - h(Q^T \| P^T) \right] \\ &\leq \frac{c - \bar{\varepsilon}}{\tau_0} - d. \end{aligned} \tag{8.5.67}$$

We will show that the satisfaction of condition (8.5.66) follows from (8.5.67).

Suppose condition (8.5.66) is not true. That is, suppose that

$$\|\mathcal{H}_{\tau_0}(s)\|_\infty > 1. \tag{8.5.68}$$

We now consider a set $\mathfrak{P}$ of deterministic power signals. According to the definition given in [271, 121], a signal $\xi(t)$, $t \in (-\infty, \infty)$ belongs to the set $\mathfrak{P}$ if the autocorrelation matrix

$$\mathcal{R}_{\xi\xi}(\theta) := \lim_{T\to\infty} \frac{1}{2T} \int_{-T}^{T} \xi(t+\theta)\xi^*(t) dt$$

exists and is finite for all θ, and also, the power spectral density function $S_{\xi\xi}(j\omega)$ exists. Here,

$$S_{\xi\xi}(j\omega) := \int_{-\infty}^{\infty} \mathcal{R}_{\xi\xi}(\theta) e^{-j\omega\theta} d\theta.$$

The power seminorm of a signal $\xi(\cdot) \in \mathfrak{P}$ is given by

$$\|\xi\|_{\mathfrak{P}} = \left[\lim_{T\to\infty} \frac{1}{2T} \int_{-T}^{T} \|\xi(t)\|^2 dt \right]^{1/2} = \sqrt{\operatorname{tr} \mathcal{R}_{\xi\xi}(0)}.$$

Furthermore, consider a set $\mathfrak{P}^+ \subset \mathfrak{P}$ which consists of input signals $\xi(t)$ such that $\xi(t) = 0$ if $t < 0$. The transfer function $\mathcal{H}_{\tau_0}(s)$ is a strictly proper real rational stable transfer function. Also, it can be shown that

$$\|\mathcal{H}_{\tau_0}\|_{\mathfrak{P}^+} = \|\mathcal{H}_{\tau_0}\|_\infty \tag{8.5.69}$$

where $\|\mathcal{H}_{\tau_0}\|_{\mathfrak{P}^+}$ denotes the induced norm of the convolution operator $\mathfrak{P}^+ \to \mathfrak{P}^+$ defined by the transfer function $\mathcal{H}_{\tau_0}(s)$; see [271]. Indeed, the proof of (8.5.69) is a minor variation[4] of a corresponding proof in [271] proving that $\|\mathcal{H}_{\tau_0}\|_{\mathfrak{P}} = \|\mathcal{H}_{\tau_0}\|_\infty$. Now consider the following state space realization of the transfer function $\mathcal{H}_{\tau_0}(s)$:

$$\begin{aligned} \frac{d\bar{x}_1}{dt} &= \bar{A}\bar{x}_1 + \bar{B}\xi(t); \\ z_1 &= \begin{bmatrix} \frac{1}{\sqrt{\tau_0}}\bar{R}^{1/2} \\ \bar{C} \\ \frac{\sqrt{\bar{\varepsilon}}}{\sqrt{\tau_0}} I \end{bmatrix} \bar{x}_1 \end{aligned} \tag{8.5.70}$$

driven by an input $\xi(\cdot) \in \mathfrak{P}^+$. Then, the fact that $\|\mathcal{H}_{\tau_0}\|_{\mathfrak{P}^+} = \|\mathcal{H}_{\tau_0}\|_\infty > 1$ leads to the following conclusion:

$$\sup_{\xi(\cdot)\in\mathfrak{P}^+} \lim_{T\to\infty} \frac{1}{T} \int_0^T \left(\|z_1(t)\|^2 dt - \|\xi(t)\|^2\right) dt = \infty. \tag{8.5.71}$$

In equation (8.5.71), $z_1(\cdot)$ is the output of the system (8.5.70) corresponding to the input $\xi(\cdot)$ and an arbitrarily chosen initial condition $\bar{x}_1(0)$[5]. That is, for any $N > 0$, there exists an uncertainty input $\xi_N(\cdot) \in \mathfrak{P}^+$ such that

$$\lim_{T\to\infty} \frac{1}{T} \int_0^T \left(\|z_1(t)\|^2 dt - \|\xi_N(t)\|^2\right) dt > N.$$

---

[4]In order to prove (8.5.69), the following modification to the proof given on pages 108–109 of [271] is used. The function $\hat{u}(t) = \sqrt{2}\sin(\omega_0 t) \in \mathfrak{P}$ is replaced by the function

$$\hat{u}(t) = \begin{cases} 2\sin(\omega_0 t) & \text{if } t \geq 0; \\ 0 & \text{if } t < 0. \end{cases}$$

Hence, $\hat{u}(t) \in \mathfrak{P}^+$. The corresponding autocorrelation function is

$$\begin{aligned} \mathcal{R}_{\hat{u}\hat{u}}(\theta) &= \lim_{T\to\infty} \frac{1}{2T} \int_{-T}^{T} \hat{u}(t+\theta)\hat{u}(t)dt \\ &= \lim_{T\to\infty} \frac{1}{T} \int_{\max\{0,-\theta\}}^{T} 2\sin(\omega_0(t+\theta))\sin(\omega_0 t)dt \\ &= \cos(\omega_0\theta). \end{aligned}$$

The remainder of the proof given in [271] remains unchanged.

[5]Note, that the limit on the left hand side of equation (8.5.71) is independent of the initial condition of the system (8.5.70).

This condition implies that for a sufficiently small $\varepsilon > 0$, there exists a constant $T(\varepsilon, N) > 0$ such that

$$\frac{1}{T}\int_0^T (\|z_1(t)\|^2 dt - \|\xi_N(t)\|^2)\, dt > N - \varepsilon \tag{8.5.72}$$

for all $T > T(\varepsilon, N)$. We now suppose the initial condition of the system (8.5.70) is a random variable $\bar{x}_0$. This system is driven by the input $\xi_N(\cdot)$. In this case, the system (8.5.70) gives rise to an $\mathcal{F}_0$-measurable stochastic process $\bar{x}_1(\cdot)$. Furthermore, for all $T > T(\varepsilon, N)$, inequality (8.5.72) holds with probability one. Now note that the signal $\xi_N(\cdot)$ is a deterministic signal. Hence, it satisfies the conditions of Lemma 2.4.1. Therefore for this process, the martingale $\zeta_N \in \mathcal{M}$, the probability measure Q_N^T and the Wiener process $\tilde{W}(\cdot)$ can be constructed as described in that lemma. Also, since $\xi_N(\cdot) \in \mathfrak{P}^+$ and is deterministic, then for any $T > 0$,

$$\mathbf{E}^{Q_N^T}\int_0^T \|\xi_N(t)\|^2 dt = \int_0^T \|\xi_N(t)\|^2 dt < \infty.$$

From this observation, it follows that $\zeta_N(\cdot) \in \mathcal{M}_\infty$ and also the random variable on the left hand side of inequality (8.5.72) has the finite expectation with respect to the probability measure Q_N^T. Furthermore, using inequality (8.5.72), one can prove that the system

$$d\bar{x} = (\bar{A}\bar{x} + \bar{B}\xi_N(t))dt + \bar{B}d\tilde{W}(t), \tag{8.5.73}$$

considered on the probability space $(\Omega, \mathcal{F}_T, Q_N^T)$, satisfies the following condition:

$$\frac{1}{T}\mathbf{E}^{Q_N^T}\int_0^T \left(\bar{x}'(t)(\frac{1}{\tau_0}\bar{R} + \bar{C}'\bar{C})\bar{x}(t) + \frac{\bar{\varepsilon}}{\tau_0}\|x(t)\|^2 - \|\xi_N(t)\|^2\right) dt > N - \varepsilon.$$

This condition can be established using the same arguments as those used in proving the corresponding fact in Section 8.4; see the proof of the second part of Theorem 8.4.2. Hence,

$$\lim_{T\to\infty}\frac{1}{T}\mathbf{E}^{Q_N^T}\int_0^T \left(\bar{x}'(t)(\frac{1}{\tau_0}\bar{R} + \bar{C}'\bar{C})\bar{x}(t) + \frac{\bar{\varepsilon}}{\tau_0}\|x(t)\|^2 - \|\xi_N(t)\|^2\right) dt \geq N. \tag{8.5.74}$$

Letting $N \to \infty$ in equation (8.5.74) and using the representation of the relative entropy $h(Q_N^T\|P^T)$, we obtain a contradiction with (8.5.67):

$$\sup_{\zeta\in\mathcal{M}_\infty}\left\{\begin{array}{l} \liminf_{T\to\infty}\frac{1}{T}\mathbf{E}^{Q^T}\int_0^T \frac{1}{2\tau_0}F(x(t),u^*(t))dt \\ +\frac{\bar{\varepsilon}}{\tau_0}\liminf_{T\to\infty}\frac{1}{2T}\mathbf{E}^{Q^T}\int_0^T \|x(s)\|^2 ds \\ +\liminf_{T\to\infty}\frac{1}{T}\left[\frac{1}{2}\mathbf{E}^{Q^T}\int_0^T \|z(s)\|^2 ds - h(Q^T\|P^T)\right] \end{array}\right\}$$

$$\geq \sup_{N>0} \left\{ \begin{array}{l} \lim_{T\to\infty} \frac{1}{T}\mathbf{E}^{Q_N^T} \int_0^T \frac{1}{2\tau_0}\bar{x}(t)'\bar{R}\bar{x}(t)dt \\ +\frac{\bar{\varepsilon}}{\tau_0} \lim_{T\to\infty} \frac{1}{2T}\mathbf{E}^{Q_N^T} \int_0^T \|x(s)\|^2 ds \\ +\lim_{T\to\infty} \frac{1}{2T}\mathbf{E}^{Q_N^T} \int_0^T (\|z(s)\|^2 - \|\xi_N(s)\|^2) ds \end{array} \right\}$$
$$= \infty.$$

Thus, condition (8.5.66) holds. As observed above, the proposition follows from this condition.

The remainder of the proof exploits a large deviation result established in the paper [170]; see Section 3.3.1. We first note that the Markov process defined by equation (8.5.3) and the corresponding invariant probability measure together satisfy the basic assumptions of Section 3.3.1. Also, it follows from Proposition 3 that the Riccati equation (8.5.65) has a positive-definite stabilizing solution. Thus, we have verified all of the conditions of Lemma 3.3.1. It follows from this lemma that

$$\lim_{T\to\infty} \frac{\tau_0}{T} \log \mathbf{E} \exp\left\{ \frac{1}{2\tau_0} \int_0^T \bar{x}'(t)(\bar{R} + \tau_0 \bar{C}'\bar{C})\bar{x}(t)dt \right\}$$
$$= \frac{1}{2} \int \left[\bar{x}'(\bar{R} + \tau_0 \bar{C}'\bar{C})\bar{x} - \tau_0 \|\phi(x)\|^2 \right] \nu^\phi(d\bar{x}) \qquad (8.5.75)$$

where $\phi(\bar{x}) = 1/\tau_0 \bar{B}'\Pi\bar{x}$ and Π is the positive-definite stabilizing solution to Riccati equation (8.5.65). On the left hand side of equation (8.5.75), $\bar{x}(\cdot)$ is the solution to equation (8.5.3) corresponding to the given controller of the form (8.5.2) and a given initial condition. It is shown in [170] that the value on both sides of equation (8.5.75) is independent of this initial condition.

For the function $\phi(\cdot)$ defined above, consider the martingale $\zeta(\cdot) \in \mathcal{M}_\infty$ and the corresponding stationary solution $\bar{x}(\cdot)$ to the system (8.5.49) with initial distribution ν^ϕ constructed as in Lemma 8.5.3. For this martingale $\zeta(\cdot)$ and the stationary solution $\bar{x}(\cdot)$, condition (8.5.75) leads to the following expression for the risk-sensitive cost:

$$\lim_{T\to\infty} \frac{\tau_0}{T} \log \mathbf{E} \exp\left\{ \frac{1}{2\tau_0} \int_0^T \bar{x}'(t)(\bar{R} + \tau_0 \bar{C}'\bar{C})\bar{x}(t)dt \right\}$$
$$= \frac{1}{2} \int \left[\bar{x}'(\bar{R} + \tau_0 \bar{C}'\bar{C})\bar{x} - \tau_0 \|\phi(x)\|^2 \right] \nu^\phi(d\bar{x})$$
$$= \liminf_{T\to\infty} \frac{1}{2T}\mathbf{E}^{Q^T} \int_0^T F(x(s), u^*(s))ds$$
$$+\tau_0 \liminf_{T\to\infty} \frac{1}{T} \left[\frac{1}{2}\mathbf{E}^{Q^T} \int_0^T \|z(s)\|^2 ds - h(Q^T \| P^T) \right] \qquad (8.5.76)$$

Also, note that the right hand side of the above equation is independent of the initial condition of the system (8.5.49). Indeed, using Ito's formula and the fact that the process $\tilde{W}(t)$ is the Wiener process which is adapted to the filtration $\{\mathcal{F}_t\}$ and is independent of the initial condition of the system (8.5.49), we obtain

$$\begin{aligned}
&\lim_{T\to\infty}\frac{1}{T}\mathbf{E}^{Q^T}\int_0^T \|\bar{x}(t)\|^2 dt \\
&= \lim_{T\to\infty}\frac{1}{T}\int_0^T e^{(\bar{A}+\frac{1}{\tau_0}\bar{B}\bar{B}'\Pi)'t}\mathbf{E}^{Q^0}\|\bar{x}_0\|^2 e^{(\bar{A}+\frac{1}{\tau_0}\bar{B}\bar{B}'\Pi)t}dt \\
&+ \lim_{T\to\infty}\frac{1}{T}\int_0^T\left[\int_0^t e^{(\bar{A}+\frac{1}{\tau_0}\bar{B}\bar{B}'\Pi)'(t-s)}\bar{B}'\bar{B}e^{(\bar{A}+\frac{1}{\tau_0}\bar{B}\bar{B}'\Pi)(t-s)}ds\right]dt.
\end{aligned}$$

Since the matrix $\bar{A}+\frac{1}{\tau_0}\bar{B}\bar{B}'\Pi$ is stable, the first term on the right hand side of the above identity is equal to zero, and the second term is independent of the initial conditions. Therefore, on the right hand side of inequality (8.5.76), the stationary process $\bar{x}(\cdot)$ can be replaced by the solution $\bar{x}(\cdot)$ to the system (8.5.49) corresponding to the given initial condition. Then, equation (8.5.76) and inequality (8.5.60) imply that

$$\lim_{T\to\infty}\frac{\tau_0}{T}\log\mathbf{E}\exp\left\{\frac{1}{2\tau_0}\int_0^T \bar{x}'(t)(\bar{R}+\tau_0\bar{C}'\bar{C})\bar{x}(t)dt\right\} \le c-\bar{\varepsilon}-\tau_0 d. \tag{8.5.77}$$

Thus,

$$\begin{aligned}
V_{\tau_0} &= \inf_{u(\cdot)}\lim_{T\to\infty}\frac{\tau_0}{T}\log\mathbf{E}\exp\left[\frac{1}{2\tau_0}\int_0^T F_\tau(x(s),u(s))ds\right] \\
&\le \lim_{T\to\infty}\frac{\tau_0}{T}\log\mathbf{E}\exp\left[\frac{1}{2\tau_0}\int_0^T \bar{x}'(t)(\bar{R}+\tau_0\bar{C}'\bar{C})\bar{x}(t)dt\right] \\
&\le c-\tau_0 d.
\end{aligned}$$

Hence the optimal value of the corresponding risk-sensitive control problem (8.5.31) is finite.

■

8.5.4 *Design of the infinite-horizon minimax optimal controller*

We now present the main result of this section. This result shows that the solution to an infinite-horizon minimax optimal control problem of the form (8.5.25) can be obtained via optimization over solutions to a scaled risk-sensitive control problem of the form (8.5.31). Therefore, this result extends the corresponding result of Section 8.4 to the case where the underlying system is considered on an infinite time interval. Note that the scaled risk-sensitive control problem considered in this subsection admits a tractable solution in terms of a pair of algebraic Riccati equations of the H^∞ type; e.g., see Section 3.3.3 and [139].

The derivation of the main result of this section makes use of time-invariant versions of the parameter dependent Riccati equations (8.4.34), (8.4.35). Let $\tau > 0$ be a constant. We consider the algebraic Riccati equations

$$\begin{aligned}&(A - B_2D_2'\Gamma^{-1}C_2)Y_\infty + Y_\infty(A - B_2D_2'\Gamma^{-1}C_2)'\\&-Y_\infty(C_2'\Gamma^{-1}C_2 - \frac{1}{\tau}R_\tau)Y_\infty + B_2(I - D_2'\Gamma^{-1}D_2)B_2' = 0;\end{aligned} \tag{8.5.78}$$

$$\begin{aligned}&X_\infty(A - B_1G_\tau^{-1}\Upsilon_\tau') + (A - B_1G_\tau^{-1}\Upsilon_\tau')'X_\infty\\&+(R_\tau - \Upsilon_\tau G_\tau^{-1}\Upsilon_\tau') - X_\infty(B_1G_\tau^{-1}B_1' - \frac{1}{\tau}B_2B_2')X_\infty = 0.\end{aligned} \tag{8.5.79}$$

Note that the Riccati equations (8.5.78) and (8.5.79) arise in connection with the infinite-horizon risk-sensitive control problem considered in Section 3.3.3. Indeed, equations (8.5.78) and (8.5.79) are obtained from equations (3.3.28) and (3.3.29) by replacing R, Υ and G with R_τ, Υ_τ and G_τ, respectively.

Recall that Section 3.3.3 and the paper [139] defined the class of admissible infinite horizon risk-sensitive controllers as those controllers of the form (8.4.2) which satisfy the causality condition given in Section 3.3.3. This causality condition was formulated in terms of corresponding martingales and ensured that the probability measure transformations required in [139] were well defined. As observed in [139], linear controllers forming the class $\mathcal{U}$, satisfy this causality condition; see also Remark 2.4.7. Furthermore, Theorem 3.3.2 of Section 3.3.3 shows that solution to the risk-sensitive optimal control problem (3.3.27) in the broader class of nonlinear output-feedback controllers satisfying such a causality condition, is attained by a linear controller which belongs to $\mathcal{U}$. This implies that the class of admissible controllers in the risk-sensitive control problem (8.5.31) can be restricted to include only linear output-feedback controllers. In the sequel, we will focus on linear output-feedback controllers of the form (8.5.2) having a controllable and observable state space realization. The class of such controllers is denoted by $\mathcal{U}_0$.

The subsequent development relies on Theorem 3.3.2; see also Theorem 3 of reference [139]. We now present a version of Theorem 3.3.2 adapted to the notation used in this section. We first note that some of the conditions of Theorem 3.3.2 are automatically satisfied in the case considered in this section. Indeed, as in Section 8.4, it is readily shown that $R_\tau - \Upsilon_\tau G_\tau^{-1}\Upsilon_\tau' \geq 0$. Also, the pair

$$(A - B_1G_\tau^{-1}\Upsilon_\tau', R_\tau - \Upsilon_\tau G_\tau^{-1}\Upsilon_\tau')$$

is detectable since the matrix

$$\begin{bmatrix} A - sI & B_1 \\ R^{1/2} & 0 \\ 0 & G^{1/2} \\ \sqrt{\tau}C_1 & \sqrt{\tau}D_1 \end{bmatrix}$$

has full column rank for all s such that $\mathrm{Re} s \geq 0$; see Lemma 3.2.2 on page 69.

Lemma 8.5.4 *Consider the risk-sensitive optimal control problem (8.5.31) with underlying system (8.5.1). Suppose the pair*

$$(A - B_2 D_2' \Gamma^{-1} C_2, B_2(I - D_2' \Gamma^{-1} D_2)) \tag{8.5.80}$$

is stabilizable. Also, suppose that there exists a constant $\tau > 0$ such that the following assumptions are satisfied:

(i) The algebraic Riccati equation (8.5.78) admits a minimal positive-definite solution Y_∞.

(ii) The algebraic Riccati equation (8.5.79) admits a minimal nonnegative-definite solution X_∞.

(iii) The matrix $I - \frac{1}{\tau} Y_\infty X_\infty$ has only positive eigenvalues; that is the spectral radius of the matrix $Y_\infty X_\infty$ satisfies the condition

$$\rho(Y_\infty X_\infty) < \tau. \tag{8.5.81}$$

If $Y_\infty \geq Y_0$, then there exists a controller solving the risk-sensitive optimal control problem (8.5.31) where the infimum is taken over the set $\mathcal{U}$. This optimal risk-sensitive controller is a controller of the form (8.5.2) with

$$\begin{aligned} K &:= -G_\tau^{-1}(B_1' X_\infty + \Upsilon_\tau'); \\ A_c &:= A + B_1 K - B_c C_2 + \frac{1}{\tau}(B_2 - B_c D_2) B_2' X_\infty; \\ B_c &:= (I - \frac{1}{\tau} Y_\infty X_\infty)^{-1}(Y_\infty C_2' + B_2 D_2')\Gamma^{-1}. \end{aligned} \tag{8.5.82}$$

The corresponding optimal value of the risk-sensitive cost is given by

$$\begin{aligned} V_\tau &:= \inf_{u \in \mathcal{U}} \lim_{T \to \infty} \frac{1}{2} \Im_{\tau,T}(u(\cdot)) = \\ &\frac{1}{2} \mathrm{tr} \begin{bmatrix} Y_\infty R_\tau + \\ (Y_\infty C_2' + B_2 D_2')\Gamma^{-1}(C_2 Y_\infty + D_2 B_2') X_\infty (I - \frac{1}{\tau} Y_\infty X_\infty)^{-1} \end{bmatrix}. \end{aligned} \tag{8.5.83}$$

Remark 8.5.1 It was shown in Chapter 3 (see Lemma 3.3.2) that the conditions of Lemma 8.5.4 also guarantee that there exists a positive-definite symmetric solution $Y(t)$ to the Riccati differential equation (8.4.34) such that $Y(t) \to Y_\infty$ as $t \to \infty$. Also, for any $T > 0$, there exists a nonnegative-definite symmetric solution $X_T(t)$ to the Riccati differential equation (8.4.35) with $M = 0$ such that $X_T(t) \to X_\infty$ as $T \to \infty$. Furthermore, the matrix $I - \frac{1}{\tau} Y(t) X_T(t)$ has only positive eigenvalues. Therefore, the assumptions of Lemma 8.5.4 are consistent with those of Theorem 3 of [139]; see Section 3.3.3 for details.

The condition $Y_\infty \geq Y_0$ required by Lemma 8.5.4 is a technical condition needed to apply the results of the reference [139] to the risk-sensitive control problem associated with the minimax optimal control problem considered in this section. However, it can be seen from Lemma 8.5.4 that the resulting optimal risk-sensitive controller is independent of the matrix Y_0. This fact is consistent with the observation that in the infinite-horizon case, the optimal risk-sensitive cost is independent of the initial condition of the system (8.5.1); see equation (8.5.75) and also Lemma 3.3.1. Therefore, the condition of Lemma 8.5.4 requiring $Y_\infty \geq Y_0$ can always be satisfied by a suitable choice of the matrix Y_0.

Reference [139] does not address the issue of stability for the closed loop system corresponding to the optimal risk-sensitive controller. However, Theorem 8.5.1 shows that the controller (8.5.2), (8.5.82) leads to a robustly stable closed loop system. This fact is consistent with results showing that risk-sensitive controllers enjoy certain robustness properties; e.g., see [54, 236]. The following result shows that the conditions of Lemma 8.5.4 are not only sufficient conditions, but also necessary conditions for the existence of a solution to the risk-sensitive optimal control problem under consideration, if such a solution is sought in the class of linear stabilizing controllers; *c.f.* [135].

Lemma 8.5.5 *Suppose the pair (8.5.80) is controllable and for some $\tau > 0$, there exists an absolutely stabilizing controller $\tilde{u}(\cdot) \in \mathcal{U}_0$ such that*

$$V_\tau^0 := \lim_{T\to\infty} \frac{1}{2}\Im_{\tau,T}(\tilde{u}) = \inf_{u\in\mathcal{U}_0} \lim_{T\to\infty} \frac{1}{2}\Im_{\tau,T}(u) < +\infty. \tag{8.5.84}$$

Then, the constant τ satisfies conditions (i)–(iii) of Lemma 8.5.4.

Furthermore, if for this τ, the corresponding pairs (A_c, B_c) and (A_c, K) defined by equations (8.5.82) are controllable and observable respectively, then

$$V_\tau^0 = \frac{1}{2}\operatorname{tr}\begin{bmatrix} Y_\infty R_\tau + \\ (Y_\infty C_2' + B_2 D_2')\Gamma^{-1}(C_2 Y_\infty + D_2 B_2')X_\infty(I - \frac{1}{\tau}Y_\infty X_\infty)^{-1} \end{bmatrix}. \tag{8.5.85}$$

In the proof of Lemma 8.5.5, the following proposition is used.

Proposition 4 *Suppose the pair (8.5.80) is controllable. Then, for any controller $u(\cdot) \in \mathcal{U}_0$, the pair $(\bar{A}, \bar{B})$ in the corresponding closed loop system is controllable and the pair $(\bar{A}, \bar{R})$ is observable.*

Proof: Note that by definition, for any controller $u(\cdot) \in \mathcal{U}_0$, the corresponding pair (A_c, B_c) is controllable and the pair (A_c, K) is observable.

To prove the controllability of the pair $(\bar{A}, \bar{B})$, we first note that the matrix

$$\begin{bmatrix} A' - sI & C_2' \\ B_2' & D_2' \end{bmatrix} \tag{8.5.86}$$

has full column rank for all $s \in \mathbb{C}$. This fact follows from Lemma 3.2.2.

Next, consider the matrix pair

$$(\bar{A}, \bar{B}) = \left(\begin{bmatrix} A & B_1K \\ B_cC_2 & A_c \end{bmatrix}, \begin{bmatrix} B_2 \\ B_cD_2 \end{bmatrix} \right). \tag{8.5.87}$$

For this matrix pair to be controllable, the equations

$$(A' - sI)x_1 + C_2'B_c'x_2 = 0; \tag{8.5.88a}$$
$$B_2'x_1 + D_2'B_c'x_2 = 0; \tag{8.5.88b}$$
$$K'B_1'x_1 + (A_c' - sI)x_2 = 0 \tag{8.5.88c}$$

must imply that $x_1 = 0$ and $x_2 = 0$ for every $s \in \mathbb{C}$. Equations (8.5.88a) and (8.5.88b) can be written as follows:

$$\begin{bmatrix} A' - sI & C_2' \\ B_2' & D_2' \end{bmatrix} \begin{bmatrix} x_1 \\ B_c'x_2 \end{bmatrix} = 0.$$

It was noted above that the matrix (8.5.86) has full column rank for all $s \in \mathbb{C}$. Hence, the above equation and equation (8.5.88c) imply that

$$x_1 = 0; \quad B_c'x_2 = 0; \quad (A_c' - sI)x_2 = 0.$$

Since the pair (A_c, B_c) is controllable, then the two last equations imply that $x_2 = 0$. Thus, the pair (8.5.87) is controllable.

In order to prove the observability of the pair $(\bar{A}, \bar{R})$, we need to show that the equations

$$(A - sI)x_1 + B_1Kx_2 = 0; \tag{8.5.89a}$$
$$B_cC_2x_1 + (A_c - sI)x_2 = 0; \tag{8.5.89b}$$
$$R^{1/2}x_1 = 0; \tag{8.5.89c}$$
$$G^{1/2}Kx_2 = 0 \tag{8.5.89d}$$

imply that $x_1 = 0$, $x_2 = 0$ for every $s \in \mathbb{C}$. Indeed, since the matrices R, G are positive-definite, then it follows from equations (8.5.89c) and (8.5.89d) that $x_1 = 0$ and $Kx_2 = 0$. Using these equations, we also obtain from (8.5.89b) that $(A_c - sI)x_2 = 0$. Since the pair (A_c, K) is observable, then this implies that $x_1 = 0$ and $x_2 = 0$. Thus, the pair $(\bar{A}, \bar{R})$ is observable.

■

Proof of Lemma 8.5.5: We prove the lemma by contradiction. Suppose that at least one of conditions (i)–(iii) of Lemma 8.5.4 does not hold. That is, either equation (8.5.78) does not admit a positive-definite stabilizing solution or equation (8.5.79) does not admit a nonnegative-definite stabilizing solution, or condition (8.5.81) fails to hold. Note that conditions (i)–(iii) of Lemma 8.5.4 are standard conditions arising in H^∞ control. Since the matrix $\bar{A}$ is stable, then it follows from standard results on H^∞ control that if at least one of conditions (i)–(iii) of Lemma 8.5.4 fails to hold then for any controller of the form (8.5.2),

$$\| \left[\tilde{C}_1 \ \tilde{D}_1 K \right] (j\omega I - \bar{A})^{-1} \bar{B} \|_\infty \geq 1; \tag{8.5.90}$$

see Theorem 3.2.3 and also Theorem 3.1 of [147]. Here, we use the notation (8.4.33) assuming that the corresponding matrices are time-invariant. It is straightforward to verify that the conditions of Theorem 3.1 of [147] are satisfied. Furthermore, the Strict Bounded Real Lemma implies that the Riccati equation

$$\bar{A}'\bar{X} + \bar{X}\bar{A} + \frac{1}{\tau}\bar{X}\bar{B}\bar{B}'\bar{X} + \bar{R} + \tau\bar{C}'\bar{C} = 0. \tag{8.5.91}$$

does not have a stabilizing positive definite solution. In this case, Lemma 8.5.2 implies that none of the controllers $u(\cdot) \in \mathcal{U}_0$ lead to an absolutely stable closed loop system. This leads to a contradiction with the assumption that an absolutely stabilizing controller exists and belongs to $\mathcal{U}_0$. This completes the proof by contradiction that the given constant τ satisfies conditions (i) - (iii) of Lemma 8.5.4.

It remains to prove equation (8.5.85). Note that Lemma 8.5.4 states that for each $\tau > 0$ satisfying the conditions of that lemma, the optimal controller solving the risk-sensitive control problem (8.5.83) is the controller (8.5.2), (8.5.82). Furthermore, it is assumed that the state-space realization of this controller is controllable and observable, and hence the optimal controller from Lemma 8.5.4 belongs to the set $\mathcal{U}_0$. Therefore,

$$V_\tau^0 = \inf_{u \in \mathcal{U}_0} \lim_{T \to \infty} \frac{1}{2} \Im_{\tau,T}(u(\cdot)) = \inf_{u \in \mathcal{U}} \lim_{T \to \infty} \frac{1}{2} \Im_{\tau,T}(u(\cdot)) = V_\tau. \tag{8.5.92}$$

From this observation, equation (8.5.85) follows.

■

We now define a set $\mathcal{T} \subset \mathbf{R}$ as the set of constants $\tau \in \mathbf{R}$ satisfying the conditions of Lemma 8.5.4. It follows from Lemma 8.5.4 that for any $\tau \in \mathcal{T}$, the controller of the form (8.5.2) with the coefficients given by equations (8.5.82) represents an optimal controller in the risk-sensitive control problem (8.5.31), which guarantees the optimal value (8.5.85).

Theorem 8.5.3 *Assume that the pair (8.5.80) is controllable.*

(i) Suppose that the set $\mathcal{T}$ is non-empty and $\tau_* \in \mathcal{T}$ attains the infimum in

$$\inf_{\tau \in \mathcal{T}} (V_\tau^0 + \tau d), \tag{8.5.93}$$

where V_τ^0 is defined in equation (8.5.85). Then, the corresponding controller $u^*(\cdot) := u_{\tau_*}(\cdot)$ of the form (8.5.2) defined by (8.5.82) with the pair (A_c, B_c) being controllable and the pair (A_c, K) being observable, is an output-feedback controller guaranteeing that

$$\inf_{u \in \mathcal{U}_0} \sup_{\zeta \in \Xi} J(u, \zeta) \leq \sup_{\zeta \in \Xi} J(u^*, \zeta) \leq \inf_{\tau \in \mathcal{T}} (V_\tau^0 + \tau d). \tag{8.5.94}$$

Furthermore, this controller is an absolutely stabilizing controller for the stochastic uncertain system (8.5.1), (8.5.6).

(ii) *Conversely, if there exists an absolutely stabilizing minimax optimal controller $\tilde{u}(\cdot) \in \mathcal{U}_0$ for the stochastic uncertain system (8.5.1), (8.5.6) such that*

$$\sup_{\zeta \in \Xi} J(\tilde{u}, \zeta) < \infty,$$

then the set $\mathcal{T}$ is non-empty. Moreover,

$$\inf_{\tau \in \mathcal{T}} (V_\tau^0 + \tau d) \leq \sup_{\zeta \in \Xi} J(\tilde{u}, \zeta). \tag{8.5.95}$$

Proof: *Part (i)* The conditions of this part of the theorem guarantee that $u^*(\cdot) \in \mathcal{U}_0$. Then, $V_{\tau_*}^0 = V_{\tau_*}$. This fact together with Theorem 8.5.1 implies that

$$\inf_{u \in \mathcal{U}_0} \sup_{\zeta \in \Xi} J(u, \zeta) \leq \sup_{\zeta \in \Xi} J(u^*, \zeta) \leq (V_{\tau_*}^0 + \tau_* d) = \inf_{\tau \in \mathcal{T}} (V_\tau^0 + \tau d). \tag{8.5.96}$$

Also from Theorem 8.5.1, the controller $u^*(\cdot)$ solving the corresponding risk-sensitive control problem is an absolutely stabilizing controller. From this observation, part (i) of the theorem follows.

Part (ii) Note that the controller $\tilde{u}(\cdot) \in \mathcal{U}_0$ satisfies the conditions of Theorem 8.5.2; see Proposition 4. Let c be a constant such that

$$\sup_{\zeta \in \Xi} J(\tilde{u}, \zeta) < c.$$

When proving Theorem 8.5.2, it was shown that there exists a constant $\tau > 0$ such that

$$\frac{1}{2} \lim_{T \to \infty} \Im_{\tau,T}(\tilde{u}) < c - \tau d < \infty; \tag{8.5.97}$$

see equation (8.5.77). Hence, $V_\tau^0 < \infty$. From this condition and using Lemma 8.5.5, we conclude that the constant τ belongs to the set $\mathcal{T}$. Thus, the set $\mathcal{T}$ is non-empty.

We now prove equation (8.5.95). Consider a sequence $\{c_i\}$, $i = 1, 2, \ldots$, such that

$$c_i \downarrow \sup_{\zeta \in \Xi} J(\tilde{u}, \zeta) \quad \text{as } i \to \infty.$$

From (8.5.97) it follows that

$$\inf_{\tau \in \mathcal{T}} (V_\tau^0 + \tau d) < c_i.$$

Hence, letting i approach infinity leads to the satisfaction of equation (8.5.95).

■

The first part of Theorem 8.5.3 provides a sufficient condition for the existence of an optimal solution to the minimax LQG control problem considered in this section. This condition is given in terms of certain Riccati equations. This makes the result useful in practical controller design since there is a wide range of software available for solving such Riccati equations.

In the control literature, there is a great deal of interest concerning the issue of conservatism in robust controller design. For example, a significant issue considered in Sections 5.3, 9.2, 8.3 is to prove that the results on the minimax optimal control considered in those sections are not conservative in that the corresponding Riccati equations fail to have stabilizing solutions if the minimax optimal controller does not exist. Thus, the conditions for the existence of a minimax optimal controller presented in those sections are necessary and sufficient conditions. The second part of Theorem 8.5.3 is analogous to the necessity results of Sections 5.3, 9.2, 8.3, 8.4. It follows from this part of Theorem 8.5.3 that the controller $u^*(\cdot)$ constructed in the first part of Theorem 8.5.3 represents a minimax optimal controller in the subclass $\mathcal{U}_{0,\mathrm{stab}} \subset \mathcal{U}_0$ of stabilizing linear output feedback controllers. This result is summarized in the following theorem.

Theorem 8.5.4 *Assume that the conditions of part (i) of Theorem 8.5.3 are satisfied. Then, the controller $u^*(\cdot)$ constructed in part (i) of Theorem 8.5.3 is the minimax optimal controller such that*

$$\inf_{u\in\mathcal{U}_{0,\mathrm{stab}}} \sup_{\zeta\in\Xi} J(u,\zeta) = \sup_{\zeta\in\Xi} J(u^*,\zeta) = \inf_{\tau\in\mathcal{T}} (V_\tau^0 + \tau d). \tag{8.5.98}$$

Proof: It was shown in part (i) of Theorem 8.5.3 that the controller $u^*(\cdot)$ belongs to the set $\mathcal{U}_{0,\mathrm{stab}}$. Hence,

$$\inf_{u\in\mathcal{U}_{0,\mathrm{stab}}} \sup_{\zeta\in\Xi} J(u,\zeta) \leq \sup_{\zeta\in\Xi} J(u^*,\zeta) \leq \inf_{\tau\in\mathcal{T}} (V_\tau^0 + \tau d). \tag{8.5.99}$$

Furthermore, condition (8.5.99) implies that

$$\inf_{u\in\mathcal{U}_{0,\mathrm{stab}}} \sup_{\zeta\in\Xi} J(u,\zeta) < \infty.$$

That is, for any sufficiently small $\varepsilon > 0$, there exists a controller $\tilde{u}(\cdot) \in \mathcal{U}_{0,\mathrm{stab}}$ such that

$$\sup_{\zeta\in\Xi} J(\tilde{u},\zeta) \leq \inf_{u\in\mathcal{U}_{0,\mathrm{stab}}} \sup_{\zeta\in\Xi} J(u,\zeta) + \varepsilon.$$

This controller satisfies the conditions of part (ii) of Theorem 8.5.3. Therefore, it follows from Theorem 8.5.3 that

$$\inf_{\tau\in\mathcal{T}} (V_\tau^0 + \tau d) \leq \inf_{u\in\mathcal{U}_{0,\mathrm{stab}}} \sup_{\zeta\in\Xi} J(u,\zeta) + \varepsilon.$$

The above inequality holds for any infinitesimally small $\varepsilon > 0$. Therefore,

$$\inf_{\tau \in \mathcal{T}} (V_\tau^0 + \tau d) \leq \inf_{u \in \mathcal{U}_{0,\mathrm{stab}}} \sup_{\zeta \in \Xi} J(u, \zeta).$$

This inequality together with (8.5.99) implies equation (8.5.98).

■

8.5.5 *Connection to H^∞ control*

It is significant to note that the Riccati equations (8.5.78), (8.5.79) are closely related to those which arise in the standard infinite-horizon output-feedback H^∞ control problem; see Theorem 3.2.3. Indeed, if we define $\bar{Y} := (1/\tau)Y_\infty$ then it follows that Riccati equation (8.5.78) is equivalent to the Riccati equation

$$\begin{aligned}&(A - B_2 D_2' \Gamma^{-1} C_2)\bar{Y} + \bar{Y}(A - B_2 D_2' \Gamma^{-1} C_2)' \\ &-\tau \bar{Y}(C_2' \Gamma^{-1} C_2 - \frac{1}{\tau} R_\tau)\bar{Y} + \frac{1}{\tau} B_2 (I - D_2' \Gamma^{-1} D_2) B_2' = 0.\end{aligned} \tag{8.5.100}$$

Furthermore, condition (8.5.81) is equivalent to the condition

$$\rho(\bar{Y} X_\infty) < 1. \tag{8.5.101}$$

Now as in Theorem 8.5.3, suppose that the pair (8.5.80) is controllable. It follows from Theorem 3.2.3 that there will exist minimal positive-definite solutions to Riccati equations (8.5.100) and (8.5.79) satisfying the condition (8.5.101) if and only if there exists a solution to the infinite-horizon output-feedback H^∞ control problem (see Section 3.2.1) defined by the system

$$\begin{aligned}\dot{x} &= Ax + B_1 u + \frac{B_2}{\sqrt{\tau}} w; \\ z &= \begin{bmatrix} R^{\frac{1}{2}} \\ 0 \\ \sqrt{\tau} C_1 \end{bmatrix} x + \begin{bmatrix} 0 \\ G^{\frac{1}{2}} \\ \sqrt{\tau} D_1 \end{bmatrix} u; \\ y &= C_2 x + \frac{D_2}{\sqrt{\tau}} w\end{aligned} \tag{8.5.102}$$

and H^∞ norm bound condition $\|\mathfrak{W}_{zw}\|_\infty < 1$. Here, x is the state, u is the control input, w is the disturbance input, z is the controlled output and y is the measured output. Note that the assumptions required by Theorem 3.2.3 follow from the assumed controllability of the pair (8.5.80), the fact that $R > 0$ and the form of the system (8.5.102).

We now introduce the scaled disturbance input $\tilde{w} := \frac{w}{\sqrt{\tau}}$ and scaled controlled output, $\tilde{z} := \frac{z}{\sqrt{\tau}}$. It is straightforward to see that the above H^∞ control problem will have a solution if and only if there exists a solution to the infinite-horizon H^∞ control problem defined by the system

$$\begin{aligned}
\dot{x} &= Ax + B_1 u + B_2 \tilde{w};\\
\tilde{z} &= \begin{bmatrix} \frac{R^{\frac{1}{2}}}{\sqrt{\tau}} \\ 0 \\ C_1 \end{bmatrix} x + \begin{bmatrix} 0 \\ \frac{G^{\frac{1}{2}}}{\sqrt{\tau}} \\ D_1 \end{bmatrix} u;\\
y &= C_2 x + D_2 \tilde{w}
\end{aligned} \tag{8.5.103}$$

and the H^∞ norm bound condition $\|\mathfrak{W}_{\tilde{z}\tilde{w}}\|_\infty < 1$. Moreover, it is easily seen that if this H^∞ control problem has a solution for some $\tau = \tau_0$, then it will have a solution for all $\tau \geq \tau_0$. Thus, we can conclude that if the Riccati equations (8.5.78) and (8.5.79) have minimal positive-definite solutions satisfying condition (8.5.81) for $\tau = \tau_0$, then they will have minimal positive-definite solutions satisfying condition (8.5.81) for all $\tau \geq \tau_0$. Also, we can conclude that the Riccati equations (8.5.78) and (8.5.79) will have minimal positive-definite solutions satisfying condition (8.5.81) for some $\tau > 0$ if and only if there exists a solution to the infinite-horizon H^∞ control problem defined by the system

$$\begin{aligned}
\dot{x} &= Ax + B_1 u + B_2 \tilde{w};\\
\tilde{z} &= C_1 x + D_1 u;\\
y &= C_2 x + D_2 \tilde{w}
\end{aligned} \tag{8.5.104}$$

and the H^∞ norm bound condition $\|\mathfrak{W}_{\tilde{z}\tilde{w}}\|_\infty < 1$. Thus, the standard H^∞ robust controller design methodology can be regarded as a special case of the minimax LQG methodology. Also note that it follows from the proof of Lemma 8.5.2 that the uncertain system (8.5.1), (8.5.6) is absolutely stabilizable if and only if there exists a solution to the infinite-horizon H^∞ control problem corresponding to the system (8.5.104).

8.5.6 *Illustrative example*

We now reconsider the tracking problem which was used as an illustrative example in Sections 5.3 and 5.4. As in Section 5.4, it is desired to design an output-feedback controller so that the controlled output of the system tracks a reference step input.

The system to be controlled is shown in Figure 2.4.2. As in Sections 5.3 and 5.4, the masses of the carts are assumed to be $m_1 = 1$ and $m_2 = 1$. Furthermore, the spring constant k is treated as an uncertain parameter subject to the bound $0.5 \leq k \leq 2.0$. From this, a corresponding uncertain system was derived in Section 5.4. This uncertain system is described by the following state equations (see also (5.4.26)):

$$\begin{aligned}\dot{x} &= \begin{bmatrix} 0 & 0\ 1\ 0 \\ 0 & 0\ 0\ 1 \\ -1.25 & 1.25\ 0\ 0 \\ 1.25 & -1.25\ 0\ 0 \end{bmatrix} x + \begin{bmatrix} 0 \\ 0 \\ 0 \\ 1 \end{bmatrix} u + \begin{bmatrix} 0 & 0\ 0 \\ 0 & 0\ 0 \\ -0.70 & 0\ 0 \\ 0.80 & 0\ 0 \end{bmatrix} \xi; \\ z &= \begin{bmatrix} 1 & -1 & 0 & 0 \end{bmatrix} x; \\ y &= \begin{bmatrix} 1\ 0\ 0\ 0 \\ 0\ 1\ 0\ 0 \end{bmatrix} x + \begin{bmatrix} 0 & 0.05 & 0 \\ 0 & 0 & 0.05 \end{bmatrix} \xi; \\ y_T &= \begin{bmatrix} 1\ 0\ 0\ 0 \end{bmatrix} x \end{aligned} \tag{8.5.105}$$

Here, the uncertainty is subject to an integral quadratic constraint which will be specified below; see also, page 164. The output y_T is the output which is required to track a step input.

The control problem solved in Section 5.4 involved finding a controller which absolutely stabilized the system and also ensured that the output y_T tracks a reference step input. In Section 5.4, the system was transformed into the form:

$$\begin{aligned}\dot{\bar{x}} &= \begin{bmatrix} 0\ 1 & 0\ 0 \\ 0\ 0 & 0\ 0 \\ 0\ 0 & 0\ 1 \\ 0\ 0 & -2.5\ 0 \end{bmatrix} \bar{x} + \begin{bmatrix} 0 \\ 0.5 \\ 0 \\ -0.5 \end{bmatrix} u + \begin{bmatrix} 0 & 0\ 0 \\ 0.05 & 0\ 0 \\ 0 & 0\ 0 \\ -0.75 & 0\ 0 \end{bmatrix} \xi; \\ z &= \begin{bmatrix} 0\ 0\ 2\ 0 \end{bmatrix} \bar{x}; \\ \bar{y} &= \begin{bmatrix} 1\ 0 & 1 & 0 \\ 1\ 0 & -1 & 0 \end{bmatrix} \bar{x} + \begin{bmatrix} 0 & 0.05 & 0 \\ 0 & 0 & 0.05 \end{bmatrix} \xi; \\ y_T - \tilde{y}_T &= \begin{bmatrix} 1\ 0\ 1\ 0 \end{bmatrix} \bar{x} \end{aligned} \tag{8.5.106}$$

where

$$\bar{y} := y - \begin{bmatrix} 1 \\ 1 \end{bmatrix} \eta.$$

Here, η denotes the state of the reference input signal model:

$$\begin{aligned} \dot{\eta} &= 0; \quad \eta(0) = 1; \\ \tilde{y}_T &= \eta. \end{aligned} \tag{8.5.107}$$

The above transformation involved the following change of variables:

$$\begin{aligned} \bar{x}_1 &= (x_1 + x_2)/2 - \eta; \\ \bar{x}_2 &= (\dot{x}_1 + \dot{x}_2)/2 = (x_3 + x_4)/2; \\ \bar{x}_3 &= (x_1 - x_2)/2; \\ \bar{x}_4 &= (\dot{x}_1 - \dot{x}_2)/2 = (x_3 - x_4)/2. \end{aligned} \tag{8.5.108}$$

To construct the required controller, the following cost function of the form (5.3.2) was used:

$$\int_0^\infty [(y_T - \tilde{y}_T)^2 + 0.1\|\bar{x}\|^2 + u^2]dt. \tag{8.5.109}$$

Hence, the matrices R and G are as defined in Section 5.4:

$$R = \begin{bmatrix} 1.1 & 0 & 1 & 0 \\ 0 & 0.1 & 0 & 0 \\ 1 & 0 & 1.1 & 0 \\ 0 & 0 & 0 & 0.1 \end{bmatrix} > 0; \qquad G = 1.$$

In equation (8.5.106), the uncertainty input $\xi(\cdot)$ has three components, $\xi(\cdot) = [\xi_1(\cdot),\ \xi_2(\cdot),\ \xi_3(\cdot)]'$. The uncertainty input $\xi_1(\cdot)$ describes the uncertainty in the spring rate. This uncertainty satisfies the following constraint

$$|\xi_1(t)| \leq |z(t)|.$$

The components ξ_2 and ξ_3 of the uncertainty input vector ξ are fictitious uncertainty inputs which were added to the system (8.5.105) in Section 5.4 in order to fit this system into the framework of the method presented in that section. Specifically, it was assumed in Section 5.4 that the uncertainty input $\xi(\cdot)$ satisfies the following integral quadratic constraint:

$$\int_0^{t_i} \|\xi(t)\|^2 dt \leq \int_0^{t_i} \|z(t)\|^2 dt + \bar{x}_0' S \bar{x}_0, \tag{8.5.110}$$

where $\{t_i\}$ is a sequence of times as discussed in Definition 5.3.1. Also, in Section 5.4, the initial condition of the system (8.5.106) was chosen to be

$$\bar{x}_0 = [-1\,0\,0\,0]'.$$

This choice of the initial condition corresponds to a zero initial condition on the system dynamics and an initial condition of $\eta(0) = 1$ on the reference input dynamics. Also, in Section 5.4, the mismatch matrix S was chosen to be

$$S = \begin{bmatrix} 0.1 & 0 & 0 & 0 \\ 0 & 0.1 & 0 & 0 \\ 0 & 0 & 0.1 & 0 \\ 0 & 0 & 0 & 0.1 \end{bmatrix} > 0.$$

The output-feedback robust controller designed in Section 5.4 was a suboptimal time-varying controller. We now apply the controller design procedure presented in this section to design a time-invariant output-feedback minimax optimal controller solving the above tracking problem. We will use the state space transformation (8.5.108) which reduces the original tracking problem to a regulator problem. However, in order to apply the results of this section to this robust control problem, we must introduce a stochastic description of the system. To satisfy this requirement, a noise input will be added to the system and the controller will be designed for the system with additive noise. That is, we replace the nominal system corresponding to (8.5.106) with $\xi(\cdot) \equiv 0$, with a stochastic system described by the following stochastic differential equation

$$
\begin{aligned}
d\bar{x} &= \left[\begin{bmatrix} 0 & 1 & 0 & 0 \\ 0 & 0 & 0 & 0 \\ 0 & 0 & 0 & 1 \\ 0 & 0 & -2.5 & 0 \end{bmatrix} \bar{x} + \begin{bmatrix} 0 \\ 0.5 \\ 0 \\ -0.5 \end{bmatrix} u \right] dt + \begin{bmatrix} 0 & 0 & 0 \\ 0.05 & 0 & 0 \\ 0 & 0 & 0 \\ -0.75 & 0 & 0 \end{bmatrix} dW(t); \\
z &= \begin{bmatrix} 0 & 0 & 2 & 0 \end{bmatrix} \bar{x}; \\
d\bar{y} &= \begin{bmatrix} 1 & 0 & 1 & 0 \\ 1 & 0 & -1 & 0 \end{bmatrix} \bar{x} dt + \begin{bmatrix} 0 & 0.05 & 0 \\ 0 & 0 & 0.05 \end{bmatrix} dW; \\
y_T - \tilde{y}_T &= \begin{bmatrix} 1 & 0 & 1 & 0 \end{bmatrix} \bar{x}
\end{aligned}
\tag{8.5.111}
$$

where $W(t) = [W_1(t),\ W_2(t),\ W_3(t)]'$ is a 3-dimensional Wiener process on a certain measurable space $(\Omega, \mathcal{F}, P)$. Here, P is the reference probability measure. Also, the uncertain system (8.5.106) is replaced by an uncertain system of the form (8.5.111) considered on an uncertain measurable space $(\Omega, \mathcal{F}_T, Q^T)$, where Q^T is the uncertainty probability measure. Such a probability measure arises as the probability distribution of an uncertain noise input as described in Chapter 2. Also, as noted in Section 2.4.2, uncertain systems of this type can be described using a stochastic differential equation of the form of equation (2.4.29). The system (8.5.111) is a system of the form (8.5.1) to which the design technique presented in this section is applicable.

Note that in this example, a robust controller is sought which stabilizes the system in the face of stochastic uncertainty. It can readily be shown using Lemma 8.5.2 that the absolute stability of the stochastic closed loop system consisting of the system (8.5.111) and this controller, implies the robust stability of the closed loop system corresponding to the deterministic system (8.5.106) driven by the same linear output-feedback controller. Indeed, Lemma 8.5.2 shows that the corresponding Riccati equation (8.5.65) has a nonnegative-definite stabilizing solution. Then, using the Strict Bounded Real Lemma and Lemma 3.4.1 leads to the conclusion that the corresponding deterministic closed loop system with norm-bounded uncertainty is quadratically stable. Also, the corresponding deterministic closed loop system with the uncertainty modeled using an integral quadratic constraint of the form (8.5.110) is absolutely stable [260]. It follows from this observation that a robust output-feedback controller designed for the uncertain stochastic system (8.5.111) also serves as a robust controller for the original uncertain system (8.5.106). Thus, a controller designed for the stochastic uncertain system (8.5.111) will solve the original tracking problem. For this example, the Wiener process W is chosen to have unity covariance matrix.

We now proceed to the derivation of a robust output-feedback controller for the system (8.5.111). We first replace the integral quadratic constraint (8.5.110) by the following stochastic uncertainty constraint: For any $T > 0$

$$\frac{1}{2T}\mathbf{E}^{Q^T}\int_0^T \|\xi(t)\|^2 dt \leq \frac{1}{2T}\mathbf{E}^{Q^T}\int_0^T \|z(t)\|^2 dt + d; \quad d = \frac{1}{2}\bar{x}_0' S \bar{x}_0. \tag{8.5.112}$$

It was shown in Section 2.4.3 that the uncertainty set defined by the constraint (8.5.112) includes all uncertainty inputs satisfying the norm-bound condition $\|\xi(\cdot)\| \leq |z(\cdot)|$. Also it was shown that the uncertainty class defined by the constraint (8.5.112) can be embedded into an uncertainty class described by the corresponding relative entropy uncertainty constraint of the form (8.5.6).

The cost functional is chosen to have the form

$$\limsup_{T\to\infty} \frac{1}{2T}\int_0^T \mathbf{E}^{Q^T}[(y_T - \tilde{y}_T)^2 + 0.1\|\bar{x}\|^2 + u^2]dt. \tag{8.5.113}$$

We are now in a position to apply the design procedure outlined in Theorem 8.5.3. For each value of $\tau > 0$, the Riccati equations (8.5.78) and (8.5.79) are solved and then a line search is carried out to find the value of $\tau > 0$ which attains the minimum of the function $V_\tau^0 + \tau d$ defined in Theorem 8.5.3. A graph

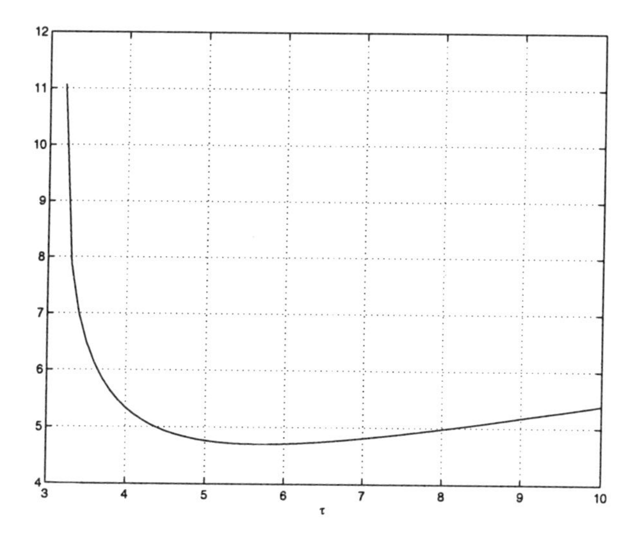

Figure 8.5.2. Cost bound $V_\tau^0 + \tau d$ versus the parameter τ

of $V_\tau^0 + \tau d$ versus τ for this example is shown in Figure 8.5.2. It was found that the optimal value of the parameter τ is $\tau = 5.6931$.

With this optimal value of τ, the following positive-definite stabilizing solutions to Riccati equations (8.5.79) and (8.5.78) were obtained:

$$X_\infty = \begin{bmatrix} 4.0028 & 6.8156 & -6.3708 & 3.8312 \\ 6.8156 & 18.3891 & -20.6541 & 9.3784 \\ -6.3708 & -20.6541 & 48.5330 & -5.8268 \\ 3.8312 & 9.3784 & -5.8268 & 12.5738 \end{bmatrix};$$

$$Y_\infty = \begin{bmatrix} 0.0007 & 0.0003 & -0.0005 & -0.0014 \\ 0.0003 & 0.0008 & -0.0017 & -0.0108 \\ -0.0005 & -0.0017 & 0.0077 & 0.0236 \\ -0.0014 & -0.0108 & 0.0236 & 0.1641 \end{bmatrix}.$$

Furthermore, a corresponding time-invariant controller of the form (8.5.2), (8.5.82) was constructed to be

$$d\hat{x} = \begin{bmatrix} -0.5868 & 1.0000 & 0.4581 & 0 \\ -1.0384 & -2.3064 & 5.7466 & 0.7202 \\ 0.4581 & 0 & -7.6627 & 1.0000 \\ 2.6530 & 3.0582 & -34.7464 & 0.3817 \end{bmatrix} \hat{x}dt + \begin{bmatrix} 0.0643 & 0.5225 \\ -0.8702 & 1.1403 \\ 3.6023 & -4.0604 \\ 13.2633 & -14.8366 \end{bmatrix} dy(t);$$

$$u = \begin{bmatrix} -1.4922 & -4.5053 & 7.4137 & 1.5977 \end{bmatrix} \hat{x}.$$

Then referring to system (8.5.105), the required tracking control system is constructed by replacing the time-varying controller in Figure 5.4.2 with the above time-invariant controller as shown in Figure 8.5.3. To verify the robust tracking properties of this control system, Figure 8.5.4 shows the step response of the system for various values of the spring constant parameter k. It can be seen from these plots that the stochastic minimax optimal control approach of this section leads to a robust tracking system which exhibits similar transient behavior to the behavior of the tracking system designed using the deterministic approach of Section 5.4. However, the controller designed using the approach of this section is time-invariant.

8.6 Conclusions

This chapter has been concerned with the existence and optimality of guaranteed cost controllers for stochastic uncertain systems subject to stochastic uncertainty constraints.

We have shown that the controller synthesis ideas developed in the previous chapters based on scaling and optimization of solutions to a parameter-dependent

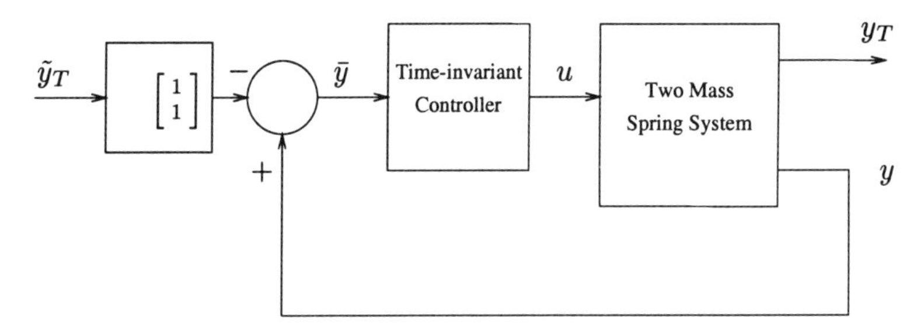

Figure 8.5.3. Tracking control system block diagram

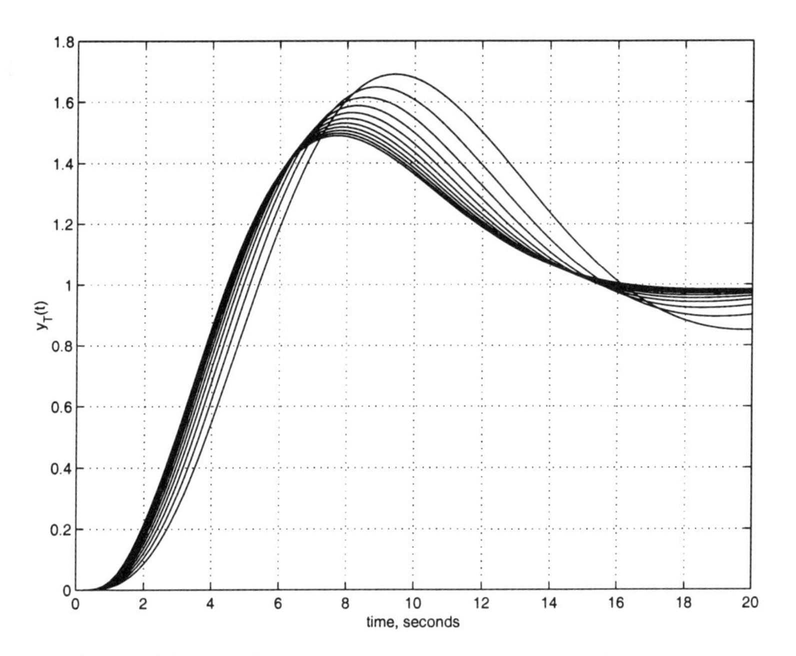

Figure 8.5.4. Control system step response for various spring constants

generalized matrix Riccati equation, can be extended to control systems involving a stochastic uncertainty description.

Of particular importance is the robust version of the LQG controller design methodology presented in this chapter. A method for the construction of a minimax optimal output-feedback controller for a stochastic uncertain system has been presented. This controller construction methodology is based on a pair of Riccati equations of the H^∞ type. This result was derived by first applying an S-procedure theorem to convert the constrained minimax optimization problem into an unconstrained one. Then the duality between stochastic dynamical games and risk-sensitive optimal control problems was used to solve the resulting unconstrained stochastic minimax optimal control problem.

9. Nonlinear versus linear control

9.1 Introduction

This chapter considers stabilization problems for uncertain linear systems containing unstructured or structured uncertainty described by integral quadratic constraints. In these problems, we will be interested in the question of whether any advantage can be obtained via the use of a nonlinear controller as opposed to a linear controller. This question is of interest since if no advantage can be obtained via the use of nonlinear controller, then there can be little motivation for the use of a complicated nonlinear controller such as adaptive controller.

To date, a number of results have appeared showing that certain classes of uncertain linear systems have the property that if they can be stabilized via a nonlinear controller then they can also be stabilized via a linear controller; see [128, 85, 8, 148, 145, 168, 103, 99, 160]. These results can be split into two categories. The results of references [128, 85, 8] concern the problem of state-feedback stabilization of uncertain systems containing no uncertainty in the input matrix. In this case, the state-feedback stabilization via linear control is achieved via the use of high gain feedback. The more interesting results are those of references [148, 145, 168, 103, 99, 160]. These results are all concerned with problems of robust stabilization in which the underlying uncertain system contains unstructured uncertainty. As mentioned in Chapter 1, the existence of these results led Khargonekar and Poolla to formulate the *Plant Uncertainty Principle*: *"In robust control problems for linear time-invariant plants, nonlinear time-varying controllers yield no advantage over linear time-invariant controllers if the plant uncertainty is unstructured"*; see [103, 168]. In [140], Petersen gave a counterexample which reinforced the connection between uncertain plants with unstructured uncertainty and the use of linear controllers. This paper presented an example of a system with *structured norm-bounded uncertainty* which can be quadratically stabilized with via a nonlinear state feedback controller but cannot be quadratically stabilized via a linear state feedback controller. Thus, from the existing results in the literature, it would appear that uncertainty structure is a most important issue relating to the question of nonlinear versus linear control of uncertain linear systems. However, the results of [185, 194] presented in this chapter indicate that the most important issue is not uncertainty structure but rather the "richness" of the class of uncertainties allowed in the uncertain system. From the results of Sections 9.2, 9.3 and 9.4 as well as Section 3.5, the reader may draw the conclusion that the class of uncertainties satisfying integral quadratic constraints is a "sufficiently rich" class of uncertainties for the following alternative to the above plant uncertainty principle to hold: *In robust control problems for linear time-invariant plants, nonlinear controllers yield no advantage over linear controllers if the class of plant uncertainties is sufficiently rich.* An important corollary to the results presented in this chapter is the fact that if an uncertain system is absolutely stabilizable via nonlinear (but time-invariant) control then it will be absolutely stabilizable via linear time-invariant control. The significance of unstructured uncertainty in the question of nonlinear versus linear control appears to be as follows: If a system with unstructured uncertainty is robustly stabilizable with a restricted class of uncertainties (such as dynamic linear time-invariant uncertainties), then it will also be robustly stabilizable with a richer class of uncertainties including nonlinear time-varying dynamic uncertainties.

However in Section 9.5, we will show that under certain circumstances, nonlinear controllers may have an advantage over linear controllers.

9.2 Nonlinear versus linear control in the absolute stabilizability of uncertain systems with structured uncertainty

9.2.1 *Problem statement*

We now consider a class of uncertain linear systems with structured uncertainty described by the following state equations of the form (2.3.3)

$$\begin{aligned}\dot{x}(t) &= Ax(t) + B_1u(t) + \sum_{j=0}^{k} B_{2j}\xi_j(t);\\ y(t) &= C_2x(t) + \sum_{j=0}^{k} D_{2j}\xi_j(t);\\ z_j(t) &= C_{1j}x(t) + D_{1j}u(t); \quad s = 0, 1, \ldots, k \end{aligned} \tag{9.2.1}$$

where $x(t) \in \mathbf{R}^n$ is the *state*, $u(t) \in \mathbf{R}^m$ is the *control input*, $y(t) \in \mathbf{R}^\ell$ is the *measured output*, $z_0(t) \in \mathbf{R}^{h_0}, \ldots, z_k(t) \in \mathbf{R}^{h_k}$ are the *uncertainty outputs* and $\xi_0(t) \in \mathbf{R}^{r_0}, \ldots, \xi_k(t) \in \mathbf{R}^{r_k}$ are the *uncertainty inputs*. The uncertainty in the above system is described by the following equations of the form (2.3.4):

$$\begin{aligned} \xi_0(t) &= \phi_0(t, x(\cdot)); \\ \xi_1(t) &= \phi_1(t, x(\cdot)); \\ &\vdots \\ \xi_k(t) &= \phi_k(t, x(\cdot)) \end{aligned} \tag{9.2.2}$$

and is assumed to satisfy the following integral quadratic constraints as defined in Chapter 2.

Definition 9.2.1 *An uncertainty of the form (9.2.2) is an admissible uncertainty for the system (9.2.1) if the following conditions hold: Given an initial condition $x(0) = x_0$ and a locally square integrable control input $u(\cdot)$ and any corresponding solution to equations (9.2.1), (9.2.2), defined on an existence interval[1] $(0, t_*)$, then there exists a sequence $\{t_i\}_{i=1}^{\infty}$ and constants $d_0 \geq 0, d_1 \geq 0, \ldots, d_k \geq 0$ such that $t_i \to t_*$, $t_i \geq 0$ and*

$$\int_0^{t_i} \|\xi_j(t)\|^2 dt \leq \int_0^{t_i} \|z_j(t)\|^2 dt + d_j \quad \forall i \text{ and } \forall j = 0, 1, \ldots, k. \tag{9.2.3}$$

Note that the sequence $\{t_i\}_{i=1}^{\infty}$ and the constants d_j may depend on $x_0, u(\cdot)$ and $\xi(\cdot)$. Also note that t_i and $t_\star$ may be equal to infinity.

In this section, we will address the problem of absolute stabilization via a nonlinear feedback controller. Both the state-feedback and output-feedback versions of this problem will be considered. A similar problem has been considered in Section 3.5. However, the form of the controllers considered here is more specific than in Section 3.5. In particular, in the problem of absolute stabilization via a nonlinear state-feedback controller, we focus on controllers of the form

$$\begin{aligned} \dot{x}_c(t) &= L(x_c(t), x(t)); \\ u(t) &= \lambda(x_c(t), x(t)) \end{aligned} \tag{9.2.4}$$

where $L(\cdot,\cdot)$ and $\lambda(\cdot,\cdot)$ are Lipschitz continuous vector functions. Also, we will consider the problem of stabilization for the system (9.2.1), (9.2.3) via a nonlinear output-feedback controller of the form

$$\begin{aligned} \dot{x}_c(t) &= L(x_c(t), y(t)); \\ u(t) &= \lambda(x_c(t), y(t)) \end{aligned} \tag{9.2.5}$$

[1] That is, t_* is the upper limit of the time interval over which the solution exists.

Note that the dimension of the state vector $x_c(t)$ may be arbitrary in both (9.2.4) and (9.2.5). The case in which controllers of a fixed order are required will be considered in the next section.

We now define the notion of absolute stabilizability for the uncertain system (9.2.1), (9.2.3).

Definition 9.2.2 *The uncertain system (9.2.1), (9.2.3) is said to be* absolutely stabilizable via nonlinear output-feedback control *if there exists a controller of the form (9.2.5) and a constant $c > 0$ such that:*

(i) For any initial condition $[x(0)\ x_c(0)] = [x_0\ x_{c0}]$ and any input $\xi(\cdot) = [\,\xi_0(\cdot)\ \ldots\ \xi_k(\cdot)\,] \in \mathbf{L}_2[0,\infty)$, the closed loop system defined by (9.2.1) and (9.2.5) has a unique solution which is defined on $[0,\infty)$.

(ii) The closed loop system (9.2.1), (9.2.5) with $\xi(t) \equiv 0$ is exponentially stable.

(iii) For any initial condition $[x(0)\ x_c(0)] = [x_0\ x_{c0}]$ and any admissible $\xi(\cdot) = [\,\xi_0(\cdot)\ \ldots\ \xi_k(\cdot)\,]$, then

$$[\,x(\cdot)\ x_c(\cdot)\ u(\cdot)\ \xi_0(\cdot)\ \ldots\ \xi_k(\cdot)\,] \in \mathbf{L}_2[0,\infty),$$

(hence, $t_\star = \infty$) and

$$\begin{aligned} &\|x(\cdot)\|_2^2 + \|x_c(\cdot)\|_2^2 + \|u(\cdot)\|_2^2 + \sum_{j=0}^{k} \|\xi_j(\cdot)\|_2^2 \\ &\qquad \leq c[\|x(0)\|^2 + \|x_c(0)\|^2 + \sum_{j=0}^{k} d_j]. \end{aligned} \tag{9.2.6}$$

(iv) For any $\varepsilon > 0$, there exists a $\delta > 0$ such that the following condition holds: For any input function $\xi(\cdot) \in \{\xi(\cdot) \in \mathbf{L}_2[0,\infty), \|\xi(\cdot)\|_2^2 \leq \delta\}$ and any vector $[x_0'\ x_{c0}']' \in \{h : \|h\|^2 \leq \delta\}$, let $[x'(t)\ x_c'(t)$ be the corresponding solution to the closed loop system defined by equations (9.2.1) and (9.2.5) with initial condition $x(0) = x_0$, $x_c(0) = x_{c0}$. Then

$$\|x(\cdot)\|_2^2 + \|x_c(\cdot)\|_2^2 + \|u(\cdot)\|_2^2 < \varepsilon.$$

The uncertain system (9.2.1), (9.2.3) is said to be absolutely stabilizable via nonlinear state-feedback control *if there exists a controller of the form (9.2.4) and a constant $c > 0$ such that conditions (i) – (iv) above are satisfied for the closed loop system defined by (9.2.1), (9.2.3), (9.2.4).*

Remark 9.2.1 The reader should note that conditions (i)–(iii) of this definition correspond directly to the conditions of absolute stabilizability via nonlinear feedback given in Section 3.5; see Definition 3.5.3 on page 97. Condition (iv) is however an additional technical stability requirement imposed on the closed loop system. This additional stability requirement is needed as a result of the presence of structured uncertainty.

Observation 9.2.1 If the uncertain system (9.2.1), (9.2.3) is absolutely stabilizable via output-feedback, then the corresponding closed loop uncertain system (9.2.1), (9.2.3), (9.2.5) will have the property that $x(t) \to 0$ as $t \to \infty$ for any admissible uncertainty $\xi(\cdot)$. Indeed, since $\{x(\cdot), u(\cdot), \xi(\cdot)\} \in \mathbf{L}_2[0, \infty)$, we can conclude from (9.2.1) that $\dot{x}(\cdot) \in \mathbf{L}_2[0, \infty)$. However, using the fact that $x(\cdot) \in \mathbf{L}_2[0, \infty)$ and $\dot{x}(\cdot) \in \mathbf{L}_2[0, \infty)$, it now follows that $x(t) \to 0$ as $t \to \infty$. Furthermore, if the controller (9.2.5) is linear, then also $x_c(t) \to 0$ as $t \to \infty$.

A similar conclusion to the above also holds if the uncertain system (9.2.1), (9.2.3) is absolutely stabilizable via state-feedback.

9.2.2 *Output-feedback nonlinear versus linear control*

We now present a result which establishes a necessary and sufficient condition for the absolute stabilizability of the uncertain linear system (9.2.1), (9.2.3) via output-feedback. The state-feedback version of this result will be given in Section 9.2.3.

This necessary and sufficient condition is given in terms of the existence of solutions to a pair of parameter dependent algebraic Riccati equations. The Riccati equations under consideration are defined as follows: Let $\tau_1 > 0, \ldots, \tau_k > 0$ be given constants and consider the Riccati equations

$$\begin{aligned}&(A - B_1 G_\tau^{-1} D_\tau' C_\tau)' X + X(A - B_1 G_\tau^{-1} D_\tau' C_\tau) + X(\tilde{B}_2 \tilde{B}_2' - B_1 G_\tau^{-1} B_1') X \\ &\quad + C_\tau' (I - D_\tau G_\tau^{-1} D_\tau') C_\tau = 0;\end{aligned} \tag{9.2.7}$$

$$\begin{aligned}&(A - \tilde{B}_2 \tilde{D}_2' \Gamma_\tau^{-1} C_2) Y + Y(A - \tilde{B}_2 \tilde{D}_2' \Gamma_\tau^{-1} C_2)' + Y(C_\tau' C_\tau - C_2' \Gamma_\tau^{-1} C_2) Y \\ &\quad + \tilde{B}_2 (I - \tilde{D}_2' \Gamma_\tau^{-1} \tilde{D}_2) \tilde{B}_2' = 0\end{aligned} \tag{9.2.8}$$

where

$$\begin{aligned}
C_\tau &:= \begin{bmatrix} C_{10} \\ \sqrt{\tau_1} C_{11} \\ \vdots \\ \sqrt{\tau_k} C_{1k} \end{bmatrix}; \quad D_\tau := \begin{bmatrix} D_{10} \\ \sqrt{\tau_1} D_{11} \\ \vdots \\ \sqrt{\tau_k} D_{1k} \end{bmatrix}; \\
\tilde{B}_2 &:= \begin{bmatrix} B_{20} & \frac{1}{\sqrt{\tau_1}} B_{21} & \ldots & \frac{1}{\sqrt{\tau_k}} B_{2k} \end{bmatrix}; \\
\tilde{D}_2 &:= \begin{bmatrix} D_{20} & \frac{1}{\sqrt{\tau_1}} D_{21} & \ldots & \frac{1}{\sqrt{\tau_k}} D_{2k} \end{bmatrix}; \\
G_\tau &:= D_\tau' D_\tau; \quad \Gamma_\tau := \tilde{D}_2 \tilde{D}_2'.
\end{aligned} \tag{9.2.9}$$

Riccati equations (9.2.7) and (9.2.8) are required in the output-feedback absolute stabilization problem. In the state-feedback absolute stabilization problem, only Riccati equation (9.2.7) will be required. In the output-feedback absolute stabilization problem we will also require the solutions to Riccati equations (9.2.7) and (9.2.8) satisfy the spectral radius condition $\rho(XY) < 1$.

The main result of this section requires that the uncertain system (9.2.1), (9.2.3) satisfies the following assumptions. Let matrices B_2, C_1, D_1, D_2, G and Γ be defined by

$$B_2 := \begin{bmatrix} B_{20} & \ldots & B_{2k} \end{bmatrix}; \quad C_1 := \begin{bmatrix} C_{10} \\ \vdots \\ C_{1k} \end{bmatrix};$$

$$D_1 := \begin{bmatrix} D_{10} \\ D_{11} \\ \vdots \\ D_{1k} \end{bmatrix}; \quad D_2 := \begin{bmatrix} D_{20} & D_{21} & \ldots & D_{2k} \end{bmatrix};$$

$$G := \sum_{j=0}^{k} D_{1j}' D_{1j}; \quad \Gamma := \sum_{j=0}^{k} D_{2j} D_{2j}'.$$

Assumption 9.2.1 *(i) The pair (A, B_1) is stabilizable.*

(ii) $G > 0$.

(iii) The pair $(A - B_1 G^{-1} D_1' C_1, (I - D_1 G^{-1} D_1') C_1)$ is observable.

Assumption 9.2.2 *(i) The pair (A, C_2) is detectable.*

(ii) $\Gamma > 0$.

(iii) The pair $(A - B_2 D_2' \Gamma^{-1} C_2, B_2 (I - D_2' \Gamma^{-1} D_2))$ is controllable.

In this subsection, we require that both Assumption 9.2.1 and Assumption 9.2.2 are satisfied. In the state-feedback case, only Assumption 9.2.1 will be needed. Note that the first conditions in Assumptions 9.2.1 and 9.2.2 are necessary for the nominal system to be stabilizable and hence can be made without loss of generality. The other conditions required in Assumptions 9.2.1 and 9.2.2 are technical assumptions needed to ensure that the underlying H^∞ problem is "non-singular"; e.g., see Section 3.2 and also [9]. This underlying "diagonally scaled" output-feedback H^∞ control problem is defined as follows: Let $\tau_1 > 0, \ldots, \tau_k > 0$ be given constants and consider the system

$$\begin{aligned} \dot{x}(t) &= Ax(t) + B_1 u(t) + \tilde{B}_2 w(t); \\ y(t) &= C_2 x(t) + \tilde{D}_2 w(t); \\ \hat{z}(t) &= C_\tau x(t) + D_\tau u(t) \end{aligned} \tag{9.2.10}$$

where $w(t)$ is the disturbance input and $\hat{z}(t)$ is the controlled output. Also, consider the following induced norm bound condition:

$$\sup_{\substack{w(\cdot) \in \mathbf{L}_2[0,\infty), \\ x(0)=0}} \frac{\|\hat{z}(\cdot)\|_2^2}{\|w(\cdot)\|_2^2} < 1. \tag{9.2.11}$$

The output-feedback H^∞ control problem corresponding to the system (9.2.10) is said to have a solution if there exists a controller mapping from y to u such that the closed loop system satisfies condition (9.2.11); see also Section 3.2. In the state-feedback case, a corresponding state-feedback H^∞ control problem is also formulated in Section 3.2. It is straightforward to verify that the system (9.2.10) satisfy the assumptions of Section 3.2. Indeed, we now show that Assumption 9.2.1 guarantees that the pair

$$\left(A - B_1 G_\tau^{-1} D_\tau' C_\tau, (I - D_\tau G_\tau^{-1} D_\tau') C_\tau\right) \tag{9.2.12}$$

is observable for any $\tau_1 > 0, \ldots, \tau_k > 0$. We need to show that the conditions

$$\begin{aligned} (A - B_1 G_\tau^{-1} D_\tau' C_\tau - sI)x &= 0; \\ (I - D_\tau G_\tau^{-1} D_\tau') C_\tau x &= 0 \end{aligned} \tag{9.2.13}$$

imply that $x = 0$ for any $s \in \mathbb{C}$. Equations (9.2.13) can be re-written as follows

$$\begin{bmatrix} A - sI & B_1 \\ C_{10} & D_{10} \\ \sqrt{\tau_1} C_{11} & \sqrt{\tau_1} D_{11} \\ \vdots & \vdots \\ \sqrt{\tau_k} C_{1k} & \sqrt{\tau_k} D_{1k} \end{bmatrix} \begin{bmatrix} x \\ -G_\tau^{-1} D_\tau' C_\tau x \end{bmatrix} = 0. \tag{9.2.14}$$

Since $\tau_1 > 0, \ldots, \tau_k > 0$, then the above equation can be transformed as follows:

$$\begin{bmatrix} A - sI & B_1 \\ C_1 & D_1 \end{bmatrix} \begin{bmatrix} x \\ - \begin{bmatrix} 1 & 0 & \ldots & 0 \\ 0 & \tau_1^{-1/2} & \ldots & 0 \\ \vdots & \vdots & \ddots & \vdots \\ 0 & 0 & \ldots & \tau_k^{-1/2} \end{bmatrix} G_\tau^{-1} D_\tau' C_\tau x \end{bmatrix} = 0. \tag{9.2.15}$$

Note that it follows from Lemma 3.2.2 that the matrix

$$\begin{bmatrix} A - sI & B_1 \\ C_1 & D_1 \end{bmatrix}$$

has full column rank for all $s \in \mathbb{C}$. Hence, equation (9.2.15) yields $x = 0$ for all $s \in \mathbb{C}$. That is, the pair (9.2.12) is observable.

Also in order to apply the results of Section 3.2 to the H^∞ control problem (9.2.10), (9.2.11), we need to ensure that the the pair

$$(A - \tilde{B}_2 \tilde{D}_2' \Gamma_\tau^{-1} C_2, \tilde{B}_2 (I - \tilde{D}_2' \Gamma_\tau^{-1} \tilde{D}_2)).$$

is controllable. This fact follows from Assumption 9.2.2 in the same manner as that of observability of the pair (9.2.12) followed from Assumption 9.2.1. Thus, the technical assumptions needed to ensure that the H^∞ control problem (9.2.10), (9.2.11) is non-singular, are satisfied.

Theorem 9.2.1 *Consider the uncertain system (9.2.1), (9.2.3) and suppose that Assumptions 9.2.1, 9.2.2 are satisfied. Then the following statements are equivalent:*

(i) The uncertain system (9.2.1), (9.2.3) is absolutely stabilizable via a nonlinear output-feedback controller of the form (9.2.5).

(ii) There exist constants $\tau_1 > 0, \ldots, \tau_k > 0$ such that the Riccati equations (9.2.7) and (9.2.8) have solutions $X > 0$ and $Y > 0$ satisfying the condition $\rho(XY) < 1$.

If condition (ii) holds, then the uncertain system (9.2.1), (9.2.3) is absolutely stabilizable via a linear output-feedback controller of the form

$$\begin{aligned} \dot{x}_c(t) &= A_c x_c(t) + B_c y(t); \\ u(t) &= C_c x_c(t) + D_c y(t) \end{aligned} \qquad (9.2.16)$$

where

$$\begin{aligned} A_c &= A + B_1 C_c - B_c C_2 + (\tilde{B}_2 - B_c \tilde{D}_2)\tilde{B}_2' X; \\ B_c &= (I - YX)^{-1}(YC_2' + \tilde{B}_2 \tilde{D}_2')\Gamma_\tau^{-1}; \\ C_c &= -G_\tau^{-1}(B_1' X + D_\tau' C_\tau); \\ D_c &= 0. \end{aligned} \qquad (9.2.17)$$

In order to prove this result, we first establish the following preliminary lemma.

Lemma 9.2.1 *Suppose that the uncertain system (9.2.1), (9.2.3) is absolutely stabilizable via a controller of the form (9.2.5). Then, there exist constants $\tau_1 > 0, \ldots, \tau_k > 0$ and $\delta > 0$ such that*

$$\begin{aligned} &\|z_0(\cdot)\|_2^2 - \|\xi_0(\cdot)\|_2^2 + \sum_{s=1}^{k} \tau_j(\|z_j(\cdot)\|_2^2 - \|\xi_j(\cdot)\|_2^2) \\ &\leq -\delta(\|x(\cdot)\|_2^2 + \|x_c(\cdot)\|_2^2 + \|u(\cdot)\|_2^2 + \sum_{j=0}^{k} \|\xi_j(\cdot)\|_2^2) \end{aligned} \qquad (9.2.18)$$

for all the solutions to the closed loop system with the initial condition $x(0) = 0$, $x_c(0) = 0$.

Proof: Suppose that the uncertain system (9.2.1), (9.2.3) is absolutely stabilizable via a controller of the form (9.2.5) and consider the corresponding closed loop system. This system is described by the state equations

$$\begin{aligned} \dot{x}(t) &= Ax(t) + B_1 \lambda(x_c(t), y(t)) + \sum_{j=0}^{k} B_{2j} \xi_j(t); \\ \dot{x}_c(t) &= L(x_c(t), y(t)). \end{aligned} \qquad (9.2.19)$$

This system can be regarded as a system of the form

$$\dot{h}(t) = \Pi(h(t), \psi(t)) \tag{9.2.20}$$

with the associations: $h(t) \sim \left[\, x(t)' \;\; x_c(t)' \,\right]'$ and $\psi(t) \sim \xi(t)$ where $\xi(\cdot) = \left[\, \xi_0(\cdot) \;\ldots\; \xi_k(\cdot) \,\right]$. For this system, consider a corresponding set $\Omega \subset \mathbf{L}_2[0,\infty)$ defined in Section 4.4: Ω is the set of vector functions $\{x(\cdot), x_c(\cdot), \xi(\cdot)\}$ such that $\xi(\cdot) \in \mathbf{L}_2[0,\infty)$ and $\left[\, x(\cdot)' \;\; x_c(\cdot)' \,\right]'$ is the corresponding solution to (9.2.1), (9.2.5) with initial condition $x(0) = 0$, $x_c(0) = 0$. Also, associated with the above system is the following set of integral functionals $\mathcal{G}_0(\cdot), \ldots, \mathcal{G}_k(\cdot)$:

$$\begin{aligned}
&\mathcal{G}_0(x(\cdot), x_c(\cdot), \xi(\cdot)) \\
&:= \int_0^\infty \left(\|\xi_0(t)\|^2 - \|C_{10}x(t) + D_{10}\lambda(x_c(t), y(t))\|^2 \right) dt \\
&\qquad - \delta_0 \int_0^\infty \left(\|x(t)\|^2 + \|x_c(t)\|^2 + \|\lambda(x_c(t), y(t))\|^2 + \|\xi(t)\|^2 \right) dt; \\
&\mathcal{G}_1(x(\cdot), x_c(\cdot), \xi(\cdot)) \\
&:= \int_0^\infty \left(\|C_{11}x(t) + D_{11}\lambda(x_c(t), y(t))\|^2 - \|\xi_1(t)\|^2 \right) dt \\
&\qquad + \delta_0 \int_0^\infty \left(\|x(t)\|^2 + \|x_c(t)\|^2 + \|\lambda(x_c(t), y(t))\|^2 + \|\xi(t)\|^2 \right) dt; \\
&\vdots \\
&\mathcal{G}_k(x(\cdot), x_c(\cdot), \xi(\cdot)) \\
&:= \int_0^\infty \left(\|C_{1k}x(t) + D_{1k}\lambda(x_c(t), y(t))\|^2 - \|\xi_k(t)\|^2 \right) dt \\
&\qquad + \delta_0 \left(\|x(t)\|^2 + \|x_c(t)\|^2 + \|\lambda(x_c(t), y(t))\|^2 + \|\xi(t)\|^2 \right) dt
\end{aligned} \tag{9.2.21}$$

where $\delta_0 = \frac{1}{2(k+1)c}$ and c is as defined in part (iii) of Definition 9.2.2. Thus using (9.2.1), we can write

$$\begin{aligned}
&\mathcal{G}_0(x(\cdot), x_c(\cdot), \xi(\cdot)) \\
&= -\left(\|z_0(\cdot)\|_2^2 - \|\xi_0(\cdot)\|_2^2 + \delta_0 \left(\|x(\cdot)\|_2^2 + \|x_c(\cdot)\|_2^2 + \|\xi(\cdot)\|_2^2 + \|u(\cdot)\|_2^2 \right) \right); \\
&\mathcal{G}_1(x(\cdot), x_c(\cdot), \xi(\cdot)) \\
&= \|z_1(\cdot)\|_2^2 - \|\xi_1(\cdot)\|_2^2 + \delta_0 \left(\|x(\cdot)\|_2^2 + \|x_c(\cdot)\|_2^2 + \|\xi(\cdot)\|_2^2 + \|u(\cdot)\|_2^2 \right); \\
&\vdots \\
&\mathcal{G}_k(x(\cdot), x_c(\cdot), \xi(\cdot)) \\
&= \|z_k(\cdot)\|_2^2 - \|\xi_k(\cdot)\|_2^2 + \delta_0 \left(\|x(\cdot)\|_2^2 + \|x_c(\cdot)\|_2^2 + \|\xi(\cdot)\|_2^2 + \|u(\cdot)\|_2^2 \right).
\end{aligned} \tag{9.2.22}$$

We wish to apply the S-procedure Theorem 4.4.1 to the system (9.2.19) and associated functionals (9.2.21). However, we must first show that Assumptions 4.4.1

– 4.4.3 on page 115 are satisfied. The satisfaction of Assumption 4.4.1 follows directly from the Lipschitz continuity of the functions $L(\cdot,\cdot)$ and $\lambda(\cdot,\cdot)$. Assumption 4.4.2 follows directly from part (iii) of Definition 9.2.2 and equations (9.2.22). To establish Assumption 4.4.3, let $\varepsilon > 0$ be given and consider an uncertainty input $\xi(\cdot) \in \mathbf{L}_2[0,\infty)$. Let $\left[x(t)' \ x_c(t)' \right]'$ be the corresponding solution to the closed loop system (9.2.19) with $[x(0) \ x_c(0)] = [x_0 \ x_{c0}]$. Now consider the quantity $|\mathcal{G}_0(x(\cdot), x_c(\cdot), \xi(\cdot))|$ where the functional $\mathcal{G}_0(\cdot,\cdot,\cdot)$ is defined in (9.2.21). Using the triangle inequality, it follows that

$$\begin{aligned}
&|\mathcal{G}_0(x(\cdot), x_c(\cdot), \xi(\cdot))| \\
&\leq \delta_0 \left(\|x(\cdot)\|_2^2 + \|x_c(\cdot)\|_2^2 + \|u(\cdot)\|_2^2 + \|\xi_0(\cdot)\|_2^2 \right) + \|z_0(\cdot)\|_2^2 + \|\xi_0(\cdot)\|_2^2 \\
&\leq \delta_0 \left(\|x(\cdot)\|_2^2 + \|x_c(\cdot)\|_2^2 + \|u(\cdot)\|_2^2 + \|\xi_0(\cdot)\|_2^2 \right) + \|\xi_0(\cdot)\|_2^2 \\
&\quad + \left\| \left[C_{10} \ D_{10} \right] \right\|^2 \left(\|x(\cdot)\|_2^2 + \|u(\cdot)\|_2^2 \right)
\end{aligned} \tag{9.2.23}$$

where $\left\| \left[C_{10} \ D_{10} \right] \right\|$ denotes the induced matrix norm of the matrix $\left[C_{10} \ D_{10} \right]$. However, it follows from part (iv) of Definition 9.2.2 that given any $\tilde{\varepsilon} > 0$, there exists a constant $\tilde{\delta} > 0$ such that $\|x_0\|^2 + \|x_{c0}\|^2 \leq \tilde{\delta}$, $\|\xi(\cdot)\|_2^2 \leq \tilde{\delta}$ implies

$$\|x(\cdot)\|_2^2 + \|x_c(\cdot)\|_2^2 + \|u(\cdot)\|_2^2 < \tilde{\varepsilon}.$$

Let $\varepsilon > 0$ be given. We choose

$$\tilde{\varepsilon} := \min_{j=0,1,\ldots,k} \left\{ \frac{\varepsilon}{2\left(\delta_0 + \left\| \left[C_{1j} \ D_{1j} \right] \right\|^2 \right)} \right\}.$$

Then, the functional $\mathcal{G}_0(x(\cdot), x_c(\cdot), \xi(\cdot))$ will satisfy Assumption 4.4.3 if the constant δ in that assumption is chosen as follows:

$$\delta = \left[\min \left\{ \tilde{\delta}, \frac{\varepsilon}{2(\delta_0 + 1)} \right\} \right]^{1/2}$$

with $\tilde{\delta}$ corresponding to the chosen value of the constant $\tilde{\varepsilon}$. Indeed, using inequality (9.2.23), it follows that

$$|\mathcal{G}_0(x(\cdot), x_c(\cdot), \xi(\cdot))| \leq \delta_0(\delta^2 + \tilde{\varepsilon}) + \left\| \left[C_{10} \ D_{10} \right] \right\|^2 \tilde{\varepsilon} + \delta^2 < \varepsilon.$$

Using the same method as above, a similar inequality can be established for the quantities $|\mathcal{G}_1(x(\cdot), x_c(\cdot), \xi(\cdot))|, \ldots, |\mathcal{G}_k(x(\cdot), x_c(\cdot), \xi(\cdot))|$. Thus, Assumption 4.4.3 is satisfied.

To apply Theorem 4.4.1, we must show that $\mathcal{G}_0((x(\cdot), x_c(\cdot), \xi(\cdot)) \geq 0$ for all $\{x(\cdot), x_c(\cdot), \xi(\cdot)\} \in \Omega$ such that $\mathcal{G}_j((x(\cdot), x_c(\cdot), \xi(\cdot)) \geq 0$ for $s = 1, \ldots, k$. If this is not true, then there exists a $\{x^0(\cdot), x_c^0(\cdot), \xi^0(\cdot)\} \in \Omega$ such that $\mathcal{G}_0(x^0(\cdot), x_c^0(\cdot), \xi^0(\cdot)) < 0$ and $\mathcal{G}_j(x^0(\cdot), x_c^0(\cdot), \xi^0(\cdot)) \geq 0$ for $j = 1, \ldots, k$. However, the input $\xi^0(\cdot)$ satisfies the integral quadratic constraint (9.2.3) with

$t_i = \infty$ and $d_j = \delta_0 \left(\|x^0(\cdot)\|_2^2 + \|x_c^0(\cdot)\|_2^2 + \|\xi^0(\cdot)\|_2^2 + \|u^0(\cdot)\|_2^2 \right)$ for $s = 0, 1, \ldots, k$. Hence, condition (iii) of Definition 9.2.2 implies that

$$\begin{aligned} &\|x^0(\cdot)\|_2^2 + \|x_c^0(\cdot)\|_2^2 + \|\xi^0(\cdot)\|_2^2 + \|u^0(\cdot)\|_2^2 \\ &\leq c \sum_{j=0}^{k} \delta_0 \left(\|x^0(\cdot)\|_2^2 + \|x_c^0(\cdot)\|_2^2 + \|\xi^0(\cdot)\|_2^2 + \|u^0(\cdot)\|_2^2 \right) \\ &\leq c(k+1)\delta_0 \left(\|x^0(\cdot)\|_2^2 + \|x_c^0(\cdot)\|_2^2 + \|\xi^0(\cdot)\|_2^2 + \|u^0(\cdot)\|_2^2 \right) \\ &\leq \frac{1}{2} \left(\|x^0(\cdot)\|_2^2 + \|x_c^0(\cdot)\|_2^2 + \|\xi^0(\cdot)\|_2^2 + \|u^0(\cdot)\|_2^2 \right) \end{aligned}$$

which yields the desired contradiction.

We now apply Theorem 4.4.1 to the system (9.2.19) and functionals (9.2.21). It follows that there exist constants $\tau_0 \geq 0, \tau_1 \geq 0, \ldots, \tau_k \geq 0$ such that $\sum_{j=0}^{k} \tau_j > 0$ and

$$\tau_0 \mathcal{G}_0 \left(x(\cdot), x_c(\cdot), \xi(\cdot) \right) \geq \sum_{s=1}^{k} \tau_j \mathcal{G}_j \left(x(\cdot), x_c(\cdot), \xi(\cdot) \right)$$

for all $\{x(\cdot), x_c(\cdot), \xi(\cdot)\} \in \Omega$. We now prove that $\tau_j > 0$ for all $j = 0, 1, \ldots, k$. If we let $\delta := \delta_0 \sum_{j=0}^{k} \tau_j > 0$, then

$$\begin{aligned} &\sum_{j=0}^{k} \tau_j (\|z_j(\cdot)\|_2^2 - \|\xi_j(\cdot)\|_2^2) \\ &\qquad \leq -\delta (\|x(\cdot)\|_2^2 + \|x_c(\cdot)\|_2^2 + \|u(\cdot)\|_2^2 + \sum_{j=0}^{k} \|\xi_j(\cdot)\|_2^2) \end{aligned} \tag{9.2.24}$$

for all $\{x(\cdot), x_c(\cdot), \xi(\cdot)\} \in \Omega$. Now if $\tau_j = 0$ for some $j \in \{0, \ldots, k\}$, then we can let $\xi_s(\cdot) \equiv 0$ for all $s \neq j$ and choose any non-zero $\xi_j(\cdot) \in \mathbf{L}_2[0, \infty)$. However, this leads to a contradiction with (9.2.24) since the left side of (9.2.24) is non-negative and the right side of (9.2.24) is negative. Therefore we must have, $\tau_j > 0$ for $j = 0, 1, \ldots, k$. Furthermore, observe that in this case, we may take $\tau_0 = 1$ without loss of generality. Moreover, with $\tau_0 = 1$, inequality (9.2.24) leads directly to (9.2.18).

■

We are now in a position to establish Theorem 9.2.1.

Proof of Theorem 9.2.1: (i)⇒(ii). Consider an uncertain system (9.2.1), (9.2.3) satisfying Assumptions 9.2.1, 9.2.2 and suppose that the system is absolutely stabilizable via a nonlinear controller of the form (9.2.5). It follows from Lemma 9.2.1 that there exist constants $\tau_1 > 0, \tau_2 > 0, \ldots, \tau_k > 0$ such that inequality (9.2.18) is satisfied. Now the system (9.2.1) is equivalent to the system (9.2.10) where

$$\begin{aligned} w(\cdot) &= [\xi_0(\cdot), \sqrt{\tau_1}\xi_1(\cdot), \ldots, \sqrt{\tau_k}\xi_k(\cdot)]; \\ \hat{z}(\cdot) &= [z_0(\cdot), \sqrt{\tau_1}z_1(\cdot), \ldots, \sqrt{\tau_k}z_k(\cdot)] \end{aligned}$$

and the matrices $\tilde{B}_2$, C_τ, D_τ and $\tilde{D}_2$ are defined as in (9.2.9). It follows from inequality (9.2.18) that there exists a constant $\delta > 0$ such that if $x(0) = 0$, then

$$\begin{aligned}
&\|\hat{z}(\cdot)\|_2^2 - \|w(\cdot)\|_2^2 \\
&= \|z_0(\cdot)\|_2^2 + \sum_{s=1}^{k} \tau_j \|z_j(\cdot)\|_2^2 - \|\xi_0(\cdot)\|_2^2 - \sum_{s=1}^{k} \tau_j \|\xi_j(\cdot)\|_2^2 \\
&\leq -\delta \sum_{j=0}^{k} \|\xi_j(\cdot)\|_2^2
\end{aligned}$$

for all $w(\cdot) \in \mathbf{L}_2[0,\infty)$. Hence, there exists a $\delta_1 > 0$ such that if $x(0) = 0$, then

$$\begin{aligned}
&\|\hat{z}(\cdot)\|_2^2 - \|w(\cdot)\|_2^2 \\
&\leq -\delta_1 \left(\|\xi_0(\cdot)\|_2^2 + \sum_{s=1}^{k} \tau_j \|\xi_j(\cdot)\|_2^2 \right) \\
&= -\delta_1 \|w(\cdot)\|_2^2
\end{aligned}$$

for all $w(\cdot) \in \mathbf{L}_2[0,\infty)$. That is, condition (9.2.11) is satisfied. Therefore, the controller (9.2.5) solves the standard output-feedback H^∞ control problem (9.2.10), (9.2.11). Furthermore, it follows from Assumptions 9.2.1, 9.2.2 that the system (9.2.10) satisfies the assumptions required by Theorem 3.2.3 on page 73. Hence using these theorems, it follows that Riccati equations (9.2.7) and (9.2.8) have solutions as described in condition (ii) of the theorem.

(ii)⇒(i). Consider an uncertain system (9.2.1), (9.2.3) satisfying Assumptions 9.2.1, 9.2.2 and suppose there exist constants $\tau_1 > 0, \tau_2 > 0, \ldots, \tau_k > 0$ such that the corresponding Riccati equations (9.2.7) and (9.2.8) have positive-definite solutions X and Y satisfying the condition $\rho(XY) < 1$. Using Theorem 3.2.3, it follows that the linear controller (9.2.16) solves a standard H^∞ control problem for the system (9.2.10) with H^∞ norm requirement (9.2.11). That is, the closed loop system defined by (9.2.10) and (9.2.16) is such that (9.2.11) is satisfied. This closed loop system is described by the state equations

$$\begin{aligned}
\dot{h}(t) &= Ph(t) + Qw(t); \\
\hat{z}(t) &= \Sigma h(t)
\end{aligned} \tag{9.2.25}$$

where

$$h = \begin{bmatrix} x \\ x_c \end{bmatrix}; \quad P = \begin{bmatrix} A & B_1 C_c \\ B_c C_2 & A_c \end{bmatrix};$$
$$Q = \begin{bmatrix} \tilde{B}_2 \\ B_c \tilde{D}_2 \end{bmatrix}; \quad \Sigma = \begin{bmatrix} C_\tau & D_\tau C_c \end{bmatrix}.$$

Since the controller (9.2.16) solves the H^∞ control problem described above, the matrix P will be stable; see Theorem 3.2.3 and also [9, 147]. Also, if we apply the controller (9.2.16) to the uncertain system (9.2.1), (9.2.3), the resulting closed loop uncertain system can be described by the state equations (9.2.25) where $w(t)$ is now interpreted as an uncertainty input and $\hat{z}(t)$ is interpreted as an uncertainty output.

We will consider an uncertain system described by the state equations (9.2.25) where the uncertainty is required to satisfy the following integral quadratic constraint analogous to (9.2.3):

$$\int_0^{t_i} \|w(t)\|^2 dt \leq \int_0^{t_i} \|\Sigma h(t)\|^2 dt + d \quad \forall i. \tag{9.2.26}$$

Note that any admissible uncertainty for the closed loop uncertain system (9.2.1), (9.2.3), (9.2.16) will also be an admissible uncertainty for the uncertain system (9.2.25), (9.2.26) with

$$d = d_0 + \sum_{s=1}^{k} \tau_j d_j \geq 0.$$

Roughly speaking, the uncertain system (9.2.25), (9.2.26) has been obtained by over-bounding the "structured" uncertainty in the closed loop uncertain system (9.2.1), (9.2.3), (9.2.16) to give an uncertain system with "unstructured" uncertainty. We can now refer to Theorem 3.5.1 which states that the uncertain system (9.2.25), (9.2.26) is absolutely stable; see Definition 3.5.2 on page 97.

We will now show that the closed loop uncertain system (9.2.1), (9.2.3), (9.2.16) satisfies the conditions for absolute stabilizability given in Definition 9.2.2. This fact is proved in a similar fashion to the proof of the corresponding result in Theorem 3.5.1. Condition (i) of Definition 9.2.2 follows directly from the absolute stability of the uncertain system(9.2.25), (9.2.26). Moreover, the absolute stability of this uncertain system implies that there exists a constant $c_0 > 0$ such that for any initial condition $x(0) = x_0$ and any admissible uncertainty input $w(\cdot)$ described by (9.2.26), then $\{x(\cdot), w(\cdot)\} \in \mathbf{L}_2[0, \infty)$ and

$$\|x(\cdot)\|_2^2 + \|w(\cdot)\|_2^2 \leq c_0[\|x_0\|^2 + d].$$

Now equation (9.2.16) implies that there exists a constant $c_1 > 0$ such that

$$\|u(t)\| \leq c_1 \|x(t)\|$$

for all solutions to the closed loop system (9.2.1), (9.2.16). Furthermore, given any admissible

$$\xi(\cdot) = [\, \xi_0(\cdot) \; \ldots \; \xi_k(\cdot) \,]$$

for the closed loop system (9.2.1), (9.2.3), (9.2.16), then

$$w(\cdot) = [\xi_0(\cdot), \sqrt{\tau_1}\xi_1(\cdot), \ldots, \sqrt{\tau_k}\xi_k(\cdot)]$$

is an admissible uncertainty input for the uncertain system (9.2.25), (9.2.26) with $d = d_0 + \sum_{s=1}^{k} \tau_j d_j$. Hence, we can conclude that

$$\begin{aligned}
&\frac{1}{2c_1}\|u(\cdot)\|_2^2 + \frac{1}{2}\|x(\cdot)\|_2^2 + \|\xi_0(\cdot)\|_2^2 + \sum_{s=1}^{k} \tau_j \|\xi_j(\cdot)\|_2^2 \\
&\leq \|x(\cdot)\|_2^2 + \|\xi_0(\cdot)\|_2^2 + \sum_{s=1}^{k} \tau_j \|\xi_j(\cdot)\|_2^2 \\
&\leq c_0 \left[\|x_0\|^2 + d_0 + \sum_{s=1}^{k} \tau_j d_j \right]. \qquad (9.2.27)
\end{aligned}$$

However, it is straightforward to verify that there exist constants $\sigma_1 > 0$ and $\sigma_2 > 0$ such that we can write

$$\begin{aligned}
&\sigma_1 \left(\|u(\cdot)\|_2^2 + \|x(\cdot)\|_2^2 + \sum_{j=0}^{k} \|\xi_j(\cdot)\|_2^2 \right) \\
&\leq \frac{1}{2c_1}\|u(\cdot)\|_2^2 + \frac{1}{2}\|x(\cdot)\|_2^2 + \|\xi_0(\cdot)\|_2^2 + \sum_{s=1}^{k} \tau_j \|\xi_j(\cdot)\|_2^2
\end{aligned}$$

and moreover,

$$c_0 \left[\|x_0\|^2 + d_0 + \sum_{s=1}^{k} \tau_j d_j \right] \leq \sigma_2 \left[\|x_0\|^2 + \sum_{j=0}^{k} d_j \right].$$

Hence, by combining these inequalities with (9.2.27), it follows that

$$\|u(\cdot)\|_2^2 + \|x(\cdot)\|_2^2 + \sum_{j=0}^{k} \|\xi_j(\cdot)\|_2^2 \leq \frac{\sigma_2}{\sigma_1} \left[\|x_0\|^2 + \sum_{j=0}^{k} d_j \right].$$

That is, condition (iii) of Definition 9.2.2 is satisfied. Conditions (ii) and (iv) of Definition 9.2.2 follow directly from the stability of the matrix P. Thus, we can now conclude that the system (9.2.1), (9.2.3) is absolutely stabilizable via the controller (9.2.16). This completes the proof of the theorem.

■

The following corollary is an immediate consequence of the above theorem.

Corollary 9.2.1 *If the uncertain system (9.2.1), (9.2.3) satisfies Assumptions 9.2.1, 9.2.2 and is absolutely stabilizable via nonlinear control, then it will be absolutely stabilizable via a linear controller of the form (9.2.16).*

9.2.3 State-feedback nonlinear versus linear control

We now present state-feedback versions of the above results.

Theorem 9.2.2 *Consider the uncertain system (9.2.1), (9.2.3) and suppose that Assumption 9.2.1 is satisfied. Then the following statements are equivalent:*

(i) The uncertain system (9.2.1), (9.2.3) is absolutely stabilizable via a nonlinear state-feedback controller of the form (9.2.4).

(ii) There exist constants $\tau_1 > 0, \ldots, \tau_k > 0$ such that Riccati equation (9.2.7) has a solution $X > 0$.

If condition (ii) holds, then the uncertain system (9.2.1), (9.2.3) is absolutely stabilizable via a linear state-feedback controller of the form

$$u(t) = -G_\tau^{-1}[B_1' X + C_\tau' D_\tau]x(t). \tag{9.2.28}$$

Proof: The proof of this theorem is analogous to the proof of Theorem 9.2.1 except that a state-feedback H^∞ result (Theorem 3.2.2) is used in place of the output-feedback H^∞ result used in the proof of Theorem 9.2.1.

■

The following corollary is an immediate consequence of the above theorem concerning the connection between nonlinear and linear state-feedback control.

Corollary 9.2.2 *If the uncertain system (9.2.1), (9.2.3) is absolutely stabilizable via a nonlinear dynamic state-feedback controller of the form (9.2.4) then it will also be absolutely stabilizable via a linear static state-feedback controller of the form (9.2.28).*

9.3 Decentralized robust state-feedback H^∞ control for uncertain large-scale systems

9.3.1 Preliminary remarks

The problem of decentralized control is one in which H^∞ control theory has proven to be useful. In this section, we consider a version of the robust decentralized control problem for a class of uncertain large-scale systems. An important feature of this problem is the fact that if it can be solved via a nonlinear dynamic state-feedback controller, then it can also be solved via a linear static state-feedback controller.

We consider a large scale system S comprising N subsystems $S_i, i = 1, \ldots, N$. Each subsystem is subject to uncertainty. The dynamics of the i-th subsystem is described by state equations of the form

$$S_i: \quad \dot{x}_i(t) = A_i x_i(t) + B_i u_i(t) + E_i \xi_i(t) + F_i w_i(t) + L_i \eta_i(t); \quad (9.3.1)$$
$$z_i(t) = C_i x_i(t) + D_i u_i(t);$$
$$\zeta_i(t) = H_i x_i(t) + G_i u_i(t);$$
$$\xi_i(t) = \phi_i(t, \zeta_i(\cdot)); \quad (9.3.2)$$
$$\eta_i(t) = \psi_i(t, \zeta_j(\cdot); j = 1, \dots, N, j \neq i) \quad (9.3.3)$$

where $x_i \in \mathbf{R}^{n_i}$ is the state, $u_i \in \mathbf{R}^{m_i}$ is the control input, $w_i \in \mathbf{R}^{p_i}$ is the disturbance input, $\xi_i \in \mathbf{R}^{r_i}$ is the local uncertainty input, $\zeta_i \in \mathbf{R}^{h_i}$ is the uncertainty output and $z_i \in \mathbf{R}^{q_i}$ is the controlled output of the subsystem. The input η_i describes the effect of the subsystem S_j on the subsystem S_i, $j \neq i$.

Apart from the exogenous disturbances $w_i(\cdot)$, the dynamics of each subsystem is affected by uncertainties which have two sources. Some of the uncertainties in the large-scale system arise from the presence of uncertain dynamics in each subsystem. Such dynamics are driven only by the uncertainty output ζ_i of the subsystem S_i. These uncertainties are described by functions ϕ_i defined on trajectories of the subsystem S_i and are referred to as local uncertainties.

A second source of uncertainties arises from the interactions between the subsystems of the large-scale system. These interactions can be described by a collection of nonlinear functionals ψ_i, each defined on trajectories of the subsystems S_j, $j \neq i$, as in equation (9.3.3).

There are two main approaches to dealing with these interactions. The first approach considers the interconnections between subsystems as undesirable effects which deteriorate the desired performance. The decentralized controllers are thus designed in order to achieve a suitable level of "interconnection rejection". Conditions on the interconnections are found under which the overall large-scale system will retain an acceptable level of performance; e.g., see [213].

A second approach is to exploit information about the structure of interconnections when designing the controller; e.g., see [66, 29, 30]. As mentioned in [242], a common feature of these two approaches is the fact that the interconnections are considered as uncertain perturbations in the given large-scale system. Also, the decomposition of a complex uncertain system into a collection of subsystems S_i will result in the uncertainty in the original system being distributed amongst the subsystems. This provides additional motivation for treating the interconnections as uncertain perturbations. In this case, the role of the functionals ψ_i is to describe the uncertainty which has not been included into a subsystem S_i when decomposing the original system.

The foregoing discussion motivates the inclusion of the inputs η_i as additional uncertainty inputs entering into the subsystem S_i. In cases where there is no uncertainty in the interconnection functionals ψ_i, this assumption may lead to conservatism. Reference [242] illustrates an alternative approach to the control of the large scale systems in which the interconnections may play a positive role in the large scale system performance. However, this approach is not pursued in this section. Instead, we will be interested in whether modeling the uncertain interactions between subsystems using IQCs leads to a tractable decentralized control problem.

For the uncertain large-scale system (9.3.1), two closely related control problems are of interest: the problem of decentralized robust stabilization and the problem of decentralized robust H^∞ control. In both problems, the class of desired controllers U_D are controllers $u = [u_1, \ldots, u_N]$ such that each local controller u_i can use only information from the local subsystem S_i. Furthermore, we restrict attention to *state-feedback* controllers in which each controller has available the state of the corresponding subsystem.

Decentralized robust stabilization. Suppose there are no exogenous disturbances entering the system. In this case, we can let $F_i = 0$ for all $i = 1, \ldots, N$. In the problem of robust decentralized stabilization, we seek a controller $u(\cdot) \in U_D$ which guarantees the robust stability of the overall system S for all admissible uncertainties: $\phi = [\phi_1, \ldots, \phi_N] \in \Xi$, $\psi = [\psi_1, \ldots, \psi_N] \in \Psi$. Here Ξ and Ψ are given sets of admissible uncertainties to be defined below.

Decentralized robust stabilization with disturbance attenuation. In this case, each subsystem is driven by an exogenous disturbance input signal $w_i(t) \in \mathbf{L}_2[0, \infty)$. Hence, $F_i \neq 0$. In the problem of robust decentralized stabilization with disturbance attenuation, we are concerned with finding a controller $u(\cdot) \in U_D$ which guarantees the stability of the overall uncertain closed loop system, and also guarantees a pre-specified bound on the level of disturbance attenuation.

In the control theory literature, a number of different approaches have been applied to the above problems. For example, the problem of decentralized quadratic stabilization for a linear interconnected system with parameter uncertainties was considered by [38]. Also, reference [269] gave sufficient conditions for the existence of a set of local controllers to achieve a required level of H^∞ performance. Furthermore, sufficient conditions for decentralized quadratic stabilizability with disturbance attenuation were derived in references [263, 270, 243]. In these references, the synthesis of a decentralized controller was reduced to the problem of designing a quadratically stabilizing controller with disturbance attenuation for each subsystem. Another approach, which involves solving a set of local algebraic Riccati equations, was considered in [37]. For systems with norm-bounded uncertainty, the results of this reference give sufficient conditions as well as necessary conditions for decentralized state-feedback stabilizability with disturbance attenuation.

As mentioned in previous chapters, the norm-bounded uncertainty model does not allow for consideration of dynamic uncertainties. In the case of large-scale systems, this limits considerably the application of such uncertainty models. Indeed, nonlinear dynamic uncertainties often arise from interconnections between subsystems of a large-scale system. The main result presented in this section shows that by allowing for a wider class of uncertainty in which perturbations and interaction between subsystems satisfy certain integral quadratic constraints, we obtain a necessary and sufficient condition for robust decentralized stabilizability with disturbance attenuation. This result, which is given in terms of a set of local algebraic Riccati equations, also allows us to conclude that if the problem can be solved using a nonlinear dynamic state-feedback controller, then it can be solved with a linear static state-feedback controller.

9.3.2 Uncertain large-scale systems

In this subsection, we present definitions of the uncertainty sets Ξ and Ψ. In particular, we assume that both the local uncertainty in each subsystem and the interconnection between subsystems of the large-scale system satisfy certain integral quadratic constraints.

The definitions that follow extend the corresponding definitions of Section 9.2 to the case of large-scale systems; see also [127, 197, 260]. The definition of the set of admissible local uncertainties Ξ presented below is given for a general case of structured uncertainties. Indeed, suppose that the uncertainty inputs $\xi_i(\cdot)$ and uncertainty outputs $\zeta_i(\cdot)$ of the subsystem S_i have the following structure:

$$\begin{aligned}
&\zeta_i = [\zeta_{i,1}, \dots, \zeta_{i,k_i}]; \qquad \xi_i = [\xi_{i,1}, \dots, \xi_{i,k_i}];\\
&\zeta_{i,\nu}(t) = H_{i,\nu}x_i(t) + G_{i,\nu}u_i(t); \qquad \nu = 1, \dots, k_i;\\
&H_i = \begin{bmatrix} H_{i,1} \\ \vdots \\ H_{i,k_i} \end{bmatrix}; \qquad G_i = \begin{bmatrix} G_{i,1} \\ \vdots \\ G_{i,k_i} \end{bmatrix};\\
&\xi_{i,1}(t) = \phi_{i,1}(t, \zeta_{i,1}(\cdot));\\
&\qquad \vdots\\
&\xi_{i,k_i}(t) = \phi_{i,k_i}(t, \zeta_{i,k_i}(\cdot)).
\end{aligned} \tag{9.3.4}$$

Definition 9.3.1 (Integral Quadratic Constraint on Local Uncertainties) *An uncertainty of the form (9.3.4) is an admissible uncertainty for the large-scale system if the following conditions hold: For each $i = 1, \dots, N$, given any locally square integrable control input $u_i(\cdot)$, locally square integrable interconnection input $\eta_i(\cdot)$ and locally square integrable disturbance input $w_i(\cdot)$, let $(0, t_*)$ be the interval of existence of the corresponding solution to the equation (9.3.1) for all i. Then there exist constants $d_{i,1} \geq 0, \dots, d_{i,k_i} \geq 0$ and a sequence $\{t_j\}_{j=1}^{\infty}$, $t_j \to t^*$, such that*

$$\int_0^{t_j} \|\xi_{i,\nu}(t)\|^2 dt \leq \int_0^{t_j} \|\zeta_{i,\nu}(t)\|^2 dt + d_{i,\nu} \quad \forall \nu = 1, \dots, k_i, \ \forall i = 1, \dots, N, \tag{9.3.5}$$

for all $j = 1, 2, \dots$. An uncertainty input $\xi_i(\cdot) = [\xi_{i,1}(\cdot), \dots, \xi_{i,k_i}(\cdot)]$ defined by an admissible uncertainty of the form (9.3.4) is referred to as an admissible uncertainty input.

Definition 9.3.2 (Integral Quadratic Constraint on Interconnections between Subsystems) *The subsystems S_i of the given large-scale system are said to have admissible interconnections of the form (9.3.3), if the following conditions hold: For each $i = 1, \dots, N$, given any locally square integrable control input $u_i(\cdot)$, locally square integrable uncertainty input $\xi_i(\cdot)$ of the form (9.3.4) and locally square integrable disturbance input $w_i(\cdot)$, let $(0, t_*)$ be the interval of existence of the corresponding solution to the equation (9.3.1) for all i. Then there exist constants $\hat{d}_1 \geq 0, \dots, \hat{d}_N \geq 0$ and a sequence $\{t_j\}_{j=1}^{\infty}$, $t_j \to t^*$, such that*

$$\int_0^{t_j} \|\eta_i(t)\|^2 dt \leq \sum_{\mu=1,\mu\neq i}^{N} \int_0^{t_j} \|\zeta_\mu(t)\|^2 dt + \hat{d}_i \qquad \forall i = 1, \dots, N, \quad (9.3.6)$$

for all $j = 1, 2, \dots$. An uncertain interconnection input $\eta_i(\cdot)$ defined by an admissible interconnection of the form (9.3.3) is also referred to as an admissible uncertain interconnection input.

Without loss of generality, we can assume that the same sequences $\{t_j\}_{j=1}^{\infty}$ are chosen in Definitions 9.3.1 and 9.3.2 whenever they correspond to the same collection of uncertainty inputs, interconnection inputs and disturbance inputs. Also note that in this case, the intervals of existence for a square integrable solution are the same in Definitions 9.3.1 and 9.3.2.

Remark 9.3.1 Definitions 9.3.1 and 9.3.2 are standard definitions of uncertainty in terms of an integral quadratic constraint; e.g., see Definition 2.3.1 on page 24. Hence, the uncertain system (9.3.1) allows for norm-bounded uncertainty as a special case. To verify this, consider the uncertain large-scale system (9.3.1) with structured norm-bounded uncertainty [243, 270], defined as follows

$$\phi_{i,\nu}(t, \zeta_{i,\nu}) = \Delta_{i,\nu}(t)\zeta_{i,\nu}; \qquad \|\Delta_{i,\nu}(t)\| \leq 1$$

and assume that the interconnections between subsystems satisfy the constraint:

$$\|\psi_i(t, \zeta_j, j \neq i)\| \leq \alpha_i(t) \sum_{\mu=1,\mu\neq i}^{N} \|\zeta_\mu\| + \beta_i(t);$$
$$0 \leq \alpha_i(t) \leq 1; \quad \beta_i(\cdot) \in \mathbf{L}_2[0, \infty);$$

c.f. [243, 270]. Then the uncertainties $\xi_i(\cdot)$ and $\eta_i(\cdot)$ satisfy conditions (9.3.5) and (9.3.6) with $d_{i,\nu} = 0$, $\hat{d}_i = \int_0^\infty \|\beta_i(t)\|^2 dt$, and any t_j.

Note that Definition 9.3.1 presents a version of the IQC uncertainty description which takes into account the fact that the uncertainty is structured. Definition 9.3.2 considers the interconnections between the subsystems of a given system as an unstructured uncertainty. To allow for structured interconnections, an obvious modification of Definition 9.3.2 is required.

For the uncertain large-scale system (9.3.1), (9.3.5), (9.3.6), we consider a problem of *decentralized absolute stabilization with disturbance attenuation* via state-feedback control. The controllers considered are decentralized dynamic state feedback controllers of the form

$$\begin{aligned} \dot{x}_{c,i} &= \mathcal{A}_{c,i}(x_{c,i}(t), x_i(t)); \\ u_i &= \mathcal{K}_{c,i}(x_{c,i}(t), x_i(t)) \end{aligned} \tag{9.3.7}$$

where $x_{c,i} \in \mathbf{R}^{n_{c,i}}$ is the ith controller state vector and $\mathcal{A}_{c,i}$, $\mathcal{K}_{c,i}$ are Lipschitz continuous matrix valued functions such that the following conditions are satisfied:

(i) For all $i = 1, \ldots, N$, if $x_i(0) = 0$, $x_{c,i}(0) = 0$, $\xi_i(\cdot) = 0$, $\eta_i(\cdot) = 0$, and $w_i(\cdot) = 0$, then $x_i(\cdot) = 0$, $x_{c,i}(\cdot) = 0$ and $u_i(\cdot) = 0$.

(ii) The closed loop system (9.3.1), (9.3.7) generates a mapping $(x_i(0), x_{c,i}(0), \xi_i(\cdot), \eta_i(\cdot), w_i(\cdot)) \to (z_i(\cdot), \zeta_i(\cdot))$, which takes elements of $\mathbf{R}^{n_i+n_{c,i}} \times \mathbf{L}_2[0,\infty)$ into $\mathbf{L}_2[0,\infty)$.

(iii) The mapping defined in (ii) is continuous in a neighborhood of the origin in the Hilbert space $\mathbf{R}^{n_i+n_{c,i}} \times \mathbf{L}_2[0,\infty)$.

We will need these conditions to guarantee that the closed loop system (9.3.1), (9.3.7) satisfies the assumptions of the S-procedure Theorem 4.4.1.

Definition 9.3.3 *The large-scale system (9.3.1), (9.3.5), (9.3.6) is said to be* absolutely stabilizable with disturbance attenuation γ via decentralized state-feedback control *if there exists a decentralized state-feedback controller (9.3.7) and constants $c_1 > 0$ and $c_2 > 0$ such that the following conditions hold:*

(i) For any initial conditions $[x_i(0), x_{c,i}(0)]$, any admissible local uncertainty inputs $\xi_i(\cdot)$, any admissible interconnection inputs $\eta_i(\cdot)$ and any disturbance inputs $w_i(\cdot) \in \mathbf{L}_2[0,\infty)$,

$$x_i(\cdot), x_{c,i}(\cdot), u_i(\cdot), \xi_i(\cdot), \eta_i(\cdot) \in \mathbf{L}_2[0,\infty),$$

and hence $t_* = \infty$. Also,

$$\begin{aligned} &\sum_{i=1}^{N} \left[\|x_i(\cdot)\|_2^2 + \|x_{c,i}(\cdot)\|_2^2 + \|u_i(\cdot)\|_2^2 + \|\xi_i(\cdot)\|_2^2 + \|\eta_i(\cdot)\|_2^2\right] \\ &\quad \leq c_1 \sum_{i=1}^{N} \left[\|x_i(0)\|^2 + \|x_{c,i}(0)\|^2 + \|w_i(\cdot)\|_2^2 + \sum_{\nu=1}^{k_i} d_{i,\nu} + \hat{d}_i\right]. \end{aligned} \tag{9.3.8}$$

(ii) The following H^∞ norm bound condition is satisfied: If $x_i(0) = 0$ and $x_{c,i} = 0$ for all $i = 1, \ldots N$, then

$$J := \sup_{\substack{w(\cdot)\in \mathbf{L}_2[0,\infty),\\ \|w(\cdot)\|_2>0}} \sup_{\substack{\xi(\cdot)\in\Xi,\\ \eta(\cdot)\in\Psi}} \frac{\|z(\cdot)\|_2^2 - c_2 \sum_{i=1}^N \left[\sum_{\nu=1}^{k_i} d_{i,\nu} + \hat{d}_i\right]}{\|w(\cdot)\|_2^2} < \gamma^2. \tag{9.3.9}$$

Remark 9.3.2 Definition 9.3.3 extends a corresponding definition of absolute stabilizability with disturbance attenuation given in reference [197] to the class of large-scale uncertain systems considered in this section. As in [197], it follows from the above definition that if the uncertain system (9.3.1), (9.3.5), (9.3.6) is absolutely stabilizable with disturbance attenuation via decentralized state-feedback control, then the corresponding closed loop system (9.3.1), (9.3.5), (9.3.6), (9.3.7) with $w_i(\cdot) \in \mathbf{L}_2[0,\infty)$, $i = 1, \ldots, N$, has the property $x_i(\cdot) \to 0$ as $t \to \infty$; see [197, Observation 2.1].

Remark 9.3.3 In the particular case of $w_i(\cdot) \equiv 0$ for $i = 1, \ldots, N$, the first condition of Definition 9.3.3 amounts to the property of absolute stability for the overall closed loop system; *c.f.* the corresponding definition in Section 9.2. Thus, any controller solving the problem of decentralized absolute stabilization with disturbance attenuation also solves a problem of decentralized absolute stabilization.

9.3.3 Decentralized controller design

In this subsection, we present the main result of the section. This result establishes a necessary and sufficient condition for the uncertain large-scale system (9.3.1), (9.3.5), (9.3.6) to be absolutely stabilizable with a given level of disturbance attenuation via decentralized state-feedback control. This result requires the following assumption.

Assumption 9.3.1 *For all $i = 1, \ldots, N$, the pair (A_i, B_i) is stabilizable, $D_i'D_i > 0$, and the pair*

$$(A_i - B_i(D_i'D_i)^{-1}D_i'C_i, (I - D_i(D_i'D_i)^{-1}D_i')C_i) \tag{9.3.10}$$

is detectable.

Let $\tau_{i,\nu} > 0$, $\nu = 1, \ldots, k_i$, and $\theta_i > 0$, $i = 1, \ldots, N$, be given constants. Also, let

$$\tau_i = [\tau_{i,1} \ \ldots \tau_{i,k_i}]'; \quad \mathbf{T}_i = \mathrm{diag}[\tau_{i,1} \ \ldots \ \tau_{i,k_i}]; \quad \bar{\theta}_i = \sum_{\substack{j=1,\\ j\neq i}}^N \theta_j.$$

We consider a collection of the game type algebraic Riccati equations

$$(A_i - B_iR_i^{-1}\bar{D}_i'\bar{C}_i)'X_i + X_i(A_i - B_iR_i^{-1}\bar{D}_i'\bar{C}_i) + \bar{C}_i'(I - \bar{D}_iR_i^{-1}\bar{D}_i')\bar{C}_i$$
$$-X_i(B_iR_i^{-1}B_i' - \bar{B}_{2,i}\bar{B}_{2,i}')X_i = 0 \qquad (9.3.11)$$

where

$$\bar{C}_i = \begin{bmatrix} C_i \\ (\tau_{i,1} + \bar{\theta}_i)^{1/2}H_{i,1} \\ \vdots \\ (\tau_{i,k_i} + \bar{\theta}_i)^{1/2}H_{i,k_i} \end{bmatrix}; \qquad \bar{D}_i = \begin{bmatrix} D_i \\ (\tau_{i,1} + \bar{\theta}_i)^{1/2}G_{i,1} \\ \vdots \\ (\tau_{i,k_i} + \bar{\theta}_i)^{1/2}G_{i,k_i} \end{bmatrix};$$
$$R_i = \bar{D}_i'\bar{D}_i; \qquad \bar{B}_{2,i} = \left[E_i\mathbf{T}_i^{-1/2} \ \theta_i^{-1/2}L_i \ \tfrac{1}{\gamma}F_i \right]. \qquad (9.3.12)$$

Also, we consider the set

$$\mathcal{T} = \left\{ \begin{array}{ll} \tau_i \in \mathbf{R}_+^{k_i}, \theta \in \mathbf{R}_+^N : & \text{For all } i = 1, \ldots, N, \text{ equation (9.3.11)} \\ (i = 1, \ldots, N) & \text{has a solution } X_i = X_i' \geq 0 \text{ such that the} \\ & \text{matrices } A_i - B_iR_i^{-1}(B_i'X_i + \bar{D}_i'\bar{C}_i) \text{ and} \\ & A_i - B_iR_i^{-1}(B_i'X_i + \bar{D}_i'\bar{C}_i) + \bar{B}_{2,i}\bar{B}_{2,i}'X_i \\ & \text{are stable.} \end{array} \right\}. \qquad (9.3.13)$$

Theorem 9.3.1 *The following statements are equivalent:*

(i) The large-scale uncertain system (9.3.1), (9.3.5), (9.3.6) is absolutely stabilizable with disturbance attenuation γ via a decentralized dynamical state-feedback controller of the form (9.3.7).

(ii) The set $\mathcal{T}$ defined in (9.3.13) is non-empty.

If condition (ii) holds, then the large-scale uncertain system (9.3.1), (9.3.5), (9.3.6) is absolutely stabilizable with disturbance attenuation γ via a linear decentralized static state-feedback controller of the form

$$u_i(t) = K_ix_i(t) \qquad (9.3.14)$$

where

$$K_i = -R_i^{-1}(B_i'X_i + \bar{D}_i'\bar{C}_i). \qquad (9.3.15)$$

Proof: The proof of this theorem follows along similar lines to the proof of the corresponding result in Section 9.2 and reference [197]. The basic idea of the proof is to use the S-procedure in order to establish a connection between the original problem and a certain associated H^∞ control problem. However, in contrast to Section 9.2 and [197], we deal with a large-scale system and a class of controllers having a particular decentralized structure. Thus, only a suitably structured solution of the algebraic Riccati equation arising from this H^∞ control problem is of interest. The main idea in obtaining the desired decentralized controller structure relies on the description of local uncertainties and interconnections between subsystems of the large-scale system (9.3.1) in terms of the integral quadratic constraints (9.3.5), (9.3.6).

(i)⇒(ii). Consider the set $\mathcal{M}$ of vector functions

$$\lambda(\cdot) = \begin{bmatrix} x_1(\cdot), \dots, x_N(\cdot), x_{c,1}(\cdot), \dots, x_{c,N}(\cdot), \xi_1(\cdot), \dots, \xi_N(\cdot), \\ \eta_1(\cdot), \dots, \eta_N(\cdot), w_1(\cdot), \dots, w_N(\cdot) \end{bmatrix}$$

in $\mathbf{L}_2[0,\infty)$ satisfying equations (9.3.1), (9.3.7), with the initial conditions $x_i(0) = 0$, $x_{c,i}(0) = 0$ and admissible uncertainty inputs $\xi_i(\cdot)$ and $\eta_i(\cdot)$. Note that

$$\begin{aligned} \|\lambda(\cdot)\|_2^2 &= \sum_{i=1}^N \left[\|x_i(\cdot)\|_2^2 + \|x_{c,i}(\cdot)\|_2^2 + \|\xi_i(\cdot)\|_2^2 + \|\eta_i(\cdot)\|_2^2 + \|w_i(\cdot)\|_2^2\right] \\ &= \|x(\cdot)\|_2^2 + \|x_c(\cdot)\|_2^2 + \|\xi(\cdot)\|_2^2 + \|\eta(\cdot)\|_2^2 + \|w(\cdot)\|_2^2. \end{aligned}$$

Condition (9.3.9) implies that there exists a constant $\delta_1 > 0$ such that

$$J < \gamma^2 - 2\delta_1. \tag{9.3.16}$$

Let

$$\delta_2 = \min\left\{ \left[2c_1\left(N + \sum_{i=1}^N k_i\right)\right]^{-1}, \delta_1\left[(c_1+1)\left(c_2\left(N + \sum_{i=1}^N k_i\right) + 1\right)\right]^{-1} \right\}$$

where c_1 and c_2 are the constants from Definitions 9.3.1 and 9.3.2. Consider the following collection of functionals $\mathcal{G}_0, \mathcal{G}_{1,1}, \dots, \mathcal{G}_{N,k_N}, \mathcal{G}_1, \dots, \mathcal{G}_N$ defined on $\mathcal{M}$:

$$\begin{aligned} \mathcal{G}_0(\lambda(\cdot)) &= -(\|z(\cdot)\|_2^2 - \gamma^2\|w(\cdot)\|_2^2 + \delta_2\|\lambda(\cdot)\|_2^2); \\ \mathcal{G}_{i,\nu}(\lambda(\cdot)) &= \|\zeta_{i,\nu}(\cdot)\|_2^2 - \|\xi_{i,\nu}(\cdot)\|_2^2 + \delta_2\|\lambda(\cdot)\|_2^2; \\ \mathcal{G}_i(\lambda(\cdot)) &= \sum_{\substack{j=1, \\ j\neq i}}^N \|\zeta_j(\cdot)\|_2^2 - \|\eta_i(\cdot)\|_2^2 + \delta_2\|\lambda(\cdot)\|_2^2; \\ &\qquad \nu = 1, \dots, k_i,\ i = 1, \dots, N. \end{aligned} \tag{9.3.17}$$

As in Section 9.2, we will prove that this collection of the functionals satisfies the requirements of the S-procedure Theorem 4.4.1; *i.e.*,

$$\begin{aligned} &\lambda(\cdot) \in \mathcal{M};\quad \mathcal{G}_{i,\nu}(\lambda(\cdot)) \geq 0;\quad \mathcal{G}_i(\lambda(\cdot)) \geq 0 \\ &\text{for all } \nu = 1, \dots, k_i,\ i = 1, \dots, N, \end{aligned} \tag{9.3.18}$$

implies $\mathcal{G}_0(\lambda(\cdot)) \geq 0$. Indeed, condition (9.3.18) implies that the vector function $\lambda(\cdot)$ satisfies the constraints (9.3.5) and (9.3.6) with $d_{i,\nu} = \hat{d}_i = \delta_2 \|\lambda(\cdot)\|_2^2$ and $t_j = \infty$. Furthermore since $\delta_2 \leq \left[2c_1(N + \sum_{i=1}^N k_i)\right]^{-1}$, then condition (i) of Definition 9.3.3 implies

$$\|\lambda(\cdot)\|_2^2 \leq 2(c_1 + 1)\|w(\cdot)\|_2^2. \tag{9.3.19}$$

Also, from condition (9.3.19) and the condition

$$\delta_2 \leq \delta_1 \left[(c_1 + 1)\left(c_2\left(N + \sum_{i=1}^N k_i\right) + 1\right)\right]^{-1},$$

it follows that $2\delta_1\|w\|_2^2 \geq (1 + c_2(N + \sum_{i=1}^N k_i))\delta_2\|\lambda(\cdot)\|_2^2$. Hence, condition (9.3.16) implies

$$\begin{aligned}
\mathcal{G}_0(\lambda(\cdot)) &= -\left(\|z(\cdot)\|_2^2 - (\gamma^2 - 2\delta_1)\|w(\cdot)\|_2^2 - c_2(N + \sum_{i=1}^N k_i)\delta_2\|\lambda(\cdot)\|_2^2\right) \\
&\quad +2\delta_1\|w\|_2^2 - (1 + c_2(N + \sum_{i=1}^N k_i))\delta_2\|\lambda(\cdot)\|_2^2 \\
&\geq -\left(\|z(\cdot)\|_2^2 - (\gamma^2 - 2\delta_1)\|w(\cdot)\|_2^2 - c_2(N + \sum_{i=1}^N k_i)\delta_2\|\lambda(\cdot)\|_2^2\right) \\
&\geq 0.
\end{aligned}$$

We now write the given large-scale system in the form

$$\dot{x} = Ax(t) + \bar{B}_1 u(t) + \hat{B}_2 \hat{w}(t) \tag{9.3.20}$$

where $x \in \mathbf{R}^{n_1+\ldots+n_N}$ is the assembled state, $u \in \mathbf{R}^{m_1+\ldots+m_N}$ is the assembled control input and $\hat{w}$ is the assembled disturbance input of the form

$$\hat{w}(\cdot) = [\xi_1(\cdot),\ \eta_1(\cdot),\ w_1(\cdot),\ \ldots\ \xi_N(\cdot),\ \eta_N(\cdot),\ w_N(\cdot)].$$

The matrices A, $\bar{B}_1$, $\hat{B}_2$ are defined as follows:

$$\begin{aligned}
A &= \operatorname{diag}[A_1\ \ldots\ A_N]; \qquad \bar{B}_1 = \operatorname{diag}[B_1\ \ldots\ B_N]; \\
\hat{B}_2 &= \begin{bmatrix} E_1 & L_1 & F_1 & \ldots & 0 & 0 & 0 \\ & \vdots & & \ddots & & \vdots & \\ 0 & 0 & 0 & \ldots & E_N & L_N & F_N \end{bmatrix}.
\end{aligned} \tag{9.3.21}$$

We can now apply Theorem 4.4.1. The fact that the closed loop system satisfies the assumptions of this theorem is a straightforward consequence of the assumptions that have been imposed on the class of controllers of the form (9.3.7). Using this theorem, it follows that there exist constants $\tau_0 \geq 0$, $\tau_{i,\nu} \geq 0$, $\theta_i \geq 0$, $\nu = 1, \ldots, k_i$, $i = 1, \ldots, N$, such that $\tau_0 + \sum_{i=1}^{N}(\theta_i + \sum_{\nu=1}^{k_i} \tau_{i,\nu}) > 0$ and

$$\tau_0 \mathcal{G}_0(\lambda(\cdot)) \geq \sum_{i=1}^{N} \left(\theta_i \mathcal{G}_i(\lambda(\cdot)) + \sum_{\nu=1}^{k_i} \tau_{i,\nu} \mathcal{G}_{i,\nu}(\lambda(\cdot)) \right) \tag{9.3.22}$$

for all $\lambda \in \mathcal{M}$. Also, as in the proof of Theorem 9.2.1, it follows from condition (9.3.22) that $\tau_0 > 0, \tau_{i,\nu} > 0, \theta_i > 0$ for all $\nu = 1, \ldots, k_i$, $i = 1, \ldots, N$. Hence, one can take $\tau_0 = 1$ in (9.3.22). Thus, from (9.3.22)

$$\begin{aligned} &\|z(\cdot)\|_2^2 - \gamma^2 \|w(\cdot)\|_2^2 + \sum_{i=1}^{N} \left[\begin{array}{l} \sum_{\nu=1}^{k_i} \tau_{i,\nu} \left(\|\zeta_{i,\nu}(\cdot)\|_2^2 - \|\xi_{i,\nu}(\cdot)\|_2^2 \right) \\ +\theta_i \left(\sum_{\substack{j=1, \\ j \neq i}}^{N} \|\zeta_j(\cdot)\|_2^2 - \|\eta_i(\cdot)\|_2^2 \right) \end{array} \right] \\ &\leq -\delta_2 \left(1 + \sum_{i=1}^{N} (\theta_i + \sum_{\nu=1}^{k_i} \tau_{i,\nu}) \right) \|\lambda(\cdot)\|_2^2. \end{aligned} \tag{9.3.23}$$

Using the fact that for each $i = 1, \ldots, N$, $\|\zeta_i\|^2 = \sum_{\nu=1}^{k_i} \|\zeta_{i,\nu}\|^2$ and also using the identity

$$\sum_{i=1}^{N} \theta_i \sum_{\substack{j=1, \\ j \neq i}}^{N} \|\zeta_j(\cdot)\|_2^2 = \sum_{i=1}^{N} \left(\sum_{\substack{j=1, \\ j \neq i}}^{N} \theta_j \right) \|\zeta_i(\cdot)\|_2^2 = \sum_{i=1}^{N} \bar{\theta}_i \|\zeta_i(\cdot)\|_2^2,$$

we can re-write (9.3.23) as

$$\sup_{\substack{\bar{w}(\cdot) \in \mathbf{L}_2[0,\infty), \\ \|\bar{w}(\cdot)\|_2 > 0}} \frac{\|\bar{z}(\cdot)\|_2^2}{\|\bar{w}(\cdot)\|_2^2} < 1 - \delta_2 \epsilon. \tag{9.3.24}$$

Here

$$\bar{w}(\cdot) = \left[\begin{array}{r} \mathbf{T}_1^{1/2} \xi_1(\cdot),\ \sqrt{\theta_1} \eta_1(\cdot),\ \gamma w_1(\cdot),\ \ldots,\ \mathbf{T}_N^{1/2} \xi_N(\cdot), \\ \sqrt{\theta_N} \eta_N(\cdot),\ \gamma w_N(\cdot) \end{array} \right] \tag{9.3.25}$$

is the disturbance input of the closed loop system corresponding to the open loop system

$$\dot{x} = Ax(t) + \bar{B}_1 u(t) + \bar{B}_2 \bar{w}(t); \qquad x(0) = 0, \tag{9.3.26}$$

$$\begin{aligned} &\bar{z} = \bar{C} x + \bar{D} u; \\ &\bar{B}_2 = \mathrm{diag}[\bar{B}_{2,1} \ \ldots \ \bar{B}_{2,N}]; \\ &\bar{C} = \mathrm{diag}[\bar{C}_1 \ \ldots \ \bar{C}_N]; \qquad \bar{D} = \mathrm{diag}[\bar{D}_1 \ \ldots \ \bar{D}_N] \end{aligned} \tag{9.3.27}$$

and the given controller (9.3.7) with initial condition $x_{c,i} = 0$ for all $i = 1, \ldots, N$. The constant ϵ in (9.3.24) is a constant satisfying the condition

$$0 < \epsilon \leq \frac{1 + \sum_{i=1}^{N} (\theta_i + \sum_{\nu=1}^{k_i} \tau_{i,\nu})}{\max_{i=1,\ldots,N} (\theta_1, \|\mathbf{T}_i\|, \gamma^2)}.$$

As a consequence, (9.3.24) implies that for each $i = 1, \dots, N$,

$$\sup_{\bar{w}_i(\cdot)\in \mathbf{L}_2[0,\infty)} \frac{\|\bar{z}_i(\cdot)\|_2^2}{\|\bar{w}_i(\cdot)\|_2^2} < 1 - \delta_2\epsilon, \tag{9.3.28}$$

where

$$\bar{w}_i(\cdot) = \left[\mathbf{T}_i^{1/2}\xi_i(\cdot),\ 0,\ \gamma w_i(\cdot)\right] \tag{9.3.29}$$

is the disturbance input for the closed loop system corresponding to the open loop system

$$\begin{aligned} \dot{x}_i &= A_i x_i(t) + B_i u_i(t) + \bar{B}_{2,i}\bar{w}_i(t); \qquad x_i(0) = 0; \\ \bar{z}_i &= \bar{C}_i x_i + \bar{D}_i u_i \end{aligned} \tag{9.3.30}$$

and the i-th entry of the given controller (9.3.7) with the initial condition $x_{c,i} = 0$. To verify this, it is sufficient to let $\bar{w}_j(\cdot) = 0, j \neq i$ in (9.3.24) and (9.3.26). Then all the entries $\bar{z}_j$, $j \neq i$, of the corresponding output vector of the system (9.3.26) will be equal to zero. Hence (9.3.28) follows from (9.3.24).

Condition (9.3.28) implies that the entry u_i of the given controller of the form (9.3.7) solves the disturbance attenuation problem associated with system (9.3.30), with a closed loop, perfect state measurement information pattern; e.g., see Section 3.2. Now, since the pairs (A_i, B_i) and (9.3.10) are assumed to be stabilizable and detectable, Lemmas 3.2.1 and 3.2.2 imply that for each $i = 1, \dots, N$, the algebraic Riccati equation (9.3.11) admits a minimal nonnegative-definite solution X_i such that the matrices $A_i - B_i R_i^{-1}(B_i' X_i + \bar{D}_i'\bar{C}_i)$ and $A_i - B_i R_i^{-1}(B_i' X_i + \bar{D}_i'\bar{C}_i) + \bar{B}_{2,i}\bar{B}_{2,i}' X_i$ are stable. That is, the collection of the constants $\tau_{i,\nu}$, θ_i determined from the S-procedure belong to the set $\mathcal{T}$. Note that Lemma 3.2.2 requires that the matrix pair

$$\left(A_i - B_i R_i^{-1}\bar{D}_i'\bar{C}_i, (I - \bar{D}_i R_i^{-1}\bar{D}_i')\bar{C}_i\right) \tag{9.3.31}$$

is detectable. It is straightforward to verify that this requirement is met since Assumption 9.3.1 holds.

(ii)⇒(i). In this part of the proof, we suppose that the set $\mathcal{T}$ is non-empty. Consider the system (9.3.26) and the matrix $X = \operatorname{diag}[X_1, \dots, X_N]$ whose entries $X_1, \dots, X_N$ satisfy condition (9.3.13). Then, letting

$$R = \operatorname{diag}[R_1, \dots, R_N],$$

it follows that the matrices

$$\begin{aligned} &A - \bar{B}_1 R^{-1}(\bar{B}_1' X + \bar{D}'\bar{C}) \text{ and} \\ &A - \bar{B}_1 R^{-1}(\bar{B}_1' X + \bar{D}'\bar{C}) + \bar{B}_2\bar{B}_2' X \end{aligned}$$

are stable and the nonnegative-definite matrix X solves the Riccati equation

$$\begin{aligned} &(A - \bar{B}_1 R^{-1}\bar{D}'\bar{C})' X + X(A - \bar{B}_1 R^{-1}\bar{D}'\bar{C}) \\ &\quad + \bar{C}'(I - \bar{D}R^{-1}\bar{D}')\bar{C} - X(\bar{B}_1 R^{-1}\bar{B}_1' - \bar{B}_2\bar{B}_2')X = 0. \end{aligned} \tag{9.3.32}$$

Since the pair $(A, \bar{B}_1)$ is stabilizable and the pair

$$(A - \bar{B}_1 R^{-1} \bar{D}' \bar{C}, (I - \bar{D} R^{-1} \bar{D}') \bar{C})$$

is detectable[2], then it follows from H^∞ control theory (e.g., see Theorem 3.2.2 and also [52, 147]), that the matrix $K = \mathrm{diag}[K_1, \ldots, K_N]$ solves a standard state-feedback H^∞ control problem defined by the system (9.3.26) and the H^∞ norm bound (9.3.24). That is, the matrix $A + \bar{B}_1 K$ is stable. Also, the closed loop system

$$\begin{aligned} \dot{x} &= (A + \bar{B}_1 K)x(t) + \bar{B}_2 \bar{w}(t), \qquad x(0) = 0, \\ \bar{z} &= (\bar{C} + \bar{D}K)x, \end{aligned} \tag{9.3.33}$$

satisfies the condition: There exists a constant $\varepsilon > 0$ such that

$$\|\bar{z}(\cdot)\|_2^2 - \|\bar{w}(\cdot)\|_2^2 < -\varepsilon \|\bar{w}(\cdot)\|_2^2 \tag{9.3.34}$$

for all $\bar{w}(\cdot) \in \mathbf{L}_2[0, \infty)$ and $x(0) = 0$. From the Strict Bounded Real Lemma 3.1.2, these conditions are equivalent to the existence of a positive-definite symmetric matrix $\tilde{X} > X$ satisfying the matrix inequality

$$\begin{aligned} (A + \bar{B}_1 K)' \tilde{X} + \tilde{X}(A + \bar{B}_1 K) + \tilde{X} \bar{B}_2 \bar{B}_2' \tilde{X} \\ + (\bar{C} + \bar{D}K)'(\bar{C} + \bar{D}K) \leq -\epsilon I \end{aligned} \tag{9.3.35}$$

for some constant $\epsilon > 0$. From this inequality and using $x' \tilde{X} x$ as a candidate Lyapunov function, it is straightforward to verify that

$$\epsilon \|x(\cdot)\|_2^2 + \|\bar{z}(\cdot)\|_2^2 \leq \|\bar{w}(\cdot)\|_2^2 + x'(0) \tilde{X} x(0) \tag{9.3.36}$$

for all $\bar{w}(\cdot) \in \mathbf{L}_2[0, \infty)$. Now, let $\{t_j\}_{j=1}^\infty$ be a sequence of times as in Definitions 9.3.1 and 9.3.2. Fix j and corresponding time t_j and choose an arbitrary collection of admissible local uncertainty inputs $\xi_1(\cdot), \ldots, \xi_N(\cdot)$ and admissible interconnections $\eta_1(\cdot), \ldots, \eta_N(\cdot)$. Based on these chosen admissible uncertainties, we define the following uncertainty input $\bar{w}^j(\cdot)$ for the system (9.3.33):

$$\bar{w}^j(\cdot) = \begin{bmatrix} \mathbf{T}_1^{1/2} \tilde{\xi}_1^j(\cdot), \ \sqrt{\theta_1} \tilde{\eta}_1^j(\cdot), \ \gamma w_1(\cdot), \ \ldots \ \mathbf{T}_N^{1/2} \tilde{\xi}_N^j(\cdot), \\ \sqrt{\theta_N} \tilde{\eta}_N^j(\cdot), \ \gamma w_N(\cdot) \end{bmatrix} \tag{9.3.37}$$

where $\tilde{\xi}_1^j(\cdot), \ldots, \tilde{\xi}_N^j(\cdot), \tilde{\eta}_1^j(\cdot), \ldots, \tilde{\eta}_N^j(\cdot)$ are obtained by extending the chosen admissible uncertainty inputs $\xi_1(\cdot), \ldots, \xi_N(\cdot)$ and interconnections $\eta_1(\cdot), \ldots, \eta_N(\cdot)$ to have a value of zero on the interval $[t_j, \infty)$. Then, $\bar{w}^j(\cdot) \in \mathbf{L}_2[0, \infty)$ and hence condition (9.3.36) holds for this particular uncertainty input. From (9.3.36), (9.3.5), (9.3.6) we have that

$$\epsilon \int_0^{t_j} \|x(t)\|^2 dt \leq \gamma^2 \|w(\cdot)\|_2^2 + x'(0) \tilde{X} x(0) + \sum_{i=1}^N (\theta_i \hat{d}_i + \sum_{\nu=1}^{k_i} \tau_{i,\nu} d_{i,\nu}) \tag{9.3.38}$$

[2] See the remark concerning the detectability of the pair (9.3.31).

for all $w(\cdot) \in \mathbf{L}_2[0,\infty)$. Here $x(\cdot)$ is the state trajectory of the closed loop system (9.3.33) driven by the input $\bar{w}^j(\cdot)$. Since $t_* = sup\{t_j : x(\cdot) \in \mathbf{L}_2[0,t_j]\}$ by definition, then (9.3.38) implies $t_* = \infty$ and hence $x(\cdot), z(\cdot), \zeta_i(\cdot) \in \mathbf{L}_2[0,\infty)$. Consequently, $\xi_i(\cdot), \eta_i(\cdot) \in \mathbf{L}_2[0,\infty)$. Then, condition (9.3.8) follows from (9.3.38), (9.3.5), (9.3.6). This proves condition (i) of Definition 9.3.3.

To see that condition (ii) of Definition 9.3.3 also holds, we refer to the H^∞ norm bound (9.3.34). If $x(0) = 0$ and the input $\bar{w}^j(\cdot)$ is defined by (9.3.37), then (9.3.34), (9.3.5), (9.3.6) imply that

$$\int_0^{t_j} \|z(t)\|^2 dt - \gamma^2 \|w(\cdot)\|_2^2 - \sum_{i=1}^{N} (\theta_i \hat{d}_i + \sum_{\nu=1}^{k_i} \tau_{i,\nu} d_{i,\nu}) < -\varepsilon \|\bar{w}^j(\cdot)\|_2^2 \leq -\varepsilon\gamma^2 \|w(\cdot)\|_2^2 \tag{9.3.39}$$

for all $w(\cdot) \in \mathbf{L}_2[0,\infty)$. In (9.3.39), $z(\cdot)$ is the controlled output of the closed loop system (9.3.33) driven by the input $\bar{w}^j(\cdot)$,

$$z(\cdot) = [(C_1 + D_1 K_1)x_1(\cdot), \ \ldots, \ (C_N + D_N K_N)x_N(\cdot)].$$

Then, condition (9.3.9) follows from (9.3.39) by letting $t_j \to \infty$, with $c_2 = \max\{\theta_i, \tau_{i,\nu};\ \nu = 1, \ldots, k_i,\ i = 1, \ldots, N\}$.

■

Corollary 9.3.1 *If the large-scale system (9.3.1), (9.3.5), (9.3.6) is absolutely stabilizable with disturbance attenuation γ via a decentralized dynamical state-feedback controller of the form (9.3.7), then this system is also absolutely stabilizable with disturbance attenuation γ via a linear decentralized static state-feedback controller of the form (9.3.14) (9.3.15).*

Using Corollary 9.3.1, one can see that nonlinear decentralized controllers do not have any advantages as compared with linear controllers if the class of interconnection uncertainties between the subsystems of a large-scale system is sufficiently rich. However, improved performance may be achievable via nonlinear control if additional information about interconnections is available.

The physical interpretation of condition (ii) of Theorem 9.3.1 is analogous to that of the corresponding condition in Sections 5.3 and 9.2. That is, in the proof of Theorem 9.3.1, the constant vectors τ_i, θ arise as vectors of scaling parameters which allow one to convert the original problem of decentralized absolute stabilization with disturbance attenuation into an intermediate problem of disturbance attenuation associated with a collection of "scaled" systems (9.3.30) which have a closed loop perfect state information pattern. This is analogous to introducing a set of scaling parameters in the "scaled" version of the small gain theorem; e.g., see [127].

In general, the search for vectors τ_i and θ satisfying condition (ii) of Theorem 9.3.1 may be a difficult computational problem. One possible practical approach to carrying out this search would be to consider the constants $\bar{\theta}_i$ and

$\bar{\tau}_i = [\bar{\tau}_{i,1} \ \dots \ \bar{\tau}_{i,k_i}]$, $\bar{\tau}_{i,\nu} := \tau_{i,\nu} + \bar{\theta}_i$ as parameters independent of τ_i, θ_i and each other. This will allow one to reduce the original problem requiring the satisfaction of condition (ii) of the theorem to a problem of determining independent constants θ_i, $\bar{\theta}_i$ and vectors τ_i, $\bar{\tau}_i$ such that the Riccati equation

$$
\begin{aligned}
&(A_i - B_i\tilde{R}_i^{-1}\tilde{D}_i'\tilde{C}_i)'X_i + X_i(A_i - B_i\tilde{R}_i^{-1}\tilde{D}_i'\tilde{C}_i) + \tilde{C}_i'(I - \tilde{D}_i\tilde{R}_i^{-1}\tilde{D}_i')\tilde{C}_i \\
&-X_i(B_i\tilde{R}_i^{-1}B_i' - \bar{B}_{2,i}\bar{B}_{2,i}')X_i = 0, \qquad (9.3.40)
\end{aligned}
$$

has nonnegative-definite stabilizing solutions for each $i = 1, \dots, N$. In equation (9.3.40), the following notation is used

$$
\tilde{C}_i = \begin{bmatrix} C_i \\ \bar{\tau}_{i,1}^{1/2} H_{i,1} \\ \vdots \\ \bar{\tau}_{i,k_i}^{1/2} H_{i,k_i} \end{bmatrix}; \quad \tilde{D}_i = \begin{bmatrix} D_i \\ \bar{\tau}_{i,1}^{1/2} G_{i,1} \\ \vdots \\ \bar{\tau}_{i,k_i}^{1/2} G_{i,k_i} \end{bmatrix}; \quad \tilde{R}_i = \tilde{D}_i'\tilde{D}_i \qquad (9.3.41)
$$

and the matrices $\bar{B}_{2,i}$ are defined as in (9.3.12). Note that the search for suitable constants θ_i, $\bar{\theta}_i$ and vectors τ_i, $\bar{\tau}_i$ can be performed individually for each Riccati equation (9.3.40). After suitable θ_i, $\bar{\theta}_i$, τ_i, $\bar{\tau}_i$ have been determined, one must extract the collections $(\theta_i, \bar{\theta}_i, \tau_i, \bar{\tau}_i)$ which satisfy the conditions

$$
\begin{aligned}
&\tau_{i,\nu} + \bar{\theta}_i = \bar{\tau}_{i,\nu}; \quad \nu = 1, \dots, k_i; \\
&\sum_{\substack{j=1, \\ j \neq i}}^{N} \theta_j = \bar{\theta}_i \qquad (9.3.42)
\end{aligned}
$$

for all $i = 1, \dots, N$. These $(\theta_i, \tau_i, \bar{\theta}_i, \bar{\tau}_i)$ then satisfy condition (ii) of the theorem.

9.4 Nonlinear versus linear control in the robust stabilizability of linear uncertain systems via a fixed-order output-feedback controller

9.4.1 *Definitions*

In this section, we consider an uncertain system of the form (3.5.1) defined as follows:

$$
\begin{aligned}
\dot{x}(t) &= Ax(t) + B_1u(t) + B_2\xi(t); \\
z(t) &= C_1x(t) + D_1u(t); \\
y(t) &= C_2x(t) + D_2\xi(t). \qquad (9.4.1)
\end{aligned}
$$

In addition, we require the following assumption.

Assumption 9.4.1

(i) $D_1' \begin{bmatrix} C_1 & D_1 \end{bmatrix} = \begin{bmatrix} 0 & I \end{bmatrix}$;

(ii) $D_2 \begin{bmatrix} B_2' & D_2' \end{bmatrix} = \begin{bmatrix} 0 & I \end{bmatrix}$.

Assumption 9.4.1 is a standard assumption in H^∞ control theory. As mentioned in Section 3.2, this assumption is not critical to the ideas presented in this section. However, it simplifies the exposition. For additional comments on this assumption, we refer the reader to the remark on page 67.

The problem considered in this section is a problem of robust stabilization for the uncertain system (9.4.1) via a nonlinear time-varying output-feedback controller of the form

$$\begin{aligned} \dot{x}_c(t) &= L(t, x_c(t), y(t)); \\ u(t) &= \lambda(t, x_c(t), y(t)) \end{aligned} \qquad (9.4.2)$$

or a linear time-invariant output-feedback controller of the form (9.2.16). Note that the controller (9.4.2) is a time-varying version of the controller (9.2.5) considered in Section 9.2. However, in contrast to Section 9.2, we assume the controllers (9.4.2) and (9.2.16) are fixed-order controllers of order k. That is, the controller state satisfies $x_c(t) \in \mathbf{R}^k$. Also, we assume that the functions $L(\cdot)$ and $\lambda(\cdot)$ in equation (9.4.2) satisfy the condition $L(\cdot, 0, 0) \equiv 0$ and $\lambda(\cdot, 0, 0) \equiv 0$.

We now introduce a notion of robust stabilizability for the uncertain system (9.4.1).

Definition 9.4.1

1. *The uncertain system (9.4.1) is said to be* robustly stabilizable with a quadratic storage function via nonlinear time-varying output-feedback controller of order k *if there exists a controller of the form (9.4.2), a matrix $H = H' > 0$ and a constant $\epsilon > 0$ such that the dimension of H is $(n+k) \times (n+k)$ and the following condition holds: Let*

$$h(t) := \begin{bmatrix} x(t) \\ x_c(t) \end{bmatrix}; \qquad V(h) := h'Hh$$

and let $\dot{V}[h(t)]$ be the derivative of $V(h(t))$ along trajectories of the closed loop system (9.4.1), (9.4.2). Then,

$$\dot{V}[h(t)] + \|z(t)\|^2 - \|\xi(t)\|^2 \leq -\epsilon(\|h(t)\|^2 + \|u(t)\|^2 + \|\xi(t)\|^2) \qquad (9.4.3)$$

for all solutions to the closed loop system (9.4.1), (9.4.2).

2. *The uncertain system (9.4.1) is said to be* robustly stabilizable with a quadratic storage function via linear time-invariant output-feedback controller of order k *if there exists a controller of the form (9.2.16), a matrix*

$H = H' > 0$ and a constant $\epsilon > 0$ such that condition (9.4.3) is satisfied with the controller (9.4.2) replaced by the linear controller (9.2.16). Note, the case of $k = 0$ corresponds to static output-feedback control.

The reader should note the connection between this notion of robust stabilizability with a quadratic storage function and the notion of absolute stability defined in Section 7.3; see Definition 7.3.5 on page 238. In particular, the system (9.4.1) is absolutely stabilizable with a quadratic storage function via linear or nonlinear control if the closed loop system corresponding to this controller is absolutely stable in terms of Definition 7.3.5. The quadratic function $V(\cdot)$ in this case is a standard storage function arising in the corresponding dissipativity problems considered in Section 7.3.

We now investigate the relationship between the concept of the robust stabilizability with a quadratic storage function and the concept of the quadratic stabilizability; e.g., see Section 3.4. Indeed, consider the following uncertain system with uncertainty satisfying a norm bound condition:

$$\begin{aligned} \dot{x}(t) &= [A + B_2\Delta(t)C_1]x(t) + [B_1 + B_2\Delta(t)D_1]u(t); \\ y(t) &= [C_2 + D_2\Delta(t)C_1]x(t) + D_2\Delta(t)D_1u(t); \\ & \quad \|\Delta(t)\| \leq 1 \end{aligned} \tag{9.4.4}$$

where $\Delta(t)$ is the uncertainty matrix; e.g., see Section 2.2 and also [143, 168, 177]. Let

$$\xi(t) = \Delta(t)[C_1x(t) + D_1u(t)] = \Delta(t)z(t).$$

Then the system (9.4.4) may be re-written in the form (9.4.1) with

$$\|\xi(t)\|^2 \leq \|z(t)\|^2. \tag{9.4.5}$$

Now suppose the uncertain system (9.4.2) is robustly stabilizable with a quadratic storage function via a controller

$$\begin{aligned} \dot{x}_c(t) &= L(t, x_c(t), y(t)); \\ u(t) &= \lambda(t, x_c(t)). \end{aligned} \tag{9.4.6}$$

It follows that there exists a matrix $H > 0$ and a constant $\epsilon > 0$ such that condition (9.4.3) holds for the corresponding closed loop system. Now consider the quadratic stability of the closed loop uncertain system (9.4.4), (9.4.6). This system is described by the state equations

$$\begin{aligned} \dot{x} &= Ax + B_1\lambda(t, x_c) + B_2\xi; \\ \dot{x}_c &= L(t, x_c, C_2x + D_2\xi) \end{aligned} \tag{9.4.7}$$

where ξ is defined as above. The Lyapunov derivative corresponding to this system and the Lyapunov function $V(h) = h'Hh$ is given by

$$\dot{V}(h) = h'H \begin{bmatrix} Ax + B_1\lambda(t, x_c) + B_2\xi \\ L(t, x_c, C_2x + D_2\xi) \end{bmatrix}$$

where $\xi = \Delta(t)z$. However, note that (9.4.7) is exactly the closed loop system (9.4.1), (9.4.6). Hence, it follows from (9.4.3) and (9.4.5) that

$$\dot{V}(h) \leq -\epsilon\|h\|^2$$

for all $\Delta(\cdot)$ such that $\|\Delta\| \leq 1$. That is the closed loop system is quadratically stable. Thus, we have established that if the system (9.4.1) is robustly stable with a quadratic storage function via the controller (9.4.6), then the corresponding uncertain system (9.4.4) is quadratically stable via the same controller.

Note that the converse of the above result is not necessarily true in general. However, if one restricts attention to linear controllers, then the notion of robust stabilizability with a quadratic storage function is equivalent to the notion of quadratic stabilizability. This fact can be established using Finsler's Theorem or the S-procedure; see Chapter 4 and also [239].

9.4.2 *Design of a fixed-order output-feedback controller*

We now present the main result of this section which shows that if the system (9.4.1) is robustly stabilizable with a quadratic storage function via a nonlinear controller of the form (9.4.2), then it is robustly stabilizable via a linear controller.

Theorem 9.4.1 *Consider the uncertain system (9.4.1) and suppose that Assumption 9.4.1 holds. Also, let $k \in \{0, 1, 2, \dots\}$ be given. Then the following statements are equivalent:*

(i) The uncertain system (9.4.1) is robustly stabilizable with a quadratic storage function via a nonlinear time-varying output-feedback controller of order k.

(ii) The uncertain system (9.4.1) is robustly stabilizable with a quadratic storage function via a linear time-invariant output-feedback controller of order k.

In order to prove this theorem, we will use the following lemma which can be found in references [64, 71, 91].

Lemma 9.4.1 *Consider the uncertain system (9.4.1) and suppose that Assumption 9.4.1 holds. Let $k \in \{0, 1, 2, \dots\}$ be given and suppose that there exist matrices $X = X' > 0$ and $Y = Y' > 0$ such that the dimension of X and Y is $n \times n$ and the following conditions hold:*

$$A'X + XA + X(B_2B_2' - B_1B_1')X + C_1'C_1 < 0; \tag{9.4.8}$$

$$AY^{-1} + Y^{-1}A' + Y^{-1}(C_1'C_1 - C_2'C_2)Y^{-1} + B_2B_2' < 0; \tag{9.4.9}$$

$$X \leq Y; \qquad \text{rank}(Y - X) \leq k. \tag{9.4.10}$$

Then, there exists a linear time-invariant controller of order k of the form (9.2.16) which solves the following H^∞ control problem:

$$\sup_{\substack{\xi(\cdot)\in \mathbf{L}_2[0,\infty),\\ x(0)=0}} \frac{\int_0^\infty \|z(t)\|^2 dt}{\int_0^\infty \|\xi(t)\|^2 dt} < 1 \tag{9.4.11}$$

for the system (9.4.1).

Proof of Theorem 9.4.1: (i)$\Rightarrow$(ii). Suppose that a controller of the form (9.4.2), a matrix $H = H' > 0$ and a constant $\epsilon > 0$ satisfy condition (9.4.3). We introduce matrices Y, Z and Ψ as follows:

$$H = \begin{bmatrix} Y & Z \\ Z' & \Psi \end{bmatrix}$$

where the dimension of Y is $n \times n$, the dimension of Z is $n \times k$ and the dimension of Ψ is $k \times k$. Then, since $H > 0$, we have $Y > 0$. Also, $\Psi > 0$ if $k > 0$. Let $X := Y - Z\Psi^{-1}Z'$ if $k > 0$ and $X := Y$ if $k = 0$. Then, conditions (9.4.10) will be satisfied. Also, the fact that $X = \begin{bmatrix} I & -Z\Psi^{-1} \end{bmatrix} H \begin{bmatrix} I & -Z\Psi^{-1} \end{bmatrix}'$ and $H > 0$ implies $X > 0$. We now prove that the matrix X satisfies condition (9.4.8). Let $V_x(x) := x'Xx$ and $\dot{V}_x[x(t)]$ be the derivative of $V_x(\cdot)$ along trajectories of the system (9.4.1). Consider the Lyapunov derivative $\dot{V}[h(t)]$ for the system (9.4.1), (9.4.2) at a point $h(t) = [x(t)' \; x_c(t)']'$ such that $x_c(t) = -\Psi^{-1}Z'x(t)$. Then

$$\begin{aligned} \dot{V}[h(t)] &= 2x(t)' \begin{bmatrix} I & -Z\Psi^{-1} \end{bmatrix} \begin{bmatrix} Y & Z \\ Z' & \Psi \end{bmatrix} \begin{bmatrix} \dot{x}(t) \\ \dot{x}_c(t) \end{bmatrix} \\ &= 2x(t)' \begin{bmatrix} X & 0 \end{bmatrix} \begin{bmatrix} \dot{x}(t) \\ \dot{x}_c(t) \end{bmatrix} \\ &= \dot{V}_x[x(t)]. \end{aligned}$$

However, since (9.4.3) must hold at all points $h(t) \in \mathbf{R}^{n+k}$, we have

$$\dot{V}_x[x(t)] + \|z(t)\|^2 - \|\xi(t)\|^2 \leq -\epsilon(\|x(t)\|^2 + \|u(t)\|^2 + \|\xi(t)\|^2) \quad (9.4.12)$$

for all solutions to the system (9.4.1) with

$$u(t) = \lambda(t, -\Psi^{-1}Z'x(t), y(t)). \quad (9.4.13)$$

Furthermore, (9.4.12) implies that

$$\begin{aligned} x(T)'Xx(T) - x(0)'Xx(0) + \int_0^T (\|z(t)\|^2 - \|\xi(t)\|^2)dt \\ \leq -\epsilon \int_0^T (\|x(t)\|^2 + \|u(t)\|^2 + \|\xi(t)\|^2)dt \end{aligned} \quad (9.4.14)$$

for any $T > 0$ and for all solutions to the closed loop system (9.4.1), (9.4.13). Also, equations (9.4.1) imply that for any $T > 0$, there exists a constant $c_T > 0$ such that

$$\|x(T)\|^2 \leq c_T \int_0^T (\|x(t)\|^2 + \|u(t)\|^2 + \|\xi(t)\|^2)dt \quad (9.4.15)$$

for all solutions to the system (9.4.1). Indeed, consider the observability gramian of the pair (A, I). Since, the pair (A, I) is observable, then there exists a constant $c_1 > 0$ such that

$$x(0)'\left[\int_0^T e^{A't}e^{At}dt\right]x(0) \geq c_1\|x(0)\|^2 \geq c_{2T}\|e^{AT}x(0)\|^2$$

for all $x(0)$; here $c_{2T} := c_1\|e^{AT}\|^{-2}$. Combining this fact with (9.4.1), we establish that

$$\begin{aligned}\|e^{AT}x(0)\|^2 &\leq c_{2T}\int_0^T \left\|x(t) - \int_0^t e^{A(t-s)}(B_1u(s) + B_2(s))ds\right\|^2 dt \\ &\leq c_{3T}\int_0^T (\|x(t)\|^2 + \|u(t)\|^2 + \|\xi(t)\|^2)dt.\end{aligned}$$

Furthermore, the above inequality and (9.4.1) together imply an inequality of the form (9.4.15).

Now we choose sufficiently small constants $\tilde{\epsilon} > 0$ and $\epsilon_0 > 0$ which satisfy the conditions

$$\tilde{\epsilon}(c_T\|X\| + \epsilon_0) < \epsilon, \quad \tilde{\epsilon}\epsilon_0 < 1.$$

With this choice of $\tilde{\epsilon}$ and ϵ_0, conditions (9.4.14) and (9.4.15) imply that

$$\begin{aligned}(1+\tilde{\epsilon})&\left[x(T)'Xx(T) + \int_0^T (\epsilon_0\|x(t)\|^2 + \|z(t)\|^2)dt\right] \\ &\leq x(0)'Xx(0) + \int_0^T \|\xi(t)\|^2 dt.\end{aligned}$$

That is, for any $T > 0$, the full information controller (9.4.13) solves the following H^∞ control problem with transients:

$$\sup_{\xi(\cdot)\in\mathbf{L}_2[0,T]} \frac{x(T)'Xx(T) + \int_0^T(\epsilon_0\|x(t)\|^2 + \|z(t)\|^2)dt}{x(0)'Xx(0) + \int_0^T \|\xi(t)\|^2 dt} < 1 \qquad (9.4.16)$$

for the system (9.4.1). Then, Theorem 3.2.4[3] on page 76 implies that the solution $X_T(\cdot)$ to the Riccati differential equation

$$\begin{aligned}-\dot{X}_T(t) &= A'X_T(t) + \hat{C}_1'\hat{C}_1 + X_T(t)A + X_T(t)(B_2B_2' - B_1B_1')X_T(t); \\ X_T(T) &= X \qquad (9.4.17)\end{aligned}$$

[3]Note that in applying Theorem 3.2.4, we identify the system (3.2.1) of Section 3.2 with our system (9.4.1). However, the output z in system (3.2.1) corresponds to the output $[z' \ \sqrt{\epsilon_0}x']'$ obtained from our system (9.4.1). Also note, the result of Theorem 3.2.4 is stated for the case of linear time-varying controllers. However, as observed in Observation 3.2.1, this theorem remains true if one allows for nonlinear time-varying controllers as well.

is defined and positive definite on $[0, T]$ and $X_T(0) < X$. Here

$$\hat{C}_1 := \left[C_1' \ \sqrt{\epsilon_0} I \right]'.$$

Furthermore,

$$X_T(0) = X_T(T) - \int_0^T \dot{X}_T(t) dt = X - \int_0^T \dot{X}_T(t) dt.$$

Hence, $\int_0^T \dot{X}_T(t) dt > 0$ for any $T > 0$. It is now clear that

$$\lim_{T \to 0} \frac{1}{T} \int_0^T \dot{X}_T(t) dt = \dot{X}(0)$$

where $X(\cdot)$ is the solution to the equation (9.4.17) with initial condition $X(0) = X$. Hence, $\dot{X}(0) \geq 0$. Therefore,

$$A'X + XA + X(B_2 B_2' - B_1 B_1')X + \hat{C}_1' \hat{C}_1 \leq 0.$$

Since $\hat{C}_1' \hat{C}_1 > C_1' C_1$, it follows that the inequality (9.4.8) holds.

We now prove that the matrix Y satisfies condition (9.4.9). Condition (9.4.3) implies that

$$h(T)'Hh(T) - h(0)'Hh(0) + \int_0^T (\|z(t)\|^2 - \|\xi(t)\|^2) dt$$
$$\leq -\epsilon \int_0^T (\|h(t)\|^2 + \|u(t)\|^2 + \|\xi(t)\|^2) dt \qquad (9.4.18)$$

for any $T > 0$ and for all solutions to the closed loop system (9.4.1), (9.4.2). Let $D_2^\perp$ be a matrix such that

$$\begin{bmatrix} D_2 \\ D_2^\perp \end{bmatrix} \begin{bmatrix} D_2 \\ D_2^\perp \end{bmatrix}' = I.$$

Such $D_2^\perp$ exists due to the second condition of Assumption 9.4.1. For any $v(\cdot) \in \mathbf{L}_2[0, T]$, define a vector function

$$\xi(t) := -D_2' C_2 x(t) + D_2^{\perp\prime} v(t).$$

Then, for the system (9.4.1), (9.4.2) with such an input $\xi(\cdot)$ and initial condition $x_c(0) = 0$, we have $y(\cdot) \equiv 0$, $u(\cdot) \equiv 0$ and condition (9.4.18) may be re-written as

$$x(T)'Yx(T) - x(0)'Yx(0) + \int_0^T (\|z(t)\|^2 - \|C_2 x(t)\|^2 - \|v(t)\|^2) dt$$
$$\leq -\epsilon \int_0^T (\|x(t)\|^2 + \|C_2 x(t)\|^2 + \|v(t)\|^2) dt \qquad (9.4.19)$$

for any $T > 0$ and for all solutions to the system

$$\dot{x}(t) = Ax(t) + B_2 D_2^{\perp'} v(t). \tag{9.4.20}$$

Also, (9.4.20) implies that for any $T > 0$, there exists a constant $c_T > 0$ such that

$$\|x(0)\|^2 \leq c_T \int_0^T (\|x(t)\|^2 + \|v(t)\|^2) dt \tag{9.4.21}$$

for all solutions to the system (9.4.20). Furthermore, conditions (9.4.19) and (9.4.21) imply that there exists a constant $\epsilon_0 > 0$ such that

$$\begin{aligned} & x(T)'Yx(T) - x(0)'Yx(0) \\ & + \int_0^T (\epsilon_0 \|x(t)\|^2 + \|z(t)\|^2 - \|C_2 x(t)\|^2 - \|v(t)\|^2) dt \\ & \leq -\epsilon_0 \left[\|x(0)\|^2 + \int_0^T (\|x(t)\|^2 + \|v(t)\|^2) dt \right] \end{aligned} \tag{9.4.22}$$

for any $T > 0$. Let

$$\begin{aligned} & F[x_T, v(\cdot)] \\ & := x(0)'Yx(0) + \int_0^T (-\epsilon_0 \|x(t)\|^2 - \|z(t)\|^2 + \|C_2 x(t)\|^2 + \|v(t)\|^2) dt \end{aligned}$$

where $x(\cdot)$ and $v(\cdot)$ satisfy the equations (9.4.20) and $x(T) = x_T$. From (9.4.22) we have that the minimum in

$$\hat{J}[x_T] := \min_{x(T)=x_T,\ v(\cdot)\in \mathbf{L}_2[0,T]} F[x_T, v(\cdot)]$$

is achieved for any $x_T \in \mathbf{R}^n$ and $\hat{J}[x_T] \geq x_T' Y x_T$; e.g., see [113]. Hence, the solution to the Riccati equation

$$\begin{aligned} -\dot{P}(t) &= P(t)A + A'P(t) + P(t)B_2 B_2' P(t) + \hat{C}_1' \hat{C}_1 - C_2' C_2; \\ P(0) &= Y \end{aligned}$$

is defined on $[0, \infty)$ and $P(T) \geq Y$ for all $T > 0$. From this we have that $\dot{P}(0) \geq 0$. Hence,

$$YA + A'Y + YB_2 B_2' Y + \hat{C}_1' \hat{C}_1 - C_2' C_2 \leq 0.$$

Since $\hat{C}_1' \hat{C}_1 > C_1' C_1$, we have that

$$YA + A'Y + YB_2 B_2' Y + C_1' C_1 - C_2' C_2 < 0.$$

Hence, condition (9.4.9) holds. Now Lemma 9.4.1 implies that there exists a linear time-invariant controller of order k such that condition (9.4.11) holds. Also, if condition (9.4.11) holds for the linear time-invariant system (9.4.1), (9.2.16), then there exist a matrix $H = H' > 0$ and a constant $\epsilon > 0$ such that condition (9.4.3) holds for the system (9.4.1), (9.2.16). This completes the proof of this part of the theorem.

(ii)⇒(i). This statement follows immediately from the definitions. ■

9.5 Simultaneous H^∞ control of a finite collection of linear plants with a single nonlinear digital controller

9.5.1 Problem statement

The main result presented in this section is a necessary and sufficient condition for the existence of a nonlinear digital output-feedback controller solving a simultaneous H^∞ control problem[4]. This condition is given in terms of the existence of suitable solutions to certain algebraic Riccati equations and a dynamic programming equation. If such solutions exist, then it is shown that they can be used to construct a corresponding controller. In the case of a single plant, the dynamic programming equation can be reduced to another algebraic Riccati equation and our controller coincides with the linear controller obtained from standard H^∞ control theory, e.g., see Section 3.2. However, in the case of several plants, the controller obtained is nonlinear.

Although the main result of this section does not address the issue of nonlinear versus linear control, it does address a related question of the required controller order in solving a robust stabilization problem.

Let k be a given integer. We consider the following k time-invariant linear systems defined on the time interval $[0, \infty)$:

$$\begin{aligned} \dot{x}_i(t) &= A_i x_i(t) + B_{1i} u(t) + B_{2i} w_i(t); \\ z_i(t) &= C_{1i} x_i(t) + D_{1i} u(t); \\ y(t) &= C_{2i} x_i(t) + v_i(t) \quad i = 1, 2, \dots, k \end{aligned} \tag{9.5.1}$$

where $x_i(t) \in \mathbf{R}^{n_i}$ is the *state*, $w_i(t) \in \mathbf{R}^{p_i}$ and $v_i(t) \in \mathbf{R}^{l_i}$ are the *disturbance inputs*, $u(t) \in \mathbf{R}^h$ is the *control input*, $z_i(t) \in \mathbf{R}^{q_i}$ are the *controlled outputs* and $y(t) \in \mathbf{R}^l$ is the *measured output*.

We suppose that the following assumption holds:

Assumption 9.5.1 *For each* $i = 1, 2, \dots, k$*:*

(i) $C_{1i}' D_{1i} = 0$ *and* $D_{1i}' D_{1i} > 0$.

(ii) The pair (A_i, C_{1i}) *is observable.*

(iii) The pair (A_i, B_{2i}) *is controllable.*

As mentioned previously, this assumption guarantees that a corresponding H^∞ control problem is non-singular.

---

[4]The main result presented in this section originally appeared in the paper [174]. The authors are grateful to Uffe Thygesen for noticing a mistake in a preliminary version of this paper and for useful suggestions which were incorporated into the final version of this result.

The digital controller

Let $\delta > 0$ be a given time. We will consider the following class of nonlinear digital output-feedback controllers which update the control input $u(t)$ at discrete times $0, \delta, 2\delta, 3\delta, \ldots$, with $u(t)$ constant between updates:

$$\begin{aligned} &u(j\delta) = U[j\delta, y(\cdot)|_0^{j\delta}] \quad \text{and} \\ &u(t) = u(j\delta) \quad \forall t \in [j\delta, (j+1)\delta) \quad \forall j = 0, 1, 2, \ldots . \end{aligned} \tag{9.5.2}$$

Also, we suppose that the following causality type assumption holds:

$$y(t) = 0 \ \forall t \in [0, j\delta] \Rightarrow u(j\delta) = 0. \tag{9.5.3}$$

Definition 9.5.1 *Consider the systems (9.5.1). A nonlinear digital output-feedback controller (9.5.2), (9.5.3) is said to solve the simultaneous H^∞ control problem for the systems (9.5.1) if there exists a constant $c > 0$ such that for all $i = 1, 2, \ldots, k$ the following conditions hold:*

(i) *For any initial condition $x_i(0) = x_i^0$ and disturbance inputs $[w_i(\cdot), v_i(\cdot)] \in \mathbf{L}_2[0, \infty)$, the closed loop system (9.5.1), (9.5.2) has a unique solution $[x_i(\cdot), u(\cdot)] \in \mathbf{L}_2[0, \infty)$.*

(ii) *The following inequality is satisfied:*

$$\sup \frac{\int_0^\infty \|z_i(t)\|^2 dt}{c\|x_i(0)\|^2 + \int_0^\infty (\|w_i(t)\|^2 + \|v_i(t)\|^2) dt} < 1 \tag{9.5.4}$$

where the supremum is taken over all $x_i(0) \in \mathbf{R}^{n_i}$ and $[w_i(\cdot), v_i(\cdot)] \in \mathbf{L}_2[0, \infty)$ such that

$$\|x_i(0)\|^2 + \int_0^\infty (\|w_i(t)\|^2 + \|v_i(t)\|^2) dt > 0.$$

Remark 9.5.1 The inequality (9.5.4) corresponds to an H^∞ control problem with non-zero initial conditions over an infinite time interval for the i−th plant; e.g., see Subsection 3.2.2 and also [100]. Hence, a nonlinear digital output-feedback controller (9.5.2), (9.5.3) is said to solve the simultaneous H^∞ control problem for the systems (9.5.1) if it solves the standard H^∞ control problem for each of the k given linear plants.

9.5.2 *The design of a digital output-feedback controller*

Let $\epsilon_0 > 0$ be a given constant. Our solution to the above problem involves the following algebraic Riccati equations

$$A_i\tilde{P}_i + \tilde{P}_iA_i' + \tilde{P}_i[(1+\epsilon_0)C_{1i}'C_{1i} - C_{2i}'C_{2i}]\tilde{P}_i + B_{2i}B_{2i}' = 0 \tag{9.5.5}$$
$$i = 1, 2, \ldots, k.$$

Also, we consider a set of state equations of the form

$$\dot{\hat{x}}(t) = A\hat{x}(t) + B_1 u(t) + B_2 y(t) \tag{9.5.6}$$

where

$$\hat{A}_i := A_i + \tilde{P}_i[(1+\epsilon_0)C_{1i}'C_{1i} - C_{2i}'C_{2i}]; \tag{9.5.7}$$

$$A := \begin{bmatrix} \hat{A}_1 & 0 & \dots & 0 \\ 0 & \hat{A}_2 & \dots & 0 \\ \dots & \dots & \dots & \dots \\ 0 & 0 & \dots & \hat{A}_k \end{bmatrix};$$

$$B_1 := \begin{bmatrix} B_{11} \\ B_{12} \\ \dots \\ B_{1k} \end{bmatrix}; \qquad B_2 := \begin{bmatrix} \tilde{P}_1 C_{21}' \\ \tilde{P}_2 C_{22}' \\ \dots \\ \tilde{P}_k C_{2k}' \end{bmatrix}. \tag{9.5.8}$$

Here $\hat{x}(t) \in \mathbf{R}^m$ where $m := n_1 + n_2 + \ldots + n_k$. Furthermore, let $\epsilon_1 > 0$ be a given constant. The vector $\hat{x}(t) \in \mathbf{R}^m$ can be partitioned as $\hat{x}(t) = \left[\, \hat{x}_1(t)' \; \hat{x}_2(t)' \; \ldots \; \hat{x}_k(t)' \,\right]'$ where $\hat{x}_i(t) \in \mathbf{R}^{n_i}$. We introduce the following cost functional

$$\begin{aligned} G^i(\epsilon_1, \hat{x}(t), u(t), y(t)) := &\ (1+\epsilon_0)\|(C_{1i}\hat{x}_i(t) + D_{1i}u(t))\|^2 \\ &+ \epsilon_1\|\hat{x}(t)\|^2 - \|(C_{2i}\hat{x}_i(t) - y(t))\|^2. \end{aligned} \tag{9.5.9}$$

Consider the following k equations

$$\dot{r}_i(t) = G^i(\epsilon_1, \hat{x}(t), u(t), y(t)) \quad i = 1, 2, \ldots, k. \tag{9.5.10}$$

Moreover, let

$$r(t) := \left[\, r_1(t) \; r_2(t) \; \ldots \; r_k(t) \,\right]', \quad z(t) := \left[\, \hat{x}(t)' \; r(t)' \,\right]'.$$

Then $r(t) \in \mathbf{R}^k$ and $z(t) \in \mathbf{R}^n$ where

$$n := m + k = n_1 + n_2 + \ldots + n_k + k.$$

Furthermore, we introduce a function $L(\cdot)$ from $\mathbf{R}^n$ to $\mathbf{R}$ as

$$L(z) := \max_{i=1,2,\ldots,k} r_i. \tag{9.5.11}$$

Let $V(\cdot)$ be a given function from $\mathbf{R}^n$ to $\mathbf{R}$, and let $u^0 \in \mathbf{R}^h$ and $z_0 \in \mathbf{R}^n$ be given vectors. Then,

$$F(\epsilon_1, z_0, u^0, V(\cdot)) := \sup_{y(\cdot)\in \mathbf{L}_2[0,\delta)} V(z(\delta)) \tag{9.5.12}$$

where the supremum is taken over all solutions to the system (9.5.6), (9.5.10) with $y(\cdot) \in \mathbf{L}_2[0,\delta)$, $u(t) \equiv u^0$ and initial condition $z(0) = z_0$.

Now we are in a position to present the main result of this section.

Theorem 9.5.1 *Consider the systems (9.5.1) and suppose that Assumption 9.5.1 holds. Then, the following statements are equivalent:*

(i) There exists a nonlinear digital output-feedback controller (9.5.2), (9.5.3) which solves the simultaneous H^∞ control problem for the systems (9.5.1).

(ii) There exist constants $\epsilon_0 > 0$ and $\epsilon_1 > 0$ and solutions $\tilde{P}_i = \tilde{P}_i' > 0$ to the algebraic Riccati equations (9.5.5) such that the matrices $\hat{A}_i$ defined by (9.5.7) are stable and the dynamic programming equation

$$V(z_0) = \inf_{u^0 \in \mathbf{R}^h} F(\epsilon_1, z_0, u^0, V(\cdot)) \tag{9.5.13}$$

has a solution $V(z_0)$ such that $V(z_0) \geq L(z_0)$ for all $z_0 \in \mathbf{R}^n$, $V(0) = 0$ and for $z_0 = 0$, the infimum in (9.5.13) is achieved at $u^0 = 0$.

Moreover, suppose that condition (ii) holds. Then there exists a function $u^0(z_0)$ such that $u^0(0) = 0$ and

$$V(z_0) \geq F(0, z_0, u^0(z_0), V(\cdot)). \tag{9.5.14}$$

Furthermore, consider a controller of the form (9.5.2) with

$$u(j\delta) = u^0(z(j\delta)) \tag{9.5.15}$$

where $z(\cdot)$ is the solution to the equations (9.5.6), (9.5.10) with initial condition $z(0) = 0$. Then any such controller solves the simultaneous H^∞ control problem for the systems (9.5.1).

In order to prove this theorem, we will use the following lemma.

Lemma 9.5.1 *Let $i \in \{1, 2, \ldots, k\}$ be given. Also let $\epsilon_0 > 0$ and $d > 0$ be given constants, $T_2 > T_1$ be given times, and $y_0(\cdot)$ and $u_0(\cdot)$ be given vector functions. Suppose that $\tilde{P}_i = \tilde{P}_i' > 0$ is a solution to the algebraic Riccati equation (9.5.5). Then, the condition*

$$\begin{aligned}&\int_{T_1}^{T_2} ((1+\epsilon_0)\|z_i(t)\|^2 - \|w_i(t)\|^2 - \|v_i(t)\|^2)dt \\ &\quad \leq x_i(T_1)'\tilde{P}_i^{-1}x_i(T_1) - d\end{aligned} \tag{9.5.16}$$

holds for all solutions to the system (9.5.1) with $y(\cdot) = y_0(\cdot)$ and $u(\cdot) = u_0(\cdot)$ if and only if

$$\int_{T_1}^{T_2} G^i(0, \hat{x}(t), u_0(t), y_0(t))dt \leq -d \tag{9.5.17}$$

for the cost functional (9.5.9). Here $\hat{x}(t)$ is the solution to the equation

$$\begin{aligned}\dot{\hat{x}}_i(t) = &\left[A_i + \tilde{P}_i(t)[(1+\epsilon_0)C_{1i}'C_{1i} - C_{2i}'C_{2i}]\right]\hat{x}_i(t) \\ &+\tilde{P}_iC_{2i}'y(t) + B_{1i}u(t)\end{aligned} \tag{9.5.18}$$

with $u(\cdot) = u_0(\cdot)$, $y(\cdot) = y_0(\cdot)$ and initial condition $\hat{x}_i(T_1) = 0$.

Proof: Given an input-output pair $[u_0(\cdot), y_0(\cdot)]$, suppose condition (9.5.16) holds for all vector functions $x_i(\cdot), w_i(\cdot)$ and $v_i(\cdot)$ satisfying equation (9.5.1) with $u(\cdot) = u_0(\cdot)$ and such that

$$y_0(t) = C_{2i}x_i(t) + v_i(t) \quad \forall t \in [T_1, T_2]. \tag{9.5.19}$$

Then, substitution of (9.5.19) into (9.5.16) implies that (9.5.16) holds if and only if $J[x_i^f, w_i(\cdot)] \geq d$ for all $w_i(\cdot) \in \mathbf{L}_2[T_1, T_2]$, $x_i^f \in \mathbf{R}^n$ where $J[x_i^f, w_i(\cdot)]$ is defined by

$$\begin{aligned} J[x_i^f, w_i(\cdot)] &:= x_i(T_1)'\tilde{P}_i^{-1}x_i(T_1) \\ &+ \int_{T_1}^{T_2} \begin{pmatrix} \|w_i(t)\|^2 + \|(y_0(t) - C_{2i}x_i(t))\|^2 \\ -(1+\epsilon_0)\|(C_{1i}x_i(t) + D_{1i}u_0(t))\|^2 \end{pmatrix} dt \end{aligned} \tag{9.5.20}$$

and $x_i(\cdot)$ is the solution to (9.5.1) with disturbance input $w_i(\cdot)$ and boundary condition $x_i(T_2) = x_i^f$.

Now consider the following minimization problem

$$\min_{w_i(\cdot) \in \mathbf{L}_2[T_1, T_2)} J[x_i^f, w_i(\cdot)] \tag{9.5.21}$$

where the minimum is taken over all $x_i(\cdot)$ and $w_i(\cdot)$ related by (9.5.1) with the boundary condition $x_i(T_2) = x_i^f$. This problem is a linear quadratic optimal tracking problem in which the system operates in reverse time.

We wish to convert the above tracking problem into a tracking problem of the form considered in [113] and [18]. In order to achieve this, first define $x_i^1(t)$ to be the solution to the state equations

$$\dot{x}_i^1(t) = A_i x_i^1(t) + B_{1i}u_0(t); \quad x_i^1(T_1) = 0. \tag{9.5.22}$$

Now let $\tilde{x}_i(t) := x_i(t) - x_i^1(t)$. Then, it follows from (9.5.1) and (9.5.22) that $\tilde{x}_i(t)$ satisfies the state equations

$$\dot{\tilde{x}}_i(t) = A_i\tilde{x}_i(t) + B_{2i}w_i(t) \tag{9.5.23}$$

where $\tilde{x}_i(T_1) = x_i(T_1)$. Furthermore, the cost functional (9.5.20) can be re-written as

$$\begin{aligned} J[x_i^f, w_i(\cdot)] &= \tilde{J}[\tilde{x}_i^f, w_i(\cdot)] \\ &= \tilde{x}_i(T_1)'\tilde{P}_i^{-1}\tilde{x}_i(T_1) \\ &+ \int_{T_1}^{T_2} \begin{pmatrix} \|w_i(t)\|^2 + \|(y_0(t) - C_{2i}[\tilde{x}_i(t) + x_i^1(t)])\|^2 \\ -(1+\epsilon_0)\|(C_{1i}[\tilde{x}_i(t) + x_i^1(t)] + D_{1i}u_0(t))\|^2 \end{pmatrix} dt \end{aligned} \tag{9.5.24}$$

where $\tilde{x}_i(T_2) = \tilde{x}_i^f = x_i^f - x_i^1(T_2)$. Equations (9.5.23) and (9.5.24) now define a tracking problem of the form considered in [113] where $y_0(\cdot)$, $u_0(\cdot)$ and $x_i^1(\cdot)$ are all treated as reference inputs. In fact, the only difference between this tracking problem and the tracking problem considered in the proof of the result of [18] is that in this section, we have a *sign-indefinite* quadratic cost function.

The solution to this tracking problem is well known; e.g., see [113]. Indeed, if the Riccati equation (9.5.5) has a positive-definite solution, then the minimum of $\tilde{J}[\tilde{x}_i^f, w_i(\cdot)]$ will be achieved for any $\tilde{x}_i^f$, $u_0(\cdot)$ and $y_0(\cdot)$. Furthermore, as in [18], we can write

$$\min_{w_i(\cdot)\in \mathbf{L}_2[T_1,T_2)} \tilde{J}[\tilde{x}_i^f, w_i(\cdot)] = (\tilde{x}_i^f - \hat{x}_i^1(T_2))'\tilde{P}_i^{-1}(\tilde{x}_i^f - \hat{x}_i^1(T_2)) - \int_{T_1}^{T_2} G^i(0, x_i^1(t) + \hat{x}_i^1(t), u_0(t), y_0(t))dt \tag{9.5.25}$$

where $\hat{x}_i^1(\cdot)$ is the solution to state equations

$$\begin{aligned}\dot{\hat{x}}_i^1(t) &= \left[A_i + \tilde{P}_i[(1+\epsilon_0)C_{1i}'C_{1i} - C_{2i}'C_{2i}]\right][x_i^1(t) + \hat{x}_i^1(t)] \\ &\quad + \tilde{P}_i C_{2i}' y_0(t) + B_{1i}u_0(t)\end{aligned}$$

with initial condition $\hat{x}_i^1(T_1) = 0$. Now let $\hat{x}_i(\cdot) := x_i^1(\cdot) + \hat{x}_i^1(\cdot)$. Using the fact that $\tilde{x}_i^f = x_i^f - x_i^1(T_2)$, it follows that (9.5.25) can be re-written as

$$\min_{w_i(\cdot)\in \mathbf{L}_2[T_1,T_2)} J[x_i^f, w_i(\cdot)] = (x_i^f - \hat{x}_i(T_2))'\tilde{P}_i^{-1}(x_i^f - \hat{x}_i(T_2)) - \int_{T_1}^{T_2} G^i(0, \hat{x}_i(t), u_0(t), y_0(t))dt$$

where $\hat{x}_i(\cdot)$ is the solution to state equations (9.5.6) with initial condition $\hat{x}_i(T_1) = 0$. From this, we can conclude that condition (9.5.16) with a given input-output pair $[u_0(\cdot), y_0(\cdot)]$ is equivalent to the inequality (9.5.17).

■

Proof of Theorem 9.5.1: (i)⇒(ii). If condition (i) holds, then there exists a constant $\epsilon_0 > 0$ such that the controller (9.5.2), (9.5.3) is a solution to the following output-feedback H^∞ control problems:

$$\sup_{\substack{x_i(0)\neq 0,\\ [w_i(\cdot),v_i(\cdot)]\in \mathbf{L}_2[0,\infty)}} \frac{\int_0^\infty ((1+\epsilon_0)\|z_i(t)\|^2 + \epsilon_0(\|u(t)\|^2 + \|y(t)\|^2))dt}{c\|x_i(0)\|^2 + \int_0^\infty (\|w_i(t)\|^2 + \|v_i(t)\|^2)dt} < 1 - \epsilon_0 \tag{9.5.26}$$

for all $i = 1, 2, \ldots, k$. Now consider the disturbance inputs $v_i(\cdot) \equiv -C_{2i}x_i(\cdot)$. Then, we have $y(\cdot) \equiv 0$ and $u(\cdot) \equiv 0$. Therefore, from (9.5.26) we have

$$\sup_{\substack{x_i(0)\neq 0,\\ w_i(\cdot)\in \mathbf{L}_2[0,\infty)}} \frac{\int_0^\infty ((1+\epsilon_0)\|C_{1i}x_i(t)\|^2 - \|C_{2i}x_i(t)\|^2)dt}{c\|x_i(0)\|^2 + \int_0^\infty (\|w_i(t)\|^2 + \|v_i(t)\|^2)dt} < 1 \quad \forall i = 1, 2, \ldots, k.$$

Hence, it follows from the proof of Theorem 4 of [136], that the solutions $P_i(\cdot)$ to the following Riccati differential equations

$$\begin{aligned}\dot{P}_i(t) &= A_i P_i(t) + P_i(t) A_i' \\ &\quad + P_i(t)[(1+\epsilon_0)C_{1i}'C_{1i} - C_{2i}'C_{2i}]P_i(t) + B_{2i}(t)B_{2i}(t)'\end{aligned} \tag{9.5.27}$$

with initial condition $P_i(0) = c^{-1}I$ are defined, positive-definite and bounded on $[0,\infty)$ for all $i = 1,2,\ldots,k$. Moreover,

$$P_i(t) \to \tilde{P}_i \;\; as \;\; t \to \infty \quad \forall i = 1,2,\ldots,k \tag{9.5.28}$$

where $\tilde{P}_i = \tilde{P}_i' > 0$ is a solution to the equation (9.5.5) such that the matrix $\hat{A}_i$ defined by (9.5.7) is stable. Since (9.5.28), we can choose a number N_0 such that

$$\|P_i^{-1}(N_0\delta)\| \leq \frac{1}{1-\epsilon_0}\|\tilde{P}_i^{-1}\| \quad \forall i = 1,2,\ldots,k. \tag{9.5.29}$$

Now let $x_i^f \in \mathbf{R}^{n_i}$ be a given vector. Then, we have that

$$\begin{aligned}&\sup_{\substack{x_i(N_0\delta)=x_i^f,\\ w_i(\cdot)\in \mathbf{L}_2[0,N_0\delta)}} \left[c\|x_i(0)\|^2 + \int_0^{N_0\delta} \begin{pmatrix} \|w_i(t)\|^2 + \|v_i(t)\|^2 - \\ (1+\epsilon_0)\|C_{1i}x_i(t)\|^2 + \|C_{2i}x_i(t)\|^2 \end{pmatrix} dt \right] \\ &= x_i^{f\prime} P_i^{-1}(N_0\delta) x_i^f\end{aligned} \tag{9.5.30}$$

for all $i = 1,2,\ldots,k$. Furthermore, consider the set of all disturbance inputs such that

$$v_i(t) = 0 \quad \forall t \in [0, N_0\delta].$$

Then, combining (9.5.30) with (9.5.26), we obtain

$$\sup \frac{\int\limits_{N_0\delta}^{N\delta} ((1+\epsilon_0)\|z_i(t)\|^2 + \epsilon_0(\|u(t)\|^2 + \|y(t)\|^2))dt}{x_i(N_0\delta)' P_i^{-1}(N_0\delta) x_i(N_0\delta) + \int\limits_{N_0\delta}^{N\delta} (\|w_i(t)\|^2 + \|v_i(t)\|^2)dt} < 1 - \epsilon_0 \tag{9.5.31}$$

for all $i = 1,2,\ldots,k$, where the supremum is taken over all $N > N_0$ and $[w_i(\cdot), v_i(\cdot)] \in \mathbf{L}_2[N_0\delta, N\delta)$. Furthermore, it follows from (9.5.31) and (9.5.29) that

$$\sup_{\substack{N>N_0,\\ [w_i(\cdot),v_i(\cdot)]\in \mathbf{L}_2[N_0\delta,N\delta)}} \frac{\int\limits_{N_0\delta}^{N\delta} ((1+\epsilon_0)\|z_i(t)\|^2 + \epsilon_0(\|u(t)\|^2 + \|y(t)\|^2))dt}{x_i(N_0\delta)' \tilde{P}_i^{-1} x_i(N_0\delta) + \int\limits_{N_0\delta}^{N\delta} (\|w_i(t)\|^2 + \|v_i(t)\|^2)dt} < 1 \tag{9.5.32}$$

for all $i = 1, 2, \ldots, k$. Condition (9.5.32) implies that condition (9.5.16) with

$$T_1 = N_0\delta, \quad T_2 = N\delta \quad \text{and} \quad d = \epsilon_0 \int_{N_0\delta}^{N\delta} (\|u(t)\|^2 + \|y(t)\|^2)dt \qquad (9.5.33)$$

holds for any $i = 1, 2, \ldots, k$ and any $u(\cdot)$ and $y(\cdot)$ related by the controller equations (9.5.2), (9.5.3). Hence, Lemma 9.5.1 implies that the inequality (9.5.17) with T_1, T_2 and d defined by (9.5.33) holds for all i. Furthermore, since the matrix $\hat{A}_i$ is stable for any i, it follows that the matrix A defined by (9.5.8) is stable. Therefore, there exists a constant $c_0 > 0$ such that

$$\int_{N_0\delta}^{N\delta} \|\hat{x}(t)\|^2 dt \le c_0 \int_{N_0\delta}^{N\delta} (\|u(t)\|^2 + \|y(t)\|^2)dt = \frac{c_0 d}{\epsilon_0}$$

for all solutions to the system (9.5.6) with $\hat{x}(N_0\delta) = 0$ and for all $N > N_0$. This and condition (9.5.17) imply that

$$\int_{N_0\delta}^{N\delta} \left[G^i(\epsilon_1, \hat{x}(t), u(t), y(t)) + \epsilon_2(\|\hat{x}(t)\|^2 + \|u(t)\|^2 + \|y(t)\|^2)\right] dt \le 0 \qquad (9.5.34)$$

for all $N > N_0$, $i = 1, 2, \ldots, k$ and $\hat{x}(t), u(t)$ and $y(t)$ related by equations (9.5.6), (9.5.2) with $\hat{x}(N_0\delta) = 0$ where $\epsilon_1 := \frac{\epsilon_0}{3c_0}$ and $\epsilon_2 := \min\left[\frac{\epsilon_0}{3c_0}, \frac{\epsilon_0}{3}\right]$.

Now let $\mathcal{U}$ be the class of all controllers of the form (9.5.2), (9.5.3). For any $N > N_0$ and $j = N_0, N_0 + 1, \ldots, N$, we define the following function

$$V_N(j, z_0) := \inf_{u(\cdot)\in\mathcal{U}} \sup_{y(\cdot)\in\mathbf{L}_2[j\delta, N\delta]} L(z(N\delta)) \qquad (9.5.35)$$

where $L(\cdot)$ is defined by (9.5.11) and the supremum is taken over all solutions to the system (9.5.18), (9.5.10) with initial condition $z(j\delta) = z_0$. Then the equation (9.5.10) and the inequality (9.5.34) imply that $V_N(j, z_0) < \infty$ for any $z_0 \in \mathbf{R}^n$. According to the theory of dynamic programming (e.g., see [9]), $V_N(\cdot, \cdot)$ satisfies the following dynamic programming equation

$$\begin{aligned} V_N(N, z_0) &= L(z_0); \\ V_N(j, z_0) &= \inf_{u_j^0 \in \mathbf{R}^h} F(\epsilon_1, z_0, u_j^0, V_N(j+1, \cdot)) \end{aligned} \qquad (9.5.36)$$

for all $j = N - 1, \ldots, N_0$. Also, if we consider all inputs $y(\cdot)$ such that $y(t) = C_{21}\hat{x}_1(t)$ for all $t \in [N\delta, M\delta]$, then it follows from (9.5.35) that

$$V_N(N_0, z_0) \le V_M(N_0, z_0) \quad \forall M \ge N. \qquad (9.5.37)$$

We now prove that there exists a function $Z(z_0)$ such that

$$V_N(N_0, z_0) \le Z(z_0) \quad \forall N > N_0 \quad \forall z_0 \in \mathbf{R}^n. \qquad (9.5.38)$$

Indeed, we have that

$$\int_{N_0\delta}^{N\delta} G^i(\epsilon_1, \hat{x}(t), u(t), y(t))dt = \int_{N_0\delta}^{N\delta} G^i(\epsilon_1, \hat{x}^0(t) + \hat{x}^1(t), u(t), y(t))dt$$

where $\hat{x}^0(\cdot)$ is the solution to the equation (9.5.6) with $\hat{x}^0(N_0\delta) = 0$, $y(\cdot), u(\cdot)$ are related by (9.5.2) and $\hat{x}^1(\cdot)$ is the solution to the equation

$$\dot{\hat{x}}^1(t) = A\hat{x}^1(t) \tag{9.5.39}$$

with $\hat{x}^1(N_0\delta) = \hat{x}_0$. Furthermore, it is clear that for any $\epsilon_2 > 0$, there exists a constant $\gamma > 0$ such that

$$\begin{aligned} &G^i(\epsilon_1, \hat{x}^0(t) + \hat{x}^1(t), u(t), y(t)) \\ &\leq G^i(\epsilon_1, \hat{x}^0(t), u(t), y(t)) + G^i(\epsilon_1, \hat{x}^1(t), 0, 0) \\ &\quad + \epsilon_2(\|\hat{x}^0(t)\|^2 + \|u(t)\|^2 + \|y(t)\|^2) + \gamma\|\hat{x}^1(t)\|^2. \end{aligned} \tag{9.5.40}$$

Now condition (9.5.38) follows immediately from (9.5.34), (9.5.40) and the stability of the system (9.5.7). Hence, from (9.5.38) and (9.5.37) we have existence of the following limit

$$V(z_0) := \lim_{N\to\infty} V_N(N_0, z_0).$$

Clearly $V(\cdot)$ is a solution to equation (9.5.13). Moreover, it follows immediately from (9.5.35) and (9.5.37) that $V(z_0) \geq L(z_0)$ for all $z_0 \in \mathbf{R}^n$. Also, it follows from (9.5.35) and (9.5.3) that $V(0) = 0$ and for $z_0 = 0$, the infimum in (9.5.13) is achieved at $u^0 = 0$. This completes the proof of this part of the theorem.

(ii)⇒(i). Suppose condition (ii) holds. We will construct a corresponding controller of the form (9.5.2), (9.5.15) satisfying (9.5.14) as follows: Given any nonzero $z_0 \in \mathbf{R}^n$, choose $\nu > 0$ such that $\nu < \epsilon_1\|z_0\|^2$. It follows from (9.5.13) that $u^0(z_0)$ can be chosen so that

$$V(z_0) \geq F(\epsilon_1, z_0, u_j^0, V(\cdot)) - \nu.$$

Since $\nu < \epsilon_1\|z_0\|^2$, equation (9.5.13) implies that this function $u^0(z_0)$ satisfies (9.5.14). Furthermore, (9.5.14) implies that for the corresponding controller (9.5.2), (9.5.15), we have

$$V(z(0)) \geq V(z(N\delta)) \geq L(z(N\delta))$$

for any solution $z(\cdot)$ to the system (9.5.6), (9.5.10) with the corresponding digital controller (9.5.2), (9.5.15). This and equations (9.5.11), (9.5.10) imply that

$$\max_{i=1,2,\ldots,k} \int_0^{N\delta} G^i(0, \hat{x}(t), u(t), y(t))dt \leq 0$$

for all solutions to the system (9.5.6), (9.5.10) with $z(0) = 0$. Furthermore, Lemma 9.5.1 implies that condition (9.5.16) holds with $d = 0$, $T_1 = 0$ and $T_2 = N\delta$. Condition (9.5.4) of Definition 9.5.1 follows immediately from inequality (9.5.16) with $d = 0$, $T_1 = 0$ and $T_2 = N\delta$. Furthermore, condition (i) of Definition 9.5.1 follows from (9.5.16) and conditions 1 and 2 of Assumption 9.5.1. This completes the proof of the theorem.

■

The following corollary can be obtained immediately from Theorem 9.5.1.

Corollary 9.5.1 *If the simultaneous H^∞ control problem for the k systems (9.5.1) of orders $n_1, n_2, \dots, n_k$ can be solved via a nonlinear digital output-feedback controller (9.5.2), (9.5.3), then this problem can be solved via a nonlinear digital output-feedback dynamic controller of order $n := n_1 + n_2 + \cdots + n_k + k$.*

Remark 9.5.2 The H^∞ control problem (9.5.4) is equivalent to a certain robust stabilization problem similar to those that arose in Sections 3.5 and 9.2, with the underlying linear system described by the state equations (9.5.1). However, in this case, $w_i(t)$ and $v_i(t)$ are the uncertainty inputs, and $z_i(t)$ is the uncertainty output. Also, the uncertainty inputs $w_i(t)$ and $v_i(t)$ are required to satisfy a certain integral quadratic constraint. Thus, Theorem 9.5.1 leads to the solution to a corresponding problem of simultaneous robust stabilization for a finite collection of uncertain systems via digital output-feedback control.

9.6 Conclusions

The objective of this chapter was to show that when using integral quadratic constraint uncertainty models, one often finds that nonlinear control has no advantage over linear control. In particular, we have considered a number of robust control problems in which the designer may face a choice between nonlinear (e.g., adaptive) and linear control schemes. The results presented show that in the situations considered, if a specified performace goal can be achieved via a nonlinear controller, then it can also be achieved via a linear controller. Furthermore, the results presented show how a suitable linear controller can be synthesized using standard Riccati equation techniques. For example, when applied to the problem of designing a decentralized controller, this conclusion means that if the class of interconnection between subsystems fits into an IQC framework, then an efficient solution can be found among linear controllers rather than complicated nonlinear controllers.

However, under certain circumstances, nonlinear controllers may have an advantage over linear controllers. One problem where a nonlinear controller gives some advantage is considered in Section 9.5.

10. Missile autopilot design via minimax optimal control of stochastic uncertain systems

10.1 Introduction

As mentioned in Chapter 1, one of the most attractive methods for designing multi-input multi-output feedback control systems is the linear quadratic Gaussian (LQG) method; e.g., see [3, 46]. In particular, the use of a quadratic cost function is well motivated in many control problems. Also, the stochastic white noise model is often a good approximation to the noise found in practical control problems. These facts, together with the fact that the LQG control problem can be solved via reliable numerical techniques based on the algebraic Riccati equation, provide good motivation for the use of the LQG method. However, the LQG design technique does not address the issue of robustness and it is known that a controller designed using the LQG technique can have arbitrarily poor robustness properties; e.g., see [50]. Since, the issue of robustness is critical in most control system design problems, we have been motivated to consider the problem of obtaining a robust version of the LQG control technique. These results were presented in Section 8.5; see also [206, 49, 134, 72, 146, 152, 230, 231]. In particular, the results of Section 8.5 generalize the LQG problem to a robust LQG problem for stochastic uncertain systems. In this chapter, we will apply the results of Section 8.5 to the problem of designing a missile autopilot.

The missile autopilot problem considered in this chapter is taken from a collection of examples for robust controller design [216]. A feature of this robust control problem is that a single controller is required to give adequate tracking performance over a range of flight conditions corresponding to different values of mass, speed and altitude. The data given in [216] consists of a finite set of state space models corresponding a different flight conditions. In order to apply the robust controller design techniques of Section 8.5 to this problem, a suitable uncertain system model must be obtained. This will be achieved by applying a constrained total least squares approach to fit the data to a norm-bounded uncertain system model; e.g., see [88, 111]. This norm-bounded uncertain system is then generalized to the class of stochastic uncertain systems considered in Section 8.5 and a controller is designed using the approach of Section 8.5.

10.2 Missile autopilot model

The controller design problem considered in this chapter is the problem of designing a controller for a rigid guided missile as described in [216]. In this section, we consider the problem of obtaining a suitable uncertain system model for the missile in order to apply the results of Section 8.5.

The model for the missile given in [216] consists of a collection of state space models of the form

$$\begin{aligned}
\begin{bmatrix} \Delta\dot{q} \\ \Delta\dot{\alpha} \\ \Delta\dot{\eta} \end{bmatrix} &= \begin{bmatrix} a_{11} & a_{12} & b_1 \\ 1 & a_{22} & b_2 \\ 0 & 0 & -30 \end{bmatrix} \begin{bmatrix} \Delta q \\ \Delta\alpha \\ \Delta\eta \end{bmatrix} + \begin{bmatrix} 0 \\ 0 \\ 30 \end{bmatrix} u; \\
\begin{bmatrix} a_z \\ \Delta q \end{bmatrix} &= \begin{bmatrix} c_{11} & c_{12} & d_1 \\ 1 & 0 & 0 \end{bmatrix} \begin{bmatrix} \Delta q \\ \Delta\alpha \\ \Delta\eta \end{bmatrix}
\end{aligned} \tag{10.2.1}$$

where the state and control variables are perturbations of the following quantities referring to Figure 10.2.1:

q	pitch rate	[rad/s]
α	angle of attack	[rad]
η	elevator deflection angle	[rad]
a_z	vertical acceleration	$[m/s^2]$

The parameters in the state space model depend on Mach number, altitude and mass. The parameter values for ten flight conditions are given in Table 10.2.1.

It is desired to design a single controller so that the normal acceleration a_z tracks a normal acceleration reference signal. The open loop system is stable but insufficiently damped at all flight conditions. It is non-minimum phase at most flight conditions.

Specifications. The following specifications for the closed loop system were given in [216]:

Step response of a_z:	overshoot	$\leq 10\ \%$
	stationary error	$\leq 5\ \%$

It is desired to meet these specifications at all ten operating points.

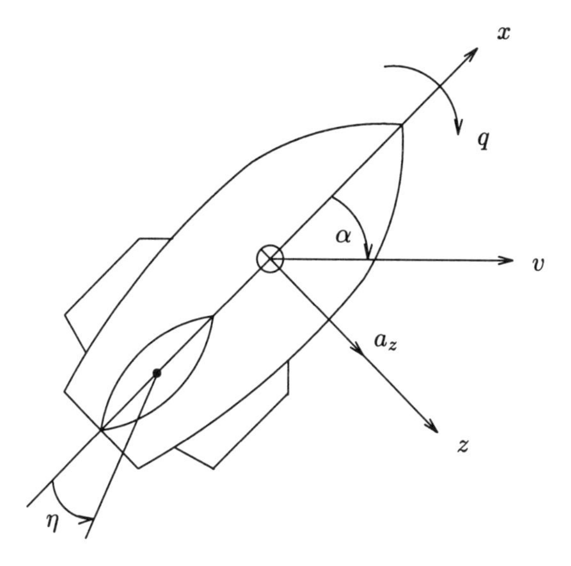

Figure 10.2.1. Missile

10.2.1 Uncertain system model

In order to apply the robust LQG controller design technique of Section 8.5, we require a suitable uncertain system model. The approach taken in this case is to construct an uncertain system with norm-bounded uncertainty which encompasses (approximately) the ten operating points defined above. The approach of Section 8.5 is then applied to the stochastic uncertain system corresponding to the uncertain system with norm-bounded uncertainty.

It is recognized that constructing a norm-bounded uncertain system model which completely encompasses the ten given state space models may lead to an excessively conservative design. Therefore, an approximate modelling technique is employed. This involves approximating the ten given operating points by points in a suitable two dimensional subspace in the parameter space of the system state space model. This procedure is represented in Figure 10.2.2.

The required two dimensional subspace is constructed via the solution to a weighted least squares problem which is solved via a quasi-Newton optimization routine.

The uncertain system with norm-bounded uncertainty to be constructed is of the following form

$$\begin{aligned}\dot{x} &= (A_0 + \Delta A)x + B_0 u;\\ y &= (C_0 + \Delta C)x\end{aligned}$$

Table 10.2.1. Missile model parameters

Altitude	0 m				
Mach number	0.8		1.6		2.5
a_{11}	-0.327	-0.391	-0.688	-0.738	-0.886
a_{12}	-63.94	-130.29	-619.27	-651.57	-1068.85
a_{22}	-1	-1.42	-2.27	-2.75	-3.38
b_1	-155.96	-186.5	-552.9	-604.18	-1004.39
b_2	-0.237	-0.337	-0.439	-0.532	-0.582
c_{11}	0.326	0.35	0.65	0.66	0.79
c_{12}	-208.5	-272.38	-651.11	-913.64	-1926.45
d_1	90.93	75.06	283.44	250.5	402.96
altitude	12000 m				
Mach number	2.0		3.0		4.0
a_{11}	-1.364	-0.333	-0.337	-0.369	-0.402
a_{12}	-92.82	-163.24	-224.03	-253.71	-277.2
a_{22}	-4.68	-0.666	-0.663	-0.8	-0.884
b_1	-128.46	-153.32	-228.72	-249.87	-419.35
b_2	-0.087	-0.124	-0.112	-0.135	-0.166
c_{11}	1.36	0.298	0.319	0.33	0.36
c_{12}	-184.26	-247.74	-375.75	-500.59	-796.18
d_1	76.43	63.77	117.4	103.76	178.59

where

$$\begin{bmatrix} \Delta A \\ \Delta C \end{bmatrix} = \begin{bmatrix} \tilde{d}_1 & 0 \\ 1 & 0 \\ 0 & 0 \\ 0 & 1 \\ 0 & 0 \end{bmatrix} \begin{bmatrix} \Delta_1 \\ \Delta_2 \end{bmatrix} [1 \; e_2 \; e_2]; \qquad \left\| \begin{bmatrix} \Delta_1 \\ \Delta_2 \end{bmatrix} \right\| \leq \mu$$

To construct a suitable uncertain system of the above form, consider the set of systems of the form (10.2.1) parameterized by the parameter vector

$$\alpha = [b_2 \; c_{11} \; c_{12} \; d_1 \; a_{11} \; a_{12} \; b_1 \; a_{21} \; a_{22}]' \in \mathbf{R}^9.$$

We first construct a two dimensional affine subspace of this nine dimensional parameter space of the following form:

$$\{\alpha \in \mathbf{R}^9 : \alpha = \alpha_0 + \alpha_1 \Delta_1 + \alpha_2 \Delta_2 : \Delta_1, \Delta_2 \in \mathbf{R}\}$$

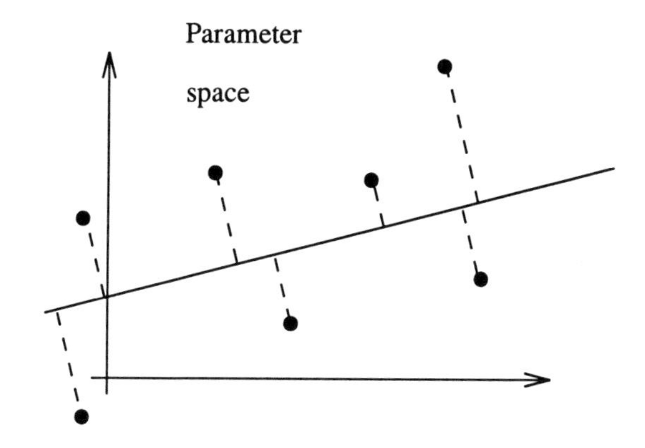

Figure 10.2.2. Subspace approximation

where the vectors α_1 and α_2 are restricted to be of the form

$$\alpha_0, \ \alpha_1 = \begin{bmatrix} e_3 \\ 0 \\ 0 \\ 0 \\ \tilde{d}_1 \\ \tilde{d}_1 e_2 \\ \tilde{d}_1 e_3 \\ 1 \\ e_2 \end{bmatrix}; \ \alpha_2 = \begin{bmatrix} 0 \\ 1 \\ e_2 \\ e_3 \\ 0 \\ 0 \\ 0 \\ 0 \\ 0 \end{bmatrix}.$$

This restriction of the vectors α_1 and α_2 corresponds to the given structure of the uncertain system (10.2.1). The parameter vectors α_0, α_1 and α_2 were determined by solving a constrained total least squares optimization problem (e.g., see [111, 88]) to minimize the weighted total least squares error for the given ten data points. The weighting was based on the maximum absolute value of each parameter. This optimization was carried out using the quasi-Newton method contained in the Matlab optimization toolbox and yielded the following results:

$$\alpha_0 = \begin{bmatrix} 0 \\ 0 \\ -973.6266 \\ 102.7566 \\ -0.0208 \\ 23.8240 \\ -304.7616 \\ 1.0024 \\ -0.2150 \end{bmatrix}; \quad \alpha_1 = 10^3 \times \begin{bmatrix} 1 \\ 0 \\ 0 \\ 0 \\ 0.0020 \\ 1.3752 \\ 0.2312 \\ 0.0000 \\ 0.0059 \end{bmatrix}; \quad \alpha_2 = 10^3 \times \begin{bmatrix} 0.6724 \\ 0.1130 \\ 0 \\ 0 \\ 0 \\ 0 \\ 0 \end{bmatrix}.$$

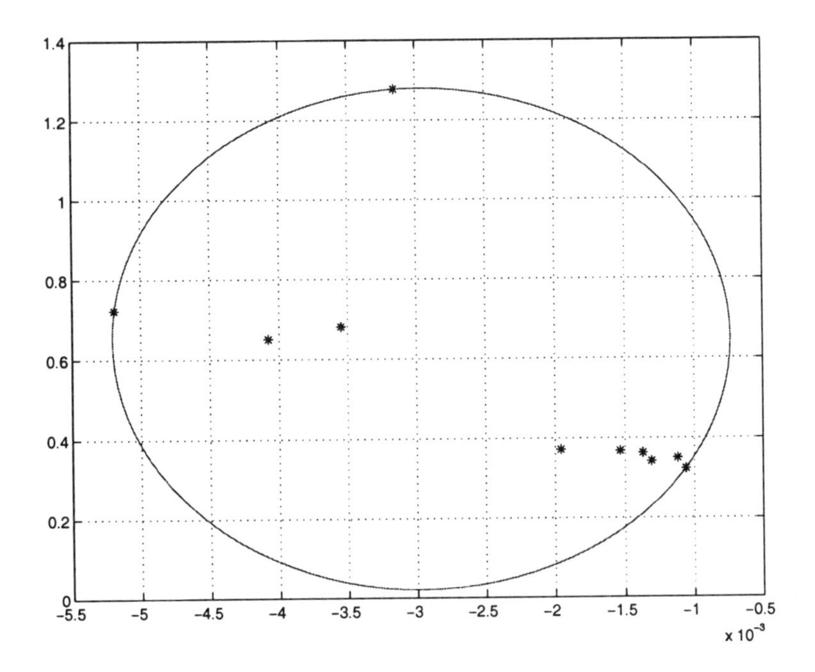

Figure 10.2.3. System data points projected onto the affine space

Once the two dimensional affine subspace was determined as above, the ten data points were projected onto this affine space and a minimal (in terms of trace norm) ellipsoid was fitted around these points. This is illustrated in Figure 10.2.3.

This then defines the final uncertain system model of the form (10.2.1) as follows:

$$\begin{aligned}
\dot{x} &= \begin{bmatrix} -0.7076 & -437.9552 & -382.3972 \\ 0.9995 & -2.2123 & -0.3358 \\ 0 & 0 & -30.0000 \end{bmatrix} x \\
&\quad + \begin{bmatrix} 0.8213 \\ 0.0036 \\ 0 \end{bmatrix} \Delta_1 [0.6295\ 423.2451\ 71.1572] x + \begin{bmatrix} 0 \\ 0 \\ 30 \end{bmatrix} u; \\
y &= \begin{bmatrix} 0.6503 & -536.3713 & 176.2693 \\ 1.0000 & 0 & 0 \end{bmatrix} x \\
&\quad + \begin{bmatrix} 1 \\ 0 \end{bmatrix} \Delta_2 [0.6295\ 423.2451\ 71.1572] x
\end{aligned} \tag{10.2.2}$$

where $\|[\Delta_1\ \Delta_2]\| \leq 1$. In the next section, we will apply the results of Section 8.5 to the stochastic uncertain system corresponding to this system.

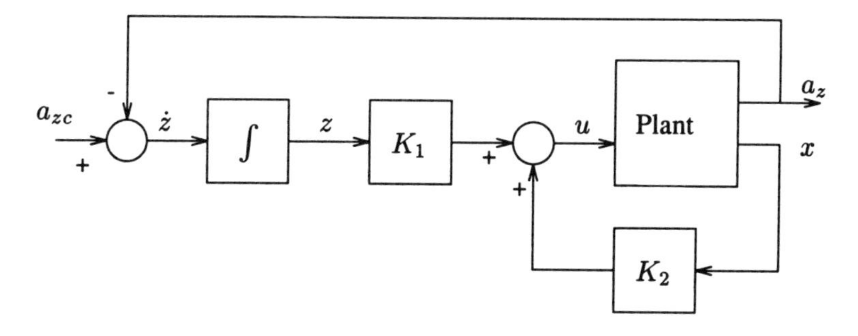

Figure 10.3.1. State-feedback tracking control system with integral action

10.3 Robust controller design

In this section, we apply the results of Section 8.5 to the uncertain system (10.2.2). Since the approach of Section 8.5 is a robust LQG approach, we proceed as is commonly done in LQG controller design problems and first consider a state-feedback version of the problem.

10.3.1 State-feedback controller design

We now consider the state-feedback case. In this case, the approach of Section 8.5 can be reduced to that of Section 5.3. In order to guarantee the specified steady state error at all operating points, a controller with integral action was chosen. This is illustrated in Figure 10.3.1. That is, we solve a robust tracking problem using a similar approach to that described in [133]. The controller gains K_1 and K_2 are chosen to minimize the worst case of the cost function

$$J = \int_0^\infty \left(z^2 + Ru^2\right) dt$$

for an initial condition of $x(0) = 0,\ z(0) = 1$. Here, the variable $z(t)$ is the integral of the tracking error as illustrated in Figure 10.3.1. This robust LQR problem was solved for various values of the control weighting R. A root locus of the nominal closed loop poles as R is varied is shown in Figure 10.3.2. From this, a weighting of $R = 2000$ was chosen to give a suitable nominal closed loop bandwidth. The step response of the state-feedback control system was then obtained for each operating point. These are shown in Figure 10.3.3.

10.3.2 Output-feedback controller design

We now consider an output-feedback controller designed using the robust LQG method of Section 8.5. As in the state-feedback case, integral action is used in the controller. This is illustrated in Figure 10.3.4.

The controller transfer function $[K_1(s)\ K_2(s)]$ is to be determined via the robust LQG method described above. The same cost function is used as in the state-

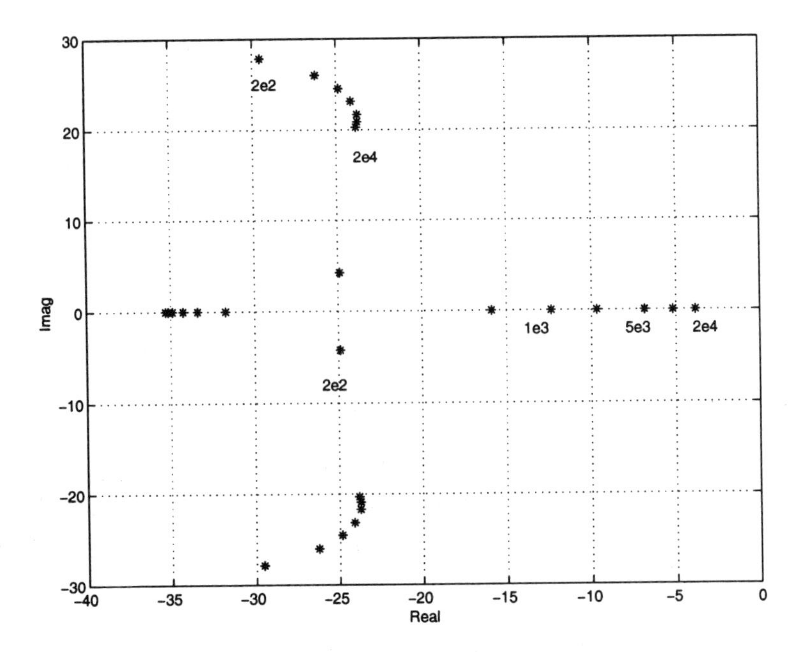

Figure 10.3.2. Root locus of nominal closed loop poles in the state-feedback case.

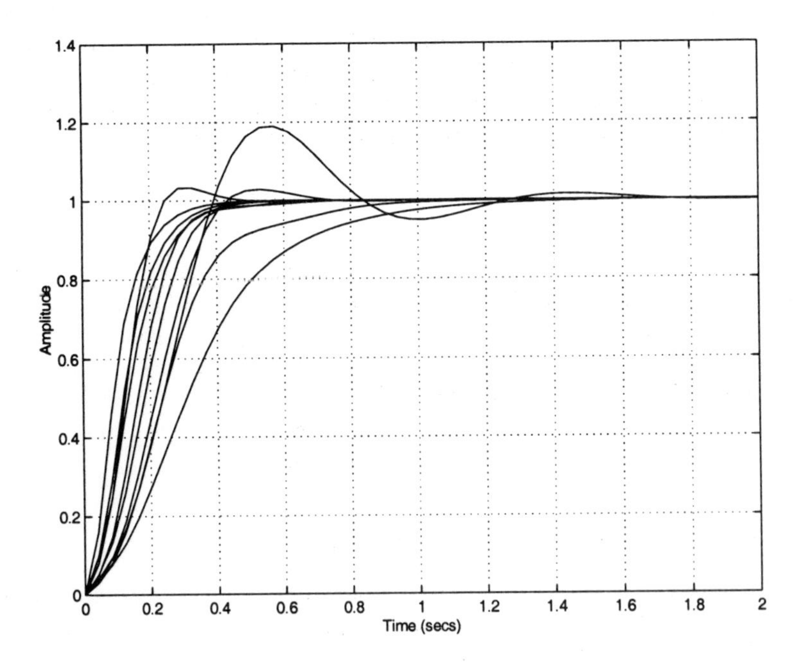

Figure 10.3.3. State-feedback closed loop step responses

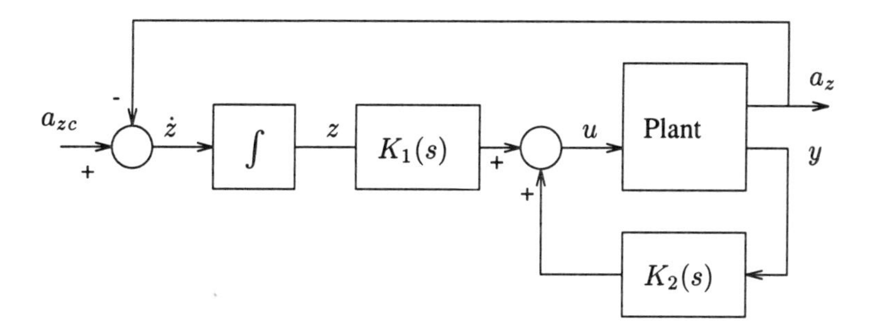

Figure 10.3.4. Output-feedback tracking controller with integral action

feedback case. The noise on the measurement of a_z is defined by the uncertainty description. However, currently the measurements of the variables q and z are noise free. Some fictitious noises are added to these measurements to ensure a finite Kalman Filter bandwidth. A root locus plot of the Kalman filter poles as this noise parameter is varied is shown in Figure 10.3.5. From this, the additional noise covariances were chosen as 0.02 to give a suitable Kalman filter bandwidth. With these parameters, the method of Section 8.5 was then applied to obtain the robust controller transfer functions. With this controller, the step response of the output-feedback control system was then obtained for each operating point. These step responses are shown in Figure 10.3.6. From these step responses, it can be seen that the robust controller designed using the robust LQG approach of Section 8.5 meets all of the required design specifications.

10.4 Conclusions

The main motivation for this section was to illustrate how the robust LQG approach of Section 8.5 might be applied to practical control system design problems. In the missile autopilot problem considered, the robust LQG technique yielded a controller which satisfied all given specifications at each of ten possible operating points. No other robustness criteria were considered. However, in practice, further robustness considerations may need to be taken into account such as robustness to unstructured uncertainty and the traditional robustness margins, gain and phase margin. To take these issues into account using the robust LQG approach of Section 8.5, the uncertain system model used would have to be modified to take into account these extra robustness requirements.

For the missile autopilot control problem under consideration, the robust LQG approach of Section 8.5 appears to have yielded a suitable controller. However, it should be realized that this controller may be overly conservative owing to the class of uncertain systems which were considered in the controller design. Also in practice, the restriction of a single linear controller may lead to excessively conservative results.

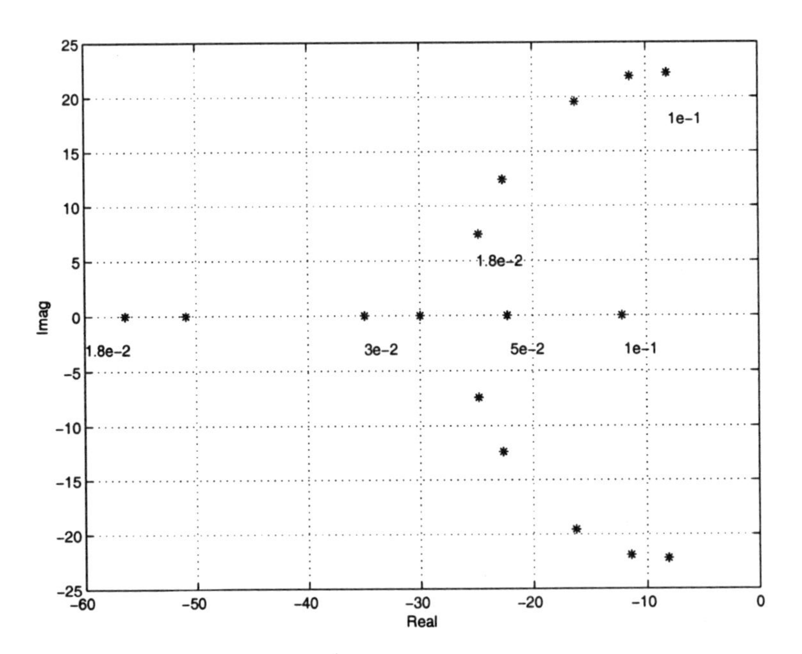

Figure 10.3.5. Root locus of nominal closed loop Kalman filter poles

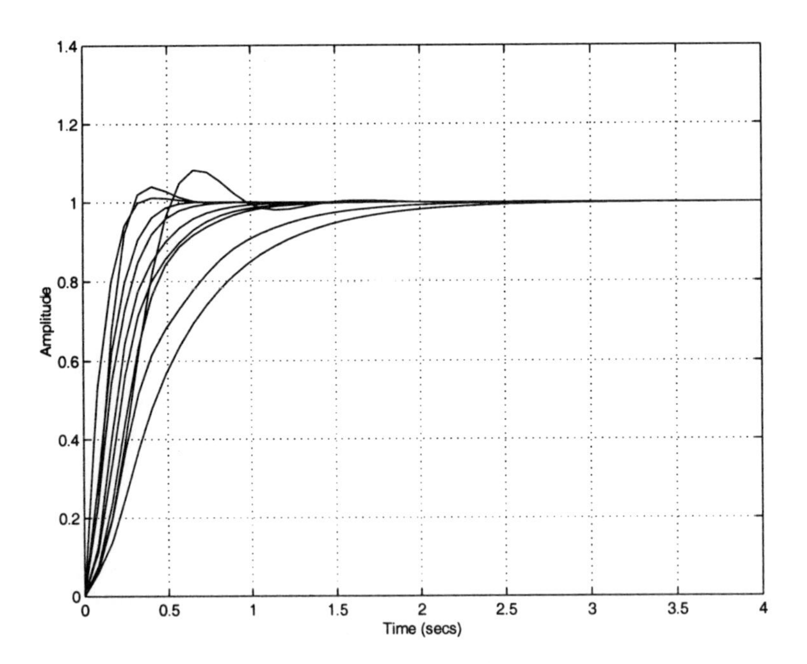

Figure 10.3.6. Output-feedback closed loop step responses

11. Robust control of acoustic noise in a duct via minimax optimal LQG control

11.1 Introduction

This chapter considers the problem of active noise control in an acoustic duct. The problem of active noise control in a duct has been the subject of a large amount of research; e.g., see [63, 86]. We consider a broadband feedback control approach to active noise control rather than the narrowband adaptive feedforward approach taken in many early papers on active noise control. The advantage of the feedback control approach is that it can handle practical noise reduction problems in which the noise source has a broad spectrum. The disadvantage of the feedback control approach is that it opens up the possibility of instability and the control system must be carefully designed to give adequate robustness. Also, it usually necessitates the construction of a model for the duct.

This chapter considers a new approach to the design of robust feedback controllers for active noise control in a duct. The basis for this approach is the minimax LQG controller design methodology developed in Sections 8.4 and 8.5; see also [152, 153, 230, 235, 231, 233].

The minimax LQG approach to the design of a controller for an acoustic duct requires a suitable model and a suitable choice for the quadratic cost functional. Our approach to constructing a suitable nominal model for the duct is a system identification approach. That is, we construct the duct model using experimental data. In our case, this experimental data consists of frequency response data measured using a swept sine spectrum analyser. A transfer function is fitted to this frequency response data using a standard least squares technique. This transfer function is then used as a nominal model for the acoustic duct.

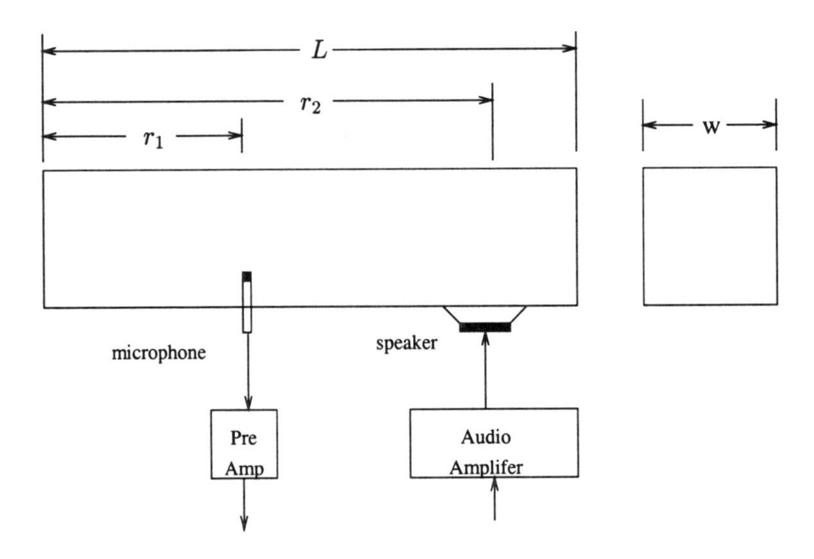

Figure 11.2.1. Acoustic Duct.

In order to apply the minimax LQG approach to any control system design problem, one must specify an uncertain system model. That is, unlike the standard LQG approach which is based on a nominal model, the minimax LQG approach is based on an uncertain system model. This leads to a controller which is robust against a specified class of uncertainties. In an active noise control problem, there are many possible sources of uncertainty in the system model. In this chapter, we concentrate on the uncertainty which arises from neglecting the higher order modes of the duct. In particular, an acoustic duct is a distributed parameter system with an infinite dimensional mathematical model. We consider a finite dimensional model which gives a good approximation to the duct behaviour over a finite frequency range. The neglected "spillover" dynamics are then treated as uncertainty. Our approach to modelling this uncertainty is a frequency weighted multiplicative uncertainty approach. This uncertainty is then overbounded by a stochastic uncertain system of the type considered in Sections 2.4.3, 8.5.

11.2 Experimental setup and modeling

11.2.1 Experimental setup

The experimental acoustic duct to be considered in this chapter is illustrated in Figure 11.2.1. This is a duct with a square cross section which is closed at both ends. A microphone is located at position r_1 and a speaker is positioned at r_2. For convenience, we assume that the acoustic noise source is at the same location as the loudspeaker. The length of the duct is L.

For the experimental duct located at the Australian Defence Force Academy (ADFA), the dimensions are given in Table 11.2.1.

Table 11.2.1. Duct dimensions

L	$3.99\,m$
w	$0.38\,m$
r_1	$1.8\,m$
r_2	$3.4\,m$

11.2.2 System identification and nominal modelling

The nominal model for the duct was obtained via a process of system identification. First frequency response data for the system was obtained using a swept sine spectrum analyser. This data was obtained in the frequency range 10 – 500 Hz (63 – 3140 rad/s). Note that the measured system included the audio amplifier driving the speakers and the microphone pre-amplifier. It also included two low pass filters (LPF) which will act as anti-aliasing and smoothing filters in the final computer control system. These filters are 8th order Bessel filters with a cutoff frequency of 2.5 kHz.

To obtain a nominal transfer function model, a 14th order transfer function $P(s)$ was fitted to the frequency response data over the frequency range 126 – 1382 rad/s. This was achieved via a standard least squares method using the matlab function **invfreqs**. Figure 11.2.2 shows the nominal model and measured frequency responses[1].

Note that it can be seen in Figure 11.2.2 that a good match has been obtained over the frequency range 126 – 1382 rad/s. Above the frequency 1382 rad/s, the model no longer attempts to match the measured frequency response data. Thus, we cannot expect our controller to attenuate noise components above this frequency. However, we must ensure that our control system is robust against the unmodelled dynamics above 1382 rad/s so that the closed loop system is stable. This is achieved by using a suitable uncertain system model.

11.2.3 Uncertainty modelling

As mentioned in the introduction, the minimax LQG approach will be used in order to design a controller for the acoustic duct. The minimax LQG approach

[1] The experimental data and the resulting nominal model can be downloaded from the book website http://routh.ee.adfa.edu.au/~irp/RCD/index.html. The state-space model of the system, the resulting minimax optimal controller, and transfer functions used for uncertainty modeling are also available from this web site.

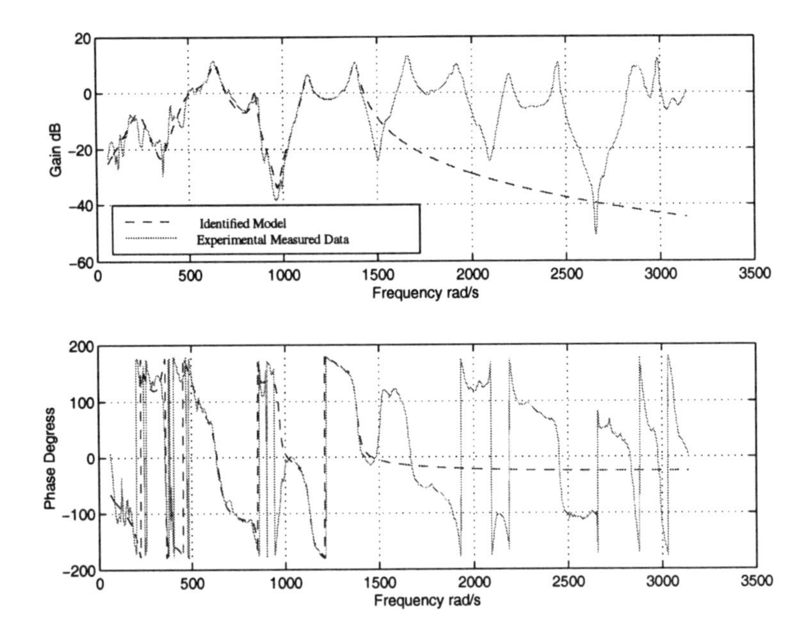

Figure 11.2.2. Model and measured duct frequency responses.

begins with an uncertain system model for the plant to be controlled. In this chapter, we will consider an uncertain system model whose primary purpose is to account for the uncertainty introduced by approximating the measured frequency response data by a finite dimensional transfer function. That is, the main uncertainty to be considered is the uncertainty arising from the spillover dynamics. This uncertainty will be represented by frequency weighted multiplicative uncertainty as shown in Figure 11.2.3. Hence, the true transfer function is assumed to be of the form

$$P_\Delta(s) = P(s)[1 + \Delta(s)W(s)] \tag{11.2.1}$$

where $P(s)$ is the nominal model transfer function obtained above. Here $W(s)$ is a suitable weighting transfer function and $\Delta(s)$ is an uncertain transfer function satisfying the H^∞ norm bound

$$||\Delta(s)||_\infty \leq 1. \tag{11.2.2}$$

This uncertainty representation is by no means the only one which could be used to represent spillover uncertainty and the consideration of different uncertainty structures is the subject of continuing research. However, it was found that this uncertainty structure was quite suitable for purposes of this example.

In order to construct a suitable uncertainty weighting transfer function $W(s)$, we note that equation (11.2.1) can be re-written as

$$\frac{P_\Delta(s) - P(s)}{P(s)} = \Delta(s)W(s).$$

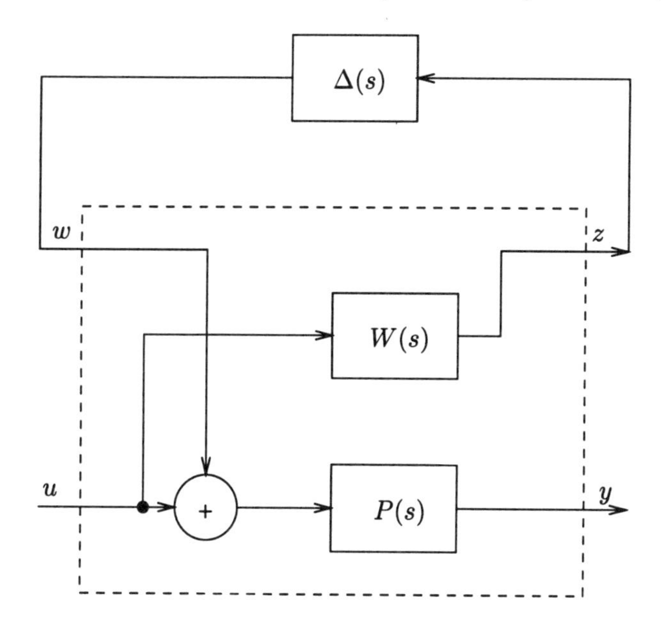

Figure 11.2.3. Uncertain system representation.

Hence, it follows from (11.2.2) that

$$\left| \frac{P_\Delta(j\omega) - P(j\omega)}{P(j\omega)} \right| \leq |W(j\omega)| \tag{11.2.3}$$

for all ω. Thus, we choose the weighting transfer function $W(s)$ so that (11.2.3) holds when $P_\Delta(j\omega)$ is the measured frequency response and $P(j\omega)$ is the model frequency response. Since, the main uncertainty arises from neglecting the duct dynamics above 1382 rad/s, $W(s)$ was chosen as a high pass filter (9th order Chebychev filter with 1 dB ripple in the pass band) with a cut off frequency of 1750 rad/s. Also, in order to enforce good gain and phase margins, the frequency response of $W(s)$ was flattened off at frequencies below 1195 rad/s. This was achieved by introducing zeros in $W(s)$ in a 9th order Chebychev pattern corresponding to a cutoff frequency of 1195 rad/s. These parameters defining $W(s)$ were chosen to achieve a suitable trade off between control system robustness and closed loop noise attenuation performance. Magnitude Bode plots of $W(j\omega)$ and $\frac{P_\Delta(j\omega)-P(j\omega)}{P(j\omega)}$ are shown in Figure 11.2.4. Note that we did not enforce condition (11.2.3) at all frequencies. In practice, it was found that it was unnecessary to enforce this condition at all frequencies in order to obtain the required level of controller robustness. Furthermore, by allowing condition (11.2.3) to be violated over a small range of frequencies, improved closed loop performance could be achieved.

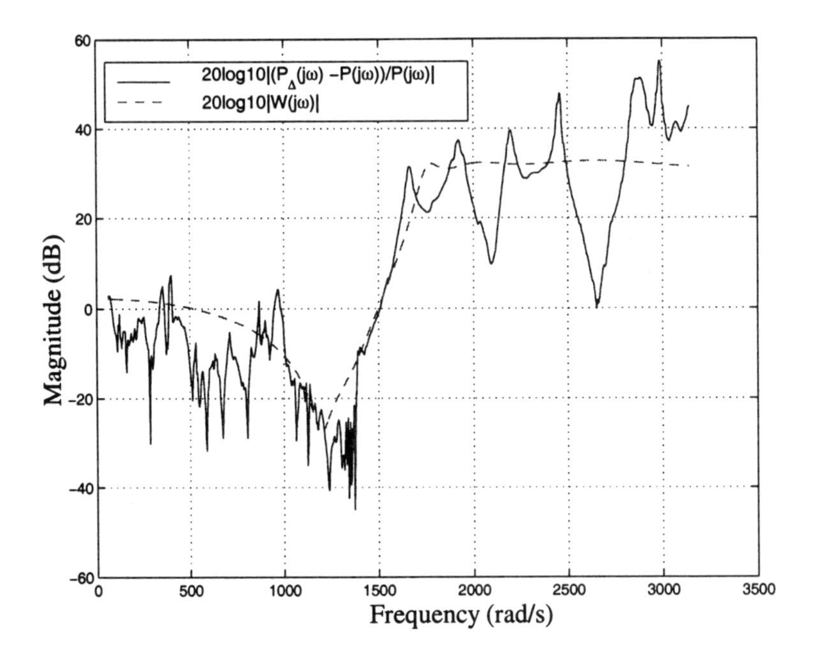

Figure 11.2.4. Multiplicative uncertainty bound.

11.3 Controller design

One of the aims of this chapter is to illustrate the efficacy of the minimax LQG method presented in Chapter 8 when applied to the problem of active noise control in an acoustic duct. Recall that in order to define the minimax LQG control problem under consideration, we consider a stochastic uncertain system defined in terms of stochastic state equations of the form of equation (2.4.29):

$$\begin{aligned} dx &= (Ax + B_1 u + B_2 \xi)dt + B_2 d\tilde{W}(t), \quad x(0) = x_0, \\ z &= C_1 x + D_1 u, \\ dy &= (C_2 x + D_2 \xi)dt + D_2 d\tilde{W}(t), \quad y(0) = 0, \end{aligned} \qquad (11.3.1)$$

As discussed in Section 2.4.3, each uncertainty input $\xi(t)$ gives rise to a corresponding martingale $\zeta(t)$. The set of admissible martingales is denoted Ξ. For each admissible martingale $\zeta(t)$, $\tilde{W}(t)$ is a standard Wiener process. This corresponds to the case in which the noise input is Gaussian white noise with unity covariance. The set of admissible uncertainties Ξ is defined in terms of a relative entropy constraint as described in Definition 8.5.1 on page 309. Recall that Definition 8.5.1 employed a constant $d > 0$ which determines the size of the uncertainty in the probability distribution of the exogenous noise signal acting on the system.

Now suppose the martingale $\zeta(t)$ is induced by a system interconnection of the form shown in Figure 11.3.1 where we identify $w(t)dt = dW(t)$ and $\tilde{w}(t)dt =$

$d\tilde{W}(t)$. In this block diagram $\Delta(s)$ is an uncertain transfer function satisfying the H^∞ norm bound (11.2.2) and $\tilde{w}(t)$ is a Gaussian white noise process with unity covariance with respect to the probability measure defined by an uncertainty martingale $\zeta(t)$. It was shown in Section 2.4.3 that the corresponding martingale $\zeta(t)$ satisfies the relative entropy constraint (8.5.6); see Definition 8.5.1. That is, the corresponding uncertain system (2.4.29), (8.5.6) allows for uncertainty of the form shown in Figure 11.3.1. In particular, the uncertain system model for the acoustic duct derived in the previous section is an uncertain system of this form.

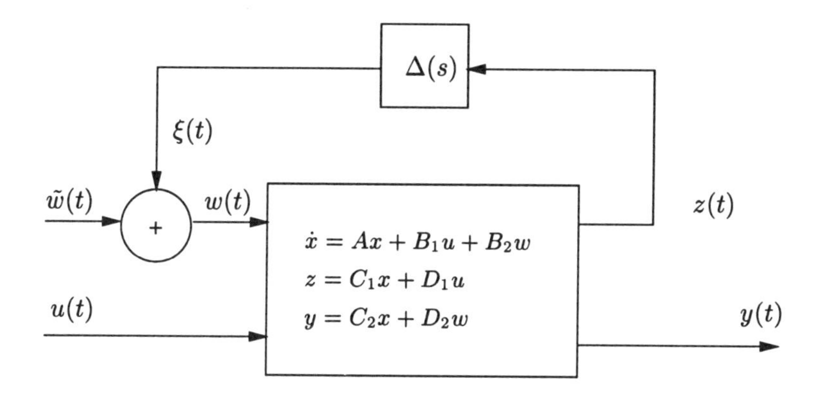

Figure 11.3.1. Stochastic uncertain system.

It is assumed in Section 8.5 that the cost functional under consideration is of the form (8.5.24). The minimax LQG control problem involves finding a controller which minimizes the maximum of this cost functional where the maximum is taken over all uncertainties satisfying the relative entropy constraint (8.5.6).

The solution to the infinite horizon minimax LQG problem given in Section 8.5 is given in terms of the parameter dependent algebraic Riccati equations (8.5.78), (8.5.79) whose solutions must satisfy the conditions of Theorem 8.5.3. It is shown in Section 8.5 that in order to obtain the minimax LQG controller, the parameter $\tau > 0$ should be chosen to minimize the quantity $V_\tau^0 + \tau d$, where V_τ^0 is given by equation (8.5.85). For this optimal value of τ, the corresponding minimax optimal controller is defined by equations (8.5.2), (8.5.82).

In order to apply the minimax LQG technique described in Section 8.5, we must first specify the stochastic uncertain system (11.3.1), (8.5.6) and the cost functional (8.5.24). As mentioned in Section 11.2.3, the main uncertainty we wish to consider arises from unmodelled spillover dynamics. Furthermore, as described in Section 11.2.3, this uncertainty can be modelled by an uncertain system of the form shown in Figure 11.2.3 where the transfer functions $P(s)$ and $W(s)$ are defined as above. Moreover, this uncertain system defines a corresponding stochastic uncertain system of the form (11.3.1), (8.5.6). In this stochastic uncertain system, the state equations (11.3.1) are obtained from a state space realization of the transfer function $P(s)$ augmented with the transfer function $W(s)$ as in Figure 11.2.3.

As mentioned in Section 2.4.3, the condition (11.2.2) guarantees the satisfaction of the relative entropy constraint (8.5.6).

In this stochastic uncertain system, the nominal noise process $w(t)$ acting on the system is a Gaussian white noise input acting through the same channel as the control input, $w(t) \equiv \tilde{w}(t)$. However, the theory of Section 8.5 requires that $D_2 D_2' > 0$ in (11.3.1). This is achieved by adding a small measurement noise to the system in addition to the process noise $w(t)$. Also, since we are mainly concerned with uncertainty in the system dynamics rather than uncertainty in the process noise probability distribution, we set the parameter d in (8.5.6) to a small value of $d = 0.001$.

Given that our nominal system model has been obtained via a process of system identification, we will chose the matrix R in the cost functional (8.5.24) as

$$R = C_2' C_2.$$

That is, the term $x(t)' R x(t)$ in the cost functional (8.5.24) corresponds to the norm squared value of the nominal system output.

The term $u' G u$ in the cost functional (8.5.24) is treated as a design parameter. This term is chosen so that gain of the resulting controller is not too large. If the gain of the controller is too large, this can lead to actuator saturation. However, in this case, it was found that the main factor limiting the controller gains was the presence of model uncertainty. Indeed setting G to the following small value

$$G = 10^{-8}.$$

did not lead to excessive controller gains.

Note that with the above choice of plant model (11.3.1) and cost functional (8.5.24), the nominal LQG problem essentially amounts to the problem of minimizing the noise energy at the microphone position when the system is subject to a white noise disturbance entering the system through the control input channel (ignoring the effect of the control weighting term $u' G u$ in the cost functional).

As mentioned in Section 8.5, the minimax LQG controller is synthesized by first choosing the parameter $\tau > 0$ to minimize the quantity $V_\tau^0 + \tau d$ defined in (8.5.85). For the uncertain system model and cost functional defined as above, a plot of $V_\tau^0 + \tau d$ versus τ is shown in Figure 11.3.2. From this plot it is found that the optimal value of τ is

$$\tau = 4.0.$$

With this value of the parameter τ the controller is constructed according to the formulae (8.5.2), (8.5.82). A magnitude Bode plot of this controller transfer function is shown in Figure 11.3.3.

As mentioned in Section 11.2.3, one of the objectives in defining the uncertainty weighting transfer function $W(s)$ was to enforce reasonable gain and phase margins. The fact that this was achieved can be seen in the control system loop gain Bode and Nichols plots given in Figures 11.3.4 – 11.3.5.

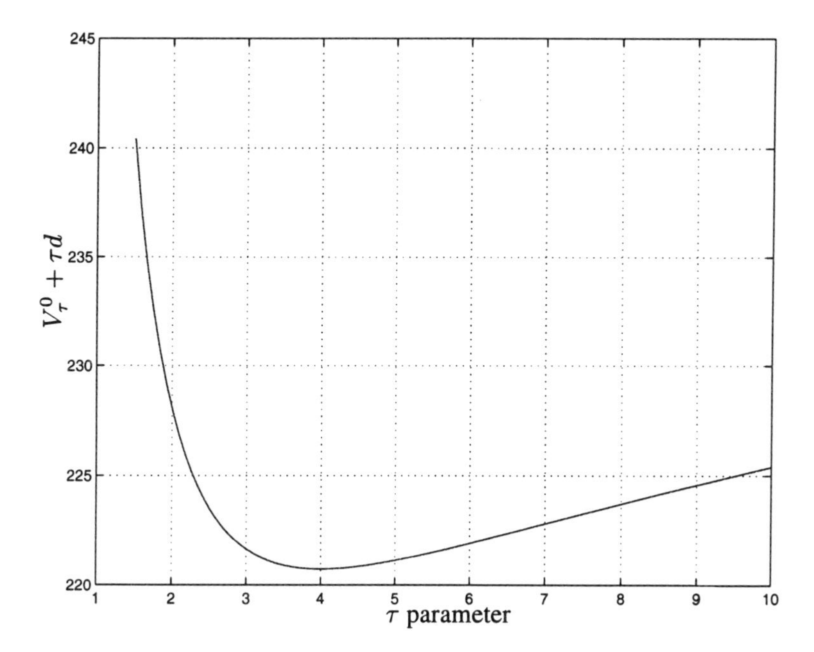

Figure 11.3.2. $V_\tau^0 + \tau d$ versus τ.

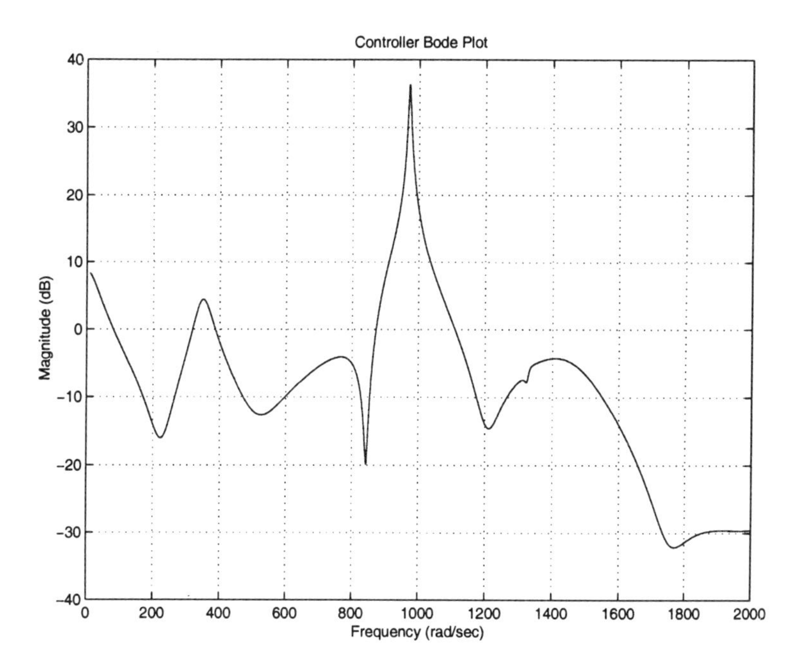

Figure 11.3.3. Controller magnitude Bode plot.

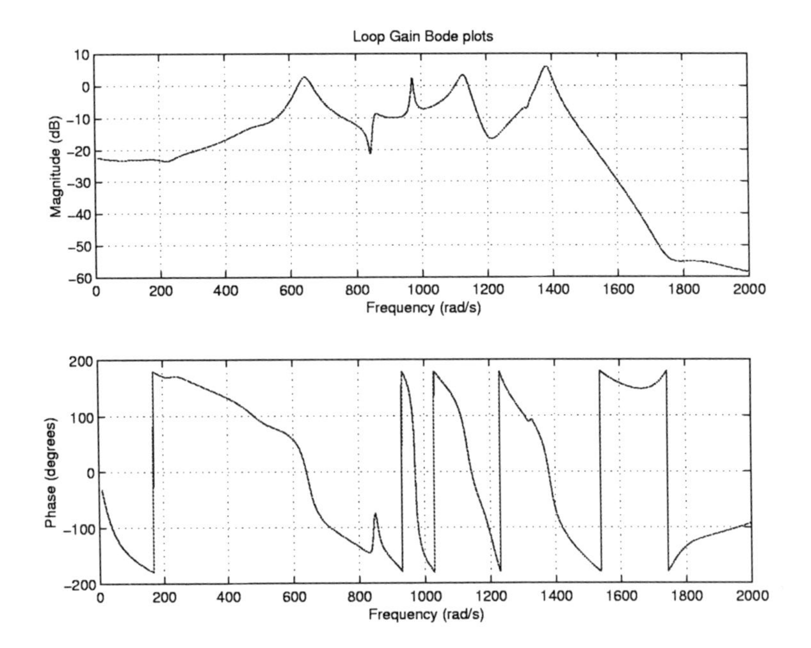

Figure 11.3.4. Loop gain Bode plots.

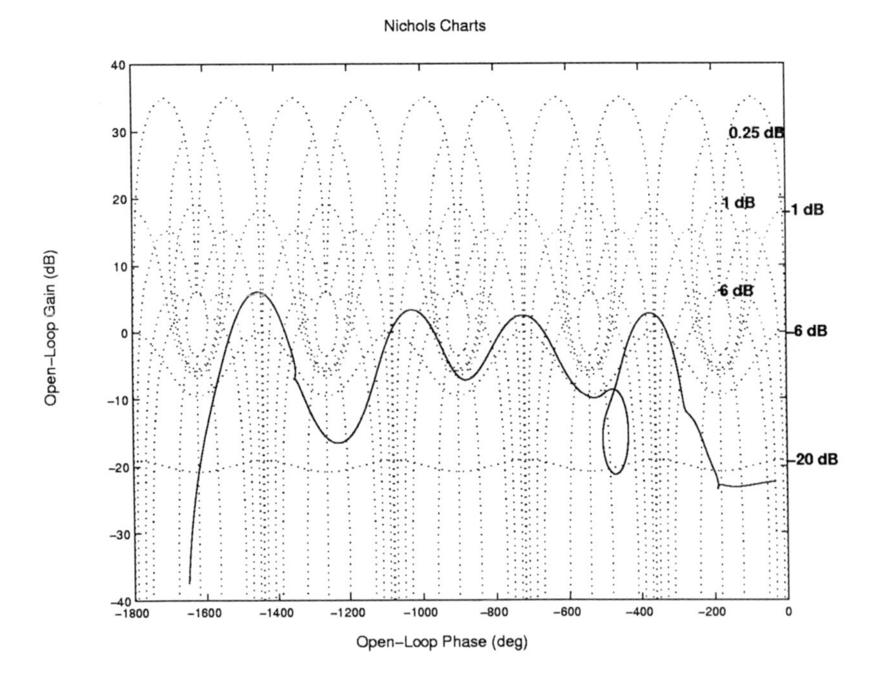

Figure 11.3.5. Nichols plot.

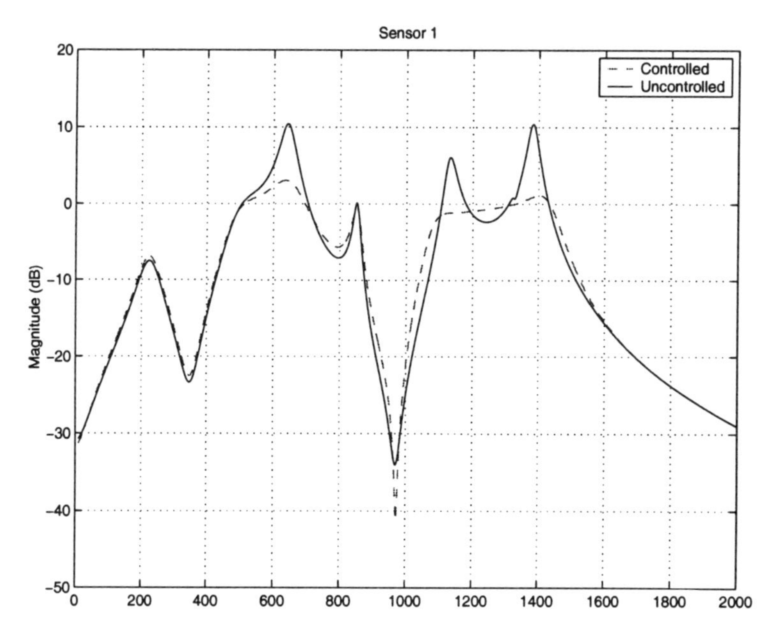

Figure 11.3.6. Simulated controlled and uncontrolled frequency responses.

Also, the disturbance attenuation performance of the controller is illustrated in Figure 11.3.6. This figure shows the frequency response of the proposed control system from the disturbance input to microphone sensor output. For the sake of comparison, this controlled frequency response is compared with the uncontrolled frequency response which is obtained from the acoustic duct model without the use of feedback control.

Remarks

In the controller design process, the main trade offs were between the control system performance on one hand and the limitations imposed by control system robustness on the other hand. In this design process, the control system robustness was controlled by the choice of the weighting transfer function $W(s)$.

In choosing the uncertainty weighting transfer function $W(s)$, the high frequency gain was determined by the presence of spillover dynamics. The low frequency gain was adjusted to achieve a desired level of robustness to account for the differences between the model frequency response and the measured frequency response. Depending on the application, it may be possible to obtain some improvement in performance by using a more accurate model and reduce the low frequency robustness requirement.

11.4 Experimental results

The controller designed in Section 11.3 was implemented on a dSPACE DSP system as shown in Figure 11.4.1. This involved first discretizing the controller state equations (8.5.2), (8.5.82) using the ZOH method with a sample period of 0.2×10^{-3} seconds. The resulting discrete time controller was then implemented on the dSPACE system with this sample period. Figure 11.4.2 shows the resulting

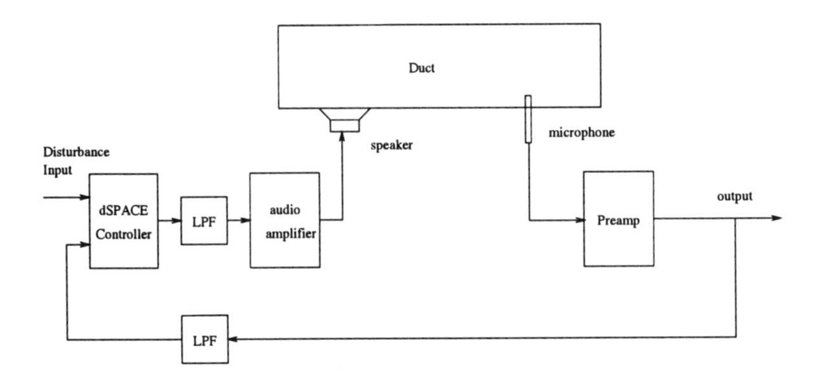

Figure 11.4.1. Feedback controller setup.

measured frequency response for the closed loop system. For the sake of comparison, this figure also shows the measured frequency response for the uncontrolled duct. In all cases, the frequency response is measured from the disturbance input (which is applied through the same channel as the control input) to the microphone sensor output. These results, which are consistent with the simulation results presented in Figure 11.3.6 , show that a significant reduction of the resonant peaks has been achieved by the use of the proposed feedback control scheme.

In order to investigate the improved damping of the resonant modes obtained via the use of the feedback controller, some transient response results were obtained for the experimental system. In the first case, a sinusoidal disturbance corresponding to the resonant frequency of 645.3 rad/s was applied to the uncontrolled duct. At time zero, this disturbance was turned off and the controller was turned on. The resulting transient response measured at the output of the microphone pre-amp is shown in Figure 11.4.3. For the sake of comparison, this figure also shows the corresponding transient responses for the uncontrolled duct.

Figure 11.4.4, shows the corresponding transient responses for the case in which the duct is excited at the resonant frequency of 1130 rad/s. Also, Figure 11.4.5 shows the corresponding transient responses for the case in which the duct is excited at the resonant frequency of 1383.7 rad/s.

The results presented in Figures 11.4.2 – 11.4.5 show that the addition of the feedback controller leads to a significantly improved damping of the resonant modes in the frequency range 126 – 1382 rad/s.

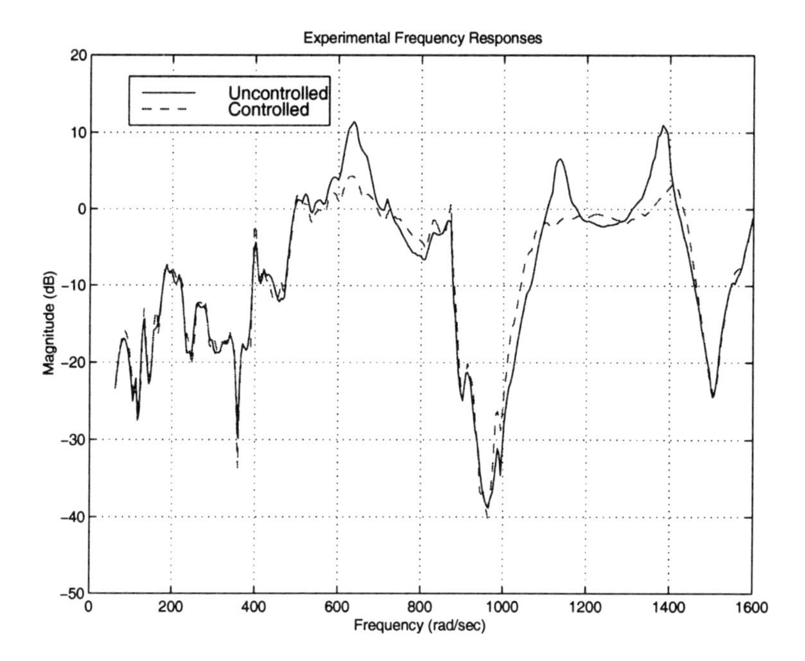

Figure 11.4.2. Experimental controlled and uncontrolled frequency responses.

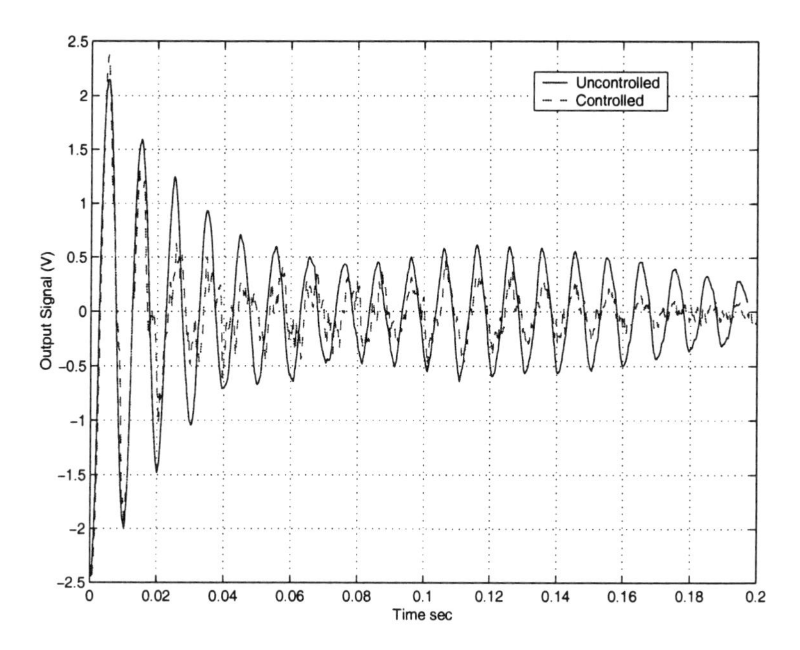

Figure 11.4.3. Controlled and uncontrolled damping of resonance at 645.3 rad/s.

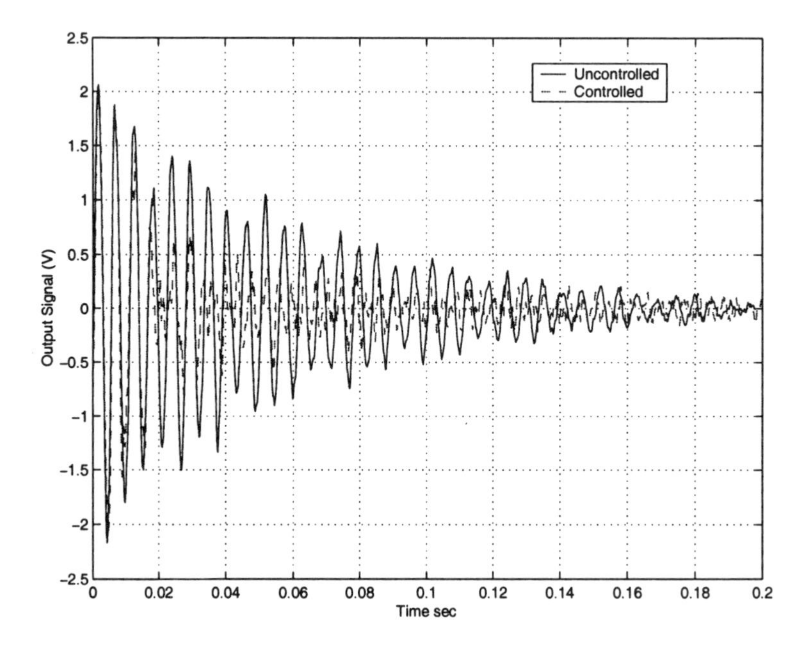

Figure 11.4.4. Controlled and uncontrolled damping of resonance at 1130 rad/s.

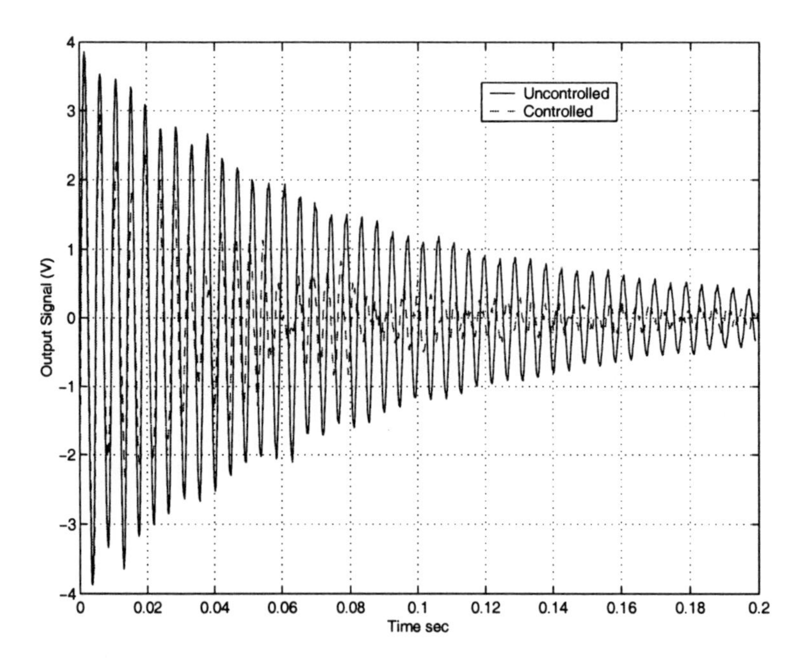

Figure 11.4.5. Controlled and uncontrolled damping of resonance at 1383.7 rad/s.

11.5 Conclusions

The results presented in this chapter indicate that the minimax LQG design methodology leads to a useful controller design methodology for the design of robust feedback controllers for acoustic ducts. In particular, the design methodology provides a systematic way of synthesizing a controller to optimize performance and satisfy given robustness specifications. The methodology has been useful in highlighting the design tradeoffs and performance limitations in the given experimental setup. Furthermore, it allows a controller to be constructed in a natural and systematic way.

One of the key advantages on the minimax LQG approach to robust controller design is the fact that it can handle in a straightforward way multi-input, multi-output control systems. In this chapter, we have considered only a single-input single-output control system. However, it is expected that a significant improvement in performance could be obtained if multiple sensors and multiple actuators were used in the control problem.

One of the critical ingredients in the proposed minimax LQG controller design methodology is the choice of uncertainty structure. In the example considered in this chapter, it was shown that a frequency weighted multiplicative uncertainty structure gave good results. However, there are many other possible uncertainty structures which might be considered and some of these might lead to an improved controller design especially in the multi-input multi-output case.

Appendix A.
Basic duality relationships for relative entropy

This appendix presents a result on the duality between free energy and relative entropy which is exploited in Chapter 8. This result is taken from the paper [39]. For a more detailed study, the reader is referred to the monographs [220, 42, 53].

Let $(\Omega, \mathcal{F})$ be a measurable space, and let $\mathcal{M}(\Omega)$ be the set of probability measures on $(\Omega, \mathcal{F})$.

Definition A.1 *Let $P \in \mathcal{M}(\Omega)$, and $\psi : \Omega \to \mathbf{R}$ be a measurable function. The quantity*

$$\mathbb{E} := \log\left(\int e^{\psi} P(d\omega)\right)$$

is called the free energy of ψ with respect to P.

Definition A.2 *Given any two probability measures $Q, P \in \mathcal{M}(\Omega)$, the relative entropy of the probability measure Q with respect to the probability measure P is defined by*

$$h(Q\|P) := \begin{cases} \int \log\left(\frac{dQ}{dP}\right) Q(d\omega) & \text{if } Q \ll P \text{ and } \log\left(\frac{dQ}{dP}\right) \in \mathbf{L}_1(\Omega, \mathcal{F}, Q), \\ +\infty & \text{otherwise.} \end{cases} \tag{A.1}$$

In the above definition, $\frac{dQ}{dP}$ is the Radon-Nikodym derivative of the probability measure Q with respect to the probability measure P. Also, the notation $Q \ll P$ denotes the fact that the probability measure Q is absolutely continuous with respect to the probability measure P. It is worth mentioning that the relative entropy

$h(Q\|P)$ can be regarded as a measure of the "distance" between the probability measure Q and the probability measure P. Note that the relative entropy is a convex, lower semicontinuous functional of Q; e.g., see [53, 44]. It is also known [53, 42] that the functions $\mathbb{E}(\psi)$ and $h(Q\|P)$ are dual with respect to a Legendre type transform as follows:

(i) For every $Q \in \mathcal{M}(\Omega)$,

$$h(Q\|P) = \sup_{\psi \in \mathcal{B}(\Omega)} \left\{ \int \psi Q(d\omega) - \mathbb{E}(\psi) \right\}; \tag{A.2}$$

(ii) For every $\psi \in \mathcal{B}(\Omega)$,

$$\mathbb{E}(\psi) = \sup_{Q \in \mathcal{M}(\Omega)} \left\{ \int \psi Q(d\omega) - h(Q\|P) \right\}. \tag{A.3}$$

In equations (A.2) and (A.3), $\mathcal{B}(\Omega)$ denotes the set of bounded $\mathcal{F}$-measurable functions on Ω.

In the applications to stochastic control of the duality relations (A.3), the function ψ is the cost function of a stochastic control problem. Even in the simple linear quadratic case such as considered in Chapter 8, this function is not bounded. It is therefore necessary to relax the assumption that $\psi \in \mathcal{B}(\Omega)$ in (A.3). This is done in [39] where the following proposition is proved:

Lemma A.1 *The following extensions of equations (A.2) and (A.3) hold:*

(i) For every $Q \in \mathcal{M}(\Omega)$,

$$h(Q\|P) = \sup_{\substack{e^{\psi} \in \mathbf{L}_1(\Omega,\mathcal{F},P), \\ \psi \text{ bounded below}}} \left\{ \int \psi Q(d\omega) - \mathbb{E}(\psi) \right\}; \tag{A.4}$$

(ii) For every ψ *bounded from below,*

$$\mathbb{E}(\psi) = \sup_{h(Q\|P) < \infty} \left\{ \int \psi Q(d\omega) - h(Q\|P) \right\}. \tag{A.5}$$

Moreover, if $\psi e^{\psi} \in \mathbf{L}_1(\Omega, \mathcal{F}, P)$, *then the supremum in (A.5) is attained at* Q^* *given by*

$$\frac{dQ^*}{dP} = \frac{e^{\psi}}{\int e^{\psi} P(d\omega)}.$$

Proof: See [39]. ∎

Appendix B.
Metrically transitive transformations

Let Ω be any space of elements ω. Let P be a probability measure defined on a Borel field $\mathcal{F}$ of ω sets. A transformation Γ taking points in Ω into points in Ω is called a one to one *measure preserving point transformation* if it is one to one, has domain and range Ω, and if it and its inverse take measurable sets into measurable sets of the same probability.

A set transformation Γ defined on the Borel field $\mathcal{F}$, taking sets of $\mathcal{F}$ into sets of $\mathcal{F}$, is called a *measure-preserving transformation* if the following conditions are satisfied:

(i) Γ is single valued, modulo sets of probability zero; *i.e.,* if $\tilde{\Lambda}$ is an image of Λ under Γ, the class of all images of Λ is the class of all measurable sets differing from $\tilde{\Lambda}$ by sets of probability zero.

(ii) $P(\Gamma\Lambda) = P(\Lambda)$.

(iii) Neglecting ω sets of probability zero,

$$\begin{aligned} \Gamma(\Lambda_1 \cup \Lambda_2) &= \Gamma\Lambda_1 \cup \Gamma\Lambda_2; \\ \Gamma(\bigcup_{j=1}^{\infty} \Lambda_j) &= \bigcup_{j=1}^{\infty} \Gamma\Lambda_j; \\ \Gamma(\Omega - \Lambda) &= \Omega - \Gamma\Lambda. \end{aligned}$$

A one-to-one measure preserving point transformation Γ induces a measure preserving set transformation.

A measurable set is called *invariant* under a measure preserving point or set transformation if it differs from its images by sets of probability zero. A measure preserving point or set transformation is called *metrically* transitive if the only invariant sets are those which have probability zero or one.

A family $\{\Gamma_s, s \geq 0\}$ of transformations taking points of Ω into points of Ω is called a *translation semigroup of measure preserving one to one point transformations* if each Γ_s is a one to one measure preserving point transformation and if

$$\Gamma_{s+t} = \Gamma_s \Gamma_t \quad \forall s, t \geq 0. \tag{B.1}$$

The transformation Γ_0 will necessarily be the identity.

A family $\{\Gamma_s, s \geq 0\}$ of set transformations is called a *translation semigroup of measure-preserving set transformations* if (B.1) is true modulo the sets of probability zero. The transformation Γ_0 will be the identity in the sense that every image of a measurable set Λ under Γ_0 will differ from Λ by at most a set of probability zero. A translation semigroup of measure preserving point or set transformations is called metrically transitive, if the only invariant sets are those which have probability zero or one.

The scalar Wiener process is an example of a process generating a metrically transitive translation semigroup of measure-preserving point transformations [45]. These transformations are generated by time shifts of strictly stationary increments of the Wiener process. Also, it is known [45] that the semigroup of shifts of a strictly stationary process is metrically transitive if and only if the semigroup of shifts for the corresponding canonical process in the coordinate space is metrically transitive. For canonical processes, shifts are point transformations. This allows one to pass from considering the Wiener process over an abstract probability space to considering its canonical version for which metrically transitive shifts have a simple construction.

The canonical representation (e.g., see [123]) for a scalar Wiener process is the probability space $(\overline{\Omega}, \overline{\mathcal{F}}, P)$ and the process $\overline{W}(t)$, $t \geq 0$ such that:

$\overline{\Omega}$ is the set of continuous functions $\omega(t)\colon [0, \infty) \to \mathbf{R}$, starting from zero at $t = 0$;

$\overline{\mathcal{F}}$ is the Borel σ-field generated by the cylindrical subsets in $\overline{\Omega}$;

P is the probability measure extending to $\overline{F}$ the function P_0 defined on cylindrical sets as follows:

$$P_0\Big(\bigcap_{k\leq n} \{\omega\colon a_k \leq \omega(t_k) \leq b_k\}$$

$$= \int_{a_1}^{b_1}\int_{a_2}^{b_2} \cdots \int_{a_n}^{b_n} \left[\frac{e^{-\frac{z_1^2}{2t_1}}}{(2\pi t_1)^{1/2}} \frac{e^{-\frac{(z_2-z_1)^2}{2(t_2-t_1)}}}{(2\pi(t_2-t_1))^{1/2}} \cdots \times \frac{e^{-\frac{(z_n-z_{n-1})^2}{2(t_n-t_{n-1})}}}{(2\pi(t_n-t_{n-1}))^{1/2}} \right] dz_1 dz_2 \dots dz_n$$

where $a_k < b_k$, $0 < t_1 < \ldots < t_n$, $n \geq 1$;

$\overline{W}(t)$ is a stochastic process defined as follows:

$$\overline{W}(t,\omega) = \omega(t) \quad (\omega \in \overline{\Omega}, t \geq 0).$$

As in [45], given any $s \geq 0$, each $\omega \in \overline{\Omega}$ has a corresponding $\omega_1 \in \overline{\Omega}$ such that

$$\overline{W}(t,\omega_1) = \overline{W}(t+s,\omega) - \overline{W}(s,\omega) \qquad \forall t \geq 0. \tag{B.2}$$

The relation $\omega_1 = \Gamma_s \omega$ generates a metrically transitive translation semigroup of measure preserving transformations.

Let $\{\overline{\mathcal{F}}_t, t \geq 0\}$ denote the Borel filtration, generated by $\overline{W}(\theta)$, $0 \leq \theta < t$.

Proposition 1 *For any $\overline{\mathcal{F}}$-measurable random variable $\nu(\omega) \in \mathbf{L}_1(\overline{\Omega}, \overline{\mathcal{F}}, P)$,*

$$\mathbf{E}\{\nu(\Gamma_s\omega)|\mathcal{F}_{t+s}\} = \mathbf{E}\{\nu(\cdot)|\mathcal{F}_t\}(\Gamma_s\omega). \tag{B.3}$$

Proof: It is sufficient to prove the proposition for $\nu = \chi_{\Lambda_0}(\omega)$, where χ_{Λ_0} is an indicator function of the set $\Lambda_0 = \{\omega : \overline{W}(\theta,\omega) \in \Delta\}$; Δ is a Borel subset of $\mathbf{R}$.

Let Λ be an arbitrary $\overline{\mathcal{F}}_{t+s}$-measurable set. Using the definition of the conditional expectation leads to the equation

$$\begin{aligned} \int_\Lambda \mathbf{E}\{\chi_{\Lambda_0}(\Gamma_s\omega)|\overline{\mathcal{F}}_{t+s}\}P(d\omega) &= \int_\Lambda \chi_{\Lambda_0}(\Gamma_s\omega)P(d\omega) \\ &= P(\Lambda \cap \{\omega : \Gamma_s\omega \in \Lambda_0\}) \end{aligned}$$

We apply equation (B.2) to the right hand side of this equation:

$$\begin{aligned} &\int_\Lambda \mathbf{E}\{\chi_{\Lambda_0}(\Gamma_s\omega)|\overline{\mathcal{F}}_{t+s}\}P(d\omega) \\ &\quad = P(\Lambda \cap \{\omega : \overline{W}(\theta+s,\omega) - \overline{W}(s,\omega) \in \Delta\}). \end{aligned} \tag{B.4}$$

Consider the following cases:

Case 1. Suppose that the set Λ satisfies an additional condition: There exists an $\overline{\mathcal{F}}_t$-measurable set $\tilde{\Lambda}$ such that $\Lambda = \Gamma_s\tilde{\Lambda}$ a.s. Then, equation (B.4) implies the following:

$$\begin{aligned} \int_\Lambda \mathbf{E}\{\chi_{\Lambda_0}(\Gamma_s\omega)|\overline{\mathcal{F}}_{t+s}\}P(d\omega) &= P(\tilde{\Lambda} \cap \{\omega : \overline{W}(\theta,\omega) \in \Delta\}) \\ &= \int_{\tilde{\Lambda}} \chi_{\Lambda_0}(\omega)P(d\omega) \\ &= \int_{\tilde{\Lambda}} \mathbf{E}\{\chi_{\Lambda_0}|\overline{\mathcal{F}}_t\}P(d\omega) \\ &= \int_\Lambda \mathbf{E}\{\chi_{\Lambda_0}|\overline{\mathcal{F}}_t\}(\Gamma_s\omega)P(d\omega). \end{aligned}$$

Case 2. Now suppose that the set $\Lambda \in \mathcal{F}_{t+s}$ is such that for any set $\tilde{\Lambda} \in \overline{\mathcal{F}}_t$, $P(\Lambda \cap \Gamma_s\tilde{\Lambda}) = 0$. Note that if $\theta \leq t$, then this property of the set Λ implies

$$P(\Lambda \cap \{\omega : \overline{W}(\theta+s,\omega) - \overline{W}(s,\omega) \in \Delta\}) = 0. \tag{B.5}$$

Also, if $\theta > t$, then the probability on the right hand side of equation (B.4) can be expressed using a complete probability integral as follows:

$$\begin{aligned} &P(\Lambda \cap \{\omega : \overline{W}(\theta + s, \omega) - \overline{W}(s, \omega) \in \Delta\}) \\ &\quad = \int P(\{\omega : \overline{W}(\theta + s, \omega) - \overline{W}(t + s, \omega) + y \in \Delta\}| \\ &\qquad\qquad\qquad \Lambda \cap \{\omega : \overline{W}(t + s, \omega) - \overline{W}(s, \omega) = y\}) \\ &\quad \times P(\Lambda \cap \{\omega : \overline{W}(t + s, \omega) - \overline{W}(s, \omega) \in dy\}). \end{aligned}$$

However, $\{\omega : \overline{W}(t + s, \omega) - \overline{W}(s, \omega) \in dy\} = \Gamma_s\{\omega : \overline{W}(t, \omega) \in dy\}$ and $\{\omega : \overline{W}(t, \omega) \in dy\} \in \overline{\mathcal{F}}_t$. Hence from the definition of the set Λ, it follows that equation (B.5) holds. Then, equations (B.5) and (B.4) imply

$$\int_\Lambda \mathbf{E}\{\chi_{\Lambda_0}(\Gamma_s\omega)|\overline{\mathcal{F}}_{t+s}\}P(d\omega) = 0. \tag{B.6}$$

Also, from the definition of the set Λ, it follows that

$$\int_\Lambda \mathbf{E}\{\chi_{\Lambda_0}|\overline{\mathcal{F}}_t\}(\Gamma_s\omega)P(d\omega) = \int xP\left(\Lambda \cap \{\omega : \mathbf{E}\{\chi_{\Lambda_0}|\overline{\mathcal{F}}_t\}(\Gamma_s\omega) \in dx\}\right) = 0.$$

Hence, the equation

$$\int_\Lambda \mathbf{E}\{\chi_{\Lambda_0}(\Gamma_s\omega)|\overline{\mathcal{F}}_{t+s}\}P(d\omega) = \int_\Lambda \mathbf{E}\{\chi_{\Lambda_0}|\overline{\mathcal{F}}_t\}(\Gamma_s\omega)P(d\omega)$$

remains true in this case as well.

Equation (B.3) now follows from the observation that an arbitrary set $\Lambda \in \mathcal{F}_{t+s}$ admits the following decomposition:

$$\Lambda = \Lambda_{\|} \cup \Lambda_{\perp}, \quad \Lambda_{\|} \cap \Lambda_{\perp} = \varnothing$$

where $\Lambda_{\|}$ is a subset of the set Λ which allows for a representation as a shifted $\mathcal{F}_t$-measurable set and $\Lambda_{\perp}$ is a complementary subset which does not allow for such a representation.

■

This proposition shows that the measure preserving transformation Γ_s induces a transformation of random variables which takes $\overline{\mathcal{F}}_t$-measurable variables into $\overline{\mathcal{F}}_{t+s}$-measurable ones. When applied to diffusion processes satisfying linear Ito equations, this fact gives the following result.

Lemma B.1 *Let $x(t, \omega)$ and $y(t+s, \omega)$, $t \geq 0$, $s > 0$, be two $\mathbf{R}^n$ valued stochastic processes being the unique solutions to the integral equations (B.7) and (B.8)* *below:*

$$x(t,\omega) = h(\omega) + \int_0^t (Ax(\theta,\omega) + Bu(\theta,\omega))d\theta + \int_0^t (Hx(\theta,\omega) + Pu(\theta,\omega))dW(\theta); \tag{B.7}$$

$$y(t+s,\omega) = g(\omega) + \int_s^{t+s} (Ay(\theta,\omega) + Bv(\theta,\omega))d\theta + \int_s^{t+s} (Hy(\theta,\omega) + Pv(\theta,\omega))dW(\theta); \tag{B.8}$$

$h \in \mathbf{L}_2(\Omega, \mathcal{F}_0, P)$, $g \in \mathbf{L}_2(\Omega, \mathcal{F}_s, P)$, $u \in \mathbf{L}_2(0, \mathbf{R}^m)$, $v \in \mathbf{L}_2(s, \mathbf{R}^m)$. *If* $u(t, \Gamma_s\omega) = v(t+s,\omega)$ *and* $g(\omega) = h(\Gamma_s\omega)$ *with probability one, then*

$$y(t+s,\omega) = x(t, \Gamma_s\omega) \quad \textit{with probability one.}$$

Proof: See [27].

■

In Section 8.2.1 and in the proof of Theorem 8.2.2, we have used another result of reference [27]. For the sake of completeness, we include this result here in a form adapted to the notation of this section. Note that in the particular case of a single input system of the form (8.3.1) with the matrix A in companion form and state dependent noise, a similar result was established in [211]. Theorem 2 in [211] dealt with an $\mathbf{L}_2$ property of solutions with probability one while the result given below concerns the mean square $\mathbf{L}_2$ property.

Lemma B.2 *Suppose Assumption 8.2.1 is satisfied. Then for any pair of inputs* $(u(\cdot), \xi(\cdot)) \in \mathbf{L}_2(s, \mathbf{R}^{m_1}) \times \mathbf{L}_2(s, \mathbf{R}^{m_2})$ *and any initial condition* $x(0) = h$, *the corresponding solution to equation (8.3.1) satisfies the condition:*

$$|||x(\cdot)|||^2 \le c_0 \left(\mathbf{E}\|h\|^2 + |||u(\cdot)|||^2 + |||\xi(\cdot)|||^2\right) \tag{B.9}$$

where $c_0 > 0$ *is a constant independent of* h, $u(\cdot)$ *and* $\xi(\cdot)$.

Proof: E.g., see Theorem 5 of [27].

■

References

[1] J. Abedor, K. Nagpal, P. P. Khargonekar, and K. Poola. Robust regulation in the presence of norm-bounded uncertainty. In *Proceedings of the 1994 American Control Conference*, pages 773–777, Baltimore, Maryland, June 1994.

[2] M. A. Aizerman and F. R. Gantmacher. *Absolute stability of regulator systems*. Holden-Day, CA, 1964.

[3] B. D. O. Anderson and J. B. Moore. *Optimal Control: Linear Quadratic Methods*. Prentice-Hall, 1990.

[4] B. D. O. Anderson and S. Vongpanitlerd. *Network Analysis and Synthesis*. Prentice Hall, Englewood Cliffs, NJ, 1973.

[5] B. R. Barmish. Stabilization of uncertain systems via linear control. *IEEE Transactions on Automatic Control*, AC-28(3):848–850, 1983.

[6] B. R. Barmish. Necessary and sufficient conditions for quadratic stabilizability of an uncertain system. *Journal of Optimization Theory and Applications*, 46(4):399–408, 1985.

[7] B. R. Barmish. *New Tools for Robustness of Linear Systems*. Macmillan, New York, 1994.

[8] B. R. Barmish, I. R. Petersen, and A. Feuer. Linear ultimate boundedness control of uncertain dynamical systems. *Automatica*, 19(5):523–532, 1983.

[9] T. Başar and P. Bernhard. *H^∞-optimal control and related minimax design problems: a dynamic game approach.* Birkhäuser, Boston, 2nd edition, 1995.

[10] A. Bensoussan. Saddle point of convex concave functionals with applications to linear quadratic differential games. In H. W. Kuhn and G. P. Szegö, editors, *Differential Games and Related Topics*, pages 177–199. North-Holland, Amsterdam, 1971.

[11] A. Bensoussan. *Stochastic control of partially observable systems.* Cambridge University Press, Cambridge, 1992.

[12] A. Bensoussan and J. H. van Schuppen. Optimal control of partially observable stochastic systems with an exponential-of-integral performance index. *SIAM Journal on Control and Optimization*, 23:599–613, 1985.

[13] D. S. Bernstein and W. M. Haddad. The optimal projection equations with Petersen-Hollot bounds: Robust stability and performance via fixed order dynamic compensation for systems with structured real valued parameter uncertainty. *IEEE Transactions on Automatic Control*, 33(6):578–582, 1988.

[14] D. S. Bernstein and W. M. Haddad. Robust stability and performance analysis for state-space systems via quadratic Lyapunov bounds. *SIAM Journal on Matrix Analysis*, 11(2):239–271, 1990.

[15] D. S. Bernstein and W. M. Haddad. Robust stability and performance via fixed-order dynamic compensation with guaranteed cost bounds. *Mathematics of Control, Signals and Systems*, 3:139–163, 1990.

[16] D. S. Bernstein and W. M. Haddad. Parameter-dependent Lyapunov functions and the Popov criterion in robust analysis and synthesis. *IEEE Trans. Autom. Contr.*, 40(3):536–543, 1995.

[17] D. S. Bernstein, W. M. Haddad, and A. G. Sparks. A Popov criterion for uncertain linear multivariable systems. *Automatica*, 31(7):175–183, 1995.

[18] D. P. Bertsekas and I. B. Rhodes. Recursive state estimation for a set-membership description of uncertainty. *IEEE Transactions on Automatic Control*, 16(2):117–128, 1971.

[19] D. P. Bertsekas and I. B. Rhodes. Sufficiently informative functions and the minimax feedback control of uncertain dynamic systems. *IEEE Transactions on Automatic Control*, AC-18(2):117–123, 1973.

[20] J. M. Bismut. Linear quadratic optimal stochastic control with random coefficients. *SIAM J. Contr. Optimization*, 14(3):419–444, 1976.

[21] R. R. Bitmead and M. Gevers. Riccati difference and differential equations, convergence, monotonicity and stability. In S. Bittanti, A. J. Laub, and J. C. Willems, editors, *The Riccati equation*, pages 264–291. Springer-Verlag, Berlin, 1991.

[22] R. R. Bitmead, M. Gevers, I. R. Petersen, and R. J. Kaye. Monotonicity and stabilizability properties of solutions of the Riccati difference equation: Propositions, lemmas, theorems, fallacious conjectures and counterexamples. *Systems and Control Letters*, 5:309–315, 1985.

[23] S. Bittani, A. J. Laub, and J. C. Willems, editors. *The Riccati Equation*. Springer-Verlag, Berlin, 1991.

[24] F. Blanchini and A. Megretski. Robust state feedback control of LTV systems: Nonlinear is better than linear. In *Proceedings of the 1997 European Control Conference*, Brussels, July 1997.

[25] V. Blondel. *Simultaneous stabilization of linear systems*. Springer-Verlag, London, 1994.

[26] S. Boyd, L. El Ghaoui, E. Feron, and V. Balakrishnan. *Linear Matrix Inequalities in System and Control Theory*. SIAM, Philadelphia, Pennsylvania, 1994.

[27] V. A. Brusin. Global stability and dichotomy of a class of nonlinear systems with random parameters. *Sibirsk. Mat. Zh.*, 22(2):57–73, 1981. English translation: *Siberian Math. J.* 22(2):210–227, 1981.

[28] V. A. Brusin. Frequency conditions for H_∞ control and absolute stabilizability. *Avtomat. i Telemekh.*, (5):17–25, 1996. English translation in *Automat. Remote Control,* 57(5, part 1):626–633, 1996.

[29] V. A. Brusin and E. Ya. Ugrinovskaya. Decentralized adaptive control with a standard model. *Avtomat. i Telemekh.*, (10):29–36, 1992. English translation: *Automat. Remote Control* 53(10, part 1):1497–1504, 1992.

[30] V. A. Brusin and E. Ya. Ugrinovskaya. On decentralized adaptive control with a reference model. *Avtomat. i Telemekh.*, (12):67–77, 1996. English translation: *Automat. Remote Control* 57(12, part 1):1743–1752, 1996.

[31] V. A. Brusin and V. A. Ugrinovskii. Investigation of stochastic stability of a class of nonlinear differential equations of Ito type. *Siberian Math. J*, 28(3):35–50, 1987. English translation: *Siberian Math. J.* 28(3):381–393, 1987.

[32] V. A. Brusin and V. A. Ugrinovskii. Absolute stability approach to stochastic stability of infinite dimensional nonlinear systems. *Automatica*, 31(10):381–393, 1995.

[33] C. I. Byrnes, A. Isidori, and J. C. Willems. Passivity, feedback equivalence, and global stabilization of minimum phase nonlinear systems. *IEEE Transactions on Automatic Control*, AC-36(11):1228–1240, 1991.

[34] R. M. Canon and E. Schmitz. Initial experiments on the endpoint control of a flexible one-link robot. *International Journal of Robotics*, 3(3):62–67, 1984.

[35] S. S. L. Chang and T. K. C. Peng. Adaptive guaranteed cost control of systems with uncertain parameters. *IEEE Transactions on Automatic Control*, AC-17(4):474–483, 1972.

[36] C. D. Charalambous. The role of information state and adjoint in relating nonlinear output feedback risk-sensitive control and dynamic games. *IEEE Trans. Automat. Control*, 42(8):1163–1170, 1997.

[37] C.-F. Cheng. Disturbances attenuation for interconnected systems by decentralized control. *International Journal of Control*, 66:213–224, 1997.

[38] C. F. Cheng, W. J. Wang, and Y. P. Lin. Quadratically decentralized stabilization for uncertain structured interconnected systems. In *Proceeding of the 31st IEEE Conference on Decision and Control*, pages 2846–2847, 1992.

[39] P. Dai Pra, L. Meneghini, and W. Runggaldier. Connections between stochastic control and dynamic games. *Mathematics of Control, Systems and Signals*, 9(4):303–326, 1996.

[40] E. J. Davison and A. Goldenberg. Robust control of a general servomechanism problem: The servo compensator. *Automatica*, 11(5):461–471, 1975.

[41] C. E. de Souza, U. Shaked, and M. Fu. Robust H^∞ filtering with parameter uncertainty and deterministic input signal. In *Proceedings of the 31st IEEE Conference on Decision and Control*, pages 2305–2310, Tucson, AZ, 1992.

[42] J.-D. Deuschel and D. W. Stroock. *Large deviations*. Academic Press Inc., Boston, MA, 1989.

[43] L. Dines. On the mapping of quadratic forms. *Bull. Amer. Math. Soc.*, 47:494–498, 1941.

[44] M. D. Donsker and S. R. S. Varadhan. Asymptotic evaluation of certain Markov process expectations for large time. IV. *Communications on Pure and Applied Mathematics*, 36:183–212, 1983.

[45] J. L. Doob. *Stochastic processes*. John Wiley, NY, 1953.

[46] P. Dorato, C. T. Abdallah, and V. Cerone. *Linear Quadratic Control : An Introduction*. MacMillan, New York, 1995.

[47] P. Dorato, R. Tempo, and G. Muscato. Bibliography on robust control. *Automatica*, 29(1):201–214, 1993.

[48] J. Doyle, A. Packard, and K. Zhou. Review of LFT's, LMI's and μ. In *Proceedings of the 30th IEEE Conference on Decision on Control*, pages 1227–1232, 1991.

[49] J. Doyle, K. Zhou, K. Glover, and B. Bodenheimer. Mixed H_2 and H_∞ performance objectives II: Optimal control. *IEEE Transactions on Automatic Control*, 39(8):1575–1586, 1994.

[50] J. C. Doyle. Guaranteed margins for LQG regulators. *IEEE Transactions on Automatic Control*, 23, 1978.

[51] J. C. Doyle. Analysis of feedback systems with structured uncertainty. *IEE Proceedings Part D*, 129(6):242–250, 1982.

[52] J. C. Doyle, K. Glover, P. P. Khargonekar, and B. Francis. State-space solutions to the standard H_2 and H_∞ control problems. *IEEE Transactions on Automatic Control*, 34(8):831–847, 1989.

[53] P. Dupuis and R. Ellis. *A Weak Convergence Approach to the Theory of Large Deviations*. Wiley, 1997.

[54] P. Dupuis, M. R. James, and I. R. Petersen. Robust properties of risk–sensitive control. In *Proceedings of the IEEE Conference on Decision and Control*, Tampa, FL, December 1998.

[55] E. B. Dynkin. *Markov processes*. Springer, Berlin, 1965.

[56] A. El Bouhtouri and A. J. Pritchard. Stability radii of linear systems with respect to stochastic perturbations. *Systems and Control Letters*, 19:29–33, 1992.

[57] L. El Ghaoui. State-feedback control of systems with multiplicative noise via linear matrix inequalities. *Syst. Contr. Letters*, 24:223–228, 1995.

[58] M. K. H. Fan, A. L. Tits, and J. C. Doyle. Robustness in the presence of mixed parametric uncertainty and unmodeled dynamics. *IEEE Trans. Automat. Contr.*, 36(1):25–38, 1991.

[59] W. H. Fleming and W. M. McEneaney. Risk-sensitive control on an infinite time horizon. *SIAM J. Control and Optim.*, 33(6):1881–1915, 1995.

[60] B. A. Francis. *A course in H_∞ control theory*, volume 88 of *Lecture Notes in Control and Information Sciences*. Springer-Verlag, New York, 1987.

[61] B. Friedland. An observer-based design of a controller for spring-coupled cars. In *Proceedings of the 1992 American Control Conference*, Chicago, Illinois, 1992.

[62] A. Friedman. *Stochastic differential equations and applications.* Academic Press, New York, 1975.

[63] C. R. Fuller and A. H. von Flotow. Active control of sound and vibration. *IEEE Control Systems Magazine*, pages 9–19, December 1995.

[64] P. Gahinet. A convex parameterization of H_∞ suboptimal controllers. In *Proceedings of the 31st IEEE Conference on Decision and Control*, pages 937–942, 1992.

[65] F. R. Gantmacher and V. A. Yakubovich. Absolute stability of nonlinear control systems. In *Proceedings of the 2nd USSR Congress on Theoretical and Applied Mechanics*. Nauka, 1965. (in Russian).

[66] D. Gavel and D. Siljak. Decentralized adaptive control. *IEEE Trans. Autom. Contr.*, AC-34(4):413–426, 1989.

[67] A. H. Gelig, G. A. Leonov, and V. A. Yakubovich. *Stability of nonlinear systems with nonunique equilibrium.* Nauka, Moscow, 1978.

[68] K. Glover. Minimum entropy and risk-sensitive control: The continuous time case. In *Proc. 28th IEEE CDC*, pages 388–391, 1989.

[69] K. Glover and J. C. Doyle. State formulae for all stabilizing controllers that satisfy an H infinity norm bound and relations to risk sensitivity. *Systems and Control Letters*, 11:167–172, 1988.

[70] M. Green and D. J. N. Limebeer. *Linear Robust Control.* Prentice-Hall, Englewood Cliffs, NJ, 1995.

[71] K. M. Grigoriadis and R. E. Skelton. Fixed-order control design for LMI control problems using alternating projection methods. In *Proceedings of the 33rd IEEE Conference on Decision and Control*, pages 2003–2008, Lake Buena Vista, FL, December 1994.

[72] W. H. Haddad, D. S. Bernstein, and D. Mustafa. Mixed-norm H_2/H_∞ regulation and estimation: The discrete-time case. *Systems and Control Letters*, 16(4):235–248, 1991.

[73] W. M. Haddad and D. S. Bernstein. Parameter-dependent Lyapunov functions, constant real parameter uncertainty, and the Popov criterion in robust analysis and synthesis. In *Proceedings of the 30th Conference on Decision and Control*, Brighton, England, December 1991.

[74] W. M. Haddad and D. S. Bernstein. Parameter-dependent Lyapunov functions and the discrete-time Popov criterion for robust analysis and synthesis. In *Proceedings of the American Control Conference*, pages 599–600, Chicago, Illinois, June 1992.

[75] W. M. Haddad and D. S. Bernstein. Explicit construction of quadratic Lyapunov functions for the small gain, positivity, circle, and Popov theorem and their application to robust stability. Part I: Continuous-time theory. *International Journal of Robust and Nonlinear Control*, 3(4):313–339, 1993.

[76] W. M. Haddad and D. S. Bernstein. The multivariable parabola criterion for robust controller synthesis: a Riccati equation approach. *Journal of Mathematical Systems, Estimation and Control*, 6(1):79–104, 1996.

[77] F. Hausdorff. Der wertvorrat einer bilinearform. *Math. Z.*, 3:314–316, 1919.

[78] J. W. Helton and M. R. James. *A General Framework for Extending H_∞ Control to Nonlinear Systems*. SIAM, Philadelphia, 1999.

[79] D. Hill and P. Moylan. The stability of nonlinear dissipative systems. *IEEE Transactions on Automatic Control*, AC-21:708–7811, 1976.

[80] D. Hill and P. Moylan. Stability results for nonlinear feedback systems. *Automatica*, 13:377–382, 1977.

[81] D. Hill and P. Moylan. Dissipative dynamical systems: Basic input-output and state properties. *Journal of the Franklin Institute*, 309:327–357, 1980.

[82] D. Hinrichsen and A. Pritchard. A Riccati equation for stochastic H^∞. In *Proceedings of 2nd Symposium on Robust Control Design*, Budapest, 1997.

[83] D. Hinrichsen and A. J. Pritchard. Stability radii of systems with stochastic uncertainty and their optimization via output feedback. *SIAM J. Contr. Opt.*, 34(6):1972–1998, 1996.

[84] D. Hinrichsen and A. J. Pritchard. Stochastic H^∞. *SIAM Journal on Control and Optimization*, 36(5):1504–1538, 1998.

[85] C. V. Hollot and B. R. Barmish. Stabilization of uncertain linear systems via a variable Lyapunov function. In *Proceedings of the 18th. Allerton Conference on Communication, Control and Computation*, Monticello, IL, 1980. University of Illinois.

[86] J. Hong, J. C. Ackers, R. Vanugopal, M. N. Lee, A. G. Sparks, P. D. Washabaugh, and D. S. Bernstein. Modeling, identification, and feedback control of noise in an acoustic duct. *IEEE Transactions on Control Systems Technology*, 4(3):283–291, 1996.

[87] I. Horowitz. *Synthesis of feedback systems*. Academic, New York, 1963.

[88] S. Van Huffel and J. Vandewalle. *The Total Least Squares Problem: Computational Aspects and Analysis*. SIAM, Philadelphia, 1991.

[89] A. Ichikawa. Dynamic programming approach to stochastic evolution equations. *SIAM J. Control and Optimization*, 17:152–174, 1979.

[90] A. Ichikawa. Quadratic games and H_∞-type problems for time varying systems. *Int. J. Contr.*, 54(5):1249–1271, 1991.

[91] T. Iwasaki and R. E. Skelton. All controllers for the general H_∞ control problem: LMI existence conditions and state space formulas. *Automatica*, 30(8):1307–1319, 1994.

[92] D. H. Jacobson. Optimal stochastic linear systems with exponential performance criteria and their relation to deterministic differential games. *IEEE Transactions on Automatic Control*, 18(2):124–131, 1973.

[93] M. R. James, J.S. Baras, and R.J. Elliott. Risk-sensitive control and dynamic games for partially observed discrete-time nonlinear systems. *IEEE Trans. Automatic Control*, 39(4):780–792, 1994.

[94] T. Kailath. *Linear Systems*. Prentice-Hall, Englewood Cliffs, NJ, 1980.

[95] R. E. Kalman. Lyapunov functions for the problem of Lur'e in automatic control. *Proc. Nat. Acad. Sci. USA*, 49(2):201–205, 1963.

[96] I. Karatzas and S. E. Shreve. *Brownian Motion and Stochastic Calculus*. Springer-Verlag, New York, 1988.

[97] M. Khammash and M. Dahleh. Time-varying control and the robust performance of systems with structured norm-bounded perturbations. *Automatica*, 28(4):819–822, 1992.

[98] M. Khammash and J. B. Pearson. Analysis and design for robust performance with structured uncertainty. *Systems and Control Letters*, 20(3):179–188, 1993.

[99] P. P. Khargonekar, T. T. Georgiou, and A. M. Pascoal. On the robust stabilizability of linear time-invariant plants with unstructured uncertainty. *IEEE Transactions on Automatic Control*, AC-32(3):201–207, 1987.

[100] P. P. Khargonekar, K. M. Nagpal, and K. R. Poolla. H_∞ control with transients. *SIAM Journal on Control and Optimization*, 29(6):1373–1393, 1991.

[101] P. P. Khargonekar, I. R. Petersen, and M. A. Rotea. H^∞ optimal control with state feedback. *IEEE Transactions on Automatic Control*, AC-33(8):786–788, 1988.

[102] P. P. Khargonekar, I. R. Petersen, and K. Zhou. Robust stabilization of uncertain systems and H^∞ optimal control. *IEEE Transactions on Automatic Control*, AC-35(3):356–361, 1990.

[103] P. P. Khargonekar and K. R. Poolla. Uniformly optimal control of linear time-invariant plants: Nonlinear time-varying controllers. *Systems and Control Letters*, 6(5):303–309, 1986.

[104] V. L. Kharitonov. Asymptotic stability of an equilibrium position of a family of systems of linear differential equations. *Differentisial'nye Uravneniya*, 14:2086–2088, 1978.

[105] R. Z. Khas'minskii. *Stochastic Stability of Differential Equations*. Sijthoff and Noordhoff, Gröningen, 1980.

[106] A. A. Krasovskii, editor. *Handbook of Automatic Control Theory*. Nauka, Moscow, 1987.

[107] B. van Kuelen. *H^∞ Control for Distributed Parameter Systems: A State-Space Approach*. Birkhäuser, Boston, 1993.

[108] H. Kwakernaak and R. Sivan. *Linear Optimal Control Systems*. Wiley, 1972.

[109] P. Lancaster and L. Rodman. *Algebraic Riccati Equations*. Oxford University Press, Oxford, 1995.

[110] G. Leitmann. Guaranteed asymptotic stability for some linear systems with bounded uncertainties. *Journal of Dynamic Systems, Measurement and Control*, 101(3):212–216, 1979.

[111] P. Lemmerling, B. De Moor, and S. Van Huffel. On the equivalence of constrained total least squares and structured total least squares. *IEEE Transactions on Signal Processing*, 44(11):2908–2911, 1996.

[112] A. M. Letov. *Stability in nonlinear control systems*. Princeton Inv. Publ., Princeton, N.J., 1961.

[113] F. L. Lewis. *Optimal Control*. Wiley, New York, New York, 1986.

[114] D. J. N. Limebeer, B. D. O. Anderson, P. P. Khargonekar, and M. Green. A game theoretic approach to H^∞ control for time-varying systems. *SIAM Journal on Control and Optimization*, 30:262–283, 1992.

[115] A. Lindquist and V. A. Yakubovich. Optimal damping of forced oscillations in discrete-time systems. *IEEE Trans. Autom. Control*, 42(6):768–802, 1997.

[116] J. L. Lions. *Contrôle optimal de systèmes gouvernés par des équations aux dérivées partielles*. Dunod Gauthier-Villars, Paris, 1968.

[117] R. S. Liptser and A. N. Shiryayev. *Statistics of Random Processes. I. General Theory*. Springer-Verlag, 1977.

[118] D. G. Luenberger. *Optimization by Vector Space Methods.* Wiley, New York, 1969.

[119] A. I. Lur'e. *Some nonlinear problems in the theory of automatic control.* H. M. Stationary Off., London, 1957.

[120] A. I. Lur'e and V. N. Postnikov. On the theory of stability of control systems. *Prikladnaya Matematica i Mehanica*, 8(3), 1944.

[121] P. M. Mäkilä, J. R. Partington, and T. Norlander. Bounded power signal spaces for robust control and modeling. *SIAM J. Control Opt.*, 37(1):92–117, 1998.

[122] A. S. Matveev and A. V. Savkin. *Qualitative Theory of Hybrid Dynamical Systems.* Birkhäuser, Boston, 2000.

[123] H. P. McKean. *Stochastic Integrals.* Academic Press, New York, 1969.

[124] A. Megretski and A. Rantzer. System analysis via integral quadratic constraints. *IEEE Trans. Autom. Contr.*, 42(6):819–830, 1997.

[125] A. Megretsky. Necessary and sufficient conditions of stability: A multiloop generalization of the circle criterion. *IEEE Transactions on Automatic Control*, AC-38(5):753–756, 1993.

[126] A. Megretsky and S. Treil. S-procedure and power distribution inequalities: a new method in optimization and robustness of uncertain systems. Preprint 1, Mittag-Leffler Institute, 1990/1991.

[127] A. Megretsky and S. Treil. Power distribution inequalities in optimization and robustness of uncertain systems. *Journal of Mathematical Systems, Estimation and Control*, 3(3):301–319, 1993.

[128] A. M. Meilakhs. Design of stable control systems subject to parametric perturbation. *Avtomat. i Telemekh.*, (10):5–16, 1978. English translation in *Automat. Remote Control,* 39(10, part 1):1409–1418, 1978.

[129] S. O. R. Moheimani, A. V. Savkin, and I. R. Petersen. A connection between H^∞ control and the absolute stabilizability of discrete-time uncertain systems. *Automatica*, 31(8):1193–1195, 1995.

[130] S. O. R. Moheimani, A. V. Savkin, and I. R. Petersen. Robust observability for a class of time-varying discrete-time uncertain systems. *Systems and Control Letters*, 27:261–266, 1996.

[131] S. O. R. Moheimani, A. V. Savkin, and I. R. Petersen. Minimax optimal control of discrete-time uncertain systems with structured uncertainty. *Dynamics and Control*, 7(1):5–24, 1997.

[132] S. O. R. Moheimani, A. V. Savkin, and I. R. Petersen. Robust filtering, prediction, smoothing and observability of uncertain systems. *IEEE Transactions on Circuits and Systems. Part 1, Fundamental Theory and Applications*, 45(4):446–457, 1998.

[133] S. O. R. Moheimani, D. Stirling, I. R. Petersen, and D. C. McFarlane. Robust control of a Sendzimir mill. In *Proceedings of the 3rd IEEE Conference on Control Applications*, Glasgow, August 1994.

[134] D. Mustafa and D. S. Bernstein. LQG bounds in discrete-time H_2/H_∞ control. *Transactions of the Institute of Measurement and Control*, 13:269–275, 1991.

[135] D. Mustafa and K. Glover. *Minimum entropy H_∞ control*. Springer-Verlag, Berlin, 1990.

[136] K. M. Nagpal and P. P. Khargonekar. Filtering and smoothing in an H^∞ setting. *IEEE Transactions on Automatic Control*, AC-36(2):152–166, 1991.

[137] K. S. Narendra and J. H. Taylor. *Frequency Domain Criteria for Absolute Stability*. Academic Press, New York, 1973.

[138] A. Packard and J. C. Doyle. Quadratic stability with real and complex perturbations. *IEEE Transactions on Automatic Control*, AC-35(2):198–201, 1990.

[139] Z. Pan and T. Başar. Model simplification and optimal control of stochastic singularly perturbed systems under exponentiated quadratic cost. *SIAM J. Contr. Opt.*, 34(5):1734–1766, 1996.

[140] I. R. Petersen. Quadratic stabilizability of uncertain linear systems: Existence of a nonlinear stabilizing control does not imply existence of a linear stabilizing control. *IEEE Transactions on Automatic Control*, AC-30(3):291–293, 1985.

[141] I. R. Petersen. Disturbance attenuation and H^∞ optimization: A design method based on the algebraic Riccati equation. *IEEE Transactions on Automatic Control*, AC-32(5):427–429, 1987.

[142] I. R. Petersen. A procedure for simultaneously stabilizing a collection of single input linear systems using nonlinear state feedback control. *Automatica*, 23(1):33–40, 1987.

[143] I. R. Petersen. A stabilization algorithm for a class of uncertain systems. *Systems and Control Letters*, 8:181–188, 1987.

[144] I. R. Petersen. Some new results on algebraic Riccati equations arising in linear quadratic differential games and the stabilization of uncertain linear systems. *Systems and Control Letters*, 10:341–348, 1988.

[145] I. R. Petersen. Stabilization of an uncertain linear system in which uncertain parameters enter into the input matrix. *SIAM Journal on Control and Optimization*, 26(6):1257–1263, 1988.

[146] I. R. Petersen. Guaranteed cost LQG control of uncertain linear systems. *IEE Proceedings, Part D*, 142(2):95–102, 1995.

[147] I. R. Petersen, B. D. O. Anderson, and E. A. Jonckheere. A first principles solution to the non-singular H^∞ control problem. *International Journal of Robust and Nonlinear Control*, 1(3):171–185, 1991.

[148] I. R. Petersen and B. R. Barmish. The stabilization of single input uncertain linear systems via linear control. In *Proceedings of the 6th International Conference on Analysis and Optimization of Systems*, pages 69–83, Berlin, 1984. Springer-Verlag.

[149] I. R. Petersen, M. Corless, and E. P. Ryan. A necessary and sufficient condition for quadratic finite time feedback controllability. In M. Mansour, S. Balemi, and W. Truöl, editors, *Robustness of Dynamic Systems with Parameter Uncertainties: Proceedings of the International Workshop on Robust Control*, pages 165–173, Ascona, Switzerland, March 1992. Birkhäuser.

[150] I. R. Petersen and C. V. Hollot. A Riccati equation approach to the stabilization of uncertain linear systems. *Automatica*, 22:397–411, 1986.

[151] I. R. Petersen and M. R. James. Performance analysis and controller synthesis for nonlinear systems with stochastic uncertainty constraints. *Automatica*, 32(7):959–972, 1996.

[152] I. R. Petersen, M. R. James, and P. Dupuis. Minimax optimal control of stochastic uncertain systems with relative entropy constraints. In *Proceedings of the 36th IEEE Conference on Decision and Control*, San Diego, 1997.

[153] I. R. Petersen, M. R. James, and P. Dupuis. Minimax optimal control of stochastic uncertain systems with relative entropy constraints. *IEEE Transactions on Automatic Control (To appear)*, 2000.

[154] I. R. Petersen and D. C. McFarlane. Optimal guaranteed cost control of uncertain linear systems. In *Proceedings of the 1992 American Control Conference*, pages 2929–2930, Chicago, Illinois, 1992.

[155] I. R. Petersen and D. C. McFarlane. Optimizing the guaranteed cost in the control of uncertain linear systems. In M. Mansour, S. Balemi, and W. Truöl, editors, *Robustness of Dynamic Systems with Parameter Uncertainties: Proceedings of the International Workshop on Robust Control*, pages 241–250, Ascona, Switzerland, March 1992. Birkhäuser.

[156] I. R. Petersen and D. C. McFarlane. Optimal guaranteed cost control and filtering for uncertain linear systems. *IEEE Transactions on Automatic Control*, 39(9):1971–1977, 1994.

[157] I. R. Petersen, D. C. McFarlane, and M. A. Rotea. Optimal guaranteed cost control of discrete-time uncertain linear systems. *International Journal of Robust and Nonlinear Control*, 8(8):649–657, 1998.

[158] I. R. Petersen and A. V. Savkin. Suboptimal guaranteed cost control of uncertain linear systems. In *Proceedings of the Hong Kong International Workshop on New Directions of Control and Manufacturing*, Hong Kong, 1994.

[159] I. R. Petersen and A. V. Savkin. *Robust Kalman Filtering for Signals and Systems with Large Uncertainties*. Birkhäuser Boston, 1999.

[160] K. R. Poolla and T. Ting. Nonlinear time-varying controllers for robust stabilization. *IEEE Transactions on Automatic Control*, AC-32(3):195–200, 1987.

[161] V. M. Popov. On absolute stability of non-linear automatic control systems. *Avtomatika i Telemekhanika*, 22(8):961–967, 1961.

[162] V. M. Popov. *Hyperstability of Control Systems*. Springer Verlag, New York, 1973.

[163] A. C. M. Ran and R. Vreugdenhil. Existence and comparison theorems for algebraic Riccati equations for continuous and discrete time systems. *Linear Algebra and its Applications*, 99:63–83, 1988.

[164] J. F. Randolph. *Basic Real and Abstract Analysis*. Academic Press, New York, London, 1968.

[165] R. Ravi, K. M. Nagpal, and P. P. Khargonekar. H^∞ control of linear time-varying systems: A state-space approach. *SIAM Journal of Control and Optimization*, 29(6):1394–1413, 1991.

[166] S. Richter, A. S. Hodel, and P.G. Pruett. Homotopy methods for the solution of general modified algebraic Riccati equations. *IEE Proceedings. D. Control Theory and Applications*, 160(6):449–454, 1993.

[167] R. T. Rockafellar. *Convex Analysis*. Princeton University Press, Princeton, NJ, 1970.

[168] M. A. Rotea and P. P. Khargonekar. Stabilization of uncertain systems with norm bounded uncertainty - A control Lyapunov approach. *SIAM Journal of Control and Optimization*, 27(6):1462–1476, 1989.

[169] E. N. Rozenvasser. On stability of nonlinear control systems. *Soviet Math. Dokl.*, 117(4), 1957.

[170] T. Runolfsson. The equivalence between infinite-horizon optimal control of stochastic systems with exponential-of-integral performance index and stochastic differential games. *IEEE Trans. Autom. Contr.*, 39(8):1551–1563, 1994.

[171] M. G. Safonov. Stability margins of diagonally perturbed multivariable feedback systems. *IEE Proceedings Part D*, 129(6):251–256, 1982.

[172] M. Sampei, T. Mita, and M. Nakamichi. An algebraic approach to H^∞ output feedback control problems. *Systems and Control Letters*, 14:13–24, 1990.

[173] A. V. Savkin. Absolute stability of nonlinear control systems with nonstationary linear part. *Avtomat. i Telemekh.*, (3):78–84, 1991. English translation in *Automat. Remote Control,* 52(3, part 1):362–367, 1991.

[174] A. V. Savkin. The problem of simultaneous H^∞ control. *Applied Mathematics Letters*, 12(1):53–56, 1999.

[175] A. V. Savkin and R. J. Evans. A new approach to robust control of hybrid systems over infinite time. *IEEE Transactions on Automatic Control,* 43(9):1292–1296, 1998.

[176] A. V. Savkin and I. R. Petersen. Nonlinear versus linear control in the absolute stabilizability of uncertain linear systems with structured uncertainty. In *Proceedings of the 32nd IEEE Conference on Decision and Control*, San Antonio, Texas, 1993.

[177] A. V. Savkin and I. R. Petersen. A connection between H^∞ control and the absolute stabilizability of uncertain systems. *Systems and Control Letters*, 23(3):197–203, 1994.

[178] A. V. Savkin and I. R. Petersen. A method for robust stabilization related to the Popov stability criterion. In *Proceedings of the IFAC Symposium on Robust Control Design*, Rio de Janeiro, Brazil, September 1994.

[179] A. V. Savkin and I. R. Petersen. Output feedback guaranteed cost control of uncertain systems on an infinite time interval. In *Proceedings of the 33rd IEEE Conference on Decision and Control*, 1994.

[180] A. V. Savkin and I. R. Petersen. Structured dissipativeness and absolute stability of nonlinear systems. In *Proceedings of the Workshop on Robust Control via Variable Structure and Lyapunov Techniques*, Benevento, Italy, September 1994.

[181] A. V. Savkin and I. R. Petersen. An uncertainty averaging approach to optimal guaranteed cost control of uncertain systems with structured uncertainty. In *Proceedings of the Workshop on Robust Control via Variable Structure and Lyapunov Techniques*, pages 132–139, Benevento, Italy, September 1994.

[182] A. V. Savkin and I. R. Petersen. An uncertainty averaging approach to output feedback optimal guaranteed cost control of uncertain systems. In *Proceedings of the 33rd IEEE Conference on Decision and Control*, Lake Buena Vista, Florida, December 1994.

[183] A. V. Savkin and I. R. Petersen. A method for robust stabilization related to the Popov stability criterion. *International Journal of Control*, 62(5):1105–1115, 1995.

[184] A. V. Savkin and I. R. Petersen. Minimax optimal control of uncertain systems with structured uncertainty. *International Journal of Robust and Nonlinear Control*, 5(2):119–137, 1995.

[185] A. V. Savkin and I. R. Petersen. Nonlinear versus linear control in the absolute stabilizability of uncertain linear systems with structured uncertainty. *IEEE Transactions on Automatic Control*, 40(1):122–127, 1995.

[186] A. V. Savkin and I. R. Petersen. Recursive state estimation for uncertain systems with an integral quadratic constraint. *IEEE Transactions on Automatic Control*, 40(6):1080–1083, 1995.

[187] A. V. Savkin and I. R. Petersen. Robust control with a terminal state constraint. In *Proceedings of the IFAC Conference on System Structure and Control*, Nantes, July 1995.

[188] A. V. Savkin and I. R. Petersen. Robust control with rejection of harmonic disturbances. In *Proceedings of the 34th IEEE Conference on Decision and Control*, New Orleans, December 1995.

[189] A. V. Savkin and I. R. Petersen. Robust control with rejection of harmonic disturbances. *IEEE Transactions on Automatic Control*, 40(11):1968–1971, 1995.

[190] A. V. Savkin and I. R. Petersen. Structured dissipativeness and absolute stability of nonlinear systems. *International Journal of Control*, 62(2):443–460, 1995.

[191] A. V. Savkin and I. R. Petersen. An uncertainty averaging approach to optimal guaranteed cost control of uncertain systems with structured uncertainty. *Automatica*, 31(11):1649–1654, 1995.

[192] A. V. Savkin and I. R. Petersen. Model validation for robust control of uncertain systems with an integral quadratic constraint. *Automatica*, 32(4):603–606, 1996.

[193] A. V. Savkin and I. R. Petersen. Nonlinear versus linear control in the robust stabilizability of linear uncertain systems via fixed-order output feedback. In *Proceedings of the 1996 Mediterranean Control Conference*, Greece, 1996.

[194] A. V. Savkin and I. R. Petersen. Nonlinear versus linear control in the robust stabilizability of linear uncertain systems via fixed-order output feedback. *IEEE Transactions on Automatic Control*, 41(9):1335–1338, 1996.

[195] A. V. Savkin and I. R. Petersen. Robust control with a terminal state constraint. *Automatica*, 32(7):1001–1005, 1996.

[196] A. V. Savkin and I. R. Petersen. Robust state estimation for uncertain systems with averaged integral quadratic constraints. *International Journal of Control*, 64(5):923–939, 1996.

[197] A. V. Savkin and I. R. Petersen. Robust H^∞ control of uncertain systems with structured uncertainty. *Journal of Mathematical Systems, Estimation and Control*, 6(4):339–342, 1996.

[198] A. V. Savkin and I. R. Petersen. Uncertainty averaging approach to output feedback optimal guaranteed cost control of uncertain systems. *Journal of Optimization Theory and Applications*, 88(2):321–337, 1996.

[199] A. V. Savkin and I. R. Petersen. Fixed-order robust filtering for linear uncertain systems. *Automatica*, 33(2):253–255, 1997.

[200] A. V. Savkin and I. R. Petersen. Output feedback guaranteed cost control of uncertain systems on an infinite time interval. *International Journal of Robust and Nonlinear Control*, 7(1):43–58, 1997.

[201] A. V. Savkin and I. R. Petersen. Robust filtering and model validation for uncertain continuous-time systems with discrete and continuous measurements. *International Journal of Control*, 69(1):163–174, 1998.

[202] A. V. Savkin and I. R. Petersen. Robust stabilization of discrete-time uncertain nonlinear systems. *J. Optimization Theory and Appl.*, 96(1):87–107, 1998.

[203] A. V. Savkin and I. R. Petersen. Robust state estimation and model validation for discrete-time uncertain systems with a deterministic description of noise and uncertainty. *Automatica*, 34(2):271–274, 1998.

[204] A. V. Savkin, I. R. Petersen, E. Skafidas, and R. J. Evans. Hybrid dynamical systems: Robust control synthesis problems. *Systems and Control Letters*, 29(2):81, 1996.

[205] A. V. Savkin, E. Skafidas, and R. J. Evans. Robust output feedback stabilizability via controller switching. *Automatica*, 35(1):69–74, 1999.

[206] C. Scherer. Mixed H_2/H_∞ control. In A. Isidori, editor, *Trends in Control: A European Perspective*, pages 173–216. Springer-Verlag, Berlin, 1995.

[207] A.F.A. Serranes, M. J. G. van den Molengraft, J. J. Kok, and L. van den Steen. H^{∞} control for suppressing stick-slip in oil well drillings. *IEEE Control Systems Magazine*, 18(2):19–30, 1998.

[208] U. Shaked and E. Soroka. On the stability robustness of the continuous time LQG optimal control. *IEEE Transactions on Automatic Control*, AC-30(9):1039–1043, 1985.

[209] J. S. Shamma. Robustness analysis for time-varying systems. In *Proceedings of the 31st IEEE Conference on Decision and Control*, Tuscon, Arizona, December 1992.

[210] J. S. Shamma. Robust stability analysis of time-varying systems using time-varying quadratic forms. *Systems and Control Letters*, 24(1):13–17, 1995.

[211] H. Shibata and S. Hata. L_2-bounded stability for stochastic systems. *Int. J. Control*, 18:1275–1280, 1973.

[212] S. V. Shil'man. *Generating Function Method in the Theory of Dynamic Systems*. Nauka, Moscow, 1978.

[213] D. D. Siljak. *Decentralized Control of Complex Systems*. Academic Press, San Diego, CA, 1991.

[214] E. Skafidas, R. J. Evans, A. V. Savkin, and I. R. Petersen. Stability results for switched controller systems. *Automatica*, 35(4):553–564, 1999.

[215] R. S. Smith and J. C. Doyle. Model validation: A connection between robust control and identification. *IEEE Transactions on Automatic Control*, 37(7):942–952, 1992.

[216] K. P. Sondergeld. A collection of plant models and design specifications for robust control. Technical report, DFVLR Institute for Flight Systems Dynamics, 1982.

[217] J. L. Speyer and U. Shaked. Minimax design for a class of linear quadratic problems with parameter uncertainty. *IEEE Transactions on Automatic Control*, AC-19(2), 1974.

[218] H. Stalford. Robust asymptotic stability of Petersen counterexample via linear static controller. *IEEE Transactions on Automatic Control*, 40(8), 1995.

[219] A. A. Stoorvogel. Nonlinear L_1 optimal controllers for linear systems. *IEEE Transactions on Automatic Control*, 40(4):694, 1995.

[220] D. W. Stroock. *An Introduction to the Theory of Large Deviations*. Springer-Verlag, New York, 1984.

[221] G. Tadmor. The standard H_∞ problem and the maximum principle: The general linear case. Technical Report 192, University of Texas in Dallas, May 1989.

[222] G. Tadmor. Worst-case design in the time domain: the maximum principle and the standard H_∞ problem. *Mathematics of Control, Signals and Systems*, 3:301–324, 1990.

[223] G. Tadmor. The standard H_∞ problem and the maximum principle: The general linear case. *SIAM Journal on Control and Optimization*, 31(4):813–846, 1993.

[224] O. Töeplitz. Das algebraische analogen zu einen satze von Fejer. *Math. Z.*, 2:187–197, 1918.

[225] O. Toker. On the order of simultaneously stabilizing compensators. *IEEE Transactions on Automatic Control*, 41(3):430–433, 1996.

[226] V. A. Ugrinovskii. Exponential stabilization of nonlinear stochastic systems. *Prikl. Mat. Mekh.*, 52(1):16–24, 1988. English translation: *J. Appl. Math. Mech.* 52(1):11–17, 1988.

[227] V. A. Ugrinovskii. Robust H_∞ control in the presence of stochastic uncertainty. *Int.J. Control*, 71(2):219–237, 1998.

[228] V. A. Ugrinovskii and I. R. Petersen. Absolute stabilization and minimax optimal control of uncertain systems with stochastic uncertainty. In *Proceedings of the Second IFAC Symposium on Robust Control Design*, Budapest, Hungary, 1997.

[229] V. A. Ugrinovskii and I. R. Petersen. Finite horizon minimax optimal control of nonlinear continuous time systems with stochastic uncertainty. In *Proceedings of the 36th IEEE Conference on Decision and Control*, San Diego, CA, 1997.

[230] V. A. Ugrinovskii and I. R. Petersen. Finite horizon minimax optimal control of stochastic partially observed time varying uncertain systems. In *Proceedings of the 36th IEEE Conference on Decision and Control*, San Diego, CA, 1997.

[231] V. A. Ugrinovskii and I. R. Petersen. Infinite-horizon minimax optimal control of stochastic partially observed uncertain systems. In *Proceedings of the Control 97 Conference*, Sydney, Australia, 1997.

[232] V. A. Ugrinovskii and I. R. Petersen. Robust stabilization of uncertain systems via Lur'e-Postnikov Lyapunov functions. In *Proceedings of the IFAC Conference on System Structure and Control*, Bucharest, Romania, October 1997.

[233] V. A. Ugrinovskii and I. R. Petersen. Time-averaged robust control of stochastic partially observed uncertain systems. In *Proceedings of the IEEE Conference on Decision and Control*, Tampa, FL, December 1998.

[234] V. A. Ugrinovskii and I. R. Petersen. Absolute stabilization and minimax optimal control of uncertain systems with stochastic uncertainty. *SIAM J. Control Optimization*, 37(4):1089–1122, 1999.

[235] V. A. Ugrinovskii and I. R. Petersen. Finite horizon minimax optimal control of stochastic partially observed time varying uncertain systems. *Mathematics of Control, Signals and Systems*, 12(1):1–23, 1999.

[236] V. A. Ugrinovskii and I. R. Petersen. Robust output feedback stabilization via risk-sensitive control. In *IEEE Conference on Decision and Control*, Phoenix, Az, 1999.

[237] V. A. Ugrinovskii and I. R. Petersen. Guaranteed cost control of uncertain systems via Lur'e-Postnikov Lyapunov functions. *Automatica*, 36:279–285, 2000.

[238] V. A. Ugrinovskii, I. R. Petersen, A. V. Savkin, and E. Ya. Ugrinovskaya. Decentralized state-feedback stabilization and robust control of uncertain large-scale systems with integrally constrained interconnections. In *Proceedings of the 1998 American Control Conference*, Philadelphia, PA, June 1998.

[239] F. Uhlig. A recurring theorem about pairs of quadratic forms and extensions: A survey. *Linear Algebra and its Applications*, 25:219–237, 1979.

[240] M. Vidyasagar. *Control System Synthesis: A Factorization Approach*. MIT Press, Cambridge, MA, 1985.

[241] A. A. Voronov. *Stability, Controllability, and Observability*. Nauka, Moscow, 1979.

[242] W.-J. Wang and Y. H. Chen. Decentralized robust control design with insufficient number of controllers. *International Journal of Control*, 65:1015–1030, 1996.

[243] Y. Wang, L. Xie, and C. E. de Souza. Robust decentralized control of interconnected uncertain linear systems. In *Proceedings of the 34nd Conference on Decision and Control*, pages 2653–2658, New Orleans, LA, 1995.

[244] P. Whittle. Risk-sensitive linear/quadratic/Gaussian control. *Advances in Applied Probability*, 13:764–777, 1981.

[245] P. Whittle. *Risk-sensitive optimal control*. Wiley, Chichester, UK, 1990.

[246] B. Wie and D. S. Bernstein. Benchmark problems for robust control design. In *Proceedings of the 1992 American Control Conference*, Chicago, Illinois, 1992.

[247] J. C. Willems. Least-squares stationary optimal control and the algebraic Riccati equation. *IEEE Trans. Automat. Contr.*, AC-16:621–634, 1971.

[248] J. C. Willems. Dissipative dynamical systems – part I: General theory. *Archive of Rational Mechanics and Analysis*, 45:321–351, 1972.

[249] W. M. Wonham. On matrix Riccati equation of stochastic control. *SIAM J. Contr. Optimization*, 6(4):681–697, 1968.

[250] W. M. Wonham. Random differential equations in control theory. In A. T. Bharucha-Reid, editor, *Probabilistic Methods in Applied Mathematics*, volume 2, pages 131–212. Academic Press, 1970.

[251] W. M. Wonham. *Linear Multivariable Control: A Geometric Approach.* Springer-Verlag, New York, 3rd edition, 1985.

[252] L. Xie and C. E. de Souza. Robust H_∞ control for linear systems with norm-bounded time-varying uncertainty. *IEEE Transactions on Automatic Control*, 37(8):1188–1191, 1992.

[253] L. Xie, M. Fu, and C.D. de Souza. H_∞ control and quadratic stabilization of systems with parameter uncertainty via output feedback. *IEEE Transactions on Automatic Control*, 37(8):1253–1256, 1992.

[254] V. A. Yakubovich. Frequency conditions of absolute stability of control systems with many nonlinearities. *Avtomatika i Telemekhanica*, 28(6):5–30, 1967.

[255] V. A. Yakubovich. S-procedure in nonlinear control theory. *Vestnik Leningrad University, Series 1*, 13(1):62–77, 1971.

[256] V. A. Yakubovich. Minimization of quadratic functionals under the quadratic constraints and the necessity of a frequency condition in the quadratic criterion for absolute stability of nonlinear control systems. *Soviet Mathematics Doklady*, 14:593–597, 1973.

[257] V. A. Yakubovich. *The Methods of Research of Nonlinear Control Systems*, chapter 2-3. Nauka, Moscow, Russia, 1975. (in Russian).

[258] V. A. Yakubovich. Abstract theory of absolute stability of nonlinear systems. *Vestnik Leningrad University, Series 1*, 41(13):99–118, 1977.

[259] V. A. Yakubovich. Absolute stability of nonlinear systems with a periodically nonstationary linear part. *Soviet Physics Doklady*, 32(1):5–7, 1988.

[260] V. A. Yakubovich. Dichotomy and absolute stability of nonlinear systems with periodically nonstationary linear part. *Systems and Control Letters*, 11(3):221–228, 1988.

[261] V. A. Yakubovich. Nonconvex optimization problem: The infinite-horizon linear-quadratic control problem with quadratic constraints. *Systems and Control Letters*, 19(1):13–22, 1992.

[262] V. A. Yakubovich. Linear quadratic problem of optimal rejection of forced oscillations under unknown harmonic external disturbances. *Doklady of Russian Academy of Sciences*, 333(2):170–172, 1993. English translation in *Phys. Dokl.*, 38(11):449–451, 1994.

[263] K. Yasuda. Decentralized quadratic stabilization of interconnected systems. In *12th IFAC World Congress*, volume 6, pages 95–98, Sydney, Australia, 1993.

[264] E. Yaz and N. Yildizbayrak. Robustness of feedback-stabilized systems in the presence of nonlinear and random perturbation. *International J. of Control*, 41(2):345–353, 1985.

[265] K. Yosida. *Functional Analysis*. Springer-Verlag, Berlin, Göttingen, Heidelberg, 1965.

[266] D. C. Youla, J. J. Bongiorno, and C. N. Lu. Single-loop feedback stabilization of linear multivariable plants. *Automatica*, 10:159–173, 1974.

[267] M. Zakai. A Lyapunov criterion for the existence of stationary probability distributions for systems perturbed by noise. *SIAM J. Contr. Optim.*, 7:390–397, 1969.

[268] G. Zames. Feedback and optimal sensitivity: Model reference transformations, multiplicative seminorms, and approximate inverses. *IEEE Trans. Automat. Control*, 26:301–320, 1981.

[269] G. Zhai and M. Ikeda. Decentralized H_∞ control of large-scale systems via output feedback. In *Proceedings of the 32nd Conference on Decision and Control*, pages 1652–1653, San Antonio, Texas, 1993.

[270] G. Zhai, K. Yasuda, and M. Ikeda. Decentralized quadratic stabilization of large-scale systems. In *Proceedings of the 33rd Conference on Decision and Control*, pages 2337–2339, Lake Buena Vista, FL, 1994.

[271] K. Zhou, J. Doyle, and K. Glover. *Robust and Optimal Control*. Prentice-Hall, Upper Saddle River, NJ, 1996.

Index